McGRAW-HILL YEARBOOK OF
Science &
Technology

2004

McGRAW-HILL YEARBOOK OF
Science &
Technology

2004

**Comprehensive coverage of recent events and research as compiled by
the staff of the McGraw-Hill Encyclopedia of Science & Technology**

McGraw-Hill

New York   Chicago   San Francisco   Lisbon   London   Madrid   Mexico City   Milan
New Delhi   San Juan   Seoul   Singapore   Sydney   Toronto

The **McGraw·Hill** Companies

Library of Congress Cataloging in Publication data

McGraw-Hill yearbook of science and technology.
1962–   . New York, McGraw-Hill.

   v. illus. 26 cm.
   Vols. for 1962–   compiled by the staff of the
McGraw-Hill encyclopedia of science and technology.
    1. Science—Yearbooks. 2. Technology—
Yearbooks. 1. McGraw-Hill encyclopedia of
science and technology.
Q1.M13   505.8   62-12028

ISBN 0-07-142784-8
ISSN 0076-2016

The following articles are excluded from McGraw-Hill Copyright:
Angiogenesis; Geologic mapping; Higgs boson; Human Genome Project;
James Webb Space Telescope (JWST); Salmonellosis in livestock;
Small tree utilization (forestry); Space flight.

1 2 3 4 5 6 7 8 9 0 DOW/DOW 0 9 8 7 6 5 4 3

This book was printed on acid-free paper.

*It was set in Garamond Book and Neue Helvetica Black Condensed by
TechBooks, Fairfax, Virginia. The art was prepared by TechBooks.
The book was printed and bound by RR Donnelley, The Lakeside Press.*

Contents

Editing, Design, and Production Staff

Consulting Editors

Dr. Ralph E. Hoffman. *Associate Professor, Yale Psychiatric Institute, Yale University School of Medicine, New Haven, Connecticut.* PSYCHIATRY.

Dr. Gary C. Hogg. *Chair, Department of Industrial Engineering, Arizona State University.* INDUSTRIAL AND PRODUCTION ENGINEERING.

Prof. Gordon Holloway. *Department of Mechanical Engineering, University of New Brunswick, Canada.* FLUID MECHANICS.

Dr. S. C. Jong. *Senior Staff Scientist and Program Director, Mycology and Protistology Program, American Type Culture Collection, Manassas, Virginia.* MYCOLOGY.

Dr. Peter M. Kareiva. *Director of Conservation and Policy Projects, Environmental Studies Institute, Santa Clara University, Santa Clara, California.* ECOLOGY AND CONSERVATION.

Prof. Gabriel N. Karpouzian. *Aerospace Engineering Department, U.S. Naval Academy, Annapolis, Maryland.* AEROSPACE ENGINEERING AND PROPULSION.

Dr. Bryan A. Kibble. *National Physical Laboratory, Teddington, Middlesex, United Kingdom.* ELECTRICITY AND ELECTROMAGNETISM.

Dr. Arnold G. Kluge. *Division of Reptiles and Amphibians, Museum of Zoology, Ann Arbor, Michigan.* SYSTEMATICS.

Prof. Robert E. Knowlton. *Department of Biological Sciences, George Washington University, Washington, DC.* INVERTEBRATE ZOOLOGY.

Dr. Cynthia Larive. *Department of Chemistry, University of Kansas, Lawrence.* INORGANIC CHEMISTRY.

Prof. Trevor Letcher. *Physical and Theoretical Chemistry Laboratory, Oxford University, United Kingdom.* THERMODYNAMICS AND HEAT.

Prof. Chao-Jun Li. *Department of Chemistry, Tulane University, New Orleans, Louisiana.* ORGANIC CHEMISTRY.

Dr. Donald W. Linzey. *Wytheville Community College, Wytheville, Virginia.* VERTEBRATE ZOOLOGY.

Dr. Philip V. Lopresti. *Retired; formerly, Engineering Research Center, AT&T Bell Laboratories, Princeton, New Jersey.* ELECTRONIC CIRCUITS.

Dr. Stuart MacLeod. *Executive Director, BC Research Institute for Children's and Women's Health, Vice President, Academic Development, Provincial Health Services Authority, and Assistant Dean, Faculty of Medicine, University of British Columbia, Canada.* PHARMACOLOGY, PHARMACY, ANTIBIOTICS.

Prof. Scott M. McLennan. *Chair, Department of Geosciences, State University of New York at Stony Brook.* GEOLOGY (PHYSICAL, HISTORICAL, AND SEDIMENTARY).

Dr. Philip L. Marston. *Department of Physics, Washington State University, Pullman.* ACOUSTICS.

Dr. Ramon A. Mata-Toledo. *Associate Professor of Computer Science, James Madison University, Harrisonburg, Virginia.* COMPUTERS.

Prof. Krzysztof Matyjaszewski. *J. C. Warner Professor of Natural Sciences, Department of Chemistry, Carnegie Mellon University, Pittsburgh, Pennsylvania.* POLYMER SCIENCE AND ENGINEERING.

Dr. Orlando J. Miller. *Center for Molecular Medicine and Genetics, Wayne State University School of Medicine, Detroit, Michigan.* GENETICS AND EVOLUTION.

Prof. Jay M. Pasachoff. *Director, Hopkins Observatory, Williams College, Williamstown, Massachusetts.* ASTRONOMY.

Prof. David J. Pegg. *Department of Physics, University of Tennessee, Knoxville.* ATOMIC, MOLECULAR, AND NUCLEAR PHYSICS.

Prof. J. Jeffrey Peirce. *Department of Civil and Environmental Engineering, Edmund T. Pratt Jr. School of Engineering, Duke University, Durham, North Carolina.* ENVIRONMENTAL ENGINEERING.

Dr. William C. Peters. *Professor Emeritus, Mining and Geological Engineering, University of Arizona, Tucson.* MINING ENGINEERING.

Prof. Arthur N. Popper. *Department of Biology, University of Maryland, College Park.* NEUROSCIENCE.

Dr. Kenneth P. H. Pritzker. *Pathologist-in-Chief and Director, Head, Connective Tissue Research Group, and Professor, Laboratory Medicine and Pathobiology, University of Toronto, Mount Sinai Hospital, Toronto, Ontario, Canada.* MEDICINE AND PATHOLOGY.

Prof. Justin Revenaugh. *Department of Geology and Geophysics, University of Minnesota, Minneapolis.* GEOPHYSICS.

Dr. Roger M. Rowell. *USDA–Forest Service, Forest Products Laboratory, Madison, Wisconsin.* FORESTRY.

Dr. Andrew P. Sage. *Founding Dean Emeritus and First American Bank Professor, University Professor, School of Information Technology and Engineering, George Mason University, Fairfax, Virginia.* CONTROL AND INFORMATION SYSTEMS.

Dr. Steven A. Slack. *Associate Vice President for Agricultural Administration, Director, Ohio Agricultural Research and Development Center, and Associate Dean for Research, College of Food, Agricultural, and Environmental Sciences, Ohio State University, Wooster.* PLANT PATHOLOGY.

Prof. Arthur A. Spector. *Department of Biochemistry, University of Iowa, Iowa City.* BIOCHEMISTRY.

Prof. Anthony P. Stanton. *Carnegie Mellon University, Pittsburgh, Pennsylvania.* GRAPHIC ARTS AND PHOTOGRAPHY.

Dr. Trent Stephens. *Department of Biological Sciences, Idaho State University, Pocatello.* DEVELOPMENTAL BIOLOGY.

Dr. John Timoney. *Department of Veterinary Science, University of Kentucky, Lexington.* VETERINARY MEDICINE.

Dr. Sally E. Walker. *Associate Professor of Geology and Marine Science, University of Georgia, Athens.* INVERTEBRATE PALEONTOLOGY.

Prof. Pao K. Wang. *Department of Atmospheric and Oceanic Sciences, University of Wisconsin–Madison.* METEOROLOGY AND CLIMATOLOGY.

Dr. Nicole Y. Weekes. *Pomona College, Claremont, California.* NEUROPSYCHOLOGY.

Prof. Mary Anne White. *Killam Research Professor in Materials Science, Department of Chemistry, Dalhousie University, Halifax, Nova Scotia, Canada.* MATERIALS SCIENCE AND METALLURGIC ENGINEERING.

Prof. Thomas A. Wikle. *Head, Department of Geography, Oklahoma State University, Stillwater.* PHYSICAL GEOGRAPHY.

Dr. Gary Wnek. *Department of Chemical Engineering, Virginia Commonwealth University, Richmond.* CHEMICAL ENGINEERING.

Dr. James C. Wyant. *University of Arizona Optical Sciences Center, Tucson.* ELECTROMAGNETIC RADIATION AND OPTICS.

Article Titles and Authors

The 2004 *McGraw-Hill Yearbook of Science & Technology* provides a broad overview of important recent developments in science, technology, and engineering as selected by our distinguished board of consulting editors. At the same time, it satisfies the nonspecialist reader's need to stay informed about important trends in research and development that will advance our knowledge in the future in fields ranging from agriculture to zoology and lead to important new practical applications. Readers of the *McGraw-Hill Encyclopedia of Science & Technology*, 9th edition (2002), also will find the *Yearbook* to be a valuable companion publication, supplementing and updating the basic information.

In the 2004 edition, important advances in the biomedical sciences are in the spotlight, with articles on topics such as RNA interference, the predictive genetics of cancer, and tissue engineering, as well as a review of the findings of the Human Genome Project. We continue to document the scientific and technical findings needed for informed decisions in such areas as environmental conservation and global climate policy, for example with articles on biodiversity hotspots and satellite observations of clouds. Noteworthy developments in materials science, engineering, and chemistry are reported, such as self-healing polymers, focused ion beam machining, and advances in green chemistry. Articles on the James Webb Space Telescope and the Wilkinson Microwave Anisotropy Probe describe major space-based projects that will contribute to the fields of astronomy and cosmology. And throughout this volume, we see the contributions made to all fields of science and technology by the application of the computational power that is now readily available.

Each contribution to the *Yearbook* is a concise yet authoritative article prepared by one or more authorities in the field. We are pleased that noted researchers have been supporting the *Yearbook* since its first edition in 1962 by taking time to share their knowledge with our readers. The topics are selected by our consulting editors in conjunction with our editorial staff based on present significance and potential applications. McGraw-Hill strives to make each article as readily understandable as possible for the nonspecialist reader through careful editing and the extensive use of graphics, much of which is prepared specially for the *Yearbook*.

Librarians, students, teachers, the scientific community, journalists and writers, and the general reader continue to find in the *McGraw-Hill Yearbook of Science & Technology* the information they need in order to follow the rapid pace of advances in science and technology and to understand the developments in these fields that will shape the world of the twenty-first century.

Mark D. Licker
PUBLISHER

A–Z

Adaptive middleware

As computer systems are being applied to more aspects of personal and professional life, the quantity and complexity of software systems are increasing considerably. At the same time, the diversity in hardware architectures remains large and is likely to grow with the deployment of embedded systems, personal digital assistants (PDAs), and portable computing devices. All these architectures will coexist with personal computers, workstations, computing servers, and supercomputers. Currently, a major goal of academic and industrial computer scientists is to develop techniques to support seamless integration of computer systems in highly heterogeneous environments. The construction of these new software systems and applications in an easy and reliable way can be achieved only through the extensive revise of software components and services.

A major obstacle in software development is the heterogeneity of computer platforms. Different operating systems, such as UNIX®, MacOS®, and Microsoft Windows®, have different programming interfaces requiring software developers to write different versions of their software for each platform. In addition, operating systems come in a number of "flavors" depending on the computing platform used, needs of users, and so forth. Windows, for example, has CE, NT, 2000, ME, and XP, among others. Each flavor may impose slightly different requirements on the application developer.

To mitigate the problem stated above, during the last 10 years software developers have created middleware technology to facilitate the development of software systems. Middleware mediates the interactions between the application and the operating system. Popular middleware technologies [including the Object Management Group's Common Object Request Broker Architecture (CORBA®), Sun's Java-based J2EE™, and Microsoft®.NET] hide the complicated details of network communication, remote invocation (activation of a remote function or program), naming, and creation of new services, easing the construction of complex distributed systems. CORBA and Java also hide the differences among the underlying software and hardware platforms,

increasing portability and facilitating maintenance as new versions of operating systems are released. A distributed CORBA or J2EE system, for example, can integrate different computers such as a PC running Linux, a PC running Windows, an iMac running MacOS, and a PDA running PalmOS.

Middleware also includes pieces of software that, in the past, were included repeatedly in many applications. These pieces, previously rewritten for each new application, are now encapsulated into the middleware so that they can be reused in various applications without large efforts.

Dynamic environments. Recent advances in distributed, mobile, and ubiquitous systems (for small computers incorporated in everyday objects, such as clothes, cars, and appliances) demand new computing environments that are characterized by a high degree of dynamism. In these environments, machines are added and removed from the system every day, with mobile computers connecting and disconnecting every minute. Variations in disk, memory and battery availability, network connectivity, and hardware and software platforms greatly affect the performance of user applications. The expected growth of ubiquitous and mobile computing in the coming years will further change the nature of the computational infrastructure, bringing a plethora of small devices and requiring customized protocols and policies in order to fulfill the user's evolving quality of service requirements.

While conventional middleware technology aids the development of distributed applications for the new computing environments, it lacks adequate support for managing the dynamic (changeable) aspects of the new computational infrastructure. Next-generation applications will require adaptive middleware, that it, middleware that can be adapted to changes in the environment and customized to fit into devices ranging from PDAs and sensors to powerful desktops and multicomputers.

Transparency versus translucency. Middleware hides the details of the underlying layers and operating-system-specific interfaces, offering a major advantage to software developers. Programmers of distributed applications can write code that looks very similar to code for centralized applications because the middleware extends local calls with networking, marshaling (converting data to a common format to eliminate differences between machine architectures in heterogeneous networks), remote dispatching (choosing which entity will process each call), and scheduling (deciding the order in which the calls received from multiple clients will be processed). Applications written for middleware are easily ported, and the programmer need not worry about either the internals of the operating system or the middleware.

On the other hand, some applications can exploit the underlying layers, both in the computational environment and in the physical environment. For example, a multimedia streaming or videoconferencing application can dramatically improve the quality of service by selecting a network transport protocol that suits the underlying network infrastructure [for example, wireless local-area network (LAN), wired LAN, or long-distance Internet] and the available bandwidth. It may also benefit from being aware of its physical context by detecting, for example, the presence of a wall display and reconfiguring the application to show the video in the larger display. An e-commerce Web site can improve its response time by examining information about resource use and dynamically changing the location of its system components, creating replicas of its most requested services or changing the middleware's request scheduling policies. An effective calendar application for ubiquitous computing may detect in what kind of hardware platform it is executing (for example, PDA, wristwatch, desktop, or wall display) so that it can provide a graphical interface that is optimized for that platform.

In other words, most applications benefit from middleware that hides the details of the underlying layers. Other applications, however, can significantly improve performance through interactions with the dynamic state of the underlying layers that adapt the middleware implementation to the application requirements. Therefore, a desirable model of middleware provides transparency to the applications that want it and translucency and fine-grain control to the applications that need it. This is achieved through reflective middleware.

Reflective middleware model. In the reflective model, the middleware is implemented as a collection of components that can be selected and configured by the application. The middleware interface remains unchanged and may support applications developed for traditional, nonadaptive middleware. In addition, system and application code may inspect the internal configuration of the middleware and, if needed, reconfigure it to adapt to changes in the environment through meta-interfaces. In this manner, it is possible to select networking protocols, security policies, encoding algorithms, and various other mechanisms to optimize system performance for different contexts and situations.

In general, reflective middleware refers to the use of a causally connected self-representation to support the inspection and adaptation of the middleware system. Self-representation means an explicit representation of the internal structure of the middleware implementation that it maintains and manipulates. In this sense, the middleware is self-aware. The self-representation is causally connected if changes in the representation lead to changes in the middleware implementation itself and, conversely, changes in the middleware implementation lead to changes in the representation.

Unlike middleware that is constructed as a monolithic black box, reflective middleware is organized as a group of collaborating components. This organization permits the configuration of very small middleware engines that are able to interoperate with traditional middleware. Conventional middleware implementations often include all of the functions that any of the applications need. However, in most cases a particular application may use only a small subset of this functionality. The current difficulties in deploying standard middleware technologies to the small devices used in ubiquitous computing do not apply to component-based middleware. While conventional CORBA and Java middleware require several megabytes of memory, component-based reflective middleware can have a memory footprint as small as a few kilobytes.

Recent projects. The construction of adaptive middleware using the reflective model has been investigated in a few academic projects. Open ORB and dynamicTAO are reflective implementations of CORBA, while mChaRM is a reflective middleware based on Java. Open ORB and dynamicTAO were developed independently on different sides of the Atlantic by people with different backgrounds using different technologies. Nevertheless, the motivations of both were the same and both projects led to similar solutions based on reflective architectures.

These projects illustrate two opposite approaches for the development of reflective systems in general and, more specifically, adaptive middleware. The development of dynamicTAO started with TAO, a complete implementation of CORBA that was modular but static. The dynamicTAO developers reused tens of thousands of lines of code that were already functional and concentrated on adding reflective features to make the system more flexible, dynamic, and customizable. Conversely, the development of Open ORB started from scratch; its designers had the opportunity to plan its architecture from the earliest stages. Therefore, while dynamicTAO focused on code reuse and on leveraging existing systems, Open ORB focused on a novel middleware architecture where all the elements are consistent with the principles of reflection.

Outlook. In recent years, existing implementations of popular middleware platforms have been incorporating some of the contributions offered by research in adaptive middleware. The CORBA standard now includes mechanisms called portable interceptors, which ease the development of adaptive systems. Commercial middleware products are starting to allow the specification of different policies for security and resource management and to support dynamic loading of new components at runtime. In the coming years, the expectation is an increasing degree of flexibility and adaptability in computer systems based on the foundations provided by adaptive middleware.

For background information *see* CLIENT-SERVER SYSTEM; DISTRIBUTED SYSTEMS (COMPUTERS); LOCAL-AREA NETWORKS; OPERATING SYSTEM; PROGRAMMING LANGUAGES; WIDE-AREA NETWORKS in the McGraw-Hill Encyclopedia of Science & Technology.

Fabio Kon

Bibliography. M. Astley, D. C. Sturman, and G. Agha, Customizable middleware for modular distributed software, *Commun. ACM*, 44(5):99–107, May 2001; F. Kon et al., The case for reflective middleware, *Commun. ACM*, 45(6):33–38, June 2002; P. Maes, Concepts and experiments in computational reflection, in *Proceedings of the ACM Conference on Object-Oriented Programming, Systems, Languages and Applications (OOPSLA'87)*, Orlando, FL, pp. 147–155, October 1987; Object Management Group, *CORBA 3.0.2 Specification*, OMG Doc. formal/2002-12-06, Framingham, MA, December 2002.

Advanced CANDU Reactor (ACR)

More than 400 nuclear power plants currently provide about 16% of the world's electricity. These plants were developed primarily for power production by regulated electric utilities. The advanced nuclear reactors now under development must meet and exceed public expectations for safety, security, and sustainability, while remaining economically competitive in the emerging deregulated electricity markets. In addition, the advanced reactors should continue to contribute to the reduction of greenhouse gases and pollution emissions resulting from electric power production.

The ACR (Advanced CANDU Reactor)™ is an evolutionary design based on more than 50 years of nuclear technology development, project experience, and feedback from CANDU® stations in Asia, Europe, and North and South America, whose combined reactor operation times exceed several hundred years. In the CANDU design, developed by Atomic Energy of Canada Limited, the reactor core consists of fuel channels within a large vessel filled with heavy-water moderator. The ACR is based on the most recently built 700 MWe (megawatts of electric power)-class CANDU 6 reactors, with enhancements developed for the 900 MWe-class CANDU 9 design and other reactor types. The strategy of developing the ACR by incorporating evolutionary advances from previous CANDU and light-water reactors has allowed the ACR to be economical to operate and build, as well as achieving enhanced safety.

Modular reactor core. The ACR design is based on replicating or adapting existing CANDU 6 compo-

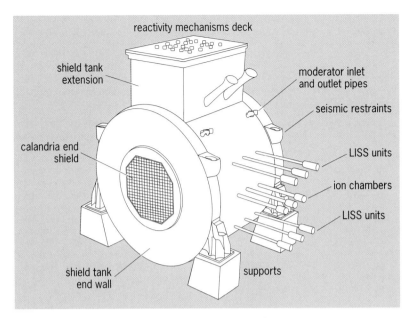

Fig. 1. ACR reactor assembly. LISS = liquid injection shutdown system.

nents for a new core design. Similar to the CANDU 6 design, the ACR reactor assembly (**Fig. 1**) contains a reactor core consisting of a horizontal cylindrical tank called the calandria that is penetrated by horizontal fuel channels that carry the fuel (**Fig. 2**).

The calandria is a low-pressure, low-temperature vessel filled with heavy water. The high-pressure–high-temperature zone of the core is limited to the horizontal fuel channels. The heavy water provides neutron moderation, and all reactivity control devices operate inside this low-temperature, low-pressure environment. This arrangement improves safety by precluding some hypothetical accident scenarios and providing an additional heat sink for protecting the fuel.

The fuel channels are symmetrically spaced in a lattice pattern in the calandria. Each fuel channel consists of a pressure tube that contains the fuel bundles surrounded by a calandria tube. Each fuel bundle weighs about 48 lb (22 kg) and has a diameter of about 4 in. (103 mm) and a length of about 20 in. (495 mm). Both ends of a pressure tube have an end fitting with a closure plug that allows access for automated on-power refueling. The pressurized

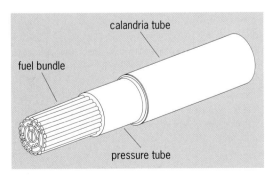

Fig. 2. ACR fuel channel.

light-water coolant flows into one end of the pressure tube, extracting heat from the fuel bundles on its way to the other end of the pressure tube. Using this arrangement, the pressurized light-water coolant is isolated from the surrounding low-pressure, low-temperature heavy-water moderator that fills the calandria.

The calandria sits inside a cylindrical shield tank, the ends of which are covered by round end shields. The tank contains light water, which provides a thermal and radiation shield around the calandria. The shield tank can also cool and retain fuel debris in the unlikely event of a severe accident during which the fuel channels disintegrate, thus providing an additional safety barrier.

The ACR modular calandria design is less expensive to build than a large pressure vessel. In addition, the thermal power output is scalable by changing the number of fuel channels. The modular calandria design is more amenable to life extension, in which only the pressure tubes may need to be replaced, rather than the entire pressure vessel (as is the case in light-water reactors).

Design evolution. The ACR design retains historical CANDU strengths, including (1) a modular reactor core that is easy to fabricate and simplifies scaling between reactor sizes; (2) a proven fuel bundle that is simple, short-length, easily handled, and inexpensive; (3) on-power refueling that contributes to high-capacity factors, optimizes fuel usage, provides versatility in outage scheduling, and reduces the need for excess reactivity in the core; (4) a low-temperature, low-pressure moderator (separated from the reactor coolant system) with back-up heat-sink capability, which has safety advantages in terms of core cooling ability under severe accident conditions; (5) a core designed for relatively low neutron absorption that ensures good uranium utilization and simplifies the use of MOX (mixed oxide) fuels; (6) proven safety systems that include two shutdown systems, an emergency core-cooling system, and containment, plus associated safety support systems; (7) a separate, water-filled shield tank surrounding the reactor that can check severe accident progression by cooling the outside of the calandria and containing any core debris; and (8) a unitized containment based on posttensioned concrete that retains radioactivity while isolating and protecting each reactor from the outside environment and from any other reactors on the site.

The ACR design incorporates innovative steps to achieve both improved economics and enhanced safety. The key innovations and their benefits relative to the CANDU 6 design include the following:

1. The ACR uses slightly enriched uranium (SEU) fuel rather than the natural uranium CANDU 6 fuel. This fuel provides operating benefits, including more efficient use of the uranium, reduced fueling-related wear on the pressure tubes and portions of the fueling machine, and a smaller volume of spent fuel per megawatt hour. It also reduces capital cost by reducing the size and frequency of use of facilities for handling the spent fuel. The incremental cost of SEU compared to natural uranium is low and stable, given the world glut of enrichment services.

2. The ACR core is more compact than that of the CANDU 6 reactor, with a reduced lattice pitch and reduced heavy-water inventory. This, along with the usage of SEU fuel, results in a highly stable core neutron flux that simplifies reactor control, improves reactor efficiency, and simplifies scaling between reactor sizes. It also reduces capital cost by reducing the volume of heavy-water moderator and the number of fuel channels required per megawatt of electric power.

3. The ACR has replaced the heavy-water primary coolant used in the CANDU 6 reactor with light water. This change eliminates many of the heavy-water interfaces in the CANDU 6 design and allows further simplification. The result is a reduction in capital cost that also benefits operating costs, environmental performance, and occupational safety. The annual cost of heavy water makeup is also reduced.

4. The ACR has moved from the positive coolant void coefficient of the CANDU 6 reactor to a negative coolant void coefficient, meaning that reactor power decreases as the amount of boiling (void fraction) increases. This simplifies reactor control and reduces the performance demands placed on the shutdown systems.

5. The pressures and temperatures in the coolant system and steam supply of the ACR are higher than those of the CANDU 6 reactor (although within the range of operating pressurized-water reactors). The steam turbine is therefore smaller and simpler than it would be if operating at CANDU 6 conditions, and the plant achieves lower capital cost.

6. The ACR design permits a much higher degree of modularization than is achieved by the CANDU 6 design. This allows more construction work to be moved from the field to the shop, reducing capital cost and enabling construction time to be significantly reduced. Larger completed modules can be shipped to the construction site, thus reducing the cost of on-site installation.

7. Passive safety is improved by the introduction of a large, elevated reserve water tank that supplies a number of diverse heat sinks in an emergency, including the heat transport system, steam generators, moderator surrounding the core, and shield tank surrounding the moderator. This reduces the likelihood of severe core damage or checks it, and increases the time available to respond to severe accidents.

8. The containment and spent-fuel bays are steel-lined, reducing the likelihood and magnitude of leaks during normal operation or accident conditions.

Safety evolution. While modern reactors have many safeguards, further safety improvements are expected to simplify licensing, reduce operating costs, and increase public acceptance. Safety in the ACR is further enhanced with improvements such as (1) use of a small, negative void coefficient; (2) a more negative power coefficient (change in reactivity per unit change in reactor power) over the operating range; (3) larger thermal margins due to the configuration of the ACR fuel bundle; (4) the ability to contain a pressure-tube failure within the calandria tube; (5) improved heat-sink reliability; (6) interconnections of water and electrical systems between units of a twin-unit plant that allow plants to support each other and hence enhance reliability of the safety support systems; (7) inherent shutdown on single-channel failure (meaning that the breaking of one of the pressure tube–calandria tube assemblies would cause the nuclear reaction within an ACR to immediately cease without any action being taken) due to the use of light-water coolant; (8) steel-lined, dry containment; (9) extended seismic qualification; (10) severe accident prevention and mitigation features; and (11) design insights from the generic CANDU Probabilistic Safety Assessment (PSA) coupled with ACR design–assist PSA.

For background information *see* NUCLEAR FUELS; NUCLEAR POWER; NUCLEAR REACTOR; REACTOR PHYSICS in the McGraw-Hill Encyclopedia of Science & Technology.   Milton Z. Caplan; Jerry Hopwood

Bibliography. Atomic Energy of Canada Limited, *Submission to the UK Energy Policy: Key Issues for Consultation*, September 2002; International Atomic Energy Agency, *Heavy Water Reactors: Status and Projected Development*, Tech. Rep. Series 407, 2002; D. F. Torgerson, The ACR-700—Raising the bar for reactor safety, performance, economics and constructability, *Nucl. News*, October 2002.

Angiogenesis

Blood vessels are composed of two basic cell types, vascular endothelial cells and peri-endothelial cells (including vascular smooth muscle cells and elongated contractile cells called pericytes, both of which support the underlying endothelial cells). The inner epithelial lining of all blood vessels, adjacent to the lumen, is a single layer of endothelial cells (**Fig. 1**). In larger blood vessels, such as arteries and veins, the inner endothelial lining, called the tunica intima, is surrounded by a medial layer, the tunica media, composed of multiple layers of vascular smooth muscle cells embedded in elastin-rich extracellular matrix. The tunica media layer is surrounded by an extracellular matrix-rich layer called the tunica adventitia. In contrast, capillary walls consist of only a single layer of endothelial cells, sometimes surrounded by pericytes.

Arteries and veins are the two fundamental types of blood vessels, carrying blood away from and toward the heart, respectively. Although both have the basic structure noted above, higher-pressure arteries generally have thicker medial smooth muscle-containing layers, whereas larger veins have thinner, more elastic walls and valves to prevent backflow. Recent studies of the process of blood vessel formation, or angiogenesis, have shown that the endothelial cells contributing to arteries and veins have distinct molecular and functional identities that are established early during development.

Blood vessel formation. Blood vessels are found only within the chordates, but within the vertebrates the anatomical architecture of the major blood vessels of the circulatory system is surprisingly well conserved, as highlighted by recent detailed characterization of the anatomy of the developing zebrafish vasculature. The first major vessels to emerge during embryonic development form by coalescence of individual mesodermal endothelial progenitor cells, or angioblasts, into cords of attached cells which then form open vascular cells, or lumenize. This process, called vasculogenesis, was thought to be restricted to early embryonic development; however, recent evidence has shown that vasculogenesis also occurs later in development and even postnatally. Most later

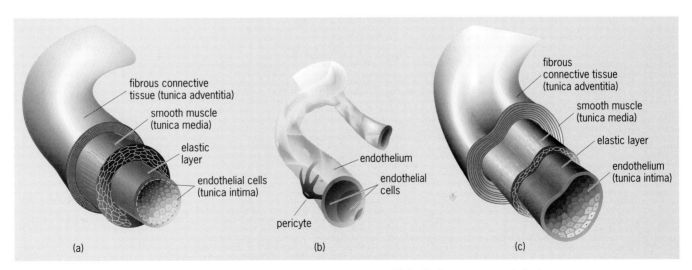

Fig. 1. Structure of blood vessels. (*a*) Arteries and (*c*) veins have the same tissue layers. (*b*) Capillaries are composed of only a single layer of endothelial cells, sometimes surrounded by elongated contractile cells known as pericytes (not to scale). (*Adapted from P. H. Raven and G. B. Johnson, Biology, 6th ed., McGraw-Hill, New York, 2002*)

developmental and postnatal blood vessel formation, however, occurs by sprouting and elongation of new vessels from preexisting vessels or remodeling of preexisting vessels, collectively known as angiogenesis.

Vessels generally form first as unlined endothelial tubes but generally become rapidly associated with supporting pericyte or vascular smooth muscle cells. A variety of studies have shown that the acquisition of these supporting cells is critical for proper morphogenesis, stability, and survival of nascent blood vessels. Although most larger blood vessels form during development and are relatively stable thereafter, significant growth and remodeling of blood vessels continues throughout adult life, for example during uterine cycling or in conjunction with gain or loss of fat or muscle mass.

Postnatal vessel growth also occurs in pathologic contexts, notably cancer. Tumors recruit new blood vessels to provide themselves with access to the host circulation and obtain oxygen and nutrients. Acquiring a vascular blood supply is essential for the progression and survival of a tumor. New tumor vessels form primarily through endothelial proliferation and growth from preexisting, adjacent host blood vessels. These vessels are usually morphologically abnormal, poorly structured, and deficient in vascular smooth muscle cells. Some tumor vascular spaces appear to lack even an endothelial lining and are surrounded only by tumor-derived cells.

Molecular regulators. Recently, interest in factors that might promote or inhibit blood vessel growth and the application of molecular methods to vascular biology and the study of vascular development have led to the discovery of many new genes encoding proteins involved in blood vessel growth and assembly. The vascular endothelial growth factor (VEGF) family of ligands and their tyrosine kinase vascular endothelial growth factor receptors (VEGFRs) expressed on the surface of endothelial cells play critical roles in the differentiation of blood vessels. VEGF-A plays a particularly central role—it is essential for the survival, proliferation, migration, and arterial differentiation of endothelial cells. Targeted disruption or "knockout" of just one of the two copies of the gene encoding this protein in mice causes (heterozygous) embryos to die early in development with a dramatic reduction in endothelial cell number. VEGFR2 is a key endothelial receptor transducing VEGF-A signals, and loss of the gene encoding this protein is also lethal (although only in homozygotes with both functional copies of the gene knocked out).

Other related VEGF ligands and VEGFR receptors have also been uncovered, and these also play important roles in modulating the development of blood vessels. VEGF-C signaling through the VEGFR3 receptor plays an essential role in the formation of lymphatic vessels. Another set of ligands, the angiopoietins, play a critical role in vascular endothelial–vascular smooth muscle cell interaction and blood vessel remodeling. Angiopoietin-1 binds to the en-

dothelially expressed Tie-2 receptor to promote the acquisition of vascular smooth muscle cells and stabilization and maturation of nascent blood vessels. Loss of either angiopoietin-1 or Tie-2 function is lethal during embryogenesis; mice with knockouts of the genes encoding these proteins die early in development with highly enlarged, poorly remodeled blood vessels deficient in vascular smooth muscle cells. Many additional genes have been identified that also play important roles in the specification, differentiation, remodeling, and maturation of blood vessels. It has become clear from the myriad of molecular studies of blood vessel formation conducted to date that regulation of vessel assembly is a highly complex and exquisitely regulated process that we have only just begun to understand.

Clinical importance. Much of the recent explosion of scientific interest in the mechanisms regulating growth and assembly of blood vessels has been driven by potential antiangiogenic (inhibiting vessel development) or proangiogenic (enhancing vessel development) therapeutic applications.

Antiangiogenic therapy. The idea that growth of tumors can be inhibited by targeting the blood vessels that supply them, rather than the tumors themselves (**Fig. 2**), grew out of observations that tumors acquire blood vessels, and that their ability to promote local angiogenesis correlates with progression from relative dormancy to rapid growth and aggressiveness (the term "angiogenic switch" has been coined to describe this transition). A variety of tumor-secreted factors mediate the angiogenic switch,

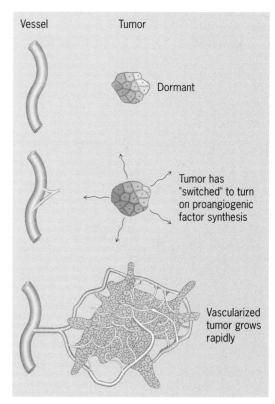

Fig. 2. Tumor angiogenesis and antiangiogenic therapy.

including the factors described above that regulate normal vessel growth and development, notably the critical vascular regulator VEGF. A worldwide search is now on for antiangiogenic compounds that can halt or reverse the growth of tumor blood vessels and thereby stop tumor growth or even cause tumor regression. One focus of attention has been on endogenously produced antiangiogenic factors, many of them normal products of the in-vivo processing of extracellular matrix or plasma proteins. More recently, some of the prominent molecular players described above (particularly VEGF) have been directly targeted for antiangiogenic cancer therapy using various means, including chemical inhibitors, interfering fragments of receptors or ligands, or interfering antibodies. Although human trials of antiangiogenic therapies are mostly in early stages, animal studies and preliminary human trials have yielded encouraging findings, including a lack of general toxicity (unlike chemotherapeutic agents traditionally used for attacking tumor cells directly, for example) and efficacy against many tumors in animal models. Fortuitously, it appears that tumor vessels are differentially sensitive to antiangiogenic therapies compared with normal host blood vessels, probably because the morphologically abnormal tumor vessels deficient in pericyte/smooth muscle cells are less stable than normal host vessels.

Proangiogenic therapy. Treatment of limb or cardiovascular ischemia (insufficient blood supply) using proangiogenic therapy is also an exciting new possibility, promoting regrowth of new vessels into tissues that because of disease and/or injury have become deficient in blood supply. To date, attempts at proangiogenic gene therapy using the genes encoding VEGF and FGFs (fibroblast growth factors) have been mixed.

Animal developmental models. One difficulty in developing effective therapies for either antiangiogenic treatment of cancer or proangiogenic treatment of ischemia is that although we have now identified many important molecular regulators of blood vessel growth and maintenance, we still have a relatively rudimentary understanding of how all of these regulators function together in the complex orchestration of blood vessel morphogenesis. A great deal of the insight into the functional roles of the key vascular signaling molecules as well as the receptors and intracellular pathways that transduce these signals has come from studies of blood vessel formation during development in animal model organisms. Targeted disruption and overexpression experiments in mice have provided definitive assays for the in-vivo functions of vascular genes during development. In most cases these genes have later been shown to play analogous functional roles in pathologic or nonpathologic angiogenesis in adults (neoangiogenesis). Studies in developing avians, *Xenopus* (frogs), and zebrafish have also begun to provide important new insights into vessel specification, differentiation, and assembly during early development.

Zebrafish. The zebrafish shows particular promise as a new model organism for understanding how blood vessels are assembled in vivo. Zebrafish are tropical fish native to Southeast Asia that have a number of advantageous attributes for developmental studies, particularly studies of vascular development. Zebrafish adults are small; fully grown animals are only about an inch long, so large numbers of fish can be housed in a small space. Zebrafish have a relatively short generation time, and adult females lay hundreds of eggs every few weeks, so large numbers of progeny can be obtained for genetic or experimental studies. The eggs are fertilized and develop externally to the mother; thus zebrafish embryos and larvae are readily accessible for noninvasive observation or experimental manipulation at all stages of their development. Early development is very rapid: Blood circulation begins within 24 hours after fertilization, larvae hatch by approximately 2.5 days, and larvae are swimming and feeding by 5–6 days.

Two additional features of zebrafish make them particularly useful for studying vascular development. First, developing zebrafish are very small—a 2-day postfertilization (dpf) embryo is just 2 mm long. The embryos are so small, in fact, that the cells and tissues of the zebrafish receive enough oxygen by passive diffusion to survive and develop in a reasonably normal fashion for the first 3 to 4 days, even in the complete absence of blood circulation. This makes it fairly straightforward to assess the cardiovascular specificity of genetic or experimental defects that affect the circulation. Second, zebrafish embryos and early larvae are virtually transparent. The embryos of zebrafish (and many other teleosts) are telolecithal; that is, yolk is contained in a single large cell separate from the embryo proper. The absence of obscuring yolk proteins gives embryos and larvae a high degree of optical clarity. Genetic variants deficient in pigment cells or pigment formation are even more transparent. This remarkable transparency is probably the most valuable feature of the fish for studying blood vessels in vivo, and has been exploited in a number of different ways.

Confocal microangiography, a technique that permits high-resolution three-dimensional visualization of fluorescent structures, has been used to study functioning blood vessels throughout the zebrafish at every stage of embryonic or early larval development. Transgenic zebrafish expressing green fluorescent proteins in blood vessels allow visualization of vascular endothelial cells and their angioblast progenitors prior to initiation of circulation through a vessel, and even before vessel assembly. A complete atlas of the anatomy of the developing vasculature throughout early development has been prepared using confocal microangiography, the first such atlas to be compiled for any vertebrate. Together these and other tools make it possible to perform in-vivo genetic and experimental dissection of the mechanisms of blood vessel formation in the zebrafish. This facility has already been used to dissect the nature of molecular signals guiding arterial-venous differentiation of endothelial cells during development, and promises to make the fish an important resource for understanding the cues and signals that guide the patterning of growing blood vessels.

For background information *see* ARTERY; BLOOD VESSELS; CANCER (MEDICINE); CAPILLARY (ANATOMY); CIRCULATORY SYSTEM; DEVELOPMENTAL BIOLOGY; TUMOR; VEIN in the McGraw-Hill Encyclopedia of Science & Technology.　　　　Brant M. Weinstein

Bibliography. E. M. Conway, D. Collen, and P. Carmeliet, Molecular mechanisms of blood vessel growth, *Cardiovasc. Res.*, 49:507–521, 2001; N. Ferrara, VEGF and the quest for tumour angiogenesis factors, *Nat. Rev. Cancer*, 2:795–803, 2002; J. M. Isner, Myocardial gene therapy, *Nature*, 415:234–239, 2002; S. Isogai, M. Horiguchi, and B. M. Weinstein, The vascular anatomy of the developing zebrafish: An atlas of embryonic and early larval development, *Dev. Biol.*, 230:278–301, 2001; R. Kerbel and J. Folkman, Clinical translation of angiogenesis inhibitors, *Nat. Rev. Cancer*, 2:727–739, 2002; N. D. Lawson and B. M. Weinstein, Arteries and veins: Making a difference with zebrafish, *Nat. Rev. Genet.*, 3:674–682, 2002; B. Weinstein, Vascular cell biology in vivo: A new piscine paradigm?, *Trends Cell Biol.*, 12:439–445, 2002.

Angiosperm genomes (plant phylogeny)

DNA C-value (the amount of deoxyribonucleic acid contained in the nuclei of unreplicated haploid cells or gametes) varies over several orders of magnitude among both higher plants and animals. Genome size in angiosperms varies about 600-fold from approximately 0.15 picogram of DNA for several species, for example, 0.16 pg [157 megabasepairs (Mbp)] in *Arabidopsis thaliana*, thale cress ($n = 5$ chromosomes) to nearly 90 pg (~88,000 Mbp) in *Fritillaria davisii*, a member of the lily family ($n = 12$ chromosomes). (Part of the range of DNA content in plants is depicted in **Fig. 1**, in which the very small chromosomes of an *Arabidopsis* relative, *Crucihimalaya himalaica*, are juxtaposed with the large chromosomes of loblolly pine, which contain 65 times more DNA.)

The developmental and evolutionary significance of the massive variation in genome size is a major enigma of evolutionary biology. The so-called DNA C-value paradox involves the lack of overall correlation between genome size and phylogenetic advancement, the common occurrence of DNA content variation of greater than two- to threefold among congeneric species (species belonging to the same genus) with the same or similar chromosome numbers, and the fact that only a small amount (often less than 1%) of the nuclear DNA has coding functions. However, the results of 40 years of research involving higher plants indicate that DNA amount may be of developmental and adaptive significance through its biophysical effects on basic cellular features.

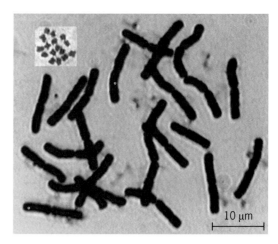

Fig. 1. Chromosomes (upper left inset) of *Crucihimalaya himalaica* (2C DNA content = 0.65 pg, $2n = 16$ chromosomes) juxtaposed with chromosomes at the same magnification of *Pinus taeda* (2C DNA content = 42.54 pg, $2n = 24$ chromosomes). (*Courtesy of G. Hodnett and M. N. Islam-Faridi*)

Developmental and adaptive significance of DNA content. The roles of varying DNA content in plant development and evolution have often been trivialized and dismissed as primarily resulting from the accumulation of junk, or selfish (noncoding), DNA. However, research suggests that DNA may play an important role by influencing plant phenotype, rate of development, and life history strategies through biophysical effects of its mass on nuclei and cells.

Effect of nucleotype on phenotype. DNA content positively correlates with chromosome volume, nuclear volume, cell volume, mitotic cycle time, and duration of meiosis. Michael Bennett has named these biophysical effects of total DNA content the nucleotype. Plant development involves a complex interaction of genetics (genotype) and environment, giving rise to observable properties (phenotype); however, the range of phenotypic expression under genetic control is limited by the plant's nucleotype.

Rate of development. Bennett presented the hypothesis that DNA content is causally correlated with the rate of plant development. He proposed that attributes allowing an annual species to differentiate rapidly in a time-limited environment require low DNA content. Thus, plants with large genomes are restricted to a perennial life form due to the growth-rate-limiting results of slow mitotic cycle time and long duration of meiosis.

Life history strategies. Patterns of DNA content variation exist at global, geographical, and climatic scales. Temperate plant families generally have larger genomes and a greater range of genome sizes compared with families restricted to tropical regions. Plants with small genomes occur in both temperate and tropical families, but large genomes are generally found in species that grow under cool temperatures. Philip Grime and colleagues found that in a British plant community herbaceous species, which undergo their main growth in the cool conditions of early spring, had larger nuclear DNA contents than species with growth occurring in the warmer late spring and summer. They suggested that climatic selection may operate on genome size through a differential effect of temperature upon cell division and cell expansion, such that at low temperature cell expansion is inhibited less than division. This theory holds that under cold conditions growth should favor expansion of larger cells with higher DNA amounts whereas under warmer conditions growth should be dominated by more rapidly dividing cells with lower DNA content.

Correlations also exist between ecological adaptation and genome size. For example, the diploid perennial species of *Microseris* and *Agoseris* (Western North American composite species related to the dandelion) occupy cooler and more mesic (having a moderate water supply) habitats than do the annual species that are adapted to warmer, more xeric (having a low or inadequate water supply), time-limited habitats. The annuals possess genomes that are only 30–60% the size of those of the perennials. A similar pattern is observed among populations of the annual *M. douglasii*, in which DNA amount varies over 20%. The higher DNA values are restricted to plants growing in more mesic sites, generally on well-developed soils. Sites with lower rainfall and/or poorly developed soils dry out more quickly than mesic sites and reduce the time available to complete the plant's life cycle. In these species, DNA amount may be subject to natural selection through its nucleotypic effects on cell size, mitotic cycle time, and the rate of development.

Repetitive DNA and plant genome evolution. The dynamics of genome size evolution are speculative. In current evolutionary theory, genetic differences among related species are viewed as resulting from evolutionary forces, for example, selection and/or genetic drift, acting on existing variability within populations. However, an analogous hypothesis maintaining that accumulation of small deletions or duplications is the primary factor in changing genome size is not supported by the distribution of DNA contents. Genome size commonly ranges severalfold among congeneric species, but well-documented examples of intraspecific variation in genome size are rare; genome size is typically very constant within a species. Since there is apparently a lack of existing variation in DNA content within species on which evolutionary forces can act, a hypothesis invoking events that generate larger discontinuous quantitative changes seems more attractive. Such events, if occurring, may be relatively rare and triggered by hybridization, environmental stress, or genetic events that destabilize the genome.

For the most part, variation in genome size is not caused by reiteration of coding genes, but is due to differences in the amount of repetitive DNA. Repetition of sequences, tandemly arranged and/or dispersed throughout the genome, comprise the bulk of eukaryotic DNA. Clusters of tandemly repeated sequences consisting of short repeats as small as a few basepairs long (microsatellites or simple sequence repeats) are dispersed throughout the genome. Longer tandemly repeated sequences, typically about 150 bp to 400 bp, are often found around centromeres of chromosomes.

Genome growth via retrotransposon accumulation. A mechanism to generate dispersed repetitive DNA sequences involves the movement and accumulation of retrotransposons, mobile DNA sequences that can move from one genomic location to another by producing ribonucleic acid (RNA) that is transcribed by reverse transcriptase into DNA that is then inserted at a new site. Members of a retrotransposon family may be several kilobasepairs long and have copy numbers up to several hundred thousand. The large copy numbers that can be attained by retrotransposons indicate that their amplification is one of the forces leading to the growth of a genome.

Genome reduction via retrotransposon crossing-over. Evidence is accumulating that indicates reductions in genome size have been common evolutionary events. Recombination in homologous regions of loops resulting from pairing of retrotransposons located in the same chromatid may be one mechanism for reducing the size of genomes that have extensively accumulated retrotransposons (**Fig. 2**). Although there is evidence from DNA sequencing that such recombination events involving retrotransposons may occur in plants, it is not known if they occur frequently enough to facilitate rapid evolutionary reductions in genome size.

Genome size and angiosperm phylogeny. It has been proposed that plants may have a "one-way ticket to genomic obesity" as a consequence of retrotransposon accumulation and polyploidy (having one or more extra sets of chromosomes). Initial evaluations of the genomic obesity hypothesis have recently been conducted by superimposing estimated ancestral genome sizes onto well-established phylogenies (branching diagrams of evolutionary history) of cotton (*Gossypium*) and its closely related genera in the cotton tribe (Gossypieae) and of the angiosperms to determine whether genome size only increases over the course of evolution or whether both increases and decreases occur.

In the cotton tribe, DNA content among diploid species varies over about a sevenfold range. Based on comparison of the well-supported phylogeny for cotton and its allies with the pattern of evolutionary history generated from an analysis of DNA content of living species and estimated ancestral genome sizes, it was inferred that both increases and decreases in genome size occurred during the evolution of this tribe. It was found that the frequency of reductions actually exceeds that of increases.

Recent advances in plant taxonomy by participants of the Angiosperm Phylogeny Group using molecular phylogenetic data, including DNA sequences, along with nonmolecular phylogenetic data, have resulted in a revised classification of the angiosperms that reflects evolutionary relationships. This phylogenetic framework and the availability of a list (currently ~3500) of angiosperm genome sizes

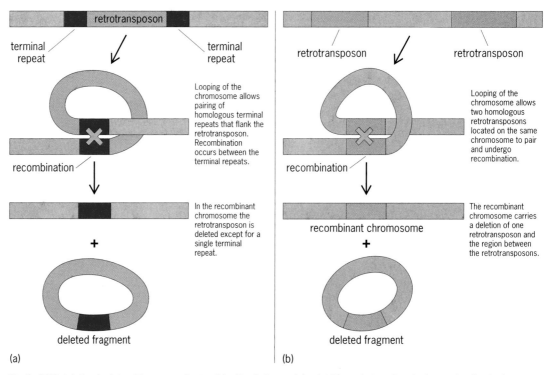

Fig. 2. DNA deletion by intrachromosomal recombination between (a) paired homologous terminal repeats of a single retrotransposon and (b) paired retrotransposons located on the same chromosome.

in the Plant DNA C-value Database are leading to analyses of evolution of genome size in relationship to angiosperm evolution. From initial broadly based analyses, it appears that the ancestral genome size for angiosperms was relatively small (≤ 1.4 pg), with the trend for the growth of genomes during the divergence of angiosperms. However, apparent decreases in genome size accompanying angiosperm phylogeny were also detected. Large genomes (≥ 35 pg) in monocots and eudicots are confined to taxa occupying derived positions within larger clades. As additional genome sizes are added to the Plant DNA C-value Database, a more detailed picture of angiosperm genome size evolution will certainly be produced. It appears that a dynamic model that includes both increases and decreases for genome size evolution in plants is emerging.

For background information *see* CROSSING-OVER (GENETICS); DEOXYRIBONUCLEIC ACID (DNA); FLOWER; GENE; MAGNOLIOPHYTA; NUCLEIC ACID; ORGANIC EVOLUTION; PLANT EVOLUTION; PLANT PHYLOGENY; TRANSPOSONS in the McGraw-Hill Encyclopedia of Science & Technology.   H. James Price

Bibliography. M. D. Bennett, Variation in genomic form in plants and its ecological implications, *New Phytol.*, 106(suppl.):177–200, 1987; J. L. Bennetzen and E. A. Kellogg, Do plants have a one way ticket to genomic obesity?, *Plant Cell*, 9:1509–1514, 1997; J. P. Grime, J. M. L. Shacklock, and S. R. Band, Nuclear DNA contents, shoot phenology and species co-existence in a limestone grassland community, *New Phytologist*, 100:435–445, 1985; S. K. Kubis, T. Schmidt, and J. S. Heslop-Harrison, Repetitive DNA elements as a major component of plant genomes, *Ann. Bot.*, 82(Suppl. A):45–55, 1998; I. J. Leitch, M. W. Chase, and M. D. Bennett, Phylogenetic analysis of DNA C-values provides evidence for a small ancestral genome size in flowering plants, *Ann. Bot.*, 82(Suppl. A):85–94, 1998; H. J. Price, DNA content variation among higher plants, *Ann. Missouri Bot. Garden*, 75:1248–1257, 1988; D. E. Soltis et al., Evolution of genome size in the angiosperms, *Amer. J. Bot.*, in press; W. F. Wendel et al., Feast and famine in plant genomes, *Genetica*, 115:37–47, 2002.

Angiotensin receptor blockers

The renin-angiotensin system plays an important role in the regulation of blood volume and is crucial in the pathogenesis of hypertension in many patients, contributing to morbidity and mortality.

Renin-angiotensin system. Renin is an enzyme released by the kidney, both locally within the kidney and into the systemic circulation. Renin acts on a substrate, angiotensinogen, synthesized both in the liver, for the systemic circulation, and locally in a number of tissues, to produce angiotensin I (AngI), a decapeptide (protein fragment consisting of 10 amino acids) with little biological activity. AngI, in turn, is converted to angiotensin II (AngII), an octapeptide (protein fragment consisting of 8 amino acids), via two pathways. Angiotensin-converting-enzyme (ACE) is the sole route for conversion of AngI to AngII in the circulation. At the tissue level, however, there is a second, non-ACE pathway for AngII generation that depends largely on another enzyme, chymase (see **illus.**).

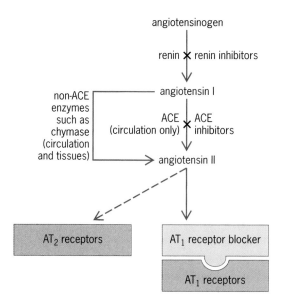

Renin-angiotensin system and possible blockades.

AngII receptors. AngII influences the tissues via two receptors that show only about 30% structural homology (similarity). The AngII receptors are divided into AT_1 and AT_2 subtypes.

AT_1 receptor. All of the clinically important, currently well-defined actions of angiotensin are mediated via the AT_1 receptor. The AT_1 receptor is responsible for all of the known actions of AngII that increase blood pressure: blood vessel constriction, aldosterone release (which increases blood pressure by promoting sodium and water retention), and activation of the sympathetic nervous system.

AT_2 receptor. Animal models have suggested that a second receptor, the AT_2 receptor, appears to have a range of actions that are the opposite of those mediated by the AT_1 receptor, including vasodilatation (an increase in blood vessel diameter), natriuresis (abnormal amounts of sodium in the urine), and growth inhibition. The AT_2 receptor is prominent during embryogenesis but becomes rapidly and progressively more sparse after birth. The role played by the AT_2 receptor in humans is speculative, as there are no direct data available from studies in humans.

Blockade options. The renin-angiotensin system can be blocked at three sites. Renin inhibitors block the interaction of renin with its substrate, angiotensinogen. ACE inhibitors block conversion of AngI to angiotensin II by competing with AngI for binding to the ACE active site. Finally, AT_1 receptor blockers can compete at the receptor site for AngII. All three classes of agents exist, and in each case have reached the clinical level.

When pharmacological interruption of the renin-angiotensin system first emerged, even the wildest enthusiast could not have predicted its success. ACE inhibitors were thought to hold promise only in hypertension associated with elevated plasma renin activity. Although this is important, since it includes patients with hypertension that is accelerated or difficult to treat, hypertension of this etiology is un-

common; thus the drugs were perceived to occupy an important but small niche. However, that niche continues to grow. A series of studies over the past two decades have identified a much broader range of efficacy. These drugs were remarkably helpful in patients with advanced heart failure, those with a large anterior myocardial infarction, and those at risk of diabetic nephropathy (kidney disease), other forms of nephropathy, and the consequences of atherosclerosis (hardening of the arteries). ACE inhibitors have gradually emerged, therefore, as the leading drug class for the treatment of not only patients with these complicated conditions but also patients with mild-to-moderate essential hypertension (cause unknown) who are free of any identifiable target organ damage.

Although there are many claims that ACE inhibitors differ in functionally important ways, such as tissue penetration and action on endothelial function, the available evidence for such differences is not persuasive, and there is no clear evidence that these differences have therapeutic implications. There are quantitatively important differences in duration of action and in metabolism, which have implications for use in patients with renal failure. However, functional differences in angiotensin receptor blocker (ARB) class may be more important.

ACE inhibition was not the product of a planned pharmacological approach, but an accidental byproduct of snake venom toxicology. A pharmacologist examining the renin cascade would first have chosen either the renin-angiotensinogen interaction or the AngII receptor site for pharmacologic intervention. In principle, blockade of a system is most effective at the rate-limiting step. In the renin cascade, the rate-limiting step is the interaction between renin and its substrate, angiotensinogen. Renin inhibitors were developed and were efficacious, but poor bioavailability and the high cost of synthesis stopped development programs. Blockade at the level of the tissue AngII receptor is also more attractive, because it inhibits both ACE-dependent and non-ACE-dependent pathways.

Angiotensin receptor blockers. The angiotensin receptor blockers (ARBs) are imidazoles, compounds which have a wide range of pharmacological activity. For this reason, no one could have anticipated how remarkably well tolerated ARBs are. In fact, they compare favorably with placebo in associated frequency of adverse reactions.

Structural differences. The first antagonists to AngII and its receptor were peptide analogs (structurally similar compounds) of angiotensin. The most widely studied, saralasin, is an octapeptide that differs from AngII in two amino acids. However, since saralasin could only be given intravenously, was a partial agonist, and was expensive to manufacture, it was abandoned. In 1982, high-throughput screening identified imidazole derivatives that bind specifically to the AngII receptor. Structure action work led to the development of losartan, and further structural modifications led to the identification of valsartan, irbesartan, candesartan, olmesartan, and telmisartan, all of which are biphenyl tetrazoles. Investigators pursued an alternative pathway to produce eprosartan, which is a nonbiphenyl, nontetrazole.

Functional differences. Eprosartan differs from the biphenyl tetrazoles in its interaction with the AT_1 receptor in that eprosartan is a pure competitive (or surmountable) antagonist (that is, eprosartan competes with AngII for binding to the active site of the AT_1 receptor; thus, increasing concentrations of AngII can overcome, or surmount, its inhibitory effects). Valsartan, irbesartan, candesartan, telmisartan, olmesartan, and the active metabolite of losartan, EXP 3174, all show noncompetitive kinetics with a nonequilibrium relation to the receptor (that is, the binding of these drugs to the AT_1 receptor is so tight that it is not overcome by an increasing concentration of AngII). One clinically relevant byproduct of this tight binding is a very long duration of action of the nonequilibrium ARBs. The half-life in plasma predicts poorly the duration of the blockade because of the long sojourn of the blocker on the receptor. The available AngII receptor antagonists also differ from one another in their oral bioavailability, metabolism, and elimination. Candesartan cilexitil is a true prodrug (needs to be activated), the active agent being candesartan, which is freed from the complex during passage through the gut wall. Bioavailability ranges from about 25% for valsartan to 80% for irbesartan. The absorption of candesartan or irbesartan is not influenced by food, whereas valsartan and losartan do show interference with absorption by the food. With the exception of candesartan, which has a 60% renal route of elimination, for most of the agents the primary excretion is biliary. Only one of the agents, losartan, has a tubular action that leads to increased uric acid excretion.

Clinical use. All of the ARBs listed earlier are currently approved for hypertension management. They have achieved rather wide use in part because they are extraordinarily well tolerated. In comparison with the ACE inhibitors, they are free of the side effect of cough, which is so common with use of these drugs (up to 20% of recipients). Second, there is accumulating evidence that ARBs share with ACE inhibitors a reduced morbidity and mortality. Two agents, irbesartan and losartan, have been approved for the management of the patient with type 2 diabetes mellitus and nephropathy. Losartan has also received approval for management of the patient with high-risk hypertension associated with left ventricular hypertrophy. Valsartan has been approved for the management of the patient with congestive heart failure. A number of ongoing studies are likely to further expand the indications for the use of these agents and the list of agents that enjoy a specific indication.

For background information *see* ENDOCRINE MECHANISMS; ENZYME INHIBITION; HEART DISORDERS; HYPERTENSION; THIRST AND SODIUM APPETITE; URINARY SYSTEM in the McGraw-Hill Encyclopedia of Science & Technology.　Norman K. Hollenberg

Bibliography. M. Burnier and H. R. Brunner, Angiotensin II receptor antagonists, *Lancet*, 355: 637–645, 2000; M. Epstein, Angiotensin II receptor antagonists: Current status, in M. Epstein and H. R. Brunner (eds.), *Angiotensin II Receptor Antagonists*, pp. 257–262, Hanley & Belfus, Philadelphia, 2001; N. K. Hollenberg, N. D. L. Fisher, and D. A. Price, Pathways for angiotensin II generation in intact human tissue: Evidence from comparative pharmacological interruption of the renin system, *Hypertension*, 32:387–392, 1998; T. H. Hostetter, Prevention of end-stage renal disease due to type 2 diabetes, *N. Engl. J. Med.*, 345:910–912, 2001; P. Timmermans, Development of nonpeptidic angiotensin II receptor antagonists, in M. Epstein and H. R. Brunner (eds.), *Angiotensin II Receptor Antagonists*, pp. 89–103, Hanley & Belfus, Philadelphia, 2001; G. Vauquelin, F. Fierens, and D. Vanderheyden, Distinction between surmountable and insurmountable angiotensin II AT_1 receptor antagonists, in M. Epstein and H. R. Brunner (eds.), *Angiotensin II Receptor Antagonists*, pp. 105–118, Hanley & Belfus, Philadelphia, 2001; R. R. Wexler et al., Nonpeptide angiotensin II receptor antagonists: The next generation in antihypertensive therapy, *J. Med. Chem.* 39:626–656, 1996.

Antenatal drug risk

Significant numbers of women of child-bearing age suffer from depression. A recent prevalence study in which 3472 pregnant women were screened for depressive symptoms using the Centre for Epidemiologic Studies depression scale found that 20% of the women surveyed scored above the cut-off score for depressive symptoms. This information and the knowledge that at least 50% of pregnancies are unplanned, suggest that a relatively large number of women will use an antidepressant during pregnancy. The prevalence during the organogenesis (organ formation) stage may be especially high, because women are often not yet aware of the pregnancy at this early stage and, therefore, have not made the decision to stop medications. It has been shown that due to a lack of information or misinformation regarding the teratogenicity (ability to cause birth defects) of these drugs, a number of women elect to abruptly discontinue needed antidepressants after learning that they are pregnant.

Maternal depression during pregnancy is more likely to be neglected than postnatal depression, even though depression during pregnancy is more prevalent and approximately 25% of postpartum depressions actually begin during pregnancy. The American Psychiatric Association has identified treatment of major depression during pregnancy as a priority issue in clinical management. Based on this recommendation, a position paper was developed describing the risk-benefit decision-making process for treatment of depression during pregnancy. The authors concluded that there is no evidence of antidepressant harm to an unborn baby, and recommended that depression should be treated as long as the benefits and possible risks are fully explained to the patient.

There is also evidence that untreated depression during pregnancy can have deleterious effects on peripartum and neonatal outcomes, including association with higher rates of cesarean section and admission to neonatal intensive care units. During pregnancy, a woman who is depressed may engage in risky behaviors, such as smoking, drinking alcohol, and not attending clinical appointments. Furthermore, it is known that a woman who is depressed may have difficulty bonding with her child after birth. Given the findings of these studies, and with the expectation that the number of women who will be treated pharmacologically for depression during pregnancy will increase, it is of the utmost importance that the fetal safety of all antidepressants be studied so that an antidepressant that is both effective for the mother and safe for the baby can be chosen.

Tricyclic antidepressants. Tricyclic antidepressants have been on the market since the late 1960s and were used by a significant number of women of childbearing age. With the advent of the selective serotonin reuptake inhibitors (SSRIs), tricyclic antidepressants are now rarely prescribed for depression. However, they continue to be prescribed for conditions other than depression, such as chronic pain syndromes, so it is conceivable that some pregnant women will still be exposed to these drugs. A meta-analysis published in 1997 encompassed all studies on tricyclic use in pregnancy for information about teratogenicity, neonatal syndromes, and postnatal behavioral development. No increased risk was identified.

SSRIs. Since 1993, there have been a substantial number of studies documenting the relative safety of the SSRI antidepressants.

Fluoxetine. Fluoxetine (Prozac®) has been studied most extensively, as it has been on the market for more than 20 years. There are now data on more than 2000 women who have used this drug in pregnancy, and there is no evidence of an increased risk for major malformations. Examination of data for postnatal complications in babies exposed during pregnancy showed only a few case reports of neonatal withdrawal syndrome. Neurodevelopmental studies of both first-trimester use and of exposure throughout pregnancy have not found any differences between babies whose mothers took fluoxetine during pregnancy and unexposed control groups. However, the relatively small numbers of children studied and the fact that followup was only to age 4 limit the generalizability of these findings.

Other studies have examined the transfer of fluoxetine across the placenta and have found lower concentrations of fluoxetine and its metabolites in babies than in the mothers. In addition, effects of maternal fluoxetine on uterine blood flow and growth were only transient.

Newer SSRIs. Much less is known about the safety of the newer SSRIs. Among these, the highest number of exposures documented in the literature is for citalopram (Celexa®). A database study recorded 365 first-trimester exposures and found no increase above the baseline rate for major malformations. Information on a total of 291 exposures to paroxetine (Paxil®) in the first trimester also showed no increased risk for major malformations. However, paroxetine appears to have the highest risk of all the SSRIs for neonatal complications. The results of a recently published study documenting neonatal complications in 55 newborn babies whose mothers took paroxetine in late pregnancy showed that 12 had complications necessitating intensive care treatment, of which the most prevalent was respiratory distress. Fortunately, the symptoms resolved within 1–2 weeks, and there were no residual adverse effects. Women who take paroxetine in late pregnancy should be advised of these possible complications and their babies watched closely postdelivery. For sertraline (Zoloft®), information on 259 exposures in the first trimester also showed no increased risk for major malformations above the baseline. Of the SSRIs, there is the least amount of safety information available for fluvoxamine (Luvox®). However, two separate reports totaling 92 cases of exposure in early pregnancy did not find an increased risk for major malformations.

Phenethylamine bicyclic derivatives. The safety of the newer, non-SSRI antidepressants during pregnancy is relatively unstudied. There is only one published study of the safety of venlafaxine (Effexor®), a phenethylamine bicyclic derivative that is chemically unrelated to other antidepressants. A prospective study of 150 women who had been exposed to venlafaxine in the first trimester found no increase in the risk of major malformations even in those who took the drug throughout pregnancy.

Phenylpiperazines. There was also no increase in the rates of major malformations found in the only published study examining the fetal safety of trazodone (Desyrel®) and nefazodone (Serzone®), phenylpiperazine antidepressants also structurally unrelated to other antidepressants. This study followed prospectively 147 women that were exposed during the first trimester (58 to trazodone and 89 to nefazodone), 52 (35%) of whom used these drugs throughout pregnancy).

Aminoketones. Bupropion is an antidepressant of the aminoketone class that is indicated for two conditions: depression (Wellbutrin®) and smoking cessation (Zyban®). To date, there are no published studies on human pregnancies. However, the manufacturer's registry documenting the outcome of 266 pregnancies during which women were exposed in the first trimester did not show an increased risk for major malformations. The study, however, lacked an appropriate control group.

Tetracyclics. Mirtazepine (Remeron®) is a tetracyclic antidepressant that enhances the release of the neurotransmitters norepinephrine and serotonin but also blocks two specific serotonin receptors (to decrease serotonin-associated side effects). There is currently no information on the safety of this antidepressant in human pregnancies; however, there were no adverse reproductive effects in animal studies, nor have there been reports of adverse postnatal effects in babies of mothers that took this drug during pregnancy.

Herbal antidepressants. St. John's wort (*Hypericum perforatum*) is a herbal antidepressant available without a prescription for the treatment of mild-to-moderate depression. Animal studies found no adverse reproductive effects; however, there is no information on its safety during human pregnancy.

Spontaneous abortions. An increase in the rates of spontaneous abortion has been reported in several studies. In a fluoxetine study, the rates of spontaneous abortions were 13.5% in the exposed group and 12% in the tricyclic group versus 7% in the general unexposed population group. With the newer SSRIs, the rates of the exposed group compared with the general population group were 12% versus 7%. More recent antidepressant studies on phenethylamine bicyclic derivatives and phenylpiperazines found spontaneous abortion rates of 12% (venlafaxine) and 13.4% (trazodone/nefazodone) in the antidepressant groups versus 7% and 8%, respectively, in the general population group. Because of the relatively small sample sizes in all of these studies, the results were not statistically significant. However, a published report of the fluoxetine pregnancy registry consisting of 796 spontaneous reports of pregnancy outcomes showed 110 (13.8%) spontaneous abortions. Taken together, these findings raise the possibility of a causative association between depression and an increase in the rates of spontaneous abortion.

Conclusion. Most evidence-based information suggests that antidepressants as a drug class do not appear to be teratogenic or cause any other significant adverse effects to a baby exposed in utero. This information can be used by women and their physicans to ensure that the benefits and risks of treating or not treating depression during pregnancy are carefully weighed. Pregnant women with depression should be treated appropriately to ensure that the mother is in optimal mental health to interact and bond with her baby after birth and beyond.

For background information *see* AFFECTIVE DISORDERS; NORADRENERGIC SYSTEM; PREGNANCY; PSYCHOPHARMACOLOGY; SEROTONIN in the McGraw-Hill Encyclopedia of Science & Technology.

Adrienne Einarson; Irena Nulman; Gideon Koren

Bibliography. A. Addis and G. Koren, Safety of fluoxetine during the first trimester of pregnancy: A meta-analytical review of epidemiological studies, *Psychol. Med.*, 30(1):89–94, 2000; L. L. Altshuler et al., Pharmacological management of psychiatric illness during pregnancy: Dilemmas and guidelines, *Amer. J. Psychiat.*, 154(5):718–719, 1997; A. M. Costei et al., Perinatal outcome following third trimester exposure to paroxetine, *Arch. Pediatr. Adolesc. Med.*, 156(11):1129–1132, 2002; A. Einarson et al., A multicentre prospective controlled study to determine the safety of trazodone and nefazodone use during pregnancy, *Can. J. Psychiat.*, 48(2):106–110, 2003; A. Einarson et al., Pregnancy outcome following gestational exposure to venlafaxine: A multicenter prospective controlled study, *Amer. J. Psychiat.*, 158:1728–1730, 2001; A. Einarson, P. Selby, and G. Koren, Abrupt discontinuation of psychotropic drugs due to fears of teratogenic risk and the impact of counseling, *J. Psychiat. Neurosci.*, 26(1):44–48, 2001; N. Kulin et al., Pregnancy outcome following maternal use of the new serotonin reuptake inhibitors: A prospective multicentre study, *JAMA*, 279:609–610, 1998; S. M. Marcus et al., Depressive symptoms among pregnant women screened in obstetric settings, *J. Women's Health*, 12(4):373–380, 2003; I. Nulman et al., Child development following exposure to tricyclic antidepressants or fluoxetine throughout fetal life: A prospective, controlled study, *Amer. J. Psychiat.*, 159(11):1889–1895, 2002.

Atmospheric water vapor

Of all the variable gases (as opposed to permanent gases such as oxygen and nitrogen) in the atmosphere, water vapor is perhaps the most important. It not only forms clouds and precipitation (the most visible weather) but also contributes to more than 80% of the greenhouse effect that warms the Earth's atmosphere. In fact, the impact of water vapor on the thermal structure of the atmosphere greatly exceeds that of carbon dioxide. Carbon dioxide is suspected to cause global warming mainly because its concentration is known to be increasing, while it is generally assumed that the water vapor concentration in the atmosphere is in a steady state and probably has no net effect on the atmospheric temperature. In reality, the effect of water vapor is not well understood, and current measurements are not conclusive enough to determine whether the steady-state assumption is valid or not. There is at least one observation indicating that midlatitude lower stratospheric water vapor concentration in the Northern Hemisphere has increased about 50% in the last three decades.

Even if its total concentration remains constant, water vapor can still cause changes in the thermodynamic and dynamic structure of the atmosphere if its distribution changes. This means that it can cause climatic change. One possible path for this occurrence is a change in the vertical distribution such that more water vapor goes from the troposphere [extending from the Earth's surface to a height of 10–16 km (6–10 mi), to the base of the stratosphere] into the stratosphere [extending to a height of nearly 50 km (30 mi)]. Because water vapor absorbs infrared radiation strongly, the stratospheric thermal structure may change and cause a dynamic or thermodynamic response in the troposphere. Recent studies show such a possibility.

Aside from a direct radiative-thermal effect, water vapor serves as the substrate for making important radicals, such as hydroxyl (OH·) and hydrogen peroxyl (HO$_2$·), that play a key role in stratospheric chemistry in general, and ozone concentration in particular. Since ozone is an absorber of solar ultraviolet (UV) radiation, heating of the upper stratosphere, due to a change in ozone concentration, may also cause changes in the stratospheric thermal structure. Again, tropospheric climatic change may follow as a response to that.

All these issues point to the urgency of monitoring the global water vapor budget, which requires knowledge of how water is transported throughout the atmosphere, particularly between the stratosphere and troposphere.

Climatological scheme. The current view of the large-scale stratosphere-troposphere exchange of water vapor is that water vapor is pumped up from the troposphere to the lower stratosphere by tropical cumulonimbus clouds. These are very tall thunderclouds that stretch almost from the ground up to the tropopause (top of the troposphere). Vigorous updrafts in these clouds presumably carry water vapor and ice crystals to the lower stratosphere. There the ice crystals evaporate, adding to the total water vapor concentration. How this upward transport works is not entirely clear.

The water vapor transported to the tropical lower stratosphere is pumped toward the middle latitudes by a suction effect created by eddies in the extratropical (that is, middle and higher latitudes) stratosphere and mesosphere, which is sometimes called the extratropical pump. The most important eddy motion that creates the suction is the breaking of large-scale atmospheric waves called Rossby or planetary waves.

Eventually the water vapor that reaches the extratropical lower stratosphere is transported downward into the troposphere, forming a global loop. The mechanisms involved in this downward transport process are also unclear, but tropopause folding (a deformation phenomenon of the tropopause) and isentropic (constant entropy) mixing with tropospheric air are possible explanations.

Anvil-top plumes. Aside from the many uncertainties, some recent discoveries about the stratospheric-tropospheric exchange of water vapor cannot be explained by the above scheme alone. One discovery is the asymmetry between the water vapor concentration in the Northern and Southern Hemispheric lower stratosphere and the seasonal variation of water vapor in the lower stratosphere in both hemispheres. Another is the anvil-top plume phenomenon observed above some severe thunderstorms in midlatitudes that has been identified as an unambiguous mechanism of transporting water vapor through the tropopause.

M. Setvak and C. A. Doswell were the first to use visible and IR (infrared) satellite images to identify clearly that there are cloud plumes above the anvils of some thunderstorms. V. Levizzani and Setvak fur-

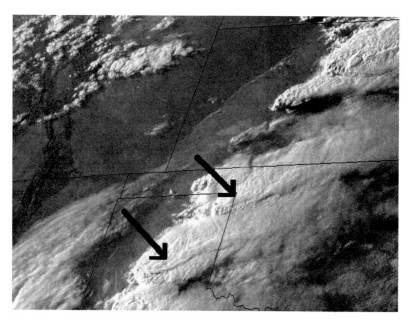

Fig. 1. Example of anvil-top plumes. The plumes on top of two thunderstorm cells (shown by arrows) over northern Texas are visible. The image in the visible channel was taken by the NOAA *GOES-8* satellite on May 6, 2001.

ther analyzed them and identified some commonly observed characteristics. The most common anvil-top plume characteristics are (1) they originate from pointlike sources usually with dimensions of a few kilometers; (2) they seem to be blown downstream by upper-level winds and form elongated chimney-like plumes; and (3) they are located above the anvils. In one of the clearer plume observations, they were able to estimate that the plume was about 3 km (1.9 mi) above the anvil. Since the thunderstorm associated with the plume was a fairly intense one and the anvil was most likely at the tropopause, the plume was about 3 km (1.9 mi) above the tropopause. In other words, it is well in the stratosphere. **Figure 1** shows an example of such plumes.

The mechanism that produces these plumes and the sources of water vapor for them were recently determined by using a three-dimensional cloud dynamic model with explicit cloud microphysics for simulations of severe thunderstorms that had occurred in the midwestern United States. The simulation results show plume formation with plume characteristics matching well with the observations summarized by Levizzani and Setvak. Analysis of the simulation results showed that the source of water vapor for the plume formation came from the thundercloud. Specifically, it came from the shell of the storm's overshooting dome, a region where strong mixing of stratospheric and tropospheric air occurs.

The mechanism by which water vapor detaches from the dome shell and enters the stratosphere is gravity-wave breaking. Gravity waves are caused by strong updrafts in the core of the thunderstorm, and the stability of the stratosphere serves as the restoring factor that results in the wave motion. These waves are called the internal gravity waves because their main energies propagate vertically. Such waves

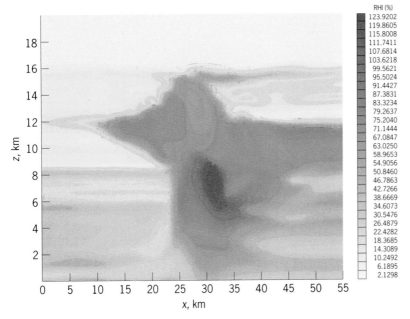

Fig. 2. Central cross section of the relative humidity (RHI) profile (with respect to ice) of the simulated Cooperative Convective Precipitation Experiment (CCOPE) supercell storm at $t = 120$ min. The actual storm event occurred in Montana on August 2, 1981. The chimney plume feature at about 15 km (9 mi) altitude (z) is clearly visible, as is its cross-sectional extent (x).

can be seen in satellite visible images and cloud model simulations. In normal wave propagation where the motion is adiabatic, no net mass transfer occurs. When the wave amplitudes become large enough or the wave speed is the same as local wind speed, wave breaking may occur (much like the breaking of surface waves over the ocean). When this occurs, materials can be transferred through the surface. This occurs for strong thunderstorms, and simulation results show that moisture becomes detached from the cloud and enters the stratosphere (**Figs. 2** and **3**). The upper-level winds play the role of carrying the vapor downstream, producing the chimney-plume-like feature.

Implications. The anvil-top plume phenomenon implies that water vapor can be transported through

the tropopause via the gravity-wave breaking mechanism at the cloud top of thunderstorms. If all storms behaved as the one simulated by P. K. Wang and if we assume that there are 2000 active storm cells globally at any time, then the net upward transport of water vapor by this mechanism would be about 500 megatons per day, a very significant source of water vapor in the lower stratosphere. Obviously not all storms behave the same and the behavior of a particular storm changes with time, so the above number only serves to illustrate the importance of this mechanism and the precise figure needs to be worked out by further research.

If water vapor can penetrate the tropopause by the plume mechanism, other chemicals (trace gases and particulates) less susceptible than water vapor to the freeze-drying effect of the cold tropopause environment can certainly be transported to the stratosphere. No conclusive observations have been made for this possibility, although model simulations similar to that by Wang have been performed and the results are positive.

For background information *see* ATMOSPHERE; CLIMATE MODIFICATION; CLOUD; CLOUD PHYSICS; DYNAMIC METEOROLOGY; ISENTROPIC SURFACES; MESOSPHERE; STRATOSPHERE; THUNDERSTORM; TROPOPAUSE; TROPOSPHERE in the McGraw-Hill Encyclopedia of Science & Technology. Pao K. Wang

Bibliography. J. R. Holton et al., Stratospheric-tropospheric exchange, *Rev. Geophys.*, 33:403–439, 1995; V. Levizzani and M. Setvak, Multispectral, high resolution satellite observations of plumes on top of convective storms, *J. Atm. Sci.*, 53:361–369, 1996; S. J. Oltmans et al., The increase in stratospheric water vapor from balloon-borne, frostpoint hygrometer measurements at Washington, D. C., and Boulder, Colorado, *Geophys. Res. Lett.*, 27:3453–3456, 2000; L. Pan et al., Hemispheric symmetries and seasonal variations of the lowermost stratospheric water vapor and ozone derived from SAGE II data, *J. Geophys. Res.*, 102:28177–28184, 1997; M. Setvak and C. A. Doswell III, The AVHRR channel 3 cloud top reflectivity of convective storms, *Mon. Weath. Rev.*, 119:841–847, 1991; P. K. Wang, Moisture plumes above thunderstorm anvils and their contributions to cross tropopause transport of water vapor in midlatitudes, *J. Geophys. Res.*, 108(D6):4194, doi:10.1029/2002JD002581, 2003.

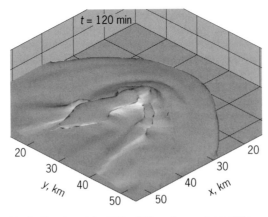

Fig. 3. Top view of the RHI = 30% contour surface of the simulated storm at $t = 120$ min, showing the plume on the cloud top. Data below 10 km (6 mi) altitude are truncated for clarity.

Atom lithography

Atom lithography designates a physical method where the forces exerted by interfering laser beams on the atoms of an atomic beam are used to steer the atoms into nanostructures fabricated on a plane surface. While atom lithography is the most frequently used term, the method is also known as light-force lithography and atomic nanofabrication (ANF).

Atom lithography is an application of a field called atom optics. In this field, atomic motion is controlled by means of forces exerted by laser beams or a combination of laser and magnetic forces. All concepts of geometric and wave optics—refraction, diffraction, and interference—are straightforwardly transferred to atomic beams. Wave properties of atomic beams are governed by the de Broglie wavelength, $\lambda_{dB} = h/mv$, where h is Planck's constant, and m and v are the mass and the speed of the atom.

Structure formation by atom lithography relies on the deposition of atoms on a surface. Typical experimental atomic beams have 1 mm^2 cross section. Taking 1 min as a reasonable exposure time, a beam with flux of the order of 10^{11} atoms per second is required to deposit a single monolayer of atoms.

Applicability. Atom lithography is applicable to any element that is able to form an atomic beam and can be manipulated by the methods of laser cooling. It is thus essential that robust and tunable laser light sources are available to excite atomic resonance lines in these atoms. Atom lithography is element-selective since the laser interaction is effective only in the vicinity of atomic resonance lines. Laser cooling is an essential preparation technique for atom optics and thus for atom lithography. It has been demonstrated to work with all alkali atoms, alkaline earths, metastable rare gases, and elements of immediate technological interest, including chromium and aluminum. Further elements such as gallium, indium, iron, and the noble metals are under investigation for this application.

Standing-wave focusing. The simplest example of atom lithography is realized using a standing-wave light field which serves as a focusing mask, free from matter, for a suitable substrate (**Fig. 1**). The energy levels of atoms are shifted as they pass through the light field. This light shift is zero at the nodes but elsewhere it is finite, so its spatial nonuniformity corresponds to a spatially varying potential whose gradient results in a force. For the case shown in Fig. 1, this force is along the direction of propagation of the laser beam that sets up the standing wave. These forces are called optical dipole forces since they rely on the interaction of the induced atomic dipole moment and the inhomogeneous light field.

The standing wave of the laser radiation acts on the incident atoms as an array of microscopic lenses with apertures $\lambda/2$, with λ denoting the wavelength of the laser light field. The lenses focus appropriately directed atoms of an incident beam toward the optical nodes or antinodes, depending on the sign of the difference between the atomic and laser frequencies. The atomic pattern generated on the surface closely resembles the pattern of the intensity distribution of the light mask. The interaction of atoms impinging onto the surface must be local in order to preserve the desired pattern; that is, atoms must not diffuse after deposition.

The standing-wave microlenses are not perfect. Like their conventional optical counterparts, they suffer from multiple aberrations. Spherical aberrations cause a more or less homogeneous background. Beam spread in the transverse direction (failure of collimation) causes broadening of the microlens

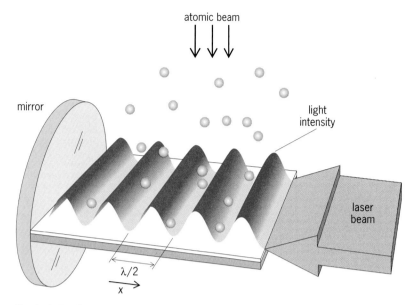

Fig. 1. A standing wave acting as an array of microlenses for an atomic beam. The atomic beam is focused into lines. The standing wave is formed by reflection from the mirror of a laser beam of wavelength λ, and the light intensity is proportional to $\sin^2 (kx)$, where $k = 2\pi/\lambda$ is the magnitude of the laser wave vector.

focus, which can be controlled by laser cooling. Longitudinal velocity spread corresponds to chromatic aberrations and broadens the microlens focus as well; it is, however, tolerable even for thermal atomic beams.

Methods. Two basic variants of atom lithography can be distinguished: In direct deposition, the atoms deposited cause a physical modification of the surface. In neutral atom lithography, atoms cause a chemical modification of a resist which is subsequently transferred to the substrate.

Direct deposition. The direct deposition method allows layers of atoms to be deposited in a two-dimensional pattern onto a substrate. While the first experiments were carried out with chemically unstable sodium atoms, it was a breakthrough when J. J. McClelland and coworkers demonstrated that atom lithography with chromium atoms could be used to produce a very large array of narrow and strictly parallel lines (**Fig. 2**). Line patterns are typically separated by $\lambda/2$, with linewidths of the order of tens

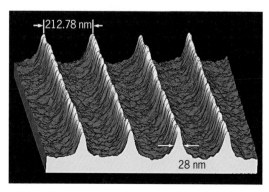

Fig. 2. Direct deposition of an array of chromium lines. (*With permission of J. J. McClelland*)

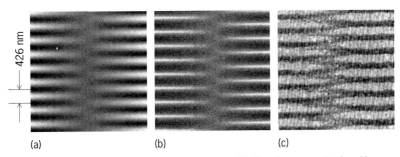

426 nm

(a)   (b)   (c)

Fig. 3. Manufacture of a nanoscale pattern, with small ($\lambda/2 = 426$ nanometers) and larger structures, by neutral atom lithography, using a cesium atomic beam and an alkane thiole resist. (*a*) Intensity distribution of the light mask. (*b*) Numerically simulated atomic pattern. (*c*) Actual structure, etched into a 30-nm gold layer.

of nanometers. Two-dimensional patterns with four- and sixfold symmetry have been manufactured. The direct deposition method offers unique promise for the manufacture of three-dimensional structures at nanometer scales through a growth process.

Neutral atom lithography. In conventional lithography, a material called resist covering a substrate is first chemically modified, for example, in optical lithography, by ultraviolet light. The structures are then prepared through a series of wet chemical processing steps. In neutral atom lithography, the surface modification is caused by reactive atoms (alkalis or metastable rare gases) chemically interacting with special resists such as alkane thioles. After wet etching, the pattern of the light mask used to steer atoms into the surface is precisely reproduced (**Fig. 3**). In comparison with optical lithography, the roles of light and matter are exchanged: Light serves as the mask determining the pattern to be formed; exposure to the atomic "radiation" causes modification of the resist. *See* EXTREME-ULTRAVIOLET LITHOGRAPHY.

Role of laser cooling. Since the forces exerted by light are small, it is absolutely essential to use only extremely well collimated atomic beams for atom lithography. If an atomic beam is collimated by geometric apertures only, atomic flux is dramatically reduced. Application of transverse laser cooling (the use of lasers to reduce the spread in the transverse components of atom velocities) is thus a necessary prerequisite to implement the methods of atom lithography. This cooling improves the brightness, brilliance, and phase-space density of atomic beams, and these improvements result in a strong enhancement (30–80-fold) of the flux density at the substrate. Transverse laser cooling is compatible with the use of any thermal atomic beam.

Longitudinal laser cooling further improves the imaging properties of atom lithography methods by reducing chromatic aberrations. However, not only does its application require costly implementation of multiple laser wavelengths, but it is also in general accompanied by significant flux reduction. Thermal atomic beams (in which the longitudinal velocity spread has not been reduced) thus seem to prove themselves as the sources of choice for atom lithography.

Technologically relevant elements. The alkali elements can be controlled with one or two simple infrared diode laser light sources, and atomic beams are easy to operate. Technologically more relevant materials typically require blue or ultraviolet laser light sources. Owing to their complicated hyperfine structure, sometimes several closely spaced laser wavelengths may have to be generated by costly instruments. Nevertheless, implementation of the methods of laser cooling and atom lithography seems realistic for the technologically relevant elements aluminum, gallium, indium, iron, chromium, copper, silver, and gold, and in some cases it is already being carried out.

Complex patterns. While large arrays of parallel lines (one-dimensional, or 1D, structures) can be straightforwardly manufactured, it remains to be seen what degree of complexity can be achieved in the atomic pattern deposited. It has been shown that two crossed standing waves result in a square array of dots, and three standing waves can generate hexagonal patterns. Superposition of multiple laser beams (up to 1000 per radian) can be achieved with volume holograms stored in photorefractive crystals.

The first step in the manufacture of a complex atomic pattern is to transfer a two-dimensional intensity pattern of interfering light beams to a substrate. If several elements are synchronously deposited, the variation of the deposition rate as well as the two-dimensional patterning can be used to extend this method to generate materials with three-dimensional modulation of their composition, which is a unique promise of atom lithography.

Applications. A natural application of atom lithography is the generation of precision rulers at nanometer scales because the laser wavelength determining line spacing is known with extremely high accuracy. A resolution of 1 in 10^6 has been demonstrated. Nanostructures produced through direct deposition of chromium in iron structures also exhibit interesting magnetic properties.

Furthermore, atom lithography promises intriguing opportunities for the generation of three-dimensional structures at nanometer scales which are unique to this method. The direct deposition concept allows growth of a material composed from several elements where the light mask selectively modulates the concentration of one element only. Hence, the periodicity of the light-field intensity distribution will induce a superlattice which could have various applications. For instance, physical properties such as the electronic band structure or the index of refraction will be subject to periodic modulation at the scale of the wavelength.

Future challenges. In a series of beautiful experiments, it has been demonstrated that atomic matter waves can successfully be employed for atomic pattern generation by reconstructing a hologram made by conventional nanolithography. Ultimate wave properties (coherence, focusability) in atom optics are promised by atomic matter waves, or so-called atom lasers. They are derived from atomic

Bose-Einstein condensates and can be treated in close analogy to coherent optical beams. However, the maximum flux achieved today is below 10^6 atoms per second, and hence their application for atom lithography, which requires deposition of at least 10^{10}–10^{11} atoms per second, seems remote at present. However, a technical breakthrough could rapidly change this situation.

For background information *see* ATOM LASER; ATOM OPTICS; ARTIFICIALLY LAYERED STRUCTURES; CRYSTAL GROWTH; HOLOGRAPHY; INTEGRATED CIRCUITS; LASER COOLING; NANOSTRUCTURE; NANOTECHNOLOGY; NONLINEAR OPTICS; OPTICAL RECORDING in the McGraw-Hill Encyclopedia of Science & Technology.   Dieter Meschede

Bibliography. C. S. Adams, M. Sigel, and J. Mlynek, Atom optics, *Phys. Rep.*, 240:143–210, 1994; J. Fujita et al., Manipulation of an atomic beam by a computer-generated hologram, *Nature*, 380:691–694, 1996; F. Lison et al., Nanoscale atomic lithography with a cesium atomic beam, *Appl. Phys. B*, 65:419–421, 1997; J. J. McClelland et al., Laser-focussed atomic deposition, *Science*, 262:877–880, 1993; D. Meschede and H. Metcalf, Atomic nanofabrication: Atomic deposition and lithography by laser and magnetic forces, *J. Phys. D: Appl. Phys.*, 36:R17–R38, 2003; H. Metcalf and P. van der Straten, *Laser Cooling and Trapping*, Springer, New York, 1999; M. Mützel et al., Atom lithography with a holographic light mask, *Phys. Rev. Lett.*, 88:083601, 2002.

Atomic Fermi gases

Researchers have developed powerful new techniques over the past two decades for cooling atoms to ultralow temperatures. Among the most significant achievements made possible by these technical developments was the creation of the long-sought Bose-Einstein condensate in 1995. In a Bose-Einstein condensate, atoms are cooled to such a low temperature that they collect in the quantum-mechanical ground state of their confinement volume. Bose-Einstein condensation is possible only in the class of particles known as bosons. Particles of the other class, fermions, are forbidden to occupy the same quantum state and are thus prevented from directly condensing.

Progress in studying ultracold fermions has lagged behind that of bosons, not because of lack of interest but because they are experimentally more difficult to cool. In fact, fermions are especially significant because they comprise the fundamental building blocks of matter: protons, neutrons, and electrons are all fermions. Most of the subject of condensed-matter physics is concerned with the behavior of fermions, especially electrons. Of particular interest is the phenomenon of Cooper pairing, where fermions, such as the electrons in a superconductor, or the atoms in helium-3 (^{3}He), form correlated pairs when cooled to sufficiently low temperature. Recent experimental progress in cooling atomic Fermi gases may enable the realization of Cooper pairing in the gas phase. Because of the inherent simplicity and tunability of the atomic interactions in an ultracold gas, this work may help clarify many troubling issues in condensed-matter physics, including the mechanisms at work in high-temperature superconductors.

Behavior of bosons and fermions. It has been known since the work of Satyendranath Bose, Albert Einstein, Enrico Fermi, and Paul Dirac in the 1920s that there are two fundamental types of particles. Bosons are particles with integer spin angular momentum ($0, \hbar, 2\hbar, \ldots$, where $\hbar$ is Planck's constant divided by 2π), and fermions are those whose spin angular momentum has half-integer values ($\frac{1}{2}\hbar$, $\frac{3}{2}\hbar, \ldots$). Since protons, neutrons, and electrons are each spin-$\frac{1}{2}$ particles, atoms made with an odd number of these constituents are themselves composite fermions. Examples are lithium-6 (^{6}Li) and potassium-40 (^{40}K), which happen to be the only stable fermionic isotopes of the alkali-metal elements. The differences between bosons and fermions are significant only when they become quantum-degenerate, that is, when the size of the atoms' quantum-mechanical de Broglie wavelength Λ, which increases as temperature decreases, is comparable to the average separation between atoms. When this condition is met, bosons undergo the Bose-Einstein condensation phase transition. Identical fermions, on the other hand, are forbidden by the Pauli exclusion principle from occupying the same quantum level (**Fig. 1**). At zero absolute temperature, they completely fill the available energy states up to some energy, which is called the Fermi energy E_F. The filling of energy levels in this manner is analogous to the way electrons fill the lowest available atomic orbitals, thus forming the periodic table of the elements. As Cooper pairs are composite bosons, superconductivity and fermionic superfluidity (which is observed in liquid helium-3) are though to be related to Bose-Einstein condensation of these pairs of fermions.

Cooling methods. The methods for cooling Fermi gases and for confining them in atom traps are similar to those used for bosons. The primary methods

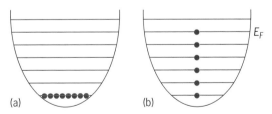

Fig. 1. Bosons and fermions at absolute temperature $T = 0$. Atoms in a magnetic trap feel a harmonic restoring potential, as indicated by the parabolic wells. The equally spaced lines in each well represent the quantized energy levels of the harmonic potential. (*a*) Bosons. At $T = 0$, they collectively occupy the ground state of the system. (*b*) Fermions. These particles must obey the Pauli exclusion principle, which forbids more than one identical fermion per state. At $T = 0$, they completely fill the energy-level ladder up to the Fermi energy E_F.

are laser cooling, atom trapping, and evaporative cooling. Laser cooling utilizes radiation pressure to slow atoms. Since temperature is related to the atom's kinetic energy, reducing their speed is equivalent to cooling. Laser cooling is typically limited to temperatures in the range of 100 millionths of a degree above absolute zero, or 100 μK. Atom traps use laser radiation pressure or magnetic fields to confine atoms without walls. Since the atom temperatures are so low, isolation from physical boundaries is absolutely necessary to prevent heating.

Even lower temperatures, of the order of 100 nK (100 billionths of a degree above absolute zero), are needed to achieve quantum degeneracy in low-density atomic gases. These temperatures, as low as any produced by any method or found anywhere in the universe, are attainable using evaporative cooling. Evaporative cooling exploits the fact that there is always a distribution of energies in a gas, including some very energetic atoms in the "tail" of the distribution, and these few atoms account for much of the total energy. By systematically removing only the hottest atoms from the trap, a large fraction of energy can be removed, while minimizing the number of atoms lost. In the actual experiments, a microwave field is tuned to a particular frequency to drive atoms with energies above a certain cutoff E_c to untrapped states (**Fig. 2**).

Effective evaporative cooling requires that the atoms in the gas undergo elastic thermalizing collisions to continuously replenish the tail of hot atoms. The main experimental difficulty encountered with fermions is that the Pauli exclusion principle prevents identical fermions from interacting, so they are unable to thermalize. Several ways around this difficulty have been developed. First, one can make use of the different projections of the nuclear and atomic spin of the atom to make subsets of atoms that are distinguishable from other subsets. For example, one can make a mixture of ^{40}K atoms in two different nuclear spin states, so that the nuclear spins of some of the atoms point in a different direction from the others. Atoms in the same spin state will not interact, but two atoms in different spin states will. Unfortunately, this method cannot be universally used, because some spin states cannot be magnetically trapped, and because collisions between some spin states can result in the creation of atoms in a third state, with a release of excess energy that sends the atoms flying from the trap.

A more useful way to cool fermions is to "sympathetically" cool them with bosons. The bosons are evaporatively cooled as usual, while the fermions are cooled by interaction with the bosons. In this way, ^{6}Li has been cooled to temperatures as low as 200 nK using either lithium-7 (^{7}Li) or sodium-23 (^{23}Na) atoms as the "refrigerant." After the fermions have been cooled, the bosons can be selectively removed from the trap by making them undergo transitions to untrapped quantum states using microwave or optical pulses. The gas is probed by shining a laser beam through the gas of atoms and imaging

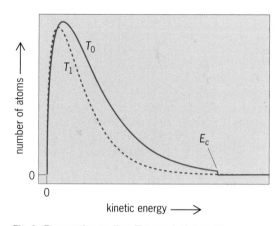

Fig. 2. Evaporative cooling. The graph shows the distributions of the kinetic energies of atoms at an initial temperature T_0 and at a lower temperature T_1. After the removal from the distribution at T_0 of all atoms with energies above E_c, collisions among the atoms cause the gas to rethermalize to the lower temperature T_1.

the shadow cast by the atoms onto a charge-coupled device (CCD) camera.

Fermi pressure. In 1926, Fermi realized that, as a result of the Pauli exclusion principle, the energy of a gas of fermions is greater than zero at zero absolute temperature (Fig. 1), and consequently it exerts a pressure. This Fermi pressure is a purely quantum-mechanical result, as the pressure p from the classical ideal gas law, $p = nkT$, vanishes as the temperature T approaches 0 (n is the number density of the gas, and k is the Boltzmann constant). The implications of this result were soon realized by S. Chandrasekhar: Under certain conditions, electron Fermi pressure could prevent the gravitational collapse of relatively cold, aging stars known as white dwarfs. This prediction was verified by observation, and the same stabilization method was later found to apply to neutron stars.

The effect of Fermi pressure has also been demonstrated for a trapped gas of ^{6}Li atoms. **Figure 3** shows a gas containing both bosons (^{7}Li) and fermions (^{6}Li), at three different temperatures. Although both types of atoms are mixed together in the trap at the same

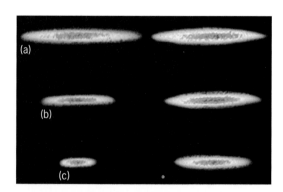

Fig. 3. Fermi pressure in a gas of trapped atoms. The images on the left are of the bosons (^{7}Li), while those on the right are of the fermions (^{6}Li), taken at progressively lower temperatures. Each pair of images corresponds to a particular temperature: (a) 810 nK, (b) 510 nK, (c) 240 nK. The ^{6}Li and ^{7}Li atoms are simultaneously trapped in the same volume, but the images are separated for clarity.

time, they can be individually imaged. The images are separated horizontally in the figure to show the difference in behavior between bosons and fermions. Figure 3*a* corresponds to the highest temperature, $T \simeq T_F$, where T_F is the Fermi temperature, which corresponds to the Fermi energy (Fig. 1) through the relation $T_F = E_F/k$. At this relatively high temperature, there is little difference in the spatial size of the fermions compared with the bosons. However, as the temperature is reduced to $\sim$0.5 T_F in Fig. 3*b*, and finally to $\sim$0.25 T_F in Fig. 3*c*, it becomes clear that the fermions occupy more volume than do the bosons. This difference is a manifestation of Fermi pressure, albeit on a much different size and energy scale than for stars.

Prospects for Cooper pairing. Cooper pairing in an ultracold atomic Fermi gas will require that the gas be cooled to even lower temperatures than have been demonstrated thus far. Furthermore, the atoms must have a strong attractive interaction. Since identical ultracold fermions cannot interact, as was the case for evaporative cooling, the experimental schemes will rely on a mixture of fermions in two spin states. A powerful technique for producing strong tunable interactions has been developed for atoms in Bose-Einstein condensates. The technique employs a phenomenon known as a Feshbach resonance, in which magnetic fields tune the energy of a bound state of the two colliding atoms, a diatomic molecule, to the same energy as the pair of unbound colliding atoms. The resulting resonance can be used to create strong or weak interatomic interactions that are either repulsive or attractive depending on the value of the magnetic field. Because of the versatility it affords, this interaction "knob" is one of the most compelling aspects of quantum gas research. Traditional superconductors and superfluid ^{3}He form pairs in the weak-coupling limit, where the size of the Cooper pairs is much larger than the average distance between fermions. This situation is very well described by the BCS (Bardeen-Cooper-Schrieffer) theory of superconductivity. On the other hand, Feshbach resonances may permit the exploration of the strongly coupled regime, for which there is yet no complete theory of Cooper pairing. Feshbach resonances in fermions are just now being explored and, although no clear demonstrations of fermion superfluidity have been made, experimentalists are beginning to reveal new aspects of strongly interacting Fermi gases.

For background information *see* BOSE-EINSTEIN CONDENSATION; EXCLUSION PRINCIPLE; FERMI-DIRAC STATISTICS; LASER COOLING; LIQUID HELIUM; NEUTRON STAR; PARTICLE TRAP; QUANTUM MECHANICS; QUANTUM STATISTICS; RESONANCE (QUANTUM MECHANICS); SUPERCONDUCTIVITY; SUPERFLUIDITY; WHITE DWARF STAR in the McGraw-Hill Encyclopedia of Science & Technology.   Randall Hulet

Bibliography. B. DeMarco and D. S. Jin, Onset of Fermi degeneracy in a trapped atomic gas, *Science*, 285:1703–1706, 1999; D. Jin, A Fermi gas of atoms, *Phys. World*, 15(4):27–32, April 2002; H. J. Metcalf and P. van der Straten, *Laser Cooling and Trapping*, Springer, 1999; C. J. Pethick and H. Smith, *Bose-Einstein Condensation in Dilute Gases*, Cambridge University Press, 2002; A. G. Truscott et al., Observation of Fermi pressure in a gas of trapped atoms, *Science*, 291:2570–2572, 2001.

Autism

Autism is a neurodevelopmental condition that impairs the way a person relates to and communicates with other people. Persons with autism can also have unusual behaviors, such as insistence on sameness, obsessions, or stereotypic behaviors (for example, hand flapping, spinning, and toe walking). The condition varies greatly in the presenting symptoms, the timing of presentation, the range and severity of symptoms, and its association with other conditions. Although recognition of autism is increasingly common, the cause of autism and the reason for its increase remain unknown. With intensive early intervention, significant improvements in a large percentage of children with autism can be achieved.

Diagnosis. Autism is a spectrum disorder that includes Asperger syndrome, Rett syndrome, childhood disintegrative disorder, and pervasive developmental disorder not otherwise specified (PDD-NOS). Leo Kanner first described autism as a condition in 1943, but descriptions matching autism can be found throughout history. Based on his observation of 11 children, Kanner identified characteristic features of "early infantile autism" that are still considered essential in making an autism diagnosis: (1) social impairments that affect how a person relates to other people, apparent early in life; (2) peculiar use of language: a failure to use language for interactive conversation; and (3) an "obsessive desire for the maintenance of sameness" associated with a "limitation in the variety of spontaneous activity." IQ ranged from 50 to 70 in classic Kanner autism, but currently autism is recognized over a broader range of IQs.

Etiology. As recently as the late 1960s, autism was attributed to disturbed mother-child relationships. The term "refrigerator mother" represents a dark period in the history of autism, when misplaced blame added to the anguish experienced by families with autistic children and delayed investigation into true causes. The recognition that 25% of children with autism developed seizures began the shift in thinking to the current view that autism is a neurodevelopmental disorder with a biological basis.

Genetics. Finding the cause(s) of autism remains elusive, but data support a strong genetic component. Families with one child with autism are more likely to have another child with autism than previously unaffected families, the risk being 3–5% in affected families compared with 0.1–0.2% in the general population. Autism occurs in males three to four times more frequently than females. Studies of families with multiply affected members have identified many chromosomes that are highly associated with autism; however, none are universally found in children with autism. Twin studies have found a concordance (both twins having autism) of 36–91% in identical twins compared with a less than 1% concordance rate in fraternal twins. Autism is more common in some conditions, such as Fragile X, neurofibromatosis, tuberous sclerosis, and Down syndrome; however, no such associated condition can be identified in most children with autism.

Prenatal and perinatal factors. Autism often manifests after the first year of life, but data from at least two sources suggest that autism often begins at or before the prenatal and perinatal periods. Analyses of neonatal blood spots taken from children later diagnosed with autism showed that 95% of a small sample of children with autism have elevated levels of four neuropeptides (short-chain proteins that can act as neurotransmitters) and neurotrophins (nerve growth factors). In this study, blood spot data from normal children and those with autism, mental retardation, or cerebral palsy were compared. Elevated neurotropin and neuropeptide levels were also found in children with mental retardation but not in normal children or those with cerebral palsy. Specifically, the damage that occurs during pregnancy resulting in cerebral palsy does not produce the changes in blood spot data that were found in autism or mental retardation. Another study found that 42% of children with autism have posteriorly rotated ears (compared with 10% of children without autism), representing changes that occur before birth in the first month of gestation. Other studies have shown neuroanatomic differences. Head sizes of autistic children are on average normal at birth, but become larger than average with age. Specific brain structures are differentially affected. For example, some structures, such as the hippocampus and amygdala, are larger than normal in autistic children, and some, such as the cerebellar vermis, are smaller. Developmental delays, including speech delays, in autistic children are often apparent early in the course of the condition. Some autistic children, however, have apparently normal development, followed by a loss of previously attained developmental milestones. Such developmental regression occurs in 20–40% of cases, leading some to suspect that postnatal factors also contribute to autism. Data showing significant increases in the autistic population over time also challenge the theory that autism is exclusively a genetic condition.

Theories. The lack of an identified cause of autism, added to an apparent increase in rates of autism, has led to the proliferation of theories for its cause and recent increase. These theories include a bowel-brain connection that is altered by some potential insult, such as food allergies, antibiotic use, and immunizations; measles-mumps-rubella or thimerosal (preservative containing 49% ethyl mercury)–containing vaccines; exposure to heavy metals and other toxic compounds such as polychlorinated biphenyls (PCBs); defective hepatic detoxification capacity; autoimmune disorder or a T-cell imbalance leading to an autoimmune disorder; and yeast overgrowth in the bowel. Some of these theories are interrelated, such as yeast overgrowth and antibiotic use, or yeast overgrowth and bowel-brain connection.

Autism spectrum disorders. Autism spectrum disorders retain many of the core features of autism.

Asperger syndrome. Deficits in communication and social interactions occur in Asperger syndrome, but to a lesser degree than in autism. Significant delay in the onset or early course of language that is seen in autism is not seen in Asperger syndrome. Overall, children with this condition have a better facility with language and more interest in social activities than children with autism; however, they usually have autistic features, such as obsessive preoccupations.

Rett syndrome. Rett syndrome, primarily affecting females, is characterized by normal early development, followed by regression some time after 5 months. Girls with Rett syndrome typically carry a single genetic marker, the *MECP2* gene, and develop unsteady gait, loss or lack of language, constant hand wringing and loss of functional use of their hands, social deficits, and mental retardation. Head sizes of affected children are on average normal at birth, but slowing of head growth during early development results in small head size. In contrast, head sizes of children with autism are about average at birth but will rapidly increase in size during early infancy.

Childhood disintegrative disorder. Children with childhood disintegrative disorder experience normal development for 2–10 years, after which their development undergoes severe regression, resulting in complete loss of speech, social interaction, and self-help skills. Recovery has been quite limited in these children.

PDD-NOS. Children with PDD-NOS have some autistic features, such as communication and social impairments and repetitive behaviors, but do not meet the criteria for full-syndrome autism and are not better described by one of the other autism spectrum disorders.

Epidemiology. Autism prevalence estimates have steadily increased over the last 20 years. Studies conducted prior to 1985 found that autism affected 4–5 per 10,000 children. However, studies conducted between 1985 and 1994 show autism prevalence rates closer to 12 per 10,000 children. More recent data suggest autism is even more common, affecting 16–40 per 10,000 children. Staffordshire (United Kingdom) preschool children in the late 1990s had an autism prevalence rate of 16.2 per 10,000. In 1998 Brick Township, New Jersey, had autism rates estimated at 40 per 10,000 in 3- to 10-year-olds. Prevalence rates of autism and Asperger syndrome in Cambridgeshire (United Kingdom) were 1 in 175 (or ~57/10,000) for 5- to 11-year-olds. Children born in the Atlanta area in 1996 had an estimated autism prevalence rate of 34 per 10,000.

Interpretation of observed increase. Autism epidemiology is mired in controversy arising from differences in interpretation of the observed increase. A 1999

California Department of Developmental Services report showed that from 1987 to 1998 autism cases in California's Regional Center System increased by 273%. This report was cited as evidence of an autism epidemic, but others challenged that the increase was due to changes in diagnostic criteria, increased awareness, increase in population, migration fluxes, earlier age at diagnosis, and problems with cross-sectional data. A followup study that compared Regional Center clients who were born between 1983 and 1985 with those born between 1993 and 1995, concluded that the increase in autism cases was not due to changes in diagnostic criteria, changes in misclassification of autism, or an influx of out-of-state autism cases. The study also found that the state's Regional Center System reliably diagnosed autism.

Although low response rates could potentially introduce biases into the followup study's results, a separate study of the California data concurred with two of its three major findings. The studies differ only in estimation of misclassification of autism (as mental retardation), but this discrepancy can be explained by differences in study methods. However, since neither study reviewed data obtained from outside the State's Regional Center System, neither examined the extent to which increased awareness and increased utilization of the Regional Center System contributed to the increase. Population changes accounted for only 10% of the observed increase. If increased awareness of autism and increased utilization were responsible for the rest of the observed increase, then two out of every three cases of autism would have been missed among children born between 1983 and 1985, assuming that all cases of autism were identified among children born from 1993 to 1995. The California Department of Developmental Services issued a 2003 report that documents a 100% increase in autism caseload since 1999 which, if valid, further weakens the increased awareness argument.

Autism spectrum disorders. Epidemiology of other autism spectrum disorders is less well established. Some studies have estimated that Asperger syndrome is three to four times more common than autism. Based on a large population study conducted prior to the identification of the *MECP2* gene, Rett syndrome was estimated to occur in about 1 in 23,000 children. Childhood disintegrative disorder is believed to be rare, occurring in less than 5 per 10,000 children. Although PDD-NOS is believed to be more common than autism, population estimates are hampered by unclear diagnostic criteria. The small-area study conducted in Brick Township estimated that autism occurs in 1 in 250 and autism spectrum disorders in 1 in 150 children.

Treatment. Early identification and early intervention with intensive behavioral programs is the cornerstone of effective treatment for autism. In the past, dismal expectations led to minimal intervention resulting in discouraging outcomes. Expectations improved when Ivar Lovaas showed that intensive early intervention programs could make significant differences in about half of the children enrolled. Subsequently, a number of programs have been developed, including Auditory Integration Training, Daily Life Therapy (Higashi), EarlyBird Programme, Lovaas/Applied Behavior Analysis (ABA), Treatment and Education of Autistic and Related Communication Handicapped Children (TEACCH), Options Son-Rise Programme, Picture Symbols (PECS) training, and Structure Positive (approaches and expectations) Empathy Low-arousal Links (SPELL). Most of these programs are time-intensive and reward mastery of individual small steps that gradually lead to mastery of larger tasks. Adjuncts to intervention programs include diet modification, music therapy, facilitated communication, and computer applications.

Although there have been no studies directly comparing these programs, some have been studied individually. ABA is best studied of the programs, having included control subjects. The original studies have shown significant improvement in about half of treated children; however, studies aimed at replicating these findings have found improvements in smaller proportions of treated children, but replication studies have failed to show as positive effects.

No one intervention is guaranteed to help every child with autism. The choice in programs is often dictated by availability and funding. Even though early intensive interventions have been shown to be effective, many families have significant difficulties in obtaining services.

There are no known medications that cure autism, but some drugs are used to ameliorate some behaviors and to improve function of the child. Families of affected children and autism researchers hope that the identification of the cause will lead to more effective treatments.

For background information *see* AUTISTIC DISORDER; DEVELOPMENTAL GENETICS; DEVELOPMENTAL PSYCHOLOGY; HUMAN GENETICS; NERVOUS SYSTEM (VERTEBRATE); PSYCHOTHERAPY in the McGraw-Hill Encyclopedia of Science & Technology.   Robert Byrd

Bibliography. S. E. Bryson and I. M. Smith, Epidemiology of autism: Prevalence, associated characteristics, and implications for research and service delivery, *MRDD Res. Rev.*, 4:97–103, 1998; P. A. Filipek et al., The screening and diagnosis of autistic spectrum disorders, *J. Autism Dev. Disorders*, 29(6): 439–484, 1999; S. Ozonoff, S. J. Rogers, and R. L. Hendren (eds.), *Autism Spectrum Disorders: A Research Review for Practitioners*, American Psychiatric Publishing, Arlington, VA, 2003; A. M. Wetherby and B. M. Prizant (eds.), *Autism Spectrum Disorders: A Transactional Developmental Perspective*, Paul H. Brookes Publishing, York, PA, 2001.

Autophagy

Autophagy, the destruction of the cell's components by its own enzymes, is a process of all eukaryotic cells. Various types of autophagy occur during normal development or as a response to various types

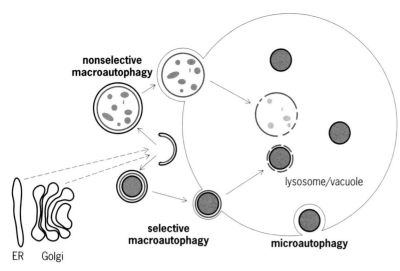

Fig. 1. Modes of autophagy. Macroautophagy involves the uptake of cytoplasm at a site away from the lysosome or vacuole. The origin of the sequestering membrane is not known but may include the endoplasmic reticulum (ER) and/or Golgi complex. During microautophagy, cytoplasm is taken up directly at the lysosome or vacuole via protrusion or invagination of the organellar membrane. Autophagy can be either nonselective, taking up bulk cytoplasm, or selective, sequestering specific biosynthetic cargo or organelles (shown here as circles) destined for degradation.

of stress, including starvation. Defects in autophagy are correlated with a number of diseases in humans. Recent advances in the study of autophagy in baker's yeast (*Saccharomyces cerevisiae*) have provided insight into the molecular basis of this process in more complex eukaryotes, including animals and humans.

Types. Autophagy involves dynamic membrane rearrangements that result in the sequestration of a portion of cytoplasm. The sequestered cytoplasm is delivered to a degradative organelle, the lysosome in animal cells or the vacuole in baker's yeast, where it

is broken down and the components recycled. Autophagy can differ with regard to the site at which it occurs and the membrane involved in the sequestration process. Microautophagy involves uptake of cytoplasm directly at the surface of the lysosome or vacuole and, as a result, uses the limiting membrane of that organelle (**Fig. 1**). Macroautophagy occurs in the cytosol and utilizes a membrane of unknown, but presumably nonlysosomal/vacuolar, origin. Relatively little is known about microautophagy; therefore this article focuses on macroautophagy, referred to as autophagy hereafter.

One additional distinction can be made regarding the specificity of autophagy. Although autophagy has long been considered to be a nonspecific process that takes up bulk cytoplasm, it is now clear that there are selective types of autophagy that utilize much of the same machinery to target specific cargo. In some cases this cargo can even be biosynthetic; for example, one type of autophagy may selectively deliver some resident enzymes to the yeast vacuole to carry out their function.

Process steps. Autophagy can be broken down into five basic steps (**Fig. 2**): (1) a signal transduction mechanism senses the environmental conditions and induces autophagy; (2) a nucleation event initiates the formation of the sequestering vesicle intermediate, a phagophore, in the cytosol; (3) the membrane expands and forms a spherical structure resulting in the formation of a double-membrane cytosolic vesicle, an autophagosome, that surrounds cytoplasm; (4) the completed autophagosome targets to, and fuses with, the lysosome or vacuole, releasing the inner single membrane vesicle, termed an autophagic body, into the organelle lumen; (5) the autophagic body is degraded, allowing breakdown and recycling of the contents. In selective autophagy, special receptors allow specific cargoes to be taken up into the autophagosome and delivered to the vacuole, where they either are degraded or, in the case of resident enzymes, carry out their function.

Role in cell growth and development. Much of the research concerning cell biology has focused on biogenesis of proteins and organelles. However, it is clear that cells operate through continual processes of both synthesis and breakdown. (After all, cellular metabolism is composed of both anabolic and catabolic processes.) Autophagy is one of the major mechanisms by which cellular components can be degraded.

Degradative pathways need to be regulated either enzymatically or spatially. A hallmark of eukaryotic cells is the presence of distinct compartments or organelles that can spatially segregate competing reactions and provide unique environments for special functions. For example, the lysosome or vacuole contains a range of hydrolytic enzymes that can degrade essentially any macromolecule, while the organelle's limiting membrane keeps these degradative enzymes away from, and protects, the rest of the cell. However, the inherent problem in using a membrane-bound organelle for degradation is that

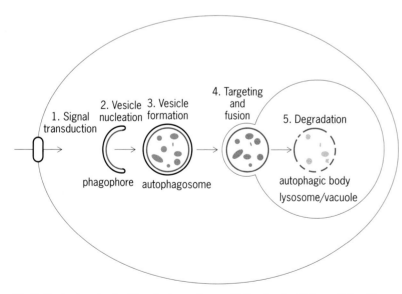

Fig. 2. Basic steps involved in autophagy, shown with the structural intermediates. The intermediate vesicle precursor is termed the phagophore. The completed vesicle, the autophagosome, has a double membrane; this structure effectively changes the topology of the sequestered cytoplasm so that it is now in the equivalent of the extracellular space. Following fusion of the autophagosome with the vacuole, the inner vesicle termed the autophagic body is now degraded to allow access to the cargo. In nonselective autophagy, the cargo is degraded and recycled.

there must be a way to bring the substrates into contact with the degradative enzymes. Autophagy has the capacity to deliver any cellular constituent, including entire organelles, from the cytoplasm into the lysosome or vacuole lumen and, as a result, is uniquely designed to meet the varying degradative needs of the cell.

Organelle removal. One of the best-characterized roles of autophagy in cellular maintenance can be seen in the elimination of superfluous or damaged organelles. For example, the peroxisome is an organelle that is capable of carrying out particular catabolic reactions that generate hydrogen peroxide; when cells are grown in the presence of certain carbon sources, such as oleic acid, peroxisome biogenesis is induced. However, it is costly to maintain organelles that are not needed. Accordingly, a process called pexophagy degrades the excess peroxisomes when cells are shifted from oleic acid to a preferred carbon source such as glucose. The machinery used in pexophagy is almost identical to that used in nonspecific autophagy, but additional proteins that are not needed for bulk autophagy ensure the specificity of the process. During pexophagy only peroxisomes, and not other organelles, are degraded.

Although pexophagy has been well documented, additional data (so far unpublished) suggest that other specific types of autophagy, for example mitophagy to remove damaged mitochondria, are possible. The removal of damaged organelles and other types of cellular debris may be important in preventing various diseases as well as in preventing premature aging; there is some evidence that caloric restriction extends lifespan and it is possible that the mechanism involves the induction of autophagy.

Programmed cell death. The process of programmed cell death, regulated self-destruction of the cell, plays a role in many developmental pathways. Apoptosis is a form of programmed cell death termed type I and is marked by cytoskeletal breakdown and condensation of cytoplasm and chromatin followed by deoxyribonucleic acid (DNA) fragmentation. Autophagy, termed type II programmed cell death, is characterized by the presence of autophagic vacuoles (autophagosomes) that sequester organelles. Although autophagic programmed cell death appears to be morphologically distinct from apoptotic programmed cell death, the two are not necessarily exclusive; both types may occur at the same time and cells may switch between them. One example of type II programmed cell death in development is seen in the death of sympathetic neurons. Although this process clearly involves apoptosis, it also appears to utilize autophagic cell death. Autophagic programmed cell death is also associated with organ morphogenesis in animals and plays a role in tissue remodeling associated with sexual reproduction.

Role in disease. Because the sequestering vesicle formed during autophagy, the autophagosome, has a nearly unlimited capacity, it may be important in removing large protein aggregates, including those that can lead to neurodegenerative diseases. Accordingly, defects in autophagy may lead to Alzheimer's or Parkinson's disease. Recent studies also show that some microbes rely on autophagy (and specifically on an ability to manipulate the environment of the autophagosome) to carry out their infective cycle and escape the host immune response. Because of its ability to bring about cell death, autophagy may play a role in tumor suppression. For example, a mammalian gene, *beclin 1*, is homologous to the yeast *APG6* gene that is required for autophagy. A correlation between cancer and autophagy is seen with the increase in tumor formation in mice having mutations in *beclin 1*.

Genetic mechanism. Autophagy has been studied in mammalian cells for over half a century. However, the difficulty of applying genetic screens in animal systems prevented the identification of many of the molecular components. In contrast, *S. cerevisiae* is ideally suited for genetic analyses. Three different screens for mutants in nonspecific or selective autophagy led to the identification of over 20 genes, termed *APG*, *AUT*, and *CVT*, which are involved in this process. Analyses of pexophagy mutants revealed that these genes largely overlapped with those required for autophagy. Many of the gene products have been characterized at the level of protein-protein interactions and subcellular localization.

The identification of the yeast autophagy genes coupled with the sequencing of increasing numbers of genomes from different organisms has allowed workers to search for homologs (genes having the same origin in different species) in various model systems. It is now clear that homologs exist in *Arabidopsis*, *Drosophila*, *Caenorhabditis elegans*, *Dictyostelium*, and mammals. In general, the corresponding gene products appear to be orthologs, carrying out the same functions in different organisms. For example, the Apg12 and Apg5 proteins in yeast undergo a series of ubiquitin-like reactions. An identical process appears to occur in human cells and probably plays an equivalent, but as yet undetermined, role in autophagy. Although it is not yet known what these proteins do in a mechanistic sense in either yeast or mammalian cells, it has been shown that Apg12 and Apg5 are needed for formation of the autophagosome in both cell types. Similarly, the human Apg8 protein is posttranslationally modified in an analogous manner to the yeast Aut7 ortholog, which appears to play a role in vesicle formation and expansion. As additional autophagy genes are analyzed in animal systems, it is likely that they will be found to carry out functions identical to their yeast counterparts.

Conclusion. Current studies are starting to reveal the molecular mechanisms underlying autophagy. Continued investigation will provide further information about this ubiquitous process, and the resulting insights may ultimately provide additional directions for research aimed at therapeutic treatment of certain genetic diseases.

For background information *see* CELL (BIOLOGY); CELL ORGANIZATION; GENE; GENETIC ENGINEERING; LYSOSOME; PEROXISOME; VACUOLE; YEAST in the McGraw-Hill Encyclopedia of Science & Technology.

Daniel J. Klionsky

Bibliography. J. Kim and D. J. Klionsky, Autophagy, cytoplasm-to-vacuole targeting pathway, and pexophagy in yeast and mammalian cells, *Annu. Rev. Biochem.*, 69:303–342, 2000; D. J. Klionsky and S. D. Emr, Autophagy as a regulated pathway of cellular degradation, *Science*, 290:1717–1721, 2000; D. J. Klionsky (ed.), *Autophagy*, Landes Bioscience, Georgetown, TX, 2003; N. Mizushima et al., Dissection of autophagosome formation using Apg5-deficient mouse embryonic stem cells, *J. Cell Biol.*, 152:657–668, 2001; T. Noda, K. Suzuki, and Y. Ohsumi, Yeast autophagosomes: de novo formation of a membrane structure, *Trends Cell Biol.*, 12:231–235, 2002; F. Reggiori and D. J. Klionsky, Autophagy in the eukaryotic cell, *Euk. Cell*, 1:11–21, 2002.

Bayesian inference (phylogeny)

When Charles Darwin drew some of the first evolutionary trees in one of his notebooks, hypothesizing how new species had evolved from existing ones, he could not have foreseen the prominent role such trees would play in biology. Today, evolutionary trees, or phylogenies, are used in most biological disciplines. For instance, phylogenies of viral strains reveal infection pathways. Gene trees (showing the evolution of individual genes) inform population geneticists about migration and selection. The tree of life, specifying the relationships among all species on Earth, forms the basis for modern classification. Tree-based reconstructions enable biologists to extend their knowledge about the history and mechanisms of evolution considerably beyond what is directly observable from paleontological and experimental evidence. *See* TREE OF LIFE.

Any inherited trait can be compared to ascertain how similar or different species are, and thus how they may be related evolutionarily. In particular, deoxyribonucleic acid (DNA) sequences, which are now easily obtained from organisms, provide an ample source of data for phylogeny reconstruction. Typically, however, DNA sequences do not point out a uniquely "true" phylogenetic tree. Instead, there are many trees that could have generated the observed sequences. Selecting among those alternative trees is a difficult statistical problem.

Classical methods. Classical statistical methods are based on the principle of maximum likelihood: they find the hypothesis with the highest probability of producing the observed data. For example, in phylogenetic inference based on DNA sequences, we would need to have a model for how DNA sequences evolve, and then we would need to find the tree with the highest probability of producing the observed sequences. In the simplest possible model, we may assume that each site in a DNA sequence evolves independently of all other sites and that all possible nucleotide substitutions occur at the same, constant rate. The probability of going from one observed sequence to another can then be calculated from the substitution rate and the length of the branches separating the two sequences (representing the amount of expected difference between them, which is related to the time they have been separated).

Finding the maximum likelihood of a single tree is a difficult computational problem, involving optimization of branch lengths. Finding the tree with the highest score among all possible trees is considerably harder. Moreover, molecular evolution is more complex than in the simple model described above. For instance, some nucleotide substitutions are more likely than others, and different sites typically evolve at different rates. Optimizing each of these variables, in addition to the branch lengths, is a daunting task. In practice, despite many smart computational tricks, biologists have been forced to work with oversimplified evolutionary models and analyze small sets of sequences at a time. Alternatively, they have had to use faster, nonstatistical methods. Unfortunately, simulations have shown that such methods are inaccurate under certain conditions. A famous example demonstrates that one method, parsimony (which assumes that the correct tree is the one that shows the fewest evolutionary changes), tends to cluster rapidly evolving lineages even if they are not closely related. Simpler methods are also less informative about the evolutionary process. For instance, parsimony systematically underestimates the rate of molecular evolution.

The introduction in the mid-1990s of Bayesian Markov chain Monte Carlo (MCMC) techniques to phylogenetic inference is rapidly changing this situation. The computational efficiency of this approach has attracted considerable attention among biologists because it allows them to examine more complex and realistic evolutionary models and to address larger sets of sequences than had been possible with statistical methods previously.

Bayesian inference. Bayesian inference differs from classical statistical inference in that it forces scientists to specify their prior beliefs. For example, if we wish to infer a phylogenetic tree, we need to specify the prior probability of all possible trees (called simply the "prior" in statisticians' shorthand). A natural and cautious solution is to place equal probability on all alternatives (**Fig. 1**) [a "diffuse" or "flat" prior], but one can also take the results of previous analyses into account. Once a prior is specified, it is updated to a so-called posterior probability distribution using observed data. The conversion is based on Bayes' rule, which first appeared in an article by the British reverend and mathematician Thomas Bayes, published posthumously in 1763.

In the phylogenetic context, Bayes' rule simply states that the posterior probability of a tree is the prior probability of the tree multiplied by the probability that the tree produced the observed data. The

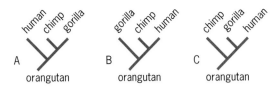

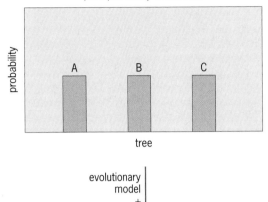

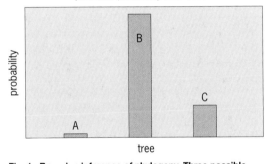

Fig. 1. Bayesian inference of phylogeny. Three possible evolutionary trees (A–C) are considered. Before the analysis, the prior beliefs about the trees must be specified; here, all trees are assigned equal probability. The prior probability distribution is updated using data (such as DNA sequences) and an evolutionary model to give a posterior probability distribution. In this case, we conclude that humans are likely most closely related to chimps.

latter is calculated using a probabilistic model of evolution, for instance a model of nucleotide base substitutions in a DNA sequence, in the same way as in maximum-likelihood inference. Let τ be the tree and X the observed data (aligned DNA sequences, for instance). Bayes' rule states that the posterior probability distribution of the tree given the data, $f(\tau|X)$, is

$$f(\tau|X) = \frac{f(\tau)f(X|\tau)}{f(X)}$$

where $f(\tau)$ is the prior probability of each tree and $f(X|\tau)$ is the probability of each tree producing the observed data. The posterior distribution $f(\tau|X)$ specifies the most likely tree and how probable it is relative to the other possible trees (Fig. 1). The denominator $f(X)$ is simply a normalizing constant ensuring that the posterior probabilities of all trees sum to 1.

It is not sufficient to consider only the relative branch (lineage) positions in an evolutionary tree. Minimally, we must also include the lengths of the branches in the tree, representing the amounts of expected evolution in each lineage. Frequently used models of DNA evolution also take into account differences in the substitution rates between nucleotides and rate variation across sites on the DNA strand. Prior probability distributions must be specified for all these parameters, which are now included together with tree topology (branching structure) in the posterior distribution. The posterior distribution thus becomes a complex multidimensional entity with both continuous and discrete variables.

With these model additions, the denominator of Bayes' rule, $f(X)$, is now a multidimensional summation and integral over all model parameters, making it impossible to derive the posterior distribution analytically (with pen and paper). This is typical of most statistical problems and had made Bayesian inference a marginal approach until the introduction of modern computers and efficient Markov chain Monte Carlo algorithms for sampling the posterior distribution.

Markov chain Monte Carlo algorithms. In the phylogenetic context, MCMC can be described as a technique for simulating evolution under different conditions. One starts with an arbitrarily chosen scenario: a tree, a set of branch lengths, and a nucleotide substitution rate, among other variables. In each step (or generation) of the algorithm, the scenario is changed slightly. For instance, a small portion of the tree may be rearranged, a branch length may be altered, or the nucleotide substitution rate may be increased or decreased. The changes are kept with a probability dependent on how likely the new scenario is relative to the old. When the algorithm has been run long enough, the MCMC procedure will be simulating evolution under each scenario in proportion to the posterior probability of the scenario. For instance, if 90% of the posterior probability falls on a particular scenario, the chain will on average spend 90% of its generations simulating evolution under that scenario.

The trick is to find out how long it takes for the chain to produce a good sample of evolutionary scenarios, or, in mathematical parlance, to converge. Typically, the chain starts from a randomly chosen scenario with low posterior probability and rapidly climbs toward more likely scenarios by successively changing one small component of the scenario at a time (**Fig. 2**). The initial climbing phase is referred to as the burn-in, and these samples are discarded because they mainly reflect the starting point of the simulations.

Often, the MCMC procedure is surprisingly efficient at finding likely evolutionary scenarios. Nevertheless, problems may occur. For instance, if there are several likely scenarios that are distinctly different, it may take a long time for the chain to move between them by randomly changing one component at a time, a problem referred to as slow mixing.

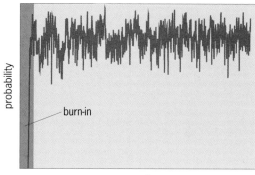

generation

Fig. 2. Generation plot for an MCMC run. The run typically starts from an evolutionary scenario tree with low probability. The probability climbs rapidly as the chain approaches better scenarios during the burn-in phase. When it starts to fluctuate around a stable value, the chain spends most of its time sampling from the most probable scenarios, as it should.

Such problems can be detected by starting independent chains from different starting points. If the final samples are similar, there is a good chance that the chains have found all the likely scenarios.

Metropolis coupling. The computational efficiency of the Bayesian MCMC approach to phylogenetic inference was demonstrated early on. However, it was not until a convergence-acceleration technique known as Metropolis coupling was introduced that the method was successfully applied to problems with hundreds of DNA sequences. Metropolis coupling was originally proposed by C. J. Geyer in 1991 and is based on the idea of using "heated" chains to accelerate mixing of the normal, "cold" chain. The heated chains change the evolutionary scenario more readily than the cold chain. At regular intervals, an attempt is made to swap the scenarios of cold and heated chains. If the swap is successful, the cold chain may jump in a single step between likely but very different scenarios. In other words, the heated chains work as pathfinders for the cold chain, helping it to find all likely scenarios. In 2001, using Metropolis-coupled MCMC analysis, convergence was demonstrated within days on a fast personal computer for a data set with 357 sequences. Only 5 years earlier, few people believed that such large statistical analyses of phylogeny would be possible in the near future.

Emerging benefits. Bayesian MCMC analysis has quickly become a standard tool in the inference of evolutionary trees from molecular data. This has resulted in many old phylogenetic questions being reexamined with new techniques. We have also seen Bayesian MCMC applied to new and more realistic models of evolution, including (1) covarion-like models, allowing the evolutionary rate at a given site on the DNA to vary over time; (2) doublet models, accommodating the fact that changes at one nucleotide site can force changes at another, distant site because the sites interact in the gene product; (3) variable selection models, recognizing that evolution at the protein level favors changes at some sites while selecting

against changes at other sites; and (4) mixed models. The mixed models can, for instance, account for the fact that mitochondrial and nuclear DNA sequences evolve at different rates and have different nucleotide composition. Previously, these differences were typically ignored when both sources of data were used simultaneously in phylogenetic inference.

Another area that has already benefited from the Bayesian MCMC approach is the dating of species divergence events. For instance, dating can be used to determine when human immunodeficiency virus (HIV) first infected humans and when the last common ancestor of humans and chimpanzees lived. Normally, evolutionary models do not distinguish between evolutionary rate and time. Untangling these factors is a difficult problem in which the advantages of the Bayesian MCMC approach have proven useful.

The hope of understanding how evolution works and using that knowledge in phylogeny reconstruction has been one of the main motivations for the development of statistical approaches to the inference of evolutionary trees. Bayesian MCMC techniques seem bound to significantly boost this research program in the near future.

For background information *see* BAYESIAN STATISTICS; MONTE CARLO METHOD; ORGANIC EVOLUTION; PHYLOGENY; STATISTICS in the McGraw-Hill Encyclopedia of Science & Technology.   Fredrik Ronquist

Bibliography. M. Holder and P. O. Lewis, Phylogeny estimation: Traditional and Bayesian approaches, *Nat. Rev. Genet.*, 4:275–284, 2003; J. P. Huelsenbeck et al., Bayesian inference of phylogeny and its impact on evolutionary biology, *Science*, 294:2310–2314, 2001; J. P. Huelsenbeck et al., Potential applications and pitfalls of Bayesian inference of phylogeny, *Systemat. Biol.*, 51:673–688, 2002; J. P. Huelsenbeck and F. Ronquist, MrBayes: Bayesian inference of phylogeny, *Bioinformatics*, 17:754–755, 2001; B. Larget and D. Simon, Markov chain Monte Carlo algorithms for the Bayesian analysis of phylogenetic trees, *Mol. Biol. Evol.*, 16:750–759, 1999; P. O. Lewis, Phylogenetic systematics turns over a new leaf, *Trends Ecol. Evol.*, 16:30–37, 2001.

Biodegradation of pollutants

In the last 100 years, humans have synthesized thousands of new chemicals, some of which have been released in large amounts into the biosphere. Other, naturally occurring pollutants such as cyanides, phenols, polycyclic aromatic compounds, or long-chain aliphatic hydrocarbons originate from coal or crude oil. These may be hazardous if accidentally or deliberately released, because they exhibit acute or chronic toxicity. However, many of these organic compounds can be eliminated by microorganisms (natural or genetically modified), if the appropriate process conditions can be established.

Xenobiotic (synthetic) compounds are rarely, if ever, found in nature and often contain structural

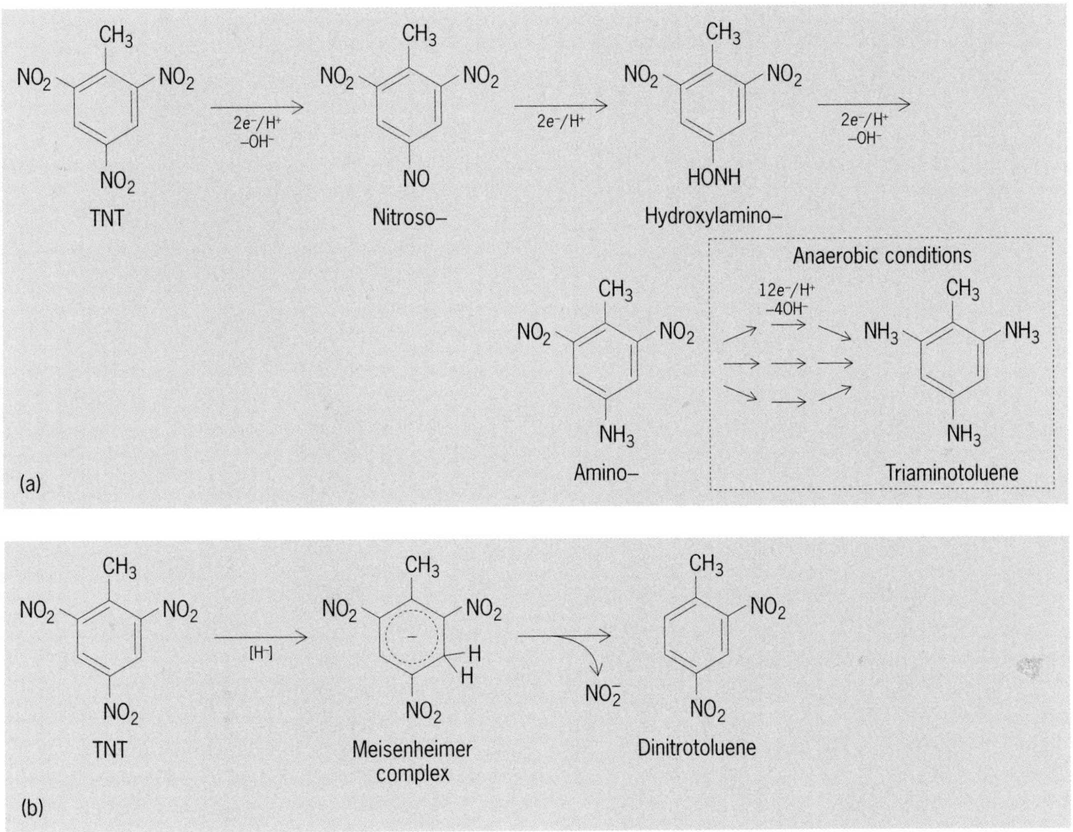

Fig. 1. TNT degradation. (a) Reduction of nitro groups in TNT. (b) Nucleophilic attack on the aromatic ring of TNT by hydride ions.

elements (for example, functional groups or stereoisomers) that cannot be synthesized by living organisms. These chemicals are often resistant to degradation because microorganisms have not been exposed to such structures during their evolution and therefore have not developed the enzymes or pathways necessary to break them down.

The limited capacity of natural ecosystems to process synthetic compounds has led to an increased interest in how these compounds can be degraded, and specifically how microorganisms can be induced to use them as substrates for growth or energy. This is especially the case for synthetic compounds such as nitroaromatics and chlorinated insecticides.

Transformation of polynitroaromatics. In industry, large quantities of nitroaromatics are used in the synthesis of herbicides, pesticides, pharmaceuticals, dyes, and explosives. The most widely used synthetic nitroaromatic compound is the explosive 2,4,6-trinitrotoluene (TNT). Its manufacture, processing, disposal, and storage have contaminated soils and ground water, leading several environmental protection agencies to declare its removal a high priority. TNT and its transformation products have been reported to be mutagenic and carcinogenic and are, in general, toxic to many microorganisms, plants, and animals, including humans.

TNT's persistence and toxicity is largely due to its chemical structure. The electron withdrawing character of the nitro groups removes the π electrons from the aromatic ring. As the number of nitro groups on the aromatic ring increases, oxidative attack on the molecule by living systems becomes more difficult. As a result, no energy-producing catabolic pathways initiated by TNT oxygenation have been identified. The nitro groups on the aromatic ring are readily reducible by reductase enzymes, which are widely distributed among living systems. Reduction of these substituents leads to nitroso, hydroxylamino, and amino derivatives in a series of two-electron transfers (**Fig. 1a**). However, under aerobic conditions nitroso and hydroxylamino derivatives can react to produce recalcitrant (nondegradable) azoxy [Ar—N=N(O)—Ar, where Ar = aryl] compounds.

Many unsuccessful attempts have been made to isolate bacteria that use TNT as a carbon source. The inability of organisms to use TNT is probably due to the unbalanced carbon-to-nitrogen ratio in the molecule, which cannot support growth. However, a few bacteria have been isolated that can use TNT as a nitrogen source.

There seems to be several mechanisms for removing nitrogen from TNT. Hydroxylamino derivatives formed from the reduction of nitroaromatic compounds have recently been identified as potential key intermediates in the metabolism of these nitroorganics. An alternative pathway suggested for eliminating nitro groups from TNT is by nucleophilic attack on the aromatic ring by hydride ions (H⁻). This seems to result in the formation of an unstable interme- diate, called the Meisenheimer complex, which, in order to reestablish its aromatic nature, may lose a nitro group as nitrite (NO_2^-) to form dinitrotoluene (Fig. 1b). This is interesting because dinitro- and mononitrotoluenes are susceptible to degradation by oxidative attack.

The microbial degradation of TNT is influenced by the type of respiration used by the microorganisms, with the use of anaerobic microorganisms receiv- ing increasing attention. Anaerobic sulfate-reducing bacteria of the genus *Desulfovibrio* reduce all three nitro groups to triaminotoluene, a product that can be formed only under low redox (below −100 mV) conditions (Fig. 1a). The exact nature of triamino- toluene metabolism remains unknown. A facultative anaerobic bacterium (capable of functioning aerobi- cally as well), *Pseudomonas* sp., can use TNT as its sole nitrogen source. This strain was found to use TNT as a nitrogen source by releasing nitro groups from the aromatic ring in the form of nitrite, which is subsequently reduced to ammonium and assimi- lated by the bacteria. The strain was shown to use TNT as a terminal electron acceptor, indicating TNT respiration.

Effective clean-up strategies to remove TNT from contaminated sites have been designed. To date, most contaminated soil or water is transported from the site and treated by various physiochemical methods, but this is expensive. Simple and cost- effective anaerobic/aerobic processes have been developed and used for the bioremediation of TNT- contaminated soil. An example is the use of a biore- actor for processing a slurry of TNT contaminated soil mixed with water and nutrients under controlled anaerobic conditions. This method removes substan- tial amounts of TNT, especially when a final aero- bic phase is introduced to remove triaminotoluene. In the bioreactor, TNT is also removed by the irre- versible binding of its derivatives to soil particles, which neutralizes their toxicity.

Increased efforts are being devoted to cleaning contaminated soil and water on-site. These include farming techniques in which contaminated soil is mixed with nutrients and moisture with periodic me- chanical turning. Another option, phytoremediation, uses plants to remove TNT from the soil. Some vari- ations include bacteria-plant combinations in which TNT-degrading bacteria in the soil region surround- ing the plant root (rhizosphere) are nourished by root exudates and carried deep underground (even below the water table) to degrade TNT.

Degradation of lindane. Hexachlorocyclohexane (HCH) is a chlorinated cyclic hydrocarbon that has been used extensively worldwide as an insecticide in agriculture, for wood protection, and for control of vectors of infectious diseases. HCH is used in two compositions: technical grade (a mixture of the α-, β-, γ-, and δ-isomers of HCH, which differ in the orientation of the chlorine atoms with respect to the carbon skeleton) and γ-HCH, also known as lindane. Lindane, the only isomer with insecticidal

activity, is also toxic to higher animals. It tends to accumulate in fatty tissues, has deleterious effects on the immune and nervous systems of humans, and is suspected to be carcinogenic. HCH has been banned in most countries because of its prolonged persistence in the environment, its resistance to bio- logical degradation, and concerns for public health. However, HCH is still used in some developing coun- tries, and HCH-contaminated sites are found world- wide.

Microorganisms play an important role in the degradation of organochlorine pesticides to less toxic metabolites and in their mineralization to car- bon dioxide and water. HCH is metabolized both under aerobic and anaerobic conditions, but the dif- ferent isomers vary in their resistance to microbial degradation. The interconversion of HCH isomers (due to physico-chemical or biological activities) may contribute to the accumulation and prolonged per- sistence of the more stable isomers in the environ- ment. Important steps in the degradation of HCH include the sequential removal of chlorine atoms, which can be achieved by reductive dechlorination (a chlorine atom is replaced with a hydrogen atom) or dehydrochlorination (the removal of both a chlo- rine and a hydrogen atom). Frequent metabolites de- rived from HCH degradation are chlorophenol and chlorobenzene, compounds that are eventually min- eralized to carbon dioxide.

γ-HCH undergoes rapid decomposition under anaerobic or oxygen-limiting conditions that are present in submerged or wetland soils. Several *Clostridium* species have been identified that de- grade γ-HCH under these conditions. An important early metabolite of anaerobic γ-HCH degradation is tetrachlorocyclohexene, which is further metab- olized to chlorobenzene. Conditions under which methane is formed (methanogenic) during the degra- dation of organic material were shown to be favor- able for the anaerobic degradation of α- and γ-HCH.

More recently, HCH isomers were found to de- grade under aerobic conditions. Bacteria able to degrade HCH were identified as *Sphingomonas paucimobilis*, *Rhodanobacter lindaniclasticus*, *Pandoraea* sp., *Anabaena* sp., and *Nostoc ellipsos- sorum*. Furthermore, white-rot fungi (for example, *Phanerochaete chrysosporium*) and microalgae (*Chlorella vulgaris*, *Chlamydomonas reinhardtii*) were reported to transform γ-HCH. In *Sphin- gomonas paucimobilis* UT26, the aerobic lindane degradation pathway and the enzymes involved have been characterized (**Fig. 2**). γ-HCH is initially dehydrochlorinated via pentachlorocyclohexene (γ-PCCH) to tetrachlorocyclohexadiene (1,4-TCDN) by the activity of the dehydrochlorinase LinA. Subse- quently, HCl is removed hydrolytically by the chloro- hydrolase LinB, resulting in the formation of dichlorocyclohexadiene-diol (2,5-DDOL). Addition- ally, trichlorobenzene (1,2,4-TCB) and dichlorophe- nol (2,5-DCP) are formed as end products. 2,5-DDOL is converted to dichlorohydroquinone (2,5-DCHQ) by the dehydrogenase LinC. A glutathione-dependent

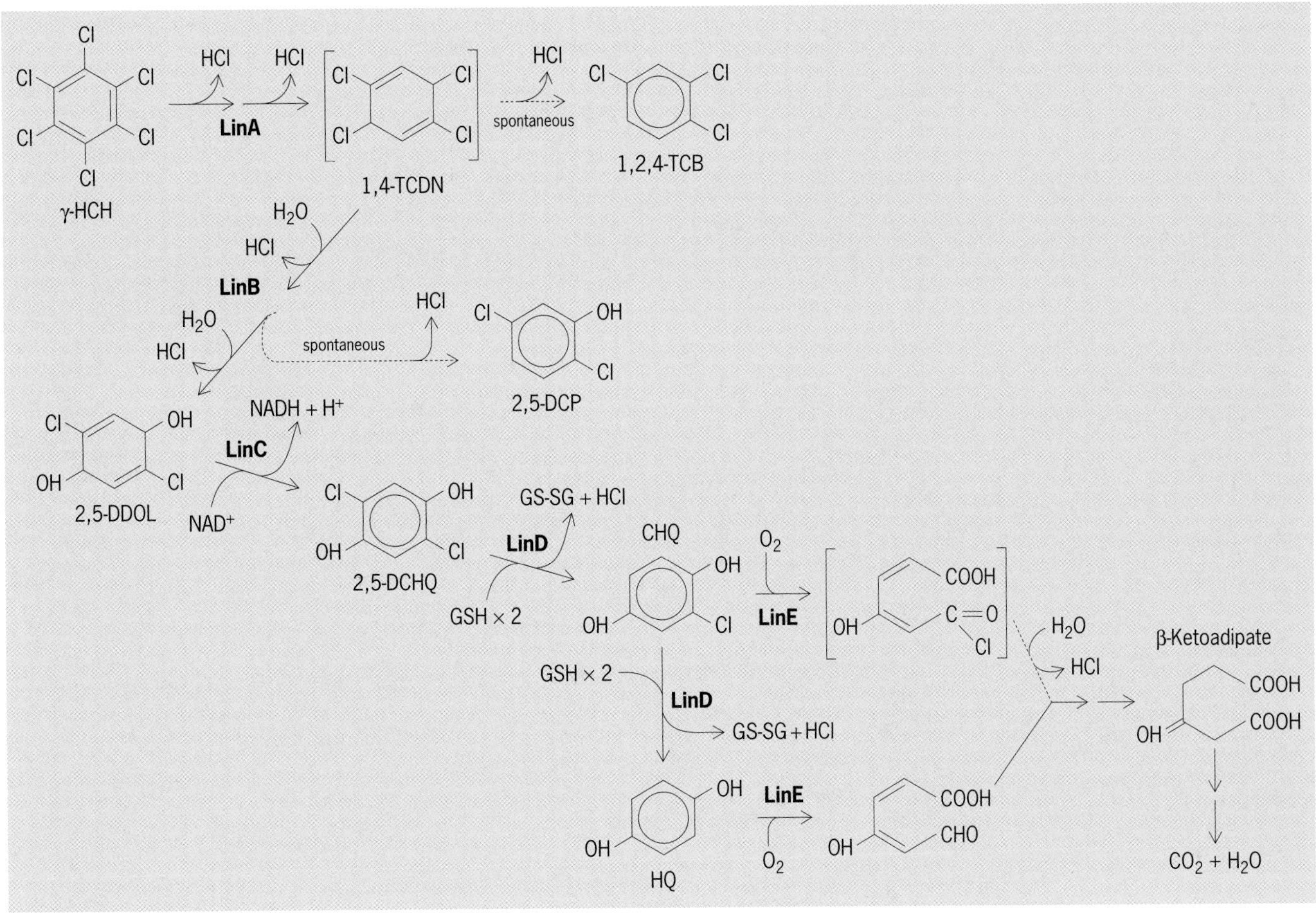

Fig. 2. Proposed aerobic degradation pathway of lindane (γ-HCH) in *Sphingomonas paucimobilis* UT26. Important reactions and metabolites are described in the text. (*Adapted from Y. Nagata, K. Miyauchi, and M. Takagi, 1999, with the authors' permission*)

dehalogenase, LinD, catalyzes the reductive dechlorination to chlorohydroquinone (CHQ) and then to hydroquinone (HQ). The (chloro) hydroquinone deoxygenase LinE mediates the cleavage of either of these metabolites, which are mineralized to carbon dioxide and water via β-ketoadipate.

Rhizoremediation. A recurring issue in TNT and lindane contamination is their persistence in soils. It now is well established that the rhizosphere of plants supports higher bacterial population numbers and greater diversity than the bulk soil, and that such rhizosphere communities can increase the degradation of pollutants in contaminated soils (rhizoremediation). Plants are useful vectors for distributing homogenously pollutant-degrading bacteria in soil and enhancing the catabolic activities against pollutants by providing nutrients in the form of root exudates. Thus, rhizoremediation allows for the in-situ treatment of chemicals such as TNT and lindane in an efficient and inexpensive manner.

For background information *see* AROMATIC HYDROCARBON; BACTERIAL PHYSIOLOGY AND METABOLISM; BIODEGRADATION; BIOSPHERE; ENVIRONMENTAL TOXICOLOGY; ENZYME; HALOGENATED HYDROCARBON; INSECTICIDE; NITRO AND NITROSO COMPOUNDS; NITROAROMATIC COMPOUND; OXIDA-

TION-REDUCTION; RHIZOSPHERE; SOIL ECOLOGY in the McGraw-Hill Encyclopedia of Science & Technology.   Juan L. Ramos; Dietmar Böltner; Pieter van Dillewijn

Bibliography. A. Esteve-Núñez, A. Caballero, and J. L. Ramos, Biological degradation of 2,4,6-trinitrotoluene, *Microbiol. Mol. Biol. Rev.*, 65:335–352, 2001; Y. Nagata, K. Miyauchi, and M. Takagi, Complete analysis of genes and enzymes for gamma-hexachlorocyclohexane degradation in *Sphingomonas paucimobilis* UT26, *J. Ind. Microbiol. Biotechnol.*, 23:380–390, 1999; B. K. Singh et al., Microbial degradation of the pesticide lindane (gamma-hexachlorocyclohexane), *Adv. Appl. Microbiol.*, 47:269–298, 2000; J. C. Spain, J. B. Hugher, and H. J. Knackmuss (eds.), *Biodegradation of Nitroaromatic Compounds and Explosives*, CRC Press, Boca Raton, FL, 1999.

Biodiversity hotspots

Land protection, including the establishment of parks, game reserves, and wildlife sanctuaries, is the major strategy used by both governments and nongovernmental organizations to counter the current extinction crisis. At present, roughly 9% of the Earth's land surface and 1% of the global marine area are protected to some degree. It is widely agreed that these levels fall well below the amounts needed to protect the variety of species and ecosystems across the world.

With limited time and money, conservation biologists must make recommendations about which regions of the planet merit top priority for conservation investment. Because species diversity is not distributed evenly across the planet, exactly which areas get selected for protection will determine how many species are protected. One major approach to establishing priorities for conservation involves the identification of biodiversity hotspots, terrestrial or marine regions that harbor large numbers of plant and animal species in relatively small geographic areas. The criteria for identifying biodiversity hotspots sometimes include the presence of a high degree of anthropogenic (human-produced) threat.

Marine conservation. In the marine realm, coral reefs provide an excellent example of biodiversity hotspots. Coral reefs cover only 0.2% of the world's oceans but are home to approximately one-third of all marine fish species. In addition, 58% of the world's reefs are currently threatened by human activities, including overfishing, pollution caused by agriculture and development, and climate change. Based on such statistics, biologists have argued that money and effort invested in coral reefs should have the highest payoff in terms of the number of marine species protected from imminent threats. More refined analyses have pinpointed exactly which areas of coral reef deserve highest priority for conservation.

Terrestrial conservation. For terrestrial conservation, identification of biodiversity hotspots has typically focused on the distribution of endemic plant species. (Endemic species are those with restricted ranges, so that if these species were lost from a single region, they would be globally extinct.) The most well-known map of terrestrial biodiversity hotspots identifies 25 regions, each of which is home to at least 1500 endemic plant species and has had at least 70% of the original vegetation destroyed. It is well known (although the possible reasons are debated) that for most groups of organisms diversity is greater at lower latitudes, that is, closer to the Equator. It is therefore no surprise that the majority of the 25 terrestrial biodiversity hotspots occur in areas of tropical forest. In addition, five of the hotspots occur in regions with Mediterranean climates, and nine are composed mainly of islands. Together, these hotspot areas cover only 1.4% of the Earth's total land area. This relatively small area is home to 44% of all plants species and 35% of all mammal, reptile, bird, and amphibian species. Based on these statistics, biologists have argued that conservation efforts should be directed primarily to these 25 hotspot regions.

Alternative identification approaches. The 25 biodiversity hotspots have received a great deal of public and scientific attention, but it is certainly not the final word on the subject of conservation priorities. Vig-

orous debate continues regarding how best to prioritize regions for conservation investment, and a number of alternative schemes have been proposed. One major issue is exactly which data should be used to assess the biological value of a region. Indeed, the number of endemic plant species is but one among a great variety of measures that might be used to compare the biological value of ecosystems. Other approaches could focus on total diversity rather than the number of endemic species, the diversity of particular animal groups rather than plants (although the data tend to be more complete for plants than for animals), or the number of genera, families, or other taxonomic levels rather than species. Beyond the wide range of potential diversity metrics, many possibilities exist for assessing the degree of threat that a region is experiencing. Besides destruction of original vegetation, analyses could focus on human population size and growth rate or the number of species at risk of extinction. Another recent scheme prioritizes areas where the least conservation work has taken place. This analysis relied on a variety of data types, including the proportion of land area that has been protected as parks or wildlife reserves; the number of genetic resource collections, zoos, botanical gardens, and other reference collections in a country; and the nation's participation in various international conventions related to the environment.

It is unlikely that biologists will reach any consensus regarding which measures are the best indicators, and the use of different combinations of metrics for biological value and threat can dramatically alter the list of top conservation priorities. Thus, it seems prudent to search for regions that "come out on top" regardless of exactly which data one chooses to analyze.

Measures of feasibility. The 25 biodiversity hotspots are based on comparisons of ecoregions rather than of nations. Ecoregions are areas that, in the judgment of biologists, represent unique and identifiable sets of ecological communities and species. The use of ecoregions as the basic unit for hot spot analysis is not without controversy. Some biologists have argued that prioritization schemes should focus on nations because protection of land is chiefly accomplished via interaction with national governments, and the success of conservation efforts is inevitably affected by national regulations and economic factors. To date, only a handful of schemes for identifying conservation priorities have focused on nations rather than ecoregions. However, none has explicitly incorporated information about how the quality of regulatory agencies, land prices, political stability, and other measures of feasibility vary across nations. An interdisciplinary analysis, incorporating information from both the social and natural sciences, would surely yield important insights regarding which regions could be most fruitfully targeted for conservation action.

Biodiversity coldspots. Every ecosystem on Earth provides a suite of products and services that are of economic value to humans. For example, tropical forests supply products such as wood, food, rubber, and medicinal plants. These same forests also provide valuable services; most notably, tropical forests contribute to regulation of atmospheric carbon dioxide via photosynthesis and respiration. Recently, a group of economists and scientists estimated the economic value per unit area for a wide variety of ecosystem types. They estimated the annual value of tropical forests to be about $2007 per hectare.

In contrast to tropical rainforests, which harbor many thousands of plant and animal species, some ecosystems are biodiversity coldspots that host just a handful of species. Certain wetlands, such as salt marshes, provide a good example of a biodiversity coldspot. Yet, despite usually consisting of only a few dozen plant species, wetlands offer a wide variety of ecosystem services such as flood regulation, waste treatment, and regulation of water supply, with an estimated annual value of $14,785 per hectare. Thus, a hectare of wetland is over seven times more valuable to the human economy than is a hectare of tropical forest. But if we were to perform a hotspot analysis based on the diversity of plant species, wetlands would rank at the very bottom of our conservation priorities.

The identification of biodiversity hotspots ignores the value of ecosystem products and services and therefore can overlook ecosystems that have low biodiversity but nonetheless provide critical ecosystem services. Given that diversity and economic value do not always go hand in hand, it makes sense to go beyond a hotspots analysis and further rank the value of various ecosystems on the basis of the products and services that each ecosystem provides.

Representation approach. Although the identification of biodiversity hotspots can help to draw attention to areas of highest concern, it would be neither prudent nor desirable for conservation organizations to invest solely in biodiversity hotspots to the neglect of other regions. All ecosystems provide valuable products and services, and each of the major ecosystems has species that can be found nowhere else. In addition, preserving the largest number of species is not the only goal for conservation. Besides protecting the diversity of species, conservation seeks to preserve the diversity of ecosystem types. Thus, many scientists and conservation organizations advocate a representation approach that seeks to ensure protection for representative examples of all major ecosystems.

Identification of conservation priorities is an interesting academic pursuit that can help direct the activities of on-the-ground conservation organizations. However, it is important to recognize that there are many different means by which one might prioritize regions for conservation. Because prioritization must rely on subjective judgments regarding which measures best indicate biological value, threat, and feasibility, there simply is no single best way to go about identifying priorities.

For background information *see* BIODIVERSITY; CONSERVATION OF RESOURCES; ENDANGERED SPE-CIES; ENVIRONMENTAL MANAGEMENT; MARINE CONSERVATION; POPULATION ECOLOGY in the McGraw-Hill Encyclopedia of Science & Technology.

Michelle A. Marvier

Bibliography. R. Constanza et al., The value of the world' ecosystem services and natural capital, *Nature*, 387:253–260, 1997; N. Myers et al., Biodiversity hotspots for conservation priorities, *Nature*, 403:853–858, 2000; D. M. Olson and E. Dinerstein, The global 200: A representation approach to conserving the Earth's most biologically valuable ecoregions, *Conserv. Biol.*, 12:502–515, 1998; W. Reid, Biodiversity hotspots, *Trend Ecol. Evol.*, 13:275–280, 1998; T. D. Sisk et al., Identifying extinction threats: Global analysis of the distribution of biodiversity and the expansion of the human enterprise, *Bioscience*, 44:592–604, 1994.

Biofuel cells

Biofuel cells produce electrical energy from organic materials using electrochemical processes that are catalyzed by biological substances (enzymes or whole microbial cells). In contrast, regular fuel cells use inorganic catalysts (such as platinum) to enhance the electrochemical processes. Most organic substrates (reactants) undergo combustion with the

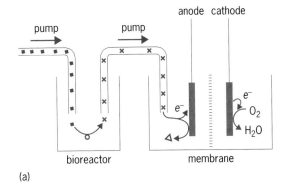

(a)

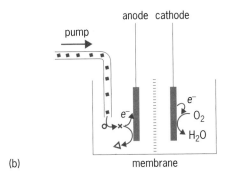

(b)

Key:

∘ microbial cell   × fuel product

▪ primary substrate   ◁ oxidized fuel

Fig. 1. Schematic configuration of a microbial biofuel cell (*a*) with a microbial bioreactor providing fuel separated from the anodic compartment of the electrochemical cell, and (*b*) with a microbial bioreactor providing fuel directly in the anodic compartment of the electrochemical cell.

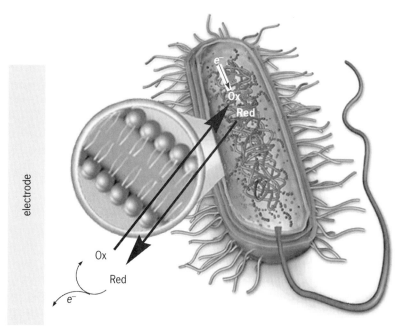

Fig. 2. Microbial cell electrically connected to a biofuel cell anode by means of a diffusional electron transfer mediator crossing the cellular membrane. Red, reduced form of the mediator generated by the reductive electron transport within the microbial cell; Ox, oxidized form of the mediator produced upon the electron transfer to the anode.

evolution of energy. The biocatalyzed oxidation of organic substrates by oxygen or other oxidizers at two electrode interfaces provides a means for converting chemical energy to electrical energy. Abundant

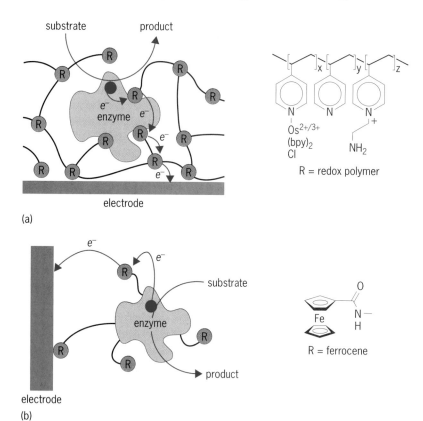

Fig. 3. Oxidative enzymes electrically connected on the anode (a) by redox polymers surrounding the enzyme molecules or (b) by redox-mediator molecules (for example, ferrocene) tethered covalently to the enzyme molecules.

organic raw materials, such as alcohols, organic acids, or sugars, can be used as substrates for the oxidation process. Molecular oxygen (O_2) or hydrogen peroxide (H_2O_2) can act as the substrate for the reduction process. The organic materials are oxidized at the electrode, called the anode, generating electrons which go through the external electrical circuit to the electrode, called the cathode, where they are accepted by the oxidizer. This process generates the electrical power (voltage and current) released by the biofuel cell. In order to produce continuous electrical power, aqueous solutions containing the organic fuel and the oxidizer are pumped to the respective electrodes in the biofuel cell, where these electrochemical processes take place.

Microbial biofuel cells. The use of entire microorganisms as microreactors in fuel cells eliminates the need to isolate individual enzymes and allows the active biomaterials to work under conditions close to their natural environment, thus at a high efficiency. However, whole microorganisms are difficult to handle because they require particular conditions to remain alive, and their direct electrochemical contacting with electrode support is virtually impossible. Thus, special means must be designed for electrical charge transport between the intercellular space and the electrode.

Microbial bioreactor. Microorganisms have the ability to produce electrochemically active substances; these may be the metabolic intermediaries or final products of anaerobic respiration. For the purpose of energy generation, fuel substances can be produced in one place and then transported to a biofuel cell. In this case, the biocatalytic microbial reactor is producing the biofuel, and the biological part of the device is not directly integrated with the electrochemical part (**Fig. 1a**). This scheme allows the electrochemical part to operate under conditions that are not compatible with the biological part of the device. The two parts can even be separated in time, operating individually. The most widely used fuel in this scheme is hydrogen gas, allowing the use of a bioreactor with well-developed and highly efficient conventional H_2/O_2 fuel cells. Various bacteria and algae, for example *Escherichia coli, Enterobacter aerogenes, Clostridium butyricum, C. acetobutylicum,* and *C. perfringens,* have been found to actively produce hydrogen under anaerobic conditions.

Microbiological fermentation. This process proceeds directly in the anodic compartment of the fuel cell, supplying the anode with the in situ–produced fermentation products (Fig. 1b). In this case, the operational conditions in the anodic compartment are dictated by the biological system, so they are significantly different from those in the conventional fuel cells. This is a real biofuel cell, not simply a bioreactor combined with a conventional fuel cell. Its configuration often is based on the biological production of hydrogen gas, but the electrochemical oxidation of H_2 occurs in the presence of the biological components under mild conditions. Other metabolic

window on the properties of the quarks that inhabit protons and neutrons.

For background information *see* BIG BANG THEORY; DEUTERON; I-SPIN; MESON; NEUTRON; NUCLEON; PARTICLE ACCELERATOR; PROTON; QUARKS; SYMMETRY BREAKING; SYMMETRY LAWS (PHYSICS) in the McGraw-Hill Encyclopedia of Science & Technology.

Edward Stephenson

Bibliography. G. A. Miller, B. M. K. Nefkens, and I. Šlaus, Charge symmetry, quarks, and mesons, *Phys. Rep.*, 194:1–116, 1990; U. van Kolck, J. A. Niskanen, and G. A. Miller, Charge symmetry violation in $pn \rightarrow d\pi^0$ and chiral effective field theory, *Phys. Lett. B*, 493:65–72, 2000.

Chemical thermodynamics modeling

The thermodynamic modeling of liquids and liquid mixtures is important to both the chemist and the chemical engineer. The chemist is interested in the physical forces that influence the behavior of liquid mixtures. The chemical engineer is interested in modeling the behavior of liquids and liquid mixtures for use in industrial applications, for example to predict the properties of multicomponent mixtures in the petrochemical, pharmaceutical, and refrigeration and air-conditioning industries. Testing the models against experimentally measured properties makes it possible to determine whether the ideas on which the models are based are correct.

The thermodynamic modeling of the liquid phase is approached from two points of view: using an equation of state or an expression for the excess Gibbs energy. The equation-of-state approach is generally used for simple liquid mixtures. These include mixtures of low- and moderate-molar-mass hydrocarbons (such as hexane and heptane). Mixtures of these compounds do not show great diversity in size, structure, and chemical nature. The excess Gibbs energy models are used when the compounds in the mixture show diversity in structure, size, and chemical nature and there are complicated intermolecular molecular phenomena such as hydrogen bonding (for example, a mixture of ethanol and diethyl ether). The excess Gibbs energy models are usually applied to mixtures that involve dissimilar chemical species and to mixtures containing sulfur, nitrogen, oxygen, or halogen compounds.

Equations of state. The ideal gas law, Eq. (1) for one mole of gaseous substance, is probably the most well-known equation of state,

$$Pv = RT \qquad (1)$$

where P is the pressure, T is the temperature, v is the molar volume, and R is the universal gas constant. An ideal gas considers all particles in a system to have zero volume, with no interaction between the particles. In essence, the particles lack any distinguishing features. Condensation and the liquid phase do not exist within the framework of the ideal gas law.

S. D. van der Waals modified the ideal gas equation by acknowledging that particles in a system interact with each other and have a finite volume. In a real system the space available to the particles (free volume) is the difference between the volume of the system and the volume of the particles. Real systems are composed not of noninteracting particles but of particles between which there are forces of repulsion and attraction. These forces affect the pressure of the system. The result of considering this more realistic behavior is the van der Waals equation, which, for one mole of a gaseous substance, is Eq. (2),

$$P = \frac{RT}{v - b} - \frac{a}{v^2} \qquad (2)$$

where a and b are constants that account for interactions between particles and the finite volume of the particles. These parameters are characteristic of the species in the system. The van der Waals equation is an example of a cubic equation of state. An important property of cubic equations of state is that two different solutions constitute the volume. These are intended to be the liquid and gas/vapor volume; hence the cubic equations of state allow for the simultaneous existence of liquid and gaseous phases. The van der Waals equation is capable of describing the formation of a liquid phase from a vapor phase, but the actual densities are not accurate.

An equation that is commonly used for quantitative calculations is the Redlich-Kwong equation of state, modified by G. Soave, shown as Eq. (3).

$$P = \frac{RT}{v - b} - \frac{a(T)}{v(v - b)} \qquad (3)$$

This equation is a modification of the van der Waals equation and is a cubic equation of state. In this case the constant a is temperature-dependent. The Redlich-Kwong equation of state is capable of accurately predicting liquid densities.

The application of equations of state to mixtures is carried out assuming the equation of state (either Eq. 2 or 3) is valid for the mixture and the constants a and b are dependent on the composition of the mixture. The constants for the mixture are determined by the application of mixing rules. A simple mixing rule for a two component mixture is shown by Eqs. (4) and (5),

$$b_{mix} = x_1^2 b_{11} + 2x_1 x_2 b_{12} + x_2^2 b_{22} \qquad (4)$$

$$a_{mix} = x_1^2 a_{11} + 2x_1 x_2 a_{12} + x_2^2 a_{22} \qquad (5)$$

where x is the mole fraction of the component in the mixture and a_{11}, a_{22}, b_{11}, and b_{22} are the pure component parameters. The parameters a_{12} and b_{12} are called binary parameters and obtained by Eqs. (6) and (7), respectively.

$$a_{12} = (a_{11} a_{22})^{1/2} (1 - k_{12}) \qquad (6)$$

$$b_{12} = \frac{1}{2}(b_{11} + b_{22})(1 - c_{12}) \qquad (7)$$

The constants k_{12} and c_{12} are characteristic of the specific interactions that exist between the different components of the mixture. These are usually determined by fitting the equation of state to experimental data such as vapor-liquid equilibrium.

Once an equation of state is known for a system, other properties such as excess enthalpies, excess volumes, activity coefficients, and fugacities can be determined using the standard equations of thermodynamics. All of these properties are used in chemical thermodynamics modeling.

A practical use of an equation of state is the prediction of bubble pressures and molar volumes of mixtures of chlorofluorocarbons. These compounds are used as refrigerants in both industrial and domestic heat pumps, such as in refrigerators and air conditioners. The equation of state is fitted to experimental vapor-liquid equilibria data to obtain constants c_{12} and k_{12}. Thereafter the bubble pressures and liquid molar volumes at any composition and temperature can be calculated. The constants a and b are pure component properties and are easily calculated from the critical temperature and critical pressure data.

Excess Gibbs energy models. The more popular approach for modeling liquids and their mixtures is an expression for the excess Gibbs energy—the difference in the values of the Gibbs energy of a real solution and an ideal solution (some defined reference state). One of the many uses of the excess Gibbs energy model is the representation of liquid-liquid equilibrium. For example, in liquid-liquid extraction (selective removal of a liquid constituent from solution) it is important to calculate the final compositions of the extract and raffinate (remaining aqueous solution) phases. This knowledge is used to determine the amount of extractive solvent that is necessary to obtain an efficient separation. Given a few experimental data points, it is possible to determine the parameters of the excess Gibbs energy model. These parameters are then used to determine the composition of the two liquid phases that will form for a mixture of any composition.

Examples of expressions for the excess Gibbs energy include the Margules equation, nonrandom two liquid (NRTL) theory, and the universal quasichemical theory (UNIQUAC). These models are commonly used in the petrochemical industry.

Margules equation. The simplest expression for the excess Gibbs energy is the two-suffix Margules equation (8),

$$g^E = Ax_1x_2 \qquad (8)$$

where g^E is the excess Gibbs energy, A an interaction parameter, and x the mole fraction. This expression is purely empirical and the parameter A has no physical significance. The excess Gibbs energy is related to the excess volume, v^E, and excess enthalpy, b^E, as shown in Eqs. (9) and (10).

$$\left(\frac{\partial g^E}{\partial P}\right)_{T,x} = v^E \qquad (9)$$

$$\left(\frac{\partial g^E/T}{\partial T}\right)_{P,x} = -\frac{b^E}{T^2} \qquad (10)$$

The activity coefficient, γ of component i, is given by Eq. (11),

$$RT \ln \gamma_i = \left(\frac{\partial n_T g^E}{\partial n_i}\right)_{T,P,n_j} \qquad (11)$$

which is the derivative of the excess Gibbs energy with respect to n_i, the number of moles of component i, when the all other components, temperature, and pressure are kept constant. The symbol n_T represents the total number of moles.

NRTL theory. The NRTL is a less empirical theory that was developed by applying G. M. Wilson's local composition concept to R. L. Scott's two-liquid theory. Scott postulated a liquid system in which two types of cells can be found. According to Scott's two-liquid theory, two cells are found in a binary mixture (see **illus.**).

Applying the local composition concept proposed by Wilson, the local mole fraction of molecules 1

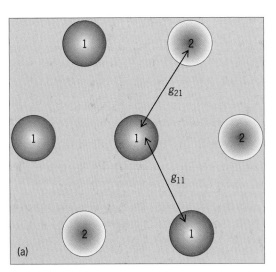 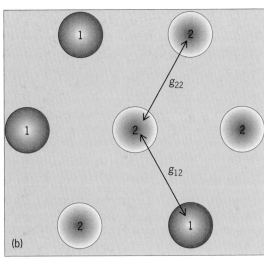

Two types of cell as described by Scott's two-liquid theory for binary mixtures. (*a*) Molecule 1 is at the center. (*b*) Molecule 2 is at the center.

surrounding a central molecule 1, x_{11}, and the local mole fraction of molecules 2 surrounding a central molecule 1, x_{21}, are related as shown in Eq. (12),

$$\frac{x_{21}}{x_{11}} = \frac{x_2}{x_1} \frac{\exp(-g_{21}/RT)}{\exp(-g_{11}/RT)} \qquad (12)$$

where g_{21} is the energy of interaction between a 1-2 pair and g_{11} is the interaction between a 1-1 pair, x_1 and x_2 are overall mole fractions of components 1 and 2 in the mixture. Using Scott's two-liquid theory, H. Renon and J. Prausnitz modified the expression for the local composition and included an additional parameter, α_{12}, which is characteristic of the nonrandomness of the mixture. Equation (13) is used for a binary mixture.

$$\frac{x_{21}}{x_{11}} = \frac{x_2}{x_1} \frac{\exp(-\alpha_{12}g_{21}/RT)}{\exp(-\alpha_{12}g_{11}/RT)} \qquad (13)$$

Using this definition for the local composition, Renon and Prausnitz developed the NRTL model, according to which the expression for the excess molar Gibbs energy, G_m^E, for a binary mixture is shown in Eq. (14),

$$\frac{G_m^E}{RT} = x_1 x_2 \left[\frac{\tau_{21}G_{21}}{x_1 + x_2 G_{21}} + \frac{\tau_{12}G_{12}}{x_2 + x_1 G_{12}} \right] \qquad (14)$$

where Eqs. 15–18 apply.

$$\tau_{12} = \frac{g_{12} - g_{22}}{RT} \qquad (15)$$

$$\tau_{21} = \frac{g_{21} - g_{11}}{RT} \qquad (16)$$

$$G_{12} = \exp(-\alpha_{12}\tau_{12}) \qquad (17)$$

$$G_{21} = \exp(-\alpha_{12}\tau_{21}) \qquad (18)$$

The term $g_{12} - g_{21}$ indicates the energy of interaction between components 1 and 2 in a cell where molecule 2 is the central species, and $g_{21} - g_{11}$ indicates the energy of interaction between molecules 1 and 2 in a cell where molecule 1 is the central species. The nonrandomness parameter, α_{12}, usually has a value in the range 0.1 to 0.3, which indicates a coordination number of 6 to 12. This relationship between the coordination number and the α_{12} does not hold under all circumstances, and α_{12} is commonly assumed to be an empirical constant. According to the NRTL equation, there are three adjustable parameters per binary pair in a mixture. The NRTL model is also applicable to multicomponent mixtures, and the equation for the excess molar Gibbs energy of a multicomponent mixture of N components is shown in Eq. (19).

$$\frac{G_m^E}{RT} = \sum_i^N x_i \frac{\sum_j \tau_{ji}G_{ji}x_j}{\sum_k^N G_{ki}x_k} \qquad (19)$$

In the case $N = 2$, Eq. (19) reduces to Eq. (14) because $\tau_{ii} = 0$ and $G_{ii} = 1$.

Universal quasichemical model. The universal quasichemical model (UNIQUAC) developed by D. S. Abrams and Prausnitz also applies the local-area fraction concept to E. A. Guggenheim's quasichemical theory. The quasichemical theory is a lattice theory and postulates that a liquid can be represented by a three-dimensional lattice. The UNIQUAC model assumes that there are two contributions to the excess Gibbs energy, G^E, namely a combinatorial contribution, $G_{(combinatorial)}^E$, which describes the entropic contribution to the excess Gibbs energy, and a residual contribution, $G_{(residual)}^E$, which is the contribution to the excess Gibbs energy of intermolecular forces. This is shown in Eq. (20).

$$G^E = G_{(combinatorial)}^E + G_{(residual)}^E \qquad (20)$$

For a mixture of N components, Eq. (21) shows how $G_{(combinatorial)}^E$ is obtained with the parameters θ_i and Φ_i, where Eqs. (22) and (23) apply.

$$\frac{G_{(combinatorial)}^E}{RT} = \sum_i^N x_i \ln \frac{\Phi_i}{x_i} + \frac{z}{2} \sum_i^N q_i x_i \ln \frac{\theta_i}{\Phi_i} \qquad (21)$$

$$\theta_i = \frac{q_i x_i}{\sum_j q_j x_j} \qquad (22)$$

$$\Phi_i = \frac{r_i x_i}{\sum_j r_j x_j} \qquad (23)$$

Equation (24) shows how $G_{(residual)}^E$ is obtained,

$$\frac{G_{(residual)}^E}{RT} = -\sum_i^N q_i x_i \ln \left(\ln \sum_j^N \theta_j \tau_{ij} \right) \qquad (24)$$

where

$$\tau_{ij} = \exp\left(-\frac{u_{ji} - u_{jj}}{RT} \right)$$

and $u_{ij} - u_{jj}$ and $u_{ji} - u_{ii}$ are adjustable binary parameters characteristic of the i-j interaction and the j-i interaction, and r_i and q_i are pure component size and surface parameters, respectively.

The UNIQUAC model is the basis for the group contribution model, universal quasichemical functional group activity coefficient (UNIFAC). The UNIFAC method is a quantitative structure property relationship method in that the constituent molecules are broken down into functional groups for which interaction parameters have already been determined. These are then used to calculate thermodynamic properties.

A difficulty associated with these excess Gibbs energy models is their failure to model the pressure dependence of the excess Gibbs energy. The pressure dependence of the excess molar Gibbs energy is expressed through the excess molar volume, as shown in Eq. (9). The excess Gibbs energy models based on a quasilattice theory predict a zero value for the excess molar volume.

For background information *see* ACTIVITY (THERMODYNAMICS); CHEMICAL EQUILIBRIUM; ENTHALPY; FREE ENERGY; FUGACITY; GIBBS FUNCTION; LIQUID; SOLUTION; THERMODYNAMICS; VAN DER WAALS

EQUATION in the McGraw-Hill Encyclopedia of Science & Technology.   Pavan K. Naicker

Bibliography. J. M. Prausnitz, R. N. Lichtenthaler, and E. Gomez de Azevedo, *Molecular Thermodynamics of Fluid-Phase Equilibria*, 3d ed., Prentice Hall, Saddle River, NJ, 1999; B. E. Poling, J. M. Prausnitz, and J. P. O'Connell, *The Properties of Gases and Liquids*, 5th ed., McGraw-Hill, New York, 2000; J. S. Rowlinson and F. L. Swinton, *Liquids in Liquid Mixtures*, 3d ed., Butterworth Scientific, London, 1982; S. I. Sandler (ed.), *Models for Thermodynamic and Phase Equilibria Calculations*, Marcel Dekker, New York, 1994; J. M. Smith and H. C. Van Ness (eds.), *Chemical Engineering Thermodynamics*, 1995; S. M. Walas, *Phase Equilibria in Chemical Engineering*, Butterworth-Heinemann, Boston, 1985.

Classification nomenclature

Charles Darwin triggered evolutionary thinking in natural history. Today, with the availability of powerful analysis methods such as parsimony, maximum likelihood, and Bayesian inference in computer software, we are well equipped to infer phylogenetic trees (evolutionary relationships among species) and to begin to unravel the details of evolutionary history and biodiversity. This increased understanding and flow of phylogenetic information sets new demands for a system of nomenclature to achieve efficient and effective communication. To a large extent, the Linnaean system of nomenclature—a system that has been around for about 250 years—is still in use. It was developed before evolution entered the limelight of biological thinking, and during the last 30 years Linnaean approaches have been increasingly criticized for not being able to handle information about phylogeny. However, there are a variety of issues involved when choosing between a traditional approach to nomenclature that has its roots in the writings of Linnaeus, and more modern approaches that aim at making phylogenetic theory the central principle in nomenclature. *See* BAYESIAN INFERENCE (PHYLOGENY).

Philosophy of systematics. A major theme in the "evolutionization" of taxonomy has been the ontological status of biological taxa. The traditional view, which can be traced back to Plato and Aristotle, holds that natural groups are classes of things—natural kinds. A natural kind has logical and necessary properties, an essence, that applies independent of space and time. To belong to a natural kind, an organism needs to possess the defining properties in question. For Linnaeus, each plant genus was a natural kind, and its specific reproductive characters constituted its essence.

Natural kind versus individual. Philosophers have long been intrigued by natural kinds because of their association with natural laws. However, natural kinds are universal and cannot evolve. At the level of particular taxa, this concept does not fit very well with evolutionary theory. This was one reason why Michael T. Ghiselin suggested that species are more beneficially treated as individuals than as natural kinds. Individuals, unlike classes, are restricted in time and space, since they have a unique beginning and a unique end. No logical and necessary properties are involved, because each individual's birth and life is contingent upon history. The individuality thesis thus breaks with tradition by arguing that taxa at all levels are individuals. Treating taxa as ontological individuals comes with benefits such as allowing for evolutionary contingencies and thus allowing taxa to evolve. A disadvantage may be the loss of association with natural laws, but these may be found elsewhere in such systems.

Phylogenetic nomenclature. The starting point for phylogenetic nomenclature is that names should be formulated in terms of evolution and common descent rather than type specimens, categorical ranks, and morphological characters. Taxon names are strictly connected to phylogenetic tree topology and evolutionary history. No traditional type specimens are used. Instead, each name is attached to a clade—a taxonomic group containing a common ancestor and all its descendants. Phylogenetic nomenclature is both a logical outcome of Ghiselin's individuality thesis and, according to some advocates, an attempt to bridge the gap between natural kind and individual. This bridging is manifested in the definitions of phylogenetic taxon names. Some believe that phylogenetic nomenclature is successful in taking the individuality thesis to its logical conclusion but that it goes astray when it blurs the distinction between class and individual.

Categorical ranks and taxa. Taxa are commonly ranked into categories. A taxon name in a Linnaean system is tightly linked to category. According to the zoological code, each specimen must be allocated to a species, each species to a genus, and each genus to a family. For instance, *Homo sapiens* is the species, *Homo* the genus, and Hominidae the family. The help of types assures this association. The botanical code uses even more mandatory categories. Linnaeus employed five categories for each kingdom and used Aristotle's logical division to allocate taxa to categories. For Linnaeus the categories were real, which can be seen in his argument that the genus provided the essence for the species. Although not seen as real anymore, categorical ranks still play pivotal roles in Linnaean systems of nomenclature. For instance, sister taxa must belong to the same rank. Besides introducing monotypic taxa and redundant names, such a procedure retains a notion of categories of the same rank being comparable units. This often takes the form of using species numbers or numbers of genera and families as measures of biodiversity. Similarly, taxa of different ranks are not considered comparable units. Another supposed advantage with ranks is the knowledge that taxa of the same rank cannot be nested within each other. The latter is true, but information on nestedness is achievable also within a rank-free system.

Phylogenetic nomenclature discards categorical ranks, recognizing that categories are not comparable units and their use is misleading in biodiversity studies. The problems with ranks become obvious if we take a look at more likely units of biodiversity—lineages and clades. Questions like "how many lineages are there?" or "how many clades are there?" become pointless, since there are no answers. These are relative concepts, illustrating the fractal nature of the tree of life and the need to let a phylogenetic hypothesis be the focus, rather than the categories, when biodiversity is quantified. Phylogenetic nomenclature helps to put focus on phylogenetic trees by offering an explicit link between names and parts of species history, that is, clades.

Traditional binomial naming. Linnaeus offered two main reasons in favor of a binomial system of names for species: a practical one for increasing the ability to memorize names, and a theoretical one. His theoretical argument had its roots in essentialism; he argued that the genus was the category in which the essence of a taxon was expressed. Accordingly, the essence of a species was to be found in its genus group name; therefore, it was important to provide a species with both a generic and a specific name. Without its generic name, a species name was scientifically empty.

In phylogenetics, the binomial name is associated with a hypothesis of relationship to each described species. This is problematic for a couple of reasons. First, it is unfortunate to make assumptions about relationships at the descriptive stage of species identification (that is, based on observation of specimens prior to phylogenetic analysis). Second, it pinpoints species as a unique category, which contradicts the idea of recognizing only clades and lineages. Uninomials seem more in line with phylogenetic thinking.

New phylogenetic alternative. To avoid the pitfalls of traditional Linnaean naming in phylogenetic nomenclature, three new methods of phylogenetic naming have been proposed: node-, stem-, and apomorphy-based. In node-based naming, taxon name A might refer to the least inclusive clade containing X and Y. In stem-based naming, A would refer to the most inclusive clade containing X and Y but not Z. In apomorphy (derived feature)–based naming, A would refer to the clade identified by a feature synapomorphic (sharing a derivation) with a feature in specimen (taxon) X. Differences between a traditional approach and these phylogenetic alternatives become obvious when the phylogenetic hypothesis changes (**Fig. 1**).

Definitions, sameness, and stability. Both traditional and phylogenetic nomenclature is built on the idea that taxon names can be defined. The traditional approach seeks stability by defining names based on types and categorical ranks, whereas the phylogenetic alternative seeks stability by defining names based on referring to the same ancestor or ancestry across phylogenetic hypotheses. Both seem to fail. The association with types and categories causes many unwanted name shifts when the discovery of

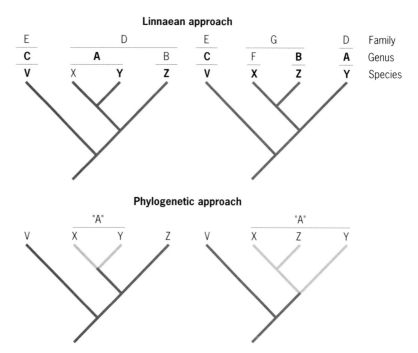

Fig. 1. Comparison between the traditional Linnaean approach to nomenclature and a phylogenetic alternative (node-based naming). Suppose that all we want to do is to give a name ("A") to a clade containing X and Y. In the Linnaean system this means that we have to introduce names for sister taxa, assigning all taxa to the categories species, genus, and family, and designate type species (boldface). No explicit reference to phylogeny is made. The phylogenetic alternative provides an explicit reference to evolutionary history, and nothing but the clade containing X and Y needs to be named. When the hypothesis of relationship changes, the phylogenetic alternative is cleaner and more explicit about what it refers to.

additional information about a species causes it to change categories, and it is questionable that phylogenetic definitions always refer to the same ancestor or ancestry across hypotheses.

Consider the sequence of phylogenetic hypotheses in **Fig. 2**. If the name *Hypothetica* is defined so as to refer to a particular clade (for example, the most recent common ancestor of taxa A and C and all its descendants), one may argue, judging exclusively from terminal taxa (A–D), that the common ancestor in tree *a* is the same as that in *b*. If we also consider internal lineages, however, the two trees differ (for example, the lineage leading to clade BC in tree *a* is not present in tree *b*), so the common ancestor referred to in *a* may not be the same as the common ancestor in *b*. In tree *c*, one additional taxon (X) is included, indicating that more gene pools are involved and therefore the common ancestor in *c* must be different than that in *a* and *b*, regardless of whether only terminal taxa or also internal lineages are considered. If we compare the three trees in terms of specific ancestry (including not only lineages but their topological relationships), all three trees represent different hypotheses of evolutionary history. For these reasons the idea of sameness across hypotheses, and therefore name stability, is hardly guaranteed.

Lately it has been questioned if stability really is a prime goal of nomenclature. Perhaps stability is better treated as a by-product of taxonomic practice. From the ontological view of individuality,

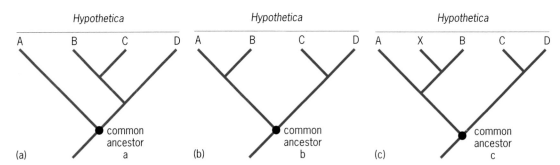

Fig. 2. Problem of sameness in phylogenetic nomenclature. In trees *a* and *b*, the taxon name *Hypothetica* refers to the same content with regard to terminal taxa (A, B, C, and D) but to different topological relations and therefore, possibly, a different common ancestor. In tree *c*, the additional taxon (X) rules out the possibility of reference to the same ancestor.

it is questionable that phylogenetic taxon names can be defined since no defining properties are involved for particular clades. Even if we accept the idea that a taxon is a natural kind with a historical essence, and thus has a defining property, there are epistemological and inferential problems of definition. Our conceptualization of phylogeny is dependent on our hypotheses. Therefore, definitions based on discarded hypotheses are problematic. Instead each new and accepted hypothesis should form the basis for a new nomenclatural conceptualization. We need to rethink nomenclature from time to time, which means that it does not necessarily fit a cumulative model of scientific progress (which assumes accumulation of knowledge).

Future of nomenclature. Biological nomenclature has taken an important step by making phylogeny its central principle. Phylogeny is general and independent of life form, which makes it ideal as a unifying principle for nomenclature. Nevertheless, there are a number of philosophical and theoretical issues to be solved.

For background information *see* CLASSIFICATION, BIOLOGICAL; PHYLOGENY; TAXON; TAXONOMIC CATEGORIES; TAXONOMY; ZOOLOGICAL NOMENCLATURE in the McGraw-Hill Encyclopedia of Science & Technology.   Mikael Härlin

Bibliography. K. de Queiroz, The Linnaean hierarchy and the evolutionization of taxonomy, with emphasis on the problem of nomenclature, *Aliso*, 15: 125–144, 1997; K. de Queiroz and J. Gauthier, Phylogenetic taxonomy, *Annu. Rev. Ecol. Sys.*, 23:449–480, 1992; M. Ereshefsky, *The Poverty of the Linnaean Hierarchy: A Philosophical Study of Biological Taxonomy*, Cambridge University Press, 2001; M. T. Ghiselin, *Metaphysics and the Origin of Species*, SUNY Press, Albany, 1997; M. Härlin, Taxonomic names and phylogenetic trees, *Zoologica Scripta*, 27:381–390, 1998.

Coda wave interferometry

An interferometer is an instrument that is sensitive to the interference of two or more waves (optical or acoustic). For example, an optical interferometer uses two interfering light beams to measure small length changes.

Coda wave interferometry is a technique for monitoring changes in media over time using acoustic or elastic waves. Sound waves that travel through a medium are scattered multiple times by heterogeneities in the medium and generate slowly decaying (late-arriving) wave trains, called coda waves. Despite their noisy and chaotic appearance, coda waves are highly repeatable such that if no change occurs in the medium over time, the waveforms are identical. If a change occurs, such as a crack in the medium, the change in the multiple scattered waves will result in an observable change in the coda waves. Coda wave interferometry uses this sensitivity to monitor temporal changes in strongly scattering media.

There are many potential applications of coda wave interferometry. In geotechnical applications, the technique can be used to monitor dams or tunnel roofs. In nondestructive testing, the technique can be used to monitor changes due to the formation of cracks or other changes in materials. In hazard monitoring, the technique can be used to monitor volcanoes, fault zones, or landslide areas. In the context of the "intelligent oilfield," coda wave interferometry can be used to monitor changes in hydrocarbon reservoirs during production.

Coda wave interferometry can be used in two different modes. In the warning mode, the technique is used to detect a change in the medium, but this change is not quantified. This mode of operation is used to prompt further action, such as more elaborate diagnostics. In the diagnostic mode, coda wave interferometry is used to quantify the change in the medium.

Volcano monitoring. The use of this technique to detect changes in a medium can be illustrated with seismological data that have been recorded on the Merapi volcano in Java by U. Wegler and coworkers. As seismic (elastic) waves pass through a volcano, they are scattered by heterogeneities (scatters) such as voids, cracks, magma bodies, and faults.

In the experiment, an air gun placed in a small water basin dug in the side of the volcano was used to generate seismic waves. (An air gun is a device that emits a bubble of compressed air in water as a source of seismic waves.) The seismic waves generated by the same source recorded at a fixed receiver at two moments in time (a year apart) are shown in **Fig. 1**.

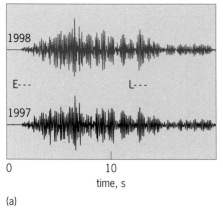
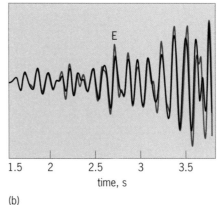
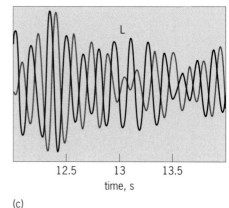

Fig. 1. Waveforms recorded at the Merapi volcano for the same source and receiver on July 1997 (black) and July 1998 (color). (*a*) Complete waveforms and definition of the early (E) and late (L) time interval. (*b*) The recorded waves in the early time interval (E). (*c*) The recorded waves in the later interval (L). (*Data courtesy of Ulrich Wegler*)

In the Fig. 1*b* and *c*, the two waveforms are shown superimposed in more detail. For the interval early in the seismogram, these waveforms are similar (Fig. 1*b*). For the later interval (coda), the waveforms are distinctly different (Fig. 1*c*), where one of the waveforms appears to be a time-shifted version of the other, indicating that the interior of the volcano had changed over time.

Note that in coda wave interferometry one needs only a single source and a single receiver, although in practice one may use more receivers to increase the reliability. This means that the hardware requirements of this technique are modest.

Theory. Suppose that a strongly scattering medium is excited by a repeatable source, and that the medium changes with time. Before the change in the medium, the unperturbed wave field $u^{(u)}(t)$ can be written in Eq. (1) as a sum of the waves that

$$u^{(u)}(t) = \sum_T A_T(t) \quad (1)$$

propagate along the multiple scattering trajectories T in the medium, where t denotes time and $A_T(t)$ is the wave that has propagated along trajectory T. When the medium changes over time, the dominant effect is a change τ_T in the arrival times of the waves that propagate along each trajectory T, so that the perturbed wave field is given by Eq. (2).

$$u^{(p)}(t) = \sum_T A_T(t - \tau_T) \quad (2)$$

The change in the waveforms as shown in Fig. 1*c* can be quantified by computing in Eq. (3) the time-

$$R(t_s) \equiv \frac{\int_{t-t_w}^{t+t_w} u^{(u)}(t') u^{(p)}(t' + t_s)\, dt'}{\left(\int_{t-t_w}^{t+t_w} u^{(u)2}(t')\, dt' \int_{t-t_w}^{t+t_w} u^{(p)2}(t')\, dt' \right)^{1/2}} \quad (3)$$

shifted cross-correlation over a time window with center time t and width $2t_w$, where t_s is the time shift of the perturbed waveform relative to the unperturbed waveform. Suppose that the waves are not perturbed. In that case, $u^{(p)}(t) = u^{(u)}(t)$ and the time-shifted cross-correlation is equal to unity for a zero lag time $R(t_s = 0) = 1$. When the perturbed wave is a time-shifted version of the original wave, then

$u^{(p)}(t) = u^{(u)}(t - \tau)$ and $R(t_s)$ attain its maximum for $t_s = \tau$.

In general, the time-shifted cross-correlation $R(t_s)$ attains its maximum at a time $t_s = t_{max}$ [Eq. (4)] when

$$t_{max} = \langle \tau \rangle \quad (4)$$

the shift time is given by the average perturbation of the travel time of the waves that arrive in the employed time window, and the value R_{max} at its maximum [Eq. (5)] is related to the variance σ_τ of the

$$R_{max}^{(t,t_w)} = 1 - \frac{1}{2}\omega^2 \sigma_\tau^2 \quad (5)$$

travel-time perturbation of the waves that arrive in the time window, where ω is the angular frequency of the waves. Given the recorded waveforms before and after the perturbation, one can readily compute the time-shifted cross-correlation and use Eqs. (4) and (5) to obtain the mean and the variance of the travel-time perturbation in the medium.

Measuring velocity change. **Figure 2*a*** shows an experiment in which ultrasound waves were propagated through a granite cylinder and recorded. The waves are complex due to the reverberations within this cylinder. With a heating coil, the temperature of the cylinder was raised 5°C (9°F). The perturbed waveforms have the same character as the unperturbed waves shown in Fig. 2*a*. The tail of the wave trains was divided in 15 nonoverlapping time intervals. For each time interval, the time shift between the perturbed and unperturbed waves was determined by computing the time-shifted cross-correlation of Eq. (3) and by picking the time for which it attains its maximum t_{max}. The relative velocity change for each time interval is given by $\delta v/v = -t_{max}/t$. This quantity is shown in Fig. 2*b* as a function of the center time of the employed time windows.

Since the employed time windows are nonoverlapping, the measurements of the velocity change in the different time windows are independent. The scatter in the different estimates of the velocity change is small; this provides a consistency check of coda wave interferometry. The variability in the measurements can be used to estimate the error in the velocity

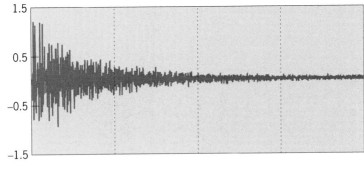

(a)

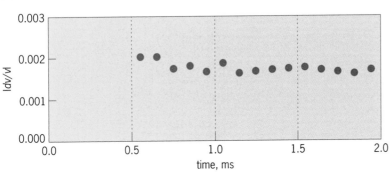

(b)

Fig. 2. Measuring velocity change. (*a*) Ultrasound waves were recorded in a small granite sample. (*b*) The relative velocity change for a 5°C (9°F) increase in temperature as a function of the center time of each time window was used to measure the velocity change.

Imprint of different type of changes on the mean $\langle \tau \rangle$ and the variance σ_τ^2 of the travel-time perturbation		
Type of change	$\langle \tau \rangle$	σ_τ^2
Change in velocity	~t	0
Movement of scatterers	0	~t
Displaced source	0	Constant

can be used to characterize the properties of the turbulent flow.

Coda wave interferometry, to a certain extent, can be used to distinguish between different perturbations. When the velocity changes, the mean travel-time perturbation is nonzero and is proportional to the total travel time. Since the waves that propagate along all paths experience the same travel-time change, the variance of the travel-time perturbation vanishes. When the location of the scatterers is perturbed randomly, the mean travel-time perturbation vanishes, because some paths are longer, while others are shorter. However, the variance of the travel-time perturbation is nonzero and can be shown to grow linearly with time. When the source is displaced, the mean travel-time perturbation vanishes, and the variance of the travel time is constant with time. These results are summarized in the **table**. The mean and the variance of the travel-time change can be computed from the unperturbed and the perturbed waveforms. Since different types of perturbations leave a different imprint on these quantities, coda wave interferometry can help determine the mechanism of the change over time.

For background information *see* ACOUSTIC EMISSION; ACOUSTICS; EARTHQUAKE; INTERFEROMETRY; SEISMOGRAPHIC INSTRUMENTATION; SEISMOLOGY; SOUND; VOLCANO; VOLCANOLOGY in the McGraw-Hill Encyclopedia of Science & Technology.

Roel Snieder

Bibliography. R. Snieder, Coda wave interferometry and the equilibration of energy in elastic media, *Phys. Rev. E*, 66:046615-1,8, 2002; R. Snieder et al., Coda wave interferometry for estimating nonlinear behavior in seismic velocity, *Science*, 295: 2253-2255, 2002; R. Snieder and J. A. Scales, Time reversed imaging as a diagnostic of wave and particle chaos, *Phys. Rev. E*, 58:5668-5675, 1998; U. Wegler, B. G. Luhr, and A. Ratdomopurbo, A repeatable seismic source for tomography at volcanoes, *Ann. Geofisica*, 42:565-571, 1999.

change. Note that the relative velocity change in this example is only about 0.16% with an error of about 0.03%. This extreme sensitivity to changes in the medium is due to the sensitivity of the multiply scattered waves to changes in the granite.

In an elastic medium such as granite, there is no single wave velocity. Compressional (*P*) waves and shear (*S*) waves propagate with different velocities v_P and v_S, respectively. The change in the velocity inferred from coda wave interferometry [Eq. (6)] is a

$$\frac{\delta v}{v} = \frac{v_S^3}{2v_P^3 + v_S^3} \frac{\delta v_P}{v_P} + \frac{2v_P^3}{2v_P^3 + v_S^3} \frac{\delta v_S}{v_S} \quad (6)$$

weighted average of the change in the velocities for the two wave types. For a Poisson medium, an elastic medium, where $v_P = \sqrt{3}v_S$, the relative velocity change is given by $\delta v/v \approx 0.09\delta v_P/v_P + 0.91\delta v_S/v_S$. This means that in practice coda wave interferometry provides a constraint on the change in the shear velocity v_S.

Other applications. Some applications may involve a change in the location of scatterers, or a change in the source position. This can be used to monitor the properties of a turbulent fluid. One can seed the fluid with neutrally buoyant particles that scatter waves. Acoustic waves that propagate through the fluid are scattered by these particles. Over time, the particles are swept along by the turbulent motion. When acoustic waves are sent into the fluid once more from the same source, the waves are scattered by particles that have been displaced by the turbulent flow. The resulting change in the recorded waves

Collaborative fixed and mobile networks

Modern information technology, supported by computers and communications, has dramatically improved productivity among individuals engaged in a wide range of tasks. Computer-supported cooperative work (CSCW) aims to provide similar improvements for multiple individuals consciously working together in the same production process or in different but related production processes. The concepts behind CSCW have led to the development of groupware (multiuser software). Groupware enables collaboration in many areas, such as document writing, virtual meetings, and distance learning and medicine (for example, telesurgery).

Increased mobile phone use, computer network capacity, and integration between telephony and computer systems are facilitating CSCW among "mobile" individuals. Some fundamental problems must be solved to achieve effective collaboration, such as those arising from the intrinsic characteristics of wireless channels (for example, lack of bandwidth and network disconnections). Moreover, different people may use different devices, such as desktop computers, personal digital assistants (PDAs), and wearable computers, and may connect through different networks, including wireless local-area network (LAN), Universal Mobile Telecommunications System (UMTS), and others.

Groupware must be able to handle heterogeneity among devices, including (1) devices with different computational power, memory, and operating systems (for example, not all applications can be used on every device); (2) devices with different display sizes, requiring that information is presented in different ways or, in some cases, filtered (for example, groupware should avoid sending large images to devices with limited display capabilities); (3) devices with different input mechanisms, such that some type of input might be absent or used with difficulty (for example, keyboard use should be avoided for users with PDAs); (4) devices with different communication bandwidth, such that some services might not be supported (for example, video streaming might not be feasible); (5) devices with different network connections and levels of reliability (for example, some networks suffer from frequent disconnections and, as a result, data might be lost or arrive with too much delay).

Collaboration models and concepts. In CSCW, a collaborative session is the activity of a group of people who exchange information and perform activities toward a common goal. Participants may play different roles in a session, and the roles can be changed dynamically. Moreover, participants may also join and leave a running session at any time.

A collaborative platform is required to provide all the facilities needed to support the dynamic nature of a collaborative session, while guaranteeing the availability of suitable media for exchanging information. In fact, the information exchange process depends on the media provided by the collaborative platform.

Collaborative activities are classified as synchronous or asynchronous, depending on the dynamism of the information exchange process. Synchronous activities are characterized by a high level of interaction among participants, with all the participants sharing a single view of the discussion and exchanging information as it becomes available. In asynchronous activities, information is transferred only when requested, lowering the degree of interaction among participants.

The collaborative platform must provide mechanisms to guarantee consistency and manage correctly the shared information. For asynchronous activities, consistency is usually managed through a data locking mechanism, where only one user at a time can update a document, although others can view it. For synchronous activities, where participants share a single view of the shared information, consistency is typically managed by floor control. In floor control, a privilege, called a modification token, is assigned to one participant at a time for modifying the shared data. Floor-control mechanisms can be classified as explicit or implicit, depending on how the collaborative platform assigns the modification token. In explicit floor control the participants are aware when the collaborative platform assigns the modification token, whereas in implicit floor control the token is assigned without the participants being aware of this action.

Collaborative platforms and services. There are two principal approaches for groupware: collaboration-aware and collaboration-unaware applications. Collaboration-aware applications, such as instant messenger and video conferencing systems, are specifically designed to support collaboration among a group of users. Collaboration-unaware applications are not specifically designed to support collaboration, but are accessed through the collaborative platform, making them collaborative. This approach not only simplifies the collaborative applications because collaboration issues are not considered directly in the design of the application, but also allows the sharing of common applications that participants customarily use in their everyday activities.

A collaborative platform should provide services to support both synchronous and asynchronous activities. In the case of synchronous activities, it should support pre- and postcollaboration activities, such as scheduling meetings, distributing documents and agenda, and preparing and distributing meeting minutes.

In terms of technology, asynchronous activities need a limited set of simple services that can be easily realized using off-the-shelf Internet technology (for example, e-mail and file transport protocols). Synchronous activities require more complex and dedicated technologies since they are centered on application sharing and on audio/video conferences. These services are compatible with Internet technology, but require additional resources for sharing collaboration-unaware applications, guaranteeing audio and video performance and usability, tracing the operations of all the participants, and supporting latecomers.

A number of collaborative platforms have been developed in the past few years. They are classified as either centralized or replicated, even if hybrid approaches are used. In a centralized architecture, the shared application is maintained in a single network node, and participants are supplied only with the

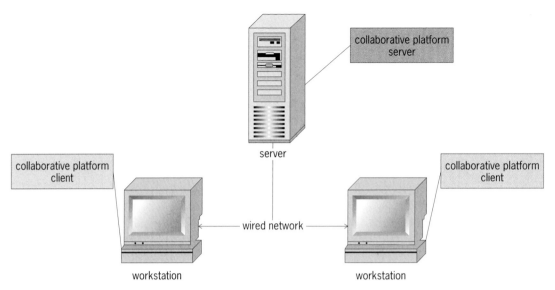

Fig. 1. Centralized platform.

output of this application (**Fig. 1**). In a replicated architecture, each participant owns an instance (copy) of the application, and the platform provides the mechanisms to synchronize the instances distributed across the network (**Fig. 2**).

Centralized platforms have some disadvantages compared to replicated platforms. A centralized platform usually requires higher network bandwidth to distribute graphical display information, while a replicated platform needs only to distribute minimal update information. Centralized platforms also require that all the participants see exactly the same view of the shared application, preventing independent work. Furthermore, centralized platforms are less responsive to user input, as each user interaction must travel to and from the central process (round-trip latency). Finally, centralized approaches are potentially less fault-tolerant than replicated approaches because the central host (a single point) is the source of possible system-wide failure. Centralized architectures do have one major advantage— they have no problem with data consistency because there are not different copies of the application data.

The principal method used to implement centralized platforms is called display broadcasting. It is based on the assumption that the participants accessing a client application have the capability of displaying its output. Centralized platforms using display broadcasting can be classified as heavyweight or

lightweight clients, based on the information the centralized application transmits to the client application. The heavyweight client approach is used mainly in collaborative platforms with windowing systems (for example, personal computers), where the display of the shared application is exported to all clients.

The lightweight client approach (also known as thin client) is used when the platform is meant to accommodate clients with very limited processing capabilities. For example, PDAs and smart phones display only a bitmap of the actual output of the centralized application. The thin client method is used by collaborative platforms based on the T-120 protocol for interactive multimedia communication [for example, Microsoft NetMeeting and virtual network computing (VNC)].

Collaboration in heterogeneous environments. The success of a collaborative platform mainly depends on user acceptance. The popularity and growth of the Web suggests that Web-based platforms are likely to be easily accepted. Moreover, the Web is assuming a central role in the way people obtain information because browsers integrate different services into a common and platform-independent user interface. As a result, the Web has been adopted as a principal means of supporting collaboration.

The basic communication support offered by the Web (that is, HTML pages and forms) does not

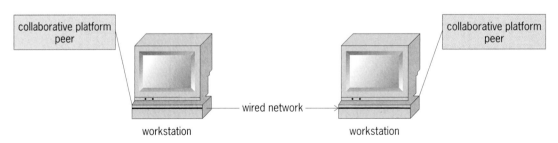

Fig. 2. Replicated platform.

support synchronous activities. Likewise, the non-interactive nature of HTTP makes it impractical to deal with even basic synchronous activity, such as a simple conversation between two participants. The Web was designed only for viewing structured documents, and does not support interactive exchanges. The possibility of integrating active components (for example, Java applets) into HTML documents will make the Web more suitable for supporting collaborative activities.

The availability of browsers on different kinds of devices and access to the Internet through different communication networks makes the Web the most suitable means for supporting collaboration. The architecture of a Web-based collaborative platform (**Fig. 3**) consists of three layers: a client layer (that is, the Web browser on the various client devices), a Web portal layer, and a service layer composed of a set of backend systems, which provide information and collaborative services through the Web portal.

The Web portal functions not only to synchronize participants and provide access to collaborative services, but also to collect and adapt the shared content for the participants. On one side, information coming from different services and participants (for example, audio and video conference streams) needs to be collected for presentation in participants' browsers. And on the other side, each participant is provided with different information, based the capability of the devices they are using. The portal does this by processing the information received from the participants. Some of the information is provided by participants' browsers (for example, information about the media types supported by the browser). Other information is provided by browser plug-ins and client-side code included in the pages sent by the Web portal to the browser.

Adaptation. In a heterogeneous environment, adaptation provides different views for different devices, different image and video quality for different communication channels and devices, and different services.

While a large display of a personal computer can present all the information coming from the active services, a small PDA display cannot. Therefore, the Web portal may send pages to a PDA containing information about a single service, as well as a menu to move from one service to another.

To be effective, remote collaboration requires that participants remain synchronized throughout all activity. Therefore, if participants are connected through lines with different bandwidth, the collaborative platform should synchronize them, reducing the image and video quality sent to participants connected through slow lines. High-quality images and videos are useful only for the devices where the size and display resolution allow their presentation without reduction.

In some cases, the information coming from some services cannot be provided to all the devices. For example, video conferencing might not be possible for devices connected through very low bandwidth

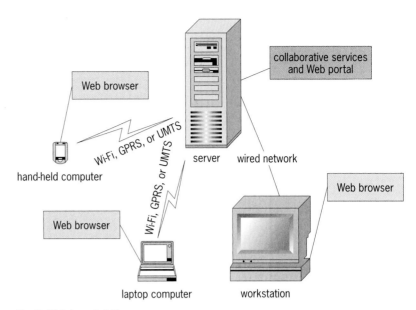

Fig. 3. **Web-based platform.**

lines. In such a case, the participant might interact with others through the audio conference services only.

In other cases, adaptation may have an effect on all the participants. For effective remote collaboration, all the participants will be required to have the same information on what is happening. Therefore, if only some of the participants have audio and video conference service, the collaborative platform might require that all the participants use an alternative service (for example, chat).

Outlook. Network requirements have been increasing constantly in terms of the connection and coordination of computer systems. At the same time, there have been significant developments in the connection infrastructure, both wired and wireless. An isolated personal computer in an office is an image of the past. The advancements in the enabling technologies of local- and wide-area networks and, more recently, of wireless networks and ad-hoc (temporary) networks for mobile or portable devices, together with improvements in interoperability and security, support the use of groupware. Soon the difference between standalone software and groupware will be insignificant because all software will be groupware, that is, it will be usable by a group of people distributed across a network. Advances in mobile communication technologies will eliminate or, at least, reduce the problems of reliability and bandwidth necessary to achieve an effective collaboration. For small devices (such as PDAs and smart phones), computation power and memory will increase to acceptable performance levels. The challenge will be to develop adequate input/output means, which will allow small devices to offer productivity comparable to that of personal computers.

For background information *see* DATA COMMUNICATIONS; DATABASE MANAGEMENT SYSTEMS; INFORMATION TECHNOLOGY; INTERNET; LOCAL-AREA

NETWORKS; SOFTWARE; WIDE-AREA NETWORKS; WORLD WIDE WEB in the McGraw-Hill Encyclopedia of Science & Technology.

Federico Bergenti; Agostino Poggi

Bibliography. M. Beaudouin-Lafon (ed.), *Computer-Supported Cooperative Work*, Trends in Software Series, Wiley, New York, 1999; F. Bergenti, A. Poggi, and M. Somacher, A collaborative platform for fixed and mobile networks, *Commun. ACM*, 45:39–44, 2002; A. Dix and R. Beale (eds.), *Remote Cooperation: CSCW Issues for Mobile and Teleworkers*, Springer-Verlag, London, 1996; I. Greif (ed.), CSCW: A Book of Readings, Morgan Kaufmann Publishers, San Mateo, CA, 1988.

Conservation paleobiology

Conservation paleobiology is a new, pragmatic, socially relevant discipline within paleobiology. Its emergence stems from the growing realization that the fossil record of the very recent past (the last centuries to millennia) can serve as an unprecedented baseline for studying rates and causes of ecological changes over time scales not accessible by studying modern conditions alone. Such a unique, multicentennial perspective can provide rigorous criteria for evaluating the current extent of ecosystem deterioration, enabling a realistic evaluation of the long-term context of anthropogenic disturbances and a critical assessment of the effectiveness of restoration efforts. For example, in a recent study of coastal ecosystems, an extensive compilation of paleontological and archeological data was used to show that ecological deterioration often has deep historical roots. The compiled studies represented a wide range of time scales, from projects reaching back only a few decades to those spanning hundreds of thousands of years. In addition, they encompassed a wide spectrum of ecosystems, from high-latitude kelp forests of Alaska to tropical reefs of the Caribbean Sea. Despite the great variety of ecosystems included in the compilation, virtually all cases revealed similar patterns: marine ecosystems were first altered by overfishing, an activity that invariably predated other anthropogenic disturbances by decades, centuries, or even millennia. Overfishing, which removes key members of marine ecosystems and often dramatically reduces total biomass, represents—most likely—a necessary precondition for various secondary disturbances such as eutrophication (overstimulation of algae and plant growth in aquatic systems due to natural or anthropogenic addition of excess nutrients), outbreaks of diseases, or species introduction.

A good example is provided by oyster reefs of Chesapeake Bay which, prior to their destruction by mechanical harvesting, may have filtered the entire water column every 3 days. Hypoxia (low levels of dissolved oxygen), anoxia (no dissolved oxygen), and other symptoms of eutrophication appeared only after the collapse of those water-purifying reefs.

Many such examples provided by the study show that the disturbances observed today in coastal ecosystems are not just a consequence of current activities, but have, in fact, a deep historical derivation that predates modern ecological investigations. Yet, the majority of restorative efforts that are being implemented today are specifically designed to deal with current events, without considering the history of the deteriorated ecosystems. The long-term ecological perspective offered by conservation paleobiology is critical for developing successful restoration strategies.

The idea of utilizing the historical ecological perspective to aid conservation biology efforts is certainly not new, but the approach has been of limited applicability until very recently. However, with rapid parallel advancements in paleontology, geochronology, and geochemistry, conservation paleobiology has been an increasingly powerful approach, especially in the study of coastal and marine ecosystems. It is now possible to use the recent fossil record of marine organisms to reconstruct the environmental history of the world's coastal and shelf regions over multicentennial or even multimillennial time scales. That is, rigorous insights into environmental, ecological, and rapid climate changes can be achieved by dating Holocene (from 10,000 years ago to the present) "fossils," analyzing them for geochemical signatures, and placing them into local biological and geological contexts.

Dating shells. The key to conservation paleobiology lies in the unexpected fact that biogenic accumulations (shells of clams, snails, and other shellfish) found on modern seashores and sea floors represent multicentennial bioarchives of the prehuman history of marine ecosystems. This outstanding record has been revealed only recently due to rapid advances in geochronological techniques that have made age-dating individual shells faster, cheaper, and more reliable. Recent age-dating efforts show that multicentennial time series of shells can be assembled for various shelled organisms from a wide spectrum of environments, from coastal mud flats to continental shelves (see **illus.**). Literature compilations and numerous case studies confirm that surficial shell accumulations almost invariably provide a continuous record of shelly fauna for recent centuries and millennia. Conservation biology can exploit such bioarchives in powerful ways.

Anthropogenic disturbance assessment. Age distributions of dated shells can be used directly to assess anthropogenic disturbances. For example, the dated shells from the human-impacted Colorado River delta are increasingly scarce in the most recent age classes (see illus.), suggesting a dramatic decline in shellfish density throughout the twentieth century. This trend coincides in time with the human-induced diversion of the Colorado River (water flow declined significantly—and eventually ceased—starting with the construction of the Hoover Dam in the 1930s).

Past bioproductivity levels. An even more valuable aspect of age distributions of dated shells is that they enable the assessment of past bioproductivity levels. For example, it was recently estimated that at least 2×10^{12} shells of dead mollusks are scattered today on the tidal flats of the Colorado Delta. The dating effort shows that virtually all these shells came from mollusks that lived in the delta between A.D. 950 and 1950. The most conservative calculation based on these numbers indicates that during the time of natural river flow, dense shellfish populations ($\sim$50 per square meter) thrived in the area. In contrast, the present abundance of shelly benthos is at most 6% of past productivity levels. This dramatic decrease in abundance not only testifies to the severe loss of benthic productivity resulting from diversion of the river's flow but also demonstrates the inadequacy of current restoration efforts (started in 1981).

The Colorado Delta study illustrates the power of the conservation paleobiology approach. By determining a quantitative baseline for various ecosystem parameters (in this example, the evaluated parameter was the average density of mollusk populations that thrived on the Colorado Delta mud flats prior to the anthropogenic diversion of the river's flow), conservation paleobiology not only can provide a yardstick for measuring the extent to which ecosystems have deteriorated, but also can put forth concrete, quantifiable goals for future restoration efforts.

Geochemical signature. Skeletal remains of marine organisms are also powerful recorders of various geochemical signals (such as element ratios, stable isotopes, and trace elements). Through the use of advanced geochemical signature analysis of biomineral solids, precise computer-guided sampling devices, and laser-ablation techniques, it is now possible to microsample small aliquots of shell on a micrometric scale. Consequently, ancient environments can be explored at high resolution by generating geochemical profiles along the growth axes of long-dead organisms. For example, stable isotopes of oxygen extracted from shells of mollusk bivalves can provide environmental data at subweekly to submonthly resolution levels.

But skeletons of single marine organisms can also provide long-term records, extending back past the time scale of ecological investigations. For example, relative changes in past levels of lead in the ocean water were estimated recently by analyzing a 223-year profile in a skeleton of a single sponge. A laser-ablation technique was used to estimate changes in lead levels through time, supplemented with microsampling methods to analyze stable isotopes of carbon (the latter data were used successfully to test the validity of the estimates against the carbon dioxide record provided by ice cores). The resulting lead curve showed that changes in lead recorded in the sponge closely matched with other records that relate to the history of lead pollution.

Dated shells can potentially provide even more powerful insights when geochronological approaches and high-resolution geochemistry are used in conjunction. That is, the advanced microsampling technology can be used to extract weekly-to-monthly

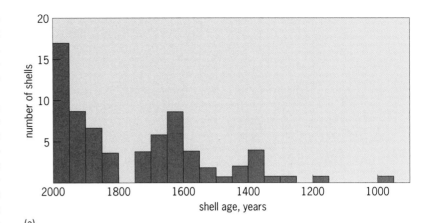

(a)

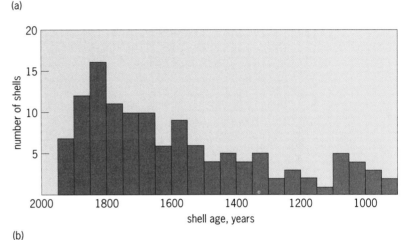

(b)

Age-frequency distributions of individually dated shells collected from the present-day sea floor. (a) Calcitic shells of terebratulid brachiopods from the inner shelf of the Southeast Brazilian Bight (western South Atlantic). **(b)** Aragonitic shells of bivalve mollusks from the macrotidal flats of the Colorado River delta (Gulf of California). (**Based on data from M. Kowalewski et al. [2000] and M. Carroll et al. [2003]**)

records of geochemical signals in shells dated by high-precision geochronological techniques. Thus it should be possible to assemble unprecedented time series of geochemical signatures of temperature, salinity, bioproductivity, erosion rates, concentration of pollutants, and other environmental and climatic parameters.

Conclusion. Conservation paleobiology is emerging as an important discipline at the interface between bio- and geosciences. Due to independent advancements in several research areas, including paleoecology, taphonomy (study of fossilization processes), geochronology, geochemistry, and sampling technology, the Holocene "fossil" record can inform us about the most recent geological past and provide critical data that complement research efforts of conservation biologists, environmental researchers, and physical anthropologists. By providing data that neither ecologists nor climate modelers can ever access, conservation paleobiology can help to establish the range of environmental and climatic variations that affect ecological systems under natural conditions, increasing our ability to evaluate their current state, forecast their future, and guide their restoration. This, in turn, will aid us in making more intelligent predictions about inherent stability of vital natural habitats and the future consequences of their intense use by humans. Thus, information about the recent geological past of ecosystems can serve as the key to their better future.

For background information *see* DATING METHODS; FOSSIL; GEOCHRONOMETRY; HOLOCENE; HUMAN ECOLOGY; MARINE ECOLOGY; MICROPALEONTOLOGY; PALEOECOLOGY in the McGraw-Hill Encyclopedia of Science & Technology.   Michal Kowalewski

Bibliography. M. Carroll et al., Quantitative estimates of time-averaging in brachiopod shell accumulations from a modern tropical shelf, *Paleobiology*, 29:382–403, 2003; D. L. Dettman and K. C. Lohmann, Microsampling carbonates for stable isotope and minor element analysis: Physical separation of samples on a 20 micrometer scale, *J. Sed. Res.*, 65(A):566–569, 1995; D. H. Goodwin et al., Cross-calibration of daily growth increments, stable isotope variation, and temperature in the Gulf of California bivalve mollusk *Chione cortezi*: Implications for paleoenvironmental analysis, *Palaios*, 16:387–398, 2001; J. B. C. Jackson et al., Historical overfishing and the recent collapse of coastal ecosystems, *Science*, 293:629–638, 2001; M. Kowalewski et al., Dead delta's former productivity: Two trillion shells at the mouth of the Colorado River, *Geology*, 28:1059–1062, 2000; C. E. Lazareth et al., Sclerosponges as a new potential record of environmental changes: Lead in *Ceratoporella nicholsoni, Geology*, 28:515–518, 2000.

Continuous casting (metallurgy)

Continuous casting is a solidification method used to mass-produce long, semifinished shapes of constant cross section from basic metals. Using this method, over 800 million tons (730 million metric tons) of steel, 20 million tons (18 million metric tons) of aluminum, 1 million tons (900,000 metric tons) of copper, and other metal products (such as nickel) are processed annually worldwide. Cross sections can be rectangular (for later rolling into a plate or sheet), square for long products, circular for wire and seamless pipes, "dog-bone" shapes for I or H beams, thin strips, or rods. Continuous casting is distinguished from other solidification processes by its steady-state appearance. That is, the molten metal freezes against the mold walls and is withdrawn from the bottom of the mold at a rate which keeps the solid/liquid interface at a constant position with time, relative to an outside observer. The process works best when all of its aspects operate in this manner.

Continuous casting is the most efficient way to solidify large volumes of metal into simple shapes with consistent quality, and has been steadily replacing older ingot casting processes since it was developed in the 1950s and 1960s. Compared with other casting processes, it generally has a higher capital cost but a lower operating cost. It produces "near-net shapes" that are closer to the final product shape, so

that they require less subsequent deformation. This makes continuous casting more energy- and cost-efficient than alternative processes.

Several different types of commercial continuous casting processes exist (**Fig. 1**). Vertical machines cast aluminum and a few special alloys. Curved machines are used for steel, which can tolerate bending or unbending of the solidifying strand. Short horizontal casters are sometimes used (most commonly in casting copper) to reduce machine height. Finally, thin-strip casting is being pioneered for steel and other metals to minimize the amount of rolling required.

Steel continuous casting. Recent innovations have transformed the continuous casting of steel into a sophisticated, high-technology process that is used to manufacture 90% of the steel currently produced worldwide. In the curved machine process, 50–200 tons (45–180 metric tons) of molten steel are supplied periodically from the steelmaking process to the caster via a ladle. A hole in the bottom of the ladle is opened to pour steel into a large, bathtub-shaped vessel, called a tundish. The tundish holds enough metal to provide a continuous flow to the mold, even while exchanging ladles. The tundish can also serve as a refining vessel to float inclusions (foreign solid particles composed of brittle oxides) to the top surface. There, the inclusions are absorbed into the slag layer, which is a thin covering of molten glass that floats on the liquid metal surface and is later discarded. Any inclusions that remain in the product may form surface defects during subsequent rolling operations, such as slivers (longitudinal streaks) that sometimes delaminate into loosely attached strips or blisters. Large inclusions also cause local internal stress concentration, which lowers the steel's strength and fatigue life (how many loading cycles can be endured before a part fails in service). To produce higher-quality product, the liquid steel must be protected from exposure to air by making sure a slag layer covers each steelmaking vessel and by using ceramic nozzles between vessels. If this is not done, oxygen in the air will react to form detrimental oxide inclusions in the steel.

A closeup of the mold region shows important phenomena in the continuous casting process (**Fig. 2**). Molten steel freezes against the water-cooled walls of a bottomless copper mold to form a solid shell. The mold is oscillated vertically in order to discourage sticking of the shell to the mold walls. Drive rolls lower in the machine continuously withdraw the shell from the mold at a rate or casting speed that matches the flow of incoming metal, so the process ideally runs in steady state. The liquid flow rate is controlled by restricting the opening in the nozzle according to a signal fed back from a level sensor in the mold.

The most critical part of the process is the meniscus, found at the junction where the top of the solidifying shell meets the mold and the liquid surface. This is where the surface of the final product is created. Defects such as surface cracks can form here, if

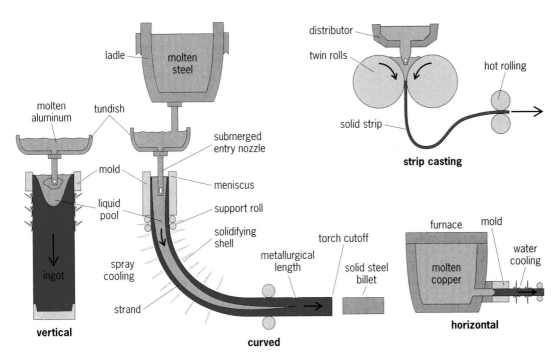

Fig. 1. Various continuous casting processes.

problems such as level fluctuations occur. To avoid this, oil or mold slag is added on top of the steel meniscus, and it then flows into the gap between the mold and shell. In addition to lubricating the contact, an optimized mold slag layer protects the steel from air, provides thermal insulation, and absorbs inclusions. It is also important to control flow across the top surface. Excessive speed can entrain mold slag, which is later trapped as inclusions, while stagnant conditions cause surface depressions and related defects.

Below the mold exit, the thin solidified shell (6–20 mm thick; 0.2–0.8 in.) acts as a container to support the remaining liquid, which makes up the interior of the strand. Water or air mist sprays cool the surface of the strand between the support rolls. The spray flow rates are adjusted to control the strand surface temperature with minimal reheating until the molten core is solid. After the center is completely solid at the "metallurgical length" of the caster, which is 10–40 m (33–131 ft) from the meniscus, the strand is cut with oxyacetylene torches into slabs or billets of any desired length (Fig. 1).

Different continuous casting machines cast cross sections of various shapes and sizes. Heavy, four-piece plate molds with rigid backing plates are used to cast large, rectangular slabs (5–25 cm thick by 0.5–2.2 m wide; 2–10 in. by 1.5–7.3 ft), which are rolled into a plate or sheet. A recent trend in the steel industry is the increased use of thin-slab casting (producing slabs 50–100 mm thick; 2–4 in.), which needs much less rolling to make sheet product than does conventional thick-slab casting; since it was first commercialized in 1985, thin-slab casting has grown to over 16 million tons (14 million metric tons) of annual capacity—almost 20% of the United States

market—and is still growing. Similar molds are used to cast relatively square blooms, which range up to 40 by 60 cm (16 by 24 in.) in cross section. Single-piece tube molds are used to cast smaller billets (10–20 cm thick; 4–8 in.) which are rolled into long products, such as bars, angles, rails, nails, and axles. The new thin-strip casting process is being developed using large rotating rolls as the mold walls to solidify an ultrathin 1–3 mm-thick (0.04–0.12 in.) steel sheet directly from the liquid; this revolutionary new technology is nearing commercialization and will likely enter the marketplace in a few years.

When casting large cross sections, such as slabs, a series of rolls must support the soft steel shell between mold exit and the metallurgical length in order to minimize bulging due to the internal liquid pressure. Extra rolls are needed to force the strand to "unbend" through the transition from the curved to the straight portion of the path shown in Fig. 1. If the roll support and alignment are not sufficient, internal cracks and segregation may result.

The process is started by plugging the bottom of the mold with a "dummy bar." After enough metal has solidified above it (a process similar to conventional casting), the dummy bar is slowly withdrawn down through the continuous casting machine, and steady-state conditions evolve. The process then operates continuously for a period of 1 h to several weeks, when the molten steel supply is stopped and the process must be restarted. A maximum casting speed of 1–8 m/min (3–26 ft/min), governed by the allowable length of the liquid core, is established to avoid quality problems, which are generally worse at higher speeds.

After the steel leaves the caster, it is reheated to a uniform temperature and rolled into sheet, bars,

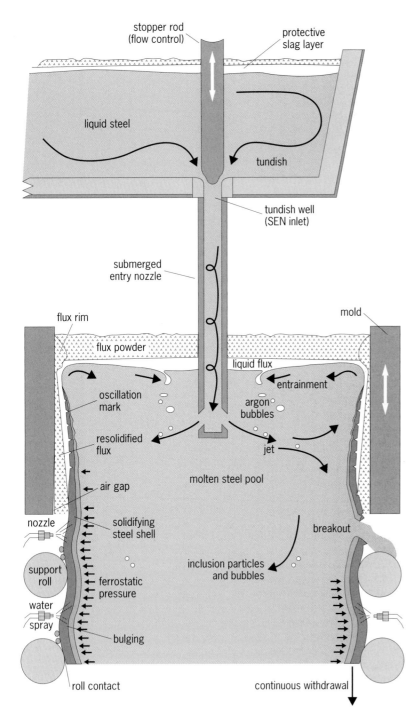

Fig. 2. Schematic of mold region used in continuous casting process for steel slabs.

(3–10 cm/min; 1–4 in./min), which is needed to avoid internal cracks and makes the metallurgical length shorter.

The two most common casting processes for aluminum are the direct-chill (DC) and electromagnetic (EM) processes, which differ in how the liquid is supported at the meniscus. The DC process uses water-cooled mold walls, similar to steel casting, while the EM process induces horizontal electromagnetic forces to suspend the metal, preventing it from ever touching the mold walls. In both processes, the solid metal surface shrinks away from the mold walls shortly below the meniscus and is cooled with spray water.

Other continuous casters. Many other continuous casting processes exist for special applications. Electroslag remelting (ESR) and vacuum arc remelting (VAR) are two forms of vertical continuous casting used for nonferrous metals, superalloys, and specialty alloy sections up to 1.5 m (5 ft) in diameter. Heat is supplied from an electrode above the melt, and the top surface is protected from air reoxidation with a thick slag layer (ESR) or a vacuum (VAR). These processes remove impurities such as sulfur to produce highly refined metal with less segregation and fewer other defects than are found in conventional continuous casting. The products are more costly but are needed for critical parts such as those used in the aerospace industry.

Challenges. Understanding and controlling the casting process is important because, if it is not done properly, defects may be introduced that persist in the final product, even after many subsequent processing steps. These defects include oxide inclusions, porosity, segregation, and cracks. In addition to expensive plant experiments, increased understanding of the process comes from physical water models and advanced computational models. Water flows in a similar manner to steel, so scale water models can visualize the flow, especially with the aid of recent tools such as particle image velocimetry. Helped by recent improvements in computer processing power and software, computational models can simulate phenomena ranging from metallurgical thermodynamics to fluid flow, heat transfer, solidification, and stress generation. The knowledge gained from models is used in designing features such as the nozzle and mold shapes. In addition, online models control the liquid steel and cooling water flows, casting speed, and other parameters. Improvements to continuous casting must be cost-efficient, owing to economic pressure from the present oversupply of primary metals worldwide. Future advances likely will rely on even better models and control systems, including more advanced computational models, more intelligent models for online process control, better process sensors, and better quality control and maintenance tools.

For background information *see* ALLOY; ALUMINUM; COPPER; METAL, MECHANICAL PROPERTIES OF; METAL CASTING; METAL FORMING; NICKEL; STAINLESS

rails, and other shapes. Modern steel plants position the rolling operations close to the caster to save on reheating energy.

Vertical semicontinuous casting. Most commercial nonferrous alloys are cast by semicontinuous vertical casting machines, typically as 5–50-cm-diameter (2–20-in.) round sections called ingots, for subsequent extrusion, forging, or rolling. This process differs from steel continuous casting in that it must be stopped periodically to remove the cast ingot. Other differences are the slower casting speed

STEEL; STEEL; STEEL MANUFACTURE in the McGraw-Hill Encyclopedia of Science & Technology.

B. G. Thomas

Bibliography. Cast shop technology, *Light Metals Conference Proceedings*, pp. 647–1095, Minerals, Metals, and Materials Society, Warrendale, PA, 1997; *Continuous Casting*, vols. 1–9, Iron and Steel Society, Warrendale, PA, 1979–1997; A. Cramb (ed.), Continuous casting, *The Making, Shaping, and Treating of Steel*, vol. 2, Association of Iron & Steel Engineers, Pittsburgh, 2003; W. R. Irving, *Continuous Casting of Steel*, Institute of Materials, London, 1993; H. F. Schrewe, *Continuous Casting of Steel: Fundamental Principles and Practice*, Stahl und Eisen, Dusseldorf, 1991; R. Wilson, *A Practical Approach to Continuous Casting of Copper Alloys and Precious Metals*, Institute of Materials, London, 1999.

Counterterrorism

Terrorism is as old as history, but the twenty-first century has brought a new form of terrorism. Terrorists now have access to technical skills that multiply their ability to wreak havoc far beyond the efforts of those in the past. However, the democracies that are most vulnerable to terrorist attack also have much greater technical resources with which to make the terrorist's task more difficult and less damaging. To understand how technical knowledge might be used to counter terrorism, one must evaluate the nature of the threat, the vulnerability of targets in civil society, and the availability of technical solutions to reduce these vulnerabilities.

Despite their small numbers, terrorists possess some advantages. Their actions (particularly those of ideological terrorists) are unpredictable, they may have cells hiding in the society they plan to attack, and they have the initiative in choosing the target and deciding when to attack. As a result, those defending against terrorism must always be alert; the terrorists need be prepared only when they choose to strike. Finally, terrorists may enjoy the technical and financial support of a nation state.

However, modern industrial societies have available a range of options in combating terrorist threats. These include the application of new technologies that can make targets less vulnerable and thus less attractive. These technologies can also limit the damage that may result from an attack, increase the speed of recovery, and provide forensic tools to identify the perpetrators. This is the main focus of a study by the National Academies of Science and of Engineering and the Institute of Medicine, released June 25, 2002. This study found that science and technology can contribute substantially to making the nation safer but cannot assure that catastrophic attacks will not take place.

Terrorist targets and weapons. Terrorists may seek to kill large numbers of people, to destroy national symbols of capitalism or of democracy, or to cripple elements of the industrial infrastructure on which normal economic and social life depend. If terrorists can obtain weapons of mass destruction—nuclear, biological, and chemical weapons—they may be expected to use them. More probably, they may use elements of the very economy they seek to attack. They may use elements of the transportation system as a weapon, as they did in the September 11, 2001, attacks. A rental truck filled with commercial fertilizer (ammonium nitrate) and fuel oil was used to blow up the Murrah Building in Oklahoma City in 1995. Terrorists might cause the release of toxic chemicals (such as railroad tank cars full of chlorine destined for municipal water supplies). They might use computers to interfere with electronic messages used to control critical industrial processes or needed by first responders to communicate when the attack occurs. This list of targets and tools to attack them is very long.

The vulnerability of civil society arises from the very efficiency of its competitive economic system. The competitive drive for commercial efficiency creates linkages and vulnerabilities in critical infrastructure industries—energy, transportation, communications, food production and distribution, public health, and financial transactions. The concentration of food packaging in a few firms, and of power production in ever bigger plants, and the production of ever larger aircraft and ships create targets that terrorists can exploit. The economy is not designed with the threat of catastrophic terrorism in mind.

Technical activities. Examples will be discussed of how technical activities might reduce either the threat of terrorism or its consequences in the areas of nuclear and radiological threats, biological threats to people and their food supply, toxic chemicals, explosive and flammable materials, energy systems, communications and information systems, transportation and borders, and cities and fixed infrastructure.

Nuclear and radiological threats. If terrorists can acquire enough highly enriched uranium, they may be able to assemble a crude but effective nuclear weapon which could kill hundreds of thousands of people. Governments with stocks of highly enriched uranium can blend them with natural uranium, leaving the material useful for power generation but not for bombs. Even more dangerous is the possible availability to terrorists of finished nuclear weapons provided by states with nuclear weapons capability. Clearly, it is extremely important to detect fissile materials entering the United States from foreign sources. The radioactivity of plutonium might be sensed, but highly enriched uranium can easily be shielded with moderate amounts of lead. Scientists are exploring the use of beams of neutrons that can penetrate steel containers and produce telltale signals from any uranium inside. Another alternative is the use of gamma rays to detect any very heavy element, including all the fissile elements.

Biological threats. Research might enable the detection of a bioterrorist attack before symptoms appear in infected people. This would also be a huge step forward in public health protection against natural agents such as SARS. Since terrorists might acquire the skill to modify deadly diseases genetically to make them resistant to known treatments and vaccines, a vigorous research effort to detect and respond is needed. Even more important is new means for detecting toxic materials or pathogens that might be inserted into the nation's food supply. While quick diagnosis and isolation of patients, as was done to quench the SARS epidemic, is the best defense today, biologists are working on networks of sensors, based on genomics and proteomics (study of the entire DNA sequences and of the entire range of proteins produced by pathogenic organisms) to detect pathogens that might be released in the air before they have time to make people ill. Since it takes years to develop vaccines for new diseases, biologists are also exploring new ways to create broad-spectrum antibiotics and antivirals for defense against a biological attack.

Toxic chemicals and dangerous materials. Dangerous chemicals in transit, such as tank cars filled with chlorine for protection of water supplies, could be blown up by terrorists with catastrophic results. These rail cars could be tracked by transponders that would report their positions using signals from the Global Positioning System and would be identified to responding fire fighters using encrypted radio messages only they can read. Sensor networks established in transportation nodes might allow attacks to be prevented, and sensors in crowded places, such as subways, theaters, and athletic facilities, could provide instant warning of dangerous chemicals that might be released there. Self-analyzing filter systems for ventilating modern office buildings might not only protect the inhabitants but detect and report the presence of dangerous materials that may be trapped in the filters. Basic research might lead to discovery of biosensors that could duplicate the ability of dogs, whose sense of smell exceeds that of humans by about 10,000 times, to smell minute traces of explosives, and perhaps toxic chemicals as well. *See* NETWORKED MICROCONTROLLERS.

Energy systems. The hazards associated with fossil fuel storage and shipment are well known. Perhaps less apparent are the vulnerabilities of a modern electric power grid. Critical points of vulnerability are the extremely high voltage, one-of-a-kind transformers that form the heart of a power distribution facility. They are hugely expensive, are imported, and have no backup. Engineers need to design reconfigurable midsized transformers, stored in a safe place off site, that could be assembled in configurations that would mimic one of these very high voltage transformers. *See* ELECTRIC POWER SUBSTATION DESIGN.

The computers that control power distribution control rooms are sensitive to cyber attack. Research to develop rigorously secure computer systems with encrypted communications might prevent attacks on these system control and data acquisition (SCADA) computers, which form the brains of the distribution system. From the perspective of a longer time frame, adaptive power grids should be developed which can automatically detect a loss of part of the system and restructure it before irreparable damage is done to the entire network.

Communications and information systems. In the United States, the most urgent issue is to provide secure communications to police, fire, and medical personnel so that they can communicate with one another and with emergency operations centers. Failure of fire fighters to hear the warnings of police to evacuate the towers greatly aggravated loss of life in the World Trade Center attack. Research is needed not only to ensure interoperability among a broad range of diverse responders to emergency but to manage priority access to the network and ensure its security against denial of service. But the main worry about cyber attacks is how to prevent terrorists from using a cyber attack to amplify the destructive effect of a conventional physical or biological attack. Greatly increased technical protection of computer communications is an urgent but difficult technical challenge. No computer operating system today is fully secure against an expert penetration effort. While the Internet and many telecommunications systems are soft targets against a cyber attack, they can be brought back into service in a matter of hours or days. But research is needed to reduce this recovery time dramatically so that information services can be available to emergency officials during an attack. *See* TELECOMMUNICATION NETWORK SECURITY.

Transportation and borders. Sensor networks for inspection of goods and passengers crossing the nation's borders will be a research priority. Off-line sensors are already used in airports to inspect trace chemicals that might indicate the presence of explosives. What is needed is a detector network that allows detection when persons or packages pass nearby. The primary technical challenge will be melding sensor networks together with data fusion and decision support software, so that false positive signals and failures to detect can be identified and unambiguous advice given to security officials for action. Biometric technology, such as devices that scan the unique patterns of each person's retina or iris, can provide more secure identification of individuals. The range of threats to the transportation networks of a modern state is very great, and careful systems analysis to identify the weak points and find the most effective and economical means for protecting them is essential.

Cities and fixed infrastructure. Much research is underway to analyze the structural characteristics of high-rise buildings that make them much more vulnerable than necessary. Buildings like the Murrah Building in Oklahoma City depend on each pillar on the ground floor remaining intact. Modifications to buildings that allow adjacent pillars to share the load and prevent catastrophic collapse are being explored. Air intakes for large buildings need to be less accessible, and equipped with better air filters and toxic chemical detectors. Instrumentation to allow first responders to detect toxic chemicals and hazardous materials, and robots to go where humans cannot, are also needed. Special provisions for protecting

executed during 2000–2001. This time the newer *Chandra X-ray Observatory* was also used to observe the nebula in x-rays at the same time that the Hubble Space Telescope was observing it at optical wavelengths (see **illus.**). The reason for using more than one wavelength was that the nebula was known to be bright at all wavelengths, and understanding the distribution of energy in the nebula, as mapped by different wavelengths of electromagnetic radiation, would provide valuable information.

Synchrotron nebula dynamics. In considering the inner nebula, it is important to build a mental picture of what is taking place there. The star itself, the size of a city, spins 32 times a second. The powerful magnetic fields are "stuck" to the star, so they are swept around at the same rate, and all the ionized particles (electrons, positrons, and protons) are swept along with them. However, the farther out from the star one goes, the faster the fields are whipped around, just like the outer edge of a merry-go-round. At some distance this speed approaches the speed of light; this distance or location is called the light cylinder. At this point the powerful magnetic fields cease to be dominant, and the particles attain enough kinetic energy to break free and flow away from the star on their own, dragging the magnetic fields along with them. This mechanism is the source of a relativistic wind of ionized particles that stream away from the pulsar as an extreme "solar wind."

The mixture of particles and fields expands away from the star mostly in the equatorial plane of the star, because that is where the rotational velocity is the maximum, and the flow eventually piles up into waves. These waves are seen as wisps from the Earth and are visible in the time-lapse images taken with the Hubble Space Telescope. The speed of these waves is about half the speed of light. As the particles flow away from the star, they radiate their energy; thus the particles closest to the star are most energetic, while those farther out are less energetic. Because of this, high-energy x-ray images show the inner nebula to be quite compact around the star, while the images taken in lower-energy optical radiation show the nebula to be much larger. However, the same wavelike structure can be seen in both optical and x-ray time-lapse images (see illus.).

Results from time-lapse images. The time-lapse images show most of the waves expanding in the equatorial plane of the star. This conclusion is based on the fact that the waves are seen as ovals centered on the star, implying that the system is being viewed from an angle. However, there are also features that are more persistent in physical location and simply change their appearance with time. Understanding how these structures are physically located in the pulsar vicinity is key to understanding their origin and persistence.

A large-scale torus-shaped region of emission is apparent in both the x-ray and optical time-lapse images. It stretches both above and below the pulsar, but is brightest above. Since the structure is centered on the star, it is interpreted as being a large-scale toroidal ring of material, again located in the equa-

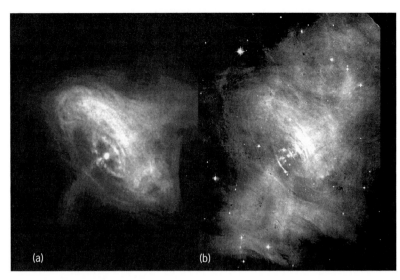

Images of the Crab Nebula made with (*a*) *Chandra X-ray Observatory* and (*b*) Hubble Space Telescope. A regular sequence of such images was made with both instruments in 2000–2001. (*NASA/Hubble Space Telescope/Chandra X-ray Observatory Center/Arizona State University/J. Hester et al.*)

torial plane of the star. The waves expanding away from the star are also in the equatorial plane, and so they expand into this toroidal ring. They keep it energized by depositing some of their energy as they expand and dissipate. Since the torus and the associated waves of energy are seen in both optical and x-ray images, they are believed to emit a continuous spectrum of electromagnetic radiation.

Above the star is a large oval-shaped ring structure that does not move but changes in brightness and illumination. This "halo" appears to be a long-lived structure. Since it is not centered on the star, it appears to be elevated away from the equatorial plane of the system and to reside at an intermediate latitude at some distance from the star. It is seen on only one side of the star. However, below the star there is a small knotlike structure that is called the sprite because it comes and goes over very short periods of time; it was seen in the earlier 1995 images as well as in the 2000–2001 images. These structures are seen mostly in the optical images and are only vaguely apparent in the x-ray images.

One final feature seen in both optical and x-ray images is a jet of material that moves quickly away from the star to the lower left. This jet follows the apparent rotational axis of the star (if the waves and torus are taken as defining the equatorial plane) and passes through the location of the sprite, which could be some kind of transition in the jet. This jet is seen to bend to one side as it expands away from the star and eventually runs into the outer wall of the nebula as defined by the filaments of ionized gas.

Overall picture. The current picture of the inner Crab Nebula is therefore as follows. At the heart of the nebula is a small, dense, fast-rotating neutron star. The intense magnetic fields associated with this star generate electron-positron pairs close to the surface that are then flung outward from the star by the rapidly rotating magnetic fields. Once the particles in this wind reach the light cylinder, they attain enough kinetic energy to dominate the magnetic fields and are free to expand physically outward, carrying the fields with them. This fast (half the speed of light) expansion produces the observed equatorial waves of energy seen in the inner synchrotron nebula, which are visible at both optical and x-ray wavelengths. As these waves expand outward, they radiate much of their energy away and deposit some into a large-scale torus of emission also located in the equatorial plane. The fast-rotating magnetic fields also generate a corkscrew jet of particles up the rotation axes, which is seen in both the optical and x-ray images. A transition in this jet marks the location of the optical feature called the sprite.

For background information *see* CRAB NEBULA; ELECTRON-POSITRON PAIR PRODUCTION; NEUTRON STAR; PULSAR; SPACE TELESCOPE, HUBBLE; SUPERNOVA; SYNCHROTRON RADIATION; X-RAY ASTRONOMY in the McGraw-Hill Encyclopedia of Science & Technology.   Paul Scowen

Bibliography. M. Bobra, The Crab's shimmering shock waves, *Sky Telesc.*, 105(3):17, March 2003; S. Mitton, *The Crab Nebula*, Scribner, 1979; R. Talcott, The Crab's inner workings, *Astron. Mag.*, 31(1):24–26, January 2003.

Cyanobacterial fossils

Until recently, the earliest chemical record of life on Earth was generally thought to be traceable to around 3.8 billion years ago. The first fossil cells and cell filaments were discovered in Archean rocks from approximately 3.5 billion years ago, at which point life is thought to have diversified rapidly, reaching perhaps to the level of oxygen-releasing photosynthesis by cyanobacteria by that time. However, this concept of an "early Eden" for life on Earth, and perhaps on Mars, is now undergoing renewed scientific review. The assumptions that led to this early Eden model were bolstered by the lack of critical investigations into the geological context (such as the field mapping, petrography, sedimentology, and geochronology) of evidence. In the future, more detailed study needs to be undertaken to distinguish abiotic (nonliving) from biotic (living) phenomena in the rock record, especially within hydrothermal systems, in which organic material can be produced by strictly inorganic processes.

Reinterpretation of Apex chert. In 2002, M. D. Brasier and colleagues began their challenge to the early Eden paradigm by questioning the Earth's oldest supposed microfossil assemblage, from the 3.46-billion-year-old Apex chert (hard, dense sedimentary rock composed of fine-grained silica). In 1993, J. W. Schopf inferred that 11 separate types of microorganisms were preserved in such cherts from Chinaman Creek near Marble Bar in Western Australia. Of these, a number were found to be morphologically similar to fossil and living cyanobacteria, with the major implication that oxygen was already being released into the atmosphere 3.46 billion

years ago. However, the biotic nature of the Apex chert assemblage came into question when the rock slices were examined in detail for the first time in the early 1990s at the Natural History Museum in London. These microfossils, which include some of the smallest fossils ever named (*Archaeotrichion* is as small as 1/3000 millimeter), were seen to intergrade with abiogenic structures resulting from recrystallization of the rock fabric. Such morphing goes against one of the cardinal rules previously established for the recognition of potentially biogenic structures.

Brasier and colleagues analyzed the Apex chert site and microstructures using new, high-resolution techniques. Geologic mapping was undertaken (to determine the type and distribution of rocks at the site) at a range of scales from kilometers to micrometers and was integrated with newly applied techniques for geochemical and microfossil morphologic analysis. Together these techniques revealed that the Apex chert microfossil site may have been produced hydrothermally by hot springs that were fed by a deep igneous heat source. The surprisingly abundant organic matter is conjectured to have been, at least in part, synthesized by something akin to strictly inorganic Fischer-Tropsch–type reactions between the highly metalliferous early crust and volcanogenic carbon dioxide and carbon monoxide to produce methane and hydrocarbons. (These processes are operating today in highly iron-rich crusts of midocean ridges.)

This major reassessment of the Apex chert and its context revolves around eight major arguments, summarized below.

1. The context for the microfossil samples is not (as previously thought) a conglomerate formed on a beach or near the mouth of a river, but part of a fissure that feeds superheated and metallic waters to a hydrothermal spring. This can be demonstrated by mapping and by analysis of fabrics (structural or spatial characteristics of a rock mass) and geochemistry.

2. The microstructures are not restricted to a distinctive class of clasts (fragments of sedimentary rock produced by physical breakdown of a larger mass), often rounded (as claimed). The structures actually occur in three layers of brecciated (angularly fragmented) rocks that were fractured and subsequently infilled a hydrothermal fissure. In places, these rocks are fused with glassy cements, indicating this fissure may have reached 500°C, a temperature ill suited for life.

3. Associated structures that were regarded as stromatolite-like clasts (resembling laminated microbial structures in carbonate rocks) are found to have a rock texture more consistent with layered fissure infilling, rather than of living cyanobacterial mats.

4. The spatial arrangements of the microfossils do not compare with those seen in the next oldest, diverse microfossil assemblage, the 1.9 billion-year-old Gunflint chert. In the Gunflint chert, filaments are wrapped around each other and clustered into layers that show clear bacterial behavioral orientation

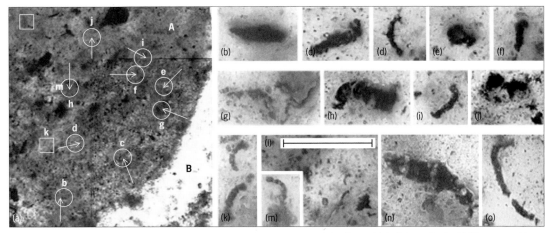

Fig. 1. A corner of the microfossiliferous, sedimentary clast reported by J. W. Schopf (1993) from the 3.46 billion-year-old Apex chert, recently reinterpreted by M. D. Brasier et al. as a shard within a subsurface hydrothermal chert dike. (a) The scatter of filamentous carbonaceous structures, originally interpreted and arrowed as microfossils, and here regarded as pseudofossils that lie within clotted, gray to black, fissure-filling fabric surrounded by white silica cement. (b–m) Detailed views of the putative microfossils shown at white arrows in (a) revealing incoherent-to-arcuate structures in which the septation, where present, is probably due to scattered quartz grains of similar size to those of the surrounding matrix. *Primaevifilum delicatulum* of Schopf is shown at *k*, with a similar structure above it. Sample NHM V.63164. (n–o) Comparable microstructures obtained by Brasier et al. from recollected sample NHM V. 647364. Brightfield transmitted light. The scale bar indicates (a) 400 μm; (l) 100 μm; (b–k, m, n, o) 40 μm.

(in which the cells line up parallel to the laminae), whereas the microfossils in the Apex chert show no coherent arrangement that might be thought consistent with biogenicity (**Fig. 1**).

5. The filaments are not all simple and unbranched, as previously thought. At least four of the important microfossils are actually branched, and all of them intergrade with other branched structures, not at all like filamentous cyanobacteria. This texture and shape is more readily consistent with abiotic formation; that is, it appears to be due to the recrystallization of hydrothermal glass (**Fig. 2**). As the glass recrystallized, it pushed carbonaceous impurities ahead of the radiating crystal fans because

they could not be incorporated in the lattice. Such a process results in forms ranging from rounded sheets of carbon, in which the impurities are abundant, to branched, dendritic, or simple arcuate (curved) filaments where the impurities become scarce (Fig. 2).

6. The presence of what appear to be septa (cell walls) and bifurcated cells (in the process of cell division) is also seen in the associated abiogenic structures (spherulitic and dendritic filaments). Both have been reinterpreted by the researchers as products of recrystallization of the glass within the fissure.

7. There is no question that the structures are made of carbonaceous matter. But this carbon is no different from that seen in the associated abiogenic

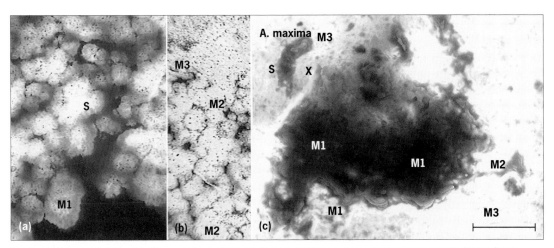

Fig. 2. Finding a pseudofossil in Precambrian hydrothermal chert. (a) Glassy silica of hydrothermal origin (Gwna Group, Precambrian of Wales) crystallizes to form spherical crystalline masses of fibrous quartz, called chalcedony. Where the impurities are abundant, they form spherical rims (M1) with septate fabric (S). (b) With decreasing impurities in the silica, these rims form a symmetry-breaking cascade of structures ranging from dendritic (M2) to arcuate (M3). (c) Digital montage of a similar spherulitic mass from the Apex chert material. Here, a pseudofossil (actually the holotype of a putative ancient cyanobacterium called *Archaeoscillatoriopsis maxima*) lies right beside a mass of spherulitic (M1) and dendritic (M2) artifacts. Smaller arcuate pseudofossils occur all around the margins of the spherulitic mass (M3). The septate appearance (S) of the arcuate pseudofossil *A. maxima* is here explained as due to crystallites that radiated from the center of a spherulitic structure (X). Sample NHM V. 63164. Brightfield transmitted light. The scale bar indicates (a, b) 400 μm, (c) 100 μm.

artifacts with which they intergrade, nor does it differ significantly (in terms of laser Raman imaging, which maps the chemical composition and structure of fossils in two dimensions) from that seen in disordered graphite or in carbonaceous meteorites. It is misleading, therefore, to use the term "kerogen" (degraded biogenic carbon) to describe the composition of the structures and, hence, to infer a biogenic origin.

8. The ratio of the light stable isotope of carbon (^{12}C) to the heavier stable isotope (^{13}C) has been used as an indication of biogenic fractionation. However, although the carbon isotope ratios are consistent with a biogenic origin for this carbon (possibly from hyperthermophilic bacteria), it is notable that a similar range of values can be produced by abiogenic Fischer-Tropsch–type synthesis, also suspected to have taken place in the Archean (approximately 3.8 to 2.5 billion years ago) dike systems.

Wider implications. Several main conclusions can be drawn from this new work on the Apex chert. First, integrated studies, in which geological context is studied at a range of scales, are crucial for the scientific study of early life on Earth as well as Mars. Until such work is undertaken and abiogenic origins ruled out, the biogenicity of all early Archean microfossils, stromatolites, and structures with carbon isotopic values (older than approximately 3 billion years) should be regarded as tentative. Second, this revised view has the potential to push the origins of oxygenic photosynthesis closer toward the first chemical signals for cyanobacteria, at approximately 2.7 billion years—much closer to oxygenation of the atmosphere at about 2.4 billion years ago. Third, it now seems possible that carbon compounds were being synthesized readily and in great abundance within the highly metalliferous early crust; however, this possibility needs to be tested. A new model of an Archean world dominated by volcanic and hydrothermal (and hyperthermophilic) systems would mean that the Apex chert dikes could provide scientists with a geological context for the synthesis of life itself. Earth's geological record may, therefore, contain much more of the early history of life than was previously thought possible.

For background information *see* ARCHEAN; BIOSPHERE; CHERT; CYANOBACTERIA; FOSSIL; GEOCHRONOMETRY; GEOLOGY; ORGANIC EVOLUTION in the McGraw-Hill Encyclopedia of Science & Technology.   Martin D. Brasier; Owen R. Green

Bibliography. M. D. Brasier et al., Questioning the evidence for Earth's oldest fossils, *Nature*, 416:76–81, 2002; C. M. Fedo and M. J. Whitehouse, Metasomatic origin of quartz-pyroxene rock, Akilia, Greenland, and its implications for Earth's earliest life, *Science*, 296:1448–1452, 2002; M. Schindlowski, Carbon isotopes as biogeochemical recorders of life over 3.8 Ga of Earth history: Evolution of a concept, *Precamb. Res.*, 106:117–134, 2001; J. W. Schopf, *Cradle of Life*, Princeton University Press, 1999; J. W. Schopf, Microfossil of the Early Archean Apex chert: New evidence for the antiquity of life, *Science*, 260:640–646, 1993; M. van Zullen et al., Reassessing the evidence for the earliest traces of life, *Nature*, 418:627–630, 2002; S. Simpson, Questioning the oldest signs of life, *Sci. Amer.*, pp. 70–77, April 2003.

Dance language of bees

The zoologist Karl von Frisch (Nobel Laureate 1973) is best known for discovering that honeybees can recruit other bees to visit a food source by performing a dance in which information on the distance and direction from the hive to the food is encoded. The communication about remote events in the dance language is an exception to the general rule that animals tend to communicate about immediate events connected with the actor and its surrounds. Furthermore, in the dance language a symbolic code is used to transmit an impressive amount of information.

Waggle dance. The dances are normally performed on the vertical wax combs in the hive. In the waggle dance, the dancer moves in a straight line (the waggle run) and circles back, alternating between a left and a right return path so that the entire dance path takes on a figure-eight shape (**Fig. 1**). During the waggle run, the dancer wags her body (all dancing workers are female) from side to side 13–15 times per second and emits 280-Hz sounds by vibrating her wings. The angle between the direction of the waggle run and the vertical indicates the angle that the recruits should later maintain between their flight path to the food and the direction to the projection of the Sun on the horizon (known as the Sun's azimuth). The distance to the food is indicated by the duration of the waggle run.

Fig. 1. Waggle dance. The bee at the center is the dancer; the four bees at the bottom are recruits receiving information from the dance to aid them in finding food.

copy of the suspect's hard drive would be examined for evidence of the address. In this case, the forensic examiner would do a keyword search and look for the words in the address. Forensic tools are software specifically designed for forensic examinations such as imaging (copying the hard drive) and restoring deleted files. Comprehensive suites of software tools are used for most forensic examinations.

Hidden data. Digital evidence may be obvious to an investigator or hidden. For example, obvious evidence might be a letter stored as a document on a hard drive, while less obvious evidence might be a word or fragment of a word that was deleted from a document but still exists on the unused space of the hard drive. Special software tools, some public and others restricted, assist forensic examiners to find this evidence. There are also forensic tools for determining when data have been deliberately hidden.

Encryption. Encryption is one way of actively hiding data. In encryption, data are encoded using an algorithm (mathematical formula), obscuring it from anyone who does not have a "secret key" to unscramble it. Once data have been encrypted, a single- or double-key decryption program is needed to read it, or else the code must be "cracked." Single-key decryption is similar to using one key to lock and unlock your home. Double-key decryption requires both a public key and a private key, where one or more people may have the public key, but only one person has the private key. Code cracking ranges from a very easy process to one that is computationally infeasible depending on the sophistication of the encryption.

Steganography. Hiding data in text, image, or audio files is called steganography. For example, a text file can be hidden in a digital picture. There are statistical analysis techniques to find such data, but they take time and effort. Commercial tools are available to detect these types of files.

Evidence authentication. A mathematical method known as hashing is used to verify the authenticity of data copied from a hard drive, disk, or file. A hash is a small mathematical summary of a data file or message. Changing one character in a file will change the numerical value of the hash. If there has been any change in the file during transmission or from tampering, the hash number will change. A hash of the hard drive is done before copying it. The hash value of a hard drive and the copy should be exactly the same number.

To ensure chain of custody, forensic scientists must sign their name to evidence. In the case of digital evidence, an authentic digital signature or a regular physical signature is required. Creating a digital signature is a simple process for verifying the integrity and originator of a message, whereby the signer calculates the hash value of the message, and then encrypts the hash value with his or her own private key. The resultant message "digest" is the digital signature. Appending it to the original message qualifies the object as being digitally signed.

Outlook. As technology changes, the nature and handling of digital evidence will remain a challenge for forensic scientists. The wireless Web will allow for "virtual evidence rooms" where evidence is stored and secured through encryption. The chain of custody will be maintained through digital signatures, and the evidence will be checked out of the virtual evidence room with a time stamp from a digital clock synchronized with the atomic clock run by the National Institute of Standards and Technology (NIST).

For background information *see* ALGORITHM; COMPUTER SECURITY; COMPUTER STORAGE TECHNOLOGY; CRIMINALISTICS; CRYPTOGRAPHY; DATA COMMUNICATIONS; DIGITAL COMPUTER; ELECTRONIC MAIL; INTERNET; SOFTWARE; WORLD WIDE WEB in the McGraw-Hill Encyclopedia of Science & Technology.

Carrie Morgan Whitcomb

Bibliography. E. Casey, *Digital Evidence and Computer Crime*, Academic Press, 2000; W. G. Kruse II and J. G. Heiser, *Computer Forensics: Incident Response Essentials*, Addison-Wesley, 2001; M. Pollitt, *Digital Evidence Standards and Principles*, Forensic Science Communication, 2000; T. Sammes, B. Jenkinson, and A. J. Sammes, *Forensic Computing: A Practitioner's Guide*, Springer, 2000; R. E. Smith, *Authentication: From Passwords to Public Keys*, Addison-Wesley, 2002.

Digital stroboscopic photography

Stroboscopic photography for depicting the changing features of subjects in motion is a technique that was enabled by one of the inventors of high-speed photography, Harold Edgerton. He applied this technique to numerous situations where a still camera (in a single shot) did not capture enough information about a subject in motion and where a motion picture camera's record was unsuitable for print reproduction.

Photographers use two types of stroboscopes. The more widely used one is a flashing light source operated in a dark environment. This is the kind that Edgerton popularized. A cheaper, simpler, mechanical alternative consists of a rotating disk with a slot cut into it. Every rotation of the disk provides a glimpse of the position of the subject at the time the slot passes in front of our eyes or a camera's lens. In either case, a stroboscope essentially allows one to view a subject on a periodic basis. This is called time-sampled image recording.

Traditional technique. Photographers have generally used a flashing light (a stroboscopic light source) to illuminate a moving subject in order to track its position over time. This is accomplished by setting up an action situation in front of the camera (usually on a tripod), opening the camera's shutter while the subject is moving and the stroboscope is flashing and, after a period, closing the shutter and terminating the exposure.

During the time the shutter is open, the moving subject is illuminated by several flashes of light. These flashes leave a superimposed sequence of images of the subject. It is often possible to gain valuable information of a subject's motion from such a record, or simply to understand the flow of motion of a given subject.

In stroboscopic photography, one is interested in photographing relatively long-duration events (1–2 s) and capturing from 10 to possibly hundreds of images during that time.

For example, to record a golf swing, one would make a record over a period of a second or so. During that time, the subject might be recorded in 20 to 100 different positions. This would require a strobe at a frequency of 20 to 100 flashes per second if the shutter was kept open for a second. An exposure time of 1 s is easy to accomplish with a regular camera, but many digital cameras have a limited maximum exposure time.

An opaque disk with a slot rotating in front of a digital camera's lens is a simpler and cheaper alternative to the flashing-light stroboscopes (**Fig. 1**). With such a device the photograph shown in **Figure 2** was made using an exposure time of 2 s. The subject was placed against a large, black velvet background. The lighting level was adjusted so that the results obtained were of an acceptable quality by making a few preliminary tests and judging the quality of the images on the liquid crystal display (LCD) screen of the camera.

For this example made over a period of 2 s, the camera recorded about 30 separate images of the action; that is, the mechanical stroboscope's disk was turning at about 15 revolutions per second. The exposure time (per image) that the disk delivers is a function of the slot size divided by 360° and multiplied by the time it takes the disk to turn once.

Fig. 1. Digital camera with a rotating-disk stroboscope attached.

Fig. 2. Multiple-exposure stroboscopic photograph made with a rotating-disk stroboscope.

As with traditional cameras, the background (which should have reproduced very dark since it was black velvet) and those parts of the subject's body that remained essentially in the same position often appear significantly overexposed because they reflect light to the same location on the camera's charge-coupled device (CCD) image sensor. The moving arms, however, were exposed in different positions on the CCD with every pass of the stroboscope disk. The arms, therefore, were exposed only once on any given area of the CCD and reproduced with less exposure than the stationary parts of the scene. One could improve slightly on the tonal range of the image by making the subject wear dark clothing while painting the moving parts with white paint. Present practice is to simply adjust the tonal range of the photograph with an image processing program such as Adobe Photoshop.

Extended time photography. If the shutter is open for too long, the sequential images will start to overlap until a point is reached where they blend into a uniform "blur" and the specific position of the subject with any given flash discharge can no longer be perceived. To record a subject's motion in detail in a film-type camera, it is possible to put the film in motion while the stroboscopic light flashes or the mechanical disk rapidly rotates in front of the camera lens.

Using a standard 35 mm camera, the film is advanced one frame at a time, with the lens covered, into the take-up chamber without exposing it. Then, by setting the camera shutter to B and locking it open during the rewinding process, the moving film records the subject's image, with the final record

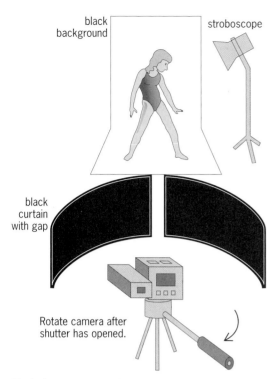

Fig. 3. Studio setup for digital time-displaced stroboscopic record photography.

appearing as a sequential set of images showing the progression of motion of the subject over time.

In digital photography, there is no possibility for moving the image receptor—the sensor in the digital camera is fixed in place. While standard, superimposed stroboscopic records can be obtained much the same as with film cameras, the visualization of motion of a subject over time by introducing motion of the sensor is not possible.

There is a method to deal with this problem if one is willing to give up a certain amount of resolution.

The approach is to put the image in motion so that it traverses the camera's field of view during the time that the shutter remains open.

One practical way to accomplish this is to move the image across the focal plane by rotating the camera. To do so, the subject remains in approximately the same location while performing some action, and the camera is aimed so that the image appears at one side of the its field of view. When the shutter is opened, the camera slowly is turned so that the subject's image moves to the other side of the viewfinder before the exposure is terminated.

With each flash of light from the stroboscope, the subject's image is recorded on a different location of the camera sensor, allowing the motion of the subject to be tracked over a significant period.

A major drawback of this technique is that the field of view of the camera (over time) is very wide and the flash, stands, and studio equipment may reflect some of the light from the stroboscope along with the light from the subject. This reflected light will appear to "bleed through" the subject's image.

To deal with this difficulty, the camera is surrounded by a curtain of black velvet set up so that there is a small open slot located between the camera, rotating on a tripod, and the subject (**Fig. 3**). This prevents the camera from seeing and recording anything that is not available to it through the gap or slot. This provides a "ghost"-free final record, although superimposition of certain parts of the moving subject is still a possibility which often just adds to the fluidity of the final motion record (**Fig. 4**).

Since the size of the camera sensor is limited, the total number of images that clearly can be recorded depends on how small the subject is made within the frame of the camera. The smaller the image size, the larger the time over which the subject, or the number of separate images of the subject, can be placed from one side of the frame to the other.

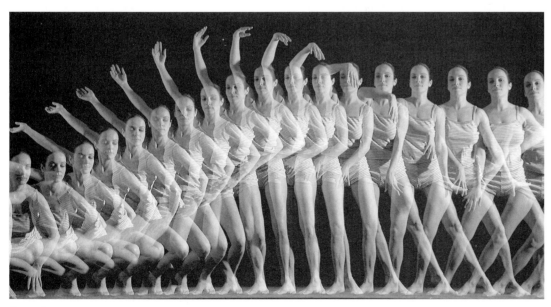

Fig. 4. Result of panning a digital camera to capture the action shown in Fig. 3.

Fig. 5. Digital stroboscopic photograph based on a comparison sampling technique.

Using digital technology, areas from each frame are isolated where the image information is different from the reference (static) view of the background. The isolated images of the action are layered, one on top of the other, in their proper spatial and temporal locations in relationship to the original, or new, background view. The final display is a time-sampled record of the position and location of a subject against a fixed background at several instances in time.

For background information *see* CAMERA; CHARGE-COUPLED DEVICES; PHOTOGRAPHY; STROBOSCOPE; STROBOSCOPIC PHOTOGRAPHY in the McGraw-Hill Encyclopedia of Science & Technology.

Andrew Davidhazy

Bibliography. H. Edgerton, *Electronic Flash, Strobe*, 3d ed., MIT Press, 1987; H. E. Edgerton and J. R. Killian, *Moments of Vision: The Stroboscopic Revolution in Photograph*, MIT Press, 1984; S. F. Ray, *Scientific Photography and Applied Imaging*, Focal Press, 1999.

Comparison sampling. A further alternative to the stroboscopic light source is a technique available only in digital photography. This is a nonintegrating, periodically updated, image output device driven by a video signal that is input from a video source such as a camera. For example, one device, which is an interface between a video camera and a video monitor, compares signal-level information associated with each point in a video frame with that of the next one. The device is set so that an increase (or a decrease) over some base signal level refreshes the display at that point. If the levels do not change in the desired direction, then a steady and unchanging image is displayed on the monitor. The frequency of update of the display can be selected from the camera's maximum capture rate (shorter time) to longer times.

Since a video camera is a time-sampling device operating at 60 frames per second, the instantaneous records of a moving subject across a dark background can be displayed at 1/60th second intervals, and the composite display looks much as if it would look if the record had been made with a standard camera (**Fig. 5**). The big difference in this case is the exposure or level of subject areas that do not move are free of unwanted "noise" (due to overexposure) in the final image.

Digital assembly. A stroboscopic image can be assembled after the fact using digital techniques that in the past would have required much too elaborate and time-consuming photographic process. Such a "stroboscopic composite" starts with a digital video record of an action captured at a given framing rate and with the camera fixed in place. A reference image (background) of the scene is made while no action takes place in front of the camera and then the action is recorded against that background.

Effectiveness of high-technology research

The effectiveness of any research, by its nature, is difficult to evaluate. There is not much agreement among managers and researchers as to what constitutes effective research. Research effectiveness depends on a number of factors such as the type of industry, the nature of the market, and the state of technology. Both financial and nonfinancial measures have been used to define effectiveness, and both have problems.

It is even more difficult to define research effectiveness in the high-technology area, since scientific and technical discoveries come at a rapid rate from many sources and research establishments. In contrast to research in mature technologies, many technical issues are not well understood, but there is pressure to introduce new products as soon as possible, utilizing the still not understood technologies. This article explores the means of achieving effective research and development (R&D) in high-technology firms and a method to evaluate research effectiveness.

Functions of research. A high-technology company relies on a constant stream of innovations and new products. As product obsolescence rates are relatively high, new products need to be brought into the market at a rapid pace, anticipating demand, in order for the company to be competitive. Effective research and development activities in a firm will lead to such a stream of innovations and successful new products and better manufacturing processes, helping the company to become stronger and more competitive.

High-technology research and development consists of two parts. Research focuses on developing new knowledge and technologies, while development focuses on the design of new products or processes based on technology already developed or being developed. The mix of effort between the two depends on the state of the industry and the firm's corporate strategy. Companies in mature technologies spend more on development than on research. In high-technology firms, the effort is either evenly split between the two or heavier on research, while in emerging-technology companies there is usually a heavy emphasis on research.

Well-managed firms spend a significant amount on research and development activities. High-technology firms and firms in emerging technologies generally spend 10–25% of their sales on research and development; and some firms, especially new ventures in emerging technologies, spend even more, mostly on research. (The national average of research and development expense for all industries is around 3.5% of sales.) It is therefore very important that this money be spent effectively.

Many research results and outputs may never find commercial applications or may not succeed when introduced as a product, as there are two major risks associated with industrial research and development: technology risk and market risk. The technology developed may not work as intended, or the demand for the product may fail to match the firm's estimates. In high technology, both risks are relatively high.

Measures of research effectiveness. No single measure captures the effectiveness of research in high-technology firms. Typical financial measures such as ROI (return on investment) and ROA (return on assets) are relatively useless, as most research and development activities bear fruit only after long gestation periods, and it is difficult to trace the profits directly to research and development efforts. Some tangible, but not financial, measures of research output, such as the number of patents generated, the number of licenses sold for the technology, and the number of technical papers presented by the researchers, are sometimes used to measure research effectiveness. Along with these, nontangible measures such as the reputation of the research staff in their respective fields are also used. All these measures, however, do not indicate how useful the research activities have been to the firm's bottom line and to its competitiveness. A composite measure of research effectiveness includes not only the output of the research and development function but also how well the research and development effort is linked to the corporate strategy and how well the research and development strategy is implemented.

Effective research. Effective research in high technology focuses resources to generate technologies and products to produce higher profits for the firm and to increase the firm's competitive position. It is the result of a number of things, of which the major parts are: designing a viable research and development strategy for the firm; organizing the research function to comply with this strategy; implementing the strategy; and managing the development and introduction of new products and processes.

Effective research is a composite of these four factors. Managing usually focuses on the last part. It is important, however, that the other parts are done well too.

Designing a viable R&D strategy. The research and development strategy should be linked to the corporate strategy of the firm. It should define the technologies that research and development should pursue that will strengthen the position of the firm in its chosen fields. It should define the allocation of resources between research activities and development activities. Failure to link research and development strategy to corporate strategy can lead to unproductive results that do not benefit the company. Linking research and development strategy to corporate strategy involves detailed discussions at the top management level and an understanding by top management of the available technologies in the firm's current and planned areas of business.

Organizing the R&D function. Research and development strategy defines the emphasis to be placed on developing new technologies and on developing new products for the existing markets and for new markets. The research and development budget is allocated to different aspects of research activities according to the emphasis.

With research and development activities becoming international, it is necessary to consider where the appropriate personnel are available. Some areas of research can be better performed abroad. For example, software research can be done more economically in countries such as India and China.

Internationalizing research activities requires good communication and coordination between different research and development activities in different locations. Communication technologies, such as the Internet and video conferencing, facilitate the operation of research and development centers abroad. Staffing and motivating research and development personnel is a complex task that is sometimes considered to be unmanageable.

Implementing R&D strategies. Deciding on specific research and development projects from among a large number of candidate proposals for budget allocation requires an analysis of the technological and market potential of the resulting products. However, choosing the most attractive proposals based on their technological and market merits alone without regard for their strategic implications for the firm is not a good approach.

A firm's portfolio of its research and development projects should reflect its strategy. The portfolio should have an appropriate mix of different product types—improvements of existing products to maintain cash flows (relatively risk-free), products for new markets (moderate risk), and products with new technologies for anticipated new markets (high risk). Recent research suggests that research and development projects and new product development projects can be classified according to whether they are aimed at existing markets or new markets, whether the technology is familiar to the firm or new, and

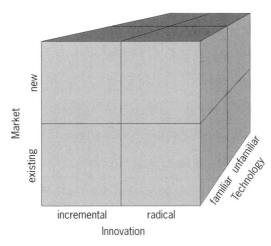

Contextual model for research and development (R&D) projects. The model has eight cells with the designations IEF (incremental innovation, existing market, familiar technology), IEU (incremental innovation, existing market, unfamiliar technology), INF (incremental innovation, new market, familiar technology), INU (incremental innovation, new market, unfamiliar technology), REF (radical innovation, existing market, familiar technology), REU (radical innovation, existing market, unfamiliar technology), RNF (radical innovation, new market, familiar technology), and RNU (radical innovation, new market, unfamiliar technology). (*Adapted from R. Balachandra and J. H. Friar, Factors for success in R&D projects and new product innovation: A contextual framework, IEEE Trans. Eng. Manag., EM-44(3):276-287, August 1997*)

whether the improvement is marginal (incremental) or large (radical) [see **illus.**]. This framework suggests eight types of projects. The portfolio should have an appropriate mix of these eight types based on the strategy of the firm. **Table 1** shows some suggested portfolio types for different corporate research and development strategies. Choosing the right portfolio of projects can help a firm in optimizing its research and development resources.

Managing R&D. Hardly one out of every 100 research projects for developing new products becomes successful in the market. The rest of them get scrapped at various stages—at the beginning before any funds are even allocated for the project, at the development stage where significant amounts of money have been invested, or even at the final market stage where the product has been introduced into the market. Carrying on a project when it is highly likely to fail is wasteful.

A number of factors can indicate in advance the likelihood of a research and development project to fail. The factors differ at different stages of a project. In the initial stages the technological factors are more important, while the market-related factors become more important toward the end of the project (just prior to introduction in the market and after initial introduction).

Different factors are also important depending on the project type (see illus.). Projects aimed for existing markets with marginal improvements have to look at market factors more closely than technological factors, while those planned for utilizing unfamiliar technology should emphasize technology factors.

The project management practices also differ for the different types of projects. A project aimed for an existing market, with familiar technology and requiring only a marginal improvement in the product, requires a more disciplined project management approach with strict time and cost schedules. On the other hand, a project attempting to create a product for a brand new market with unfamiliar technology needs a more flexible management style with loose time and cost schedules (**Table 2**).

In spite of favorable conditions, some projects may have to be terminated either because the market did not materialize as anticipated or the technological problems were too difficult to solve within the time and budget constraints. Though they do not add to

TABLE 1. Suggested R&D portfolios for different corporate strategies

Corporate strategy	Suggested portfolio of R&D projects*
Maintain and improve strength in current markets.	Focus on projects in the IEF cell (80%). Have a few in the INF cell (10%). Have one or two in the RNU cell (10%).
Develop and move to new markets.	Focus on the INF cell (60%). Have a few in the IEF cell (10%). Have a number of projects in the RNU and INU cells (30%).
Be in the forefront of technology.	Focus on the IEU cell (60%). Distribute remaining budget in the INU and RNU cells.

*Refer to Fig. 1 for cell designations.

TABLE 2. Management approaches for different project types*

Cell no.	Cell code[†]	Organization factors
1	IEF	Very important. Functional organization. Strict budgets and schedules.
2	IEU	Important. Provide for better interfacing between R&D and manufacturing and for greater flexibility in budgeting and scheduling.
3	INF	Very important. Functional organization. Strict budgets and schedules.
4	INU	Important. Provide for better interfacing between R&D and manufacturing and for greater flexibility in budgeting and scheduling.
5	REF	Important. Provide for some flexibility in budgets and scheduling.
6	REU	Important. Provide for better interfacing between marketing, R&D, and manufacturing; and greater flexibility in budgeting and scheduling.
7	RNF	Important. Provide for some flexibility in budgets and scheduling.
8	RNU	Important. Provide for better interfacing between marketing, R&D, and manufacturing; and greater flexibility in budgeting and scheduling.

*Adapted from R. Balachandra and J. H. Friar, Managing new product developed processes the right way, *Information • Knowledge • System Management*, vol. 1, no. 1, 1999.
†Refer to Fig. 1 for cell designations.

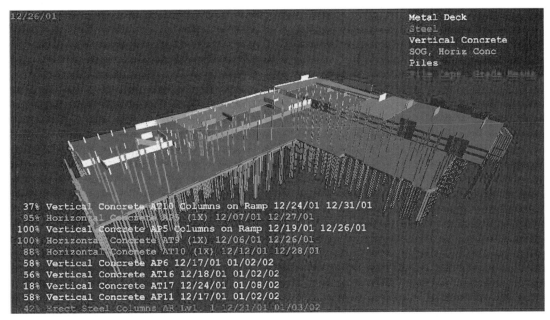

Fig. 1. Example of a 4D model. (*Courtesy of M. Fischer, Common Point Technologies, Inc., and DPR Construction, Inc.*)

business information. This enhanced access to data improves communication and collaboration among project participants. The project database can be continuously updated and is available on a real-time basis to owners, architects, engineers, and subcontractors, providing immediate access to project information such as change orders, drawing modifications, planning notes, and meeting minutes. Such a system supports project tracking and allows for real-time decision-making. In addition, a wide variety of Web-based applications are entering the marketplace to help implement e-commerce systems for materials procurement and purchasing.

4D systems. Virtual reality and three-dimensional (3D) modeling have impacted the way in which facilities are designed, constructed, and put in service. Various simulation methods have been developed to better study and analyze work site processes for saving time and effort. So-called four-dimensional (4D) systems assist in studying construction processes and improving resource allocation.

Visual 4D models combine 3D computer-aided design (CAD) models with simulations of construction activities to display the progression of construction over time, the fourth dimension (**Fig. 1**). These models enable project stakeholders to understand the relationship between construction activities and facility operation for retrofit (upgrade) projects (such as the remodeling and expansion of hospital wards). The models also help site personnel gain a better visual understanding of work sequences such as access and staging during construction.

4D models are used for analyzing construction processes to achieve lean production. The objective of lean production is to improve the flow of resources (for example, trucks, cranes, and crews) and reduce wasted effort so that a high-quality product can be produced at acceptable cost. The adaptation of lean production concepts is called lean construction.

Analysis using construction simulation. Dragados y Construcciones, one of Europe's largest construction companies, uses schematic (rather than three- or four-dimensional) simulation to analyze construction operations. Many of their field processes have been redesigned using lean production concepts based on extensive simulation studies. Processes such as the casting of large bridge girders, installation of

Fig. 2. Floating caisson constructed using PROSIDYC simulation. (*Courtesy of Dragados y Construcciones*)

Productivity obtained for each alternative based on 100% or 90% time availability (60 min/h versus 54 min/h) during the pouring period		
Time efficiency	100%	90%
Original situation	34 m³/h	31 m³/h
Alternative 1	42 m³/h	37 m³/h
Alternative 2	47 m³/h	42 m³/h

utility tunnels, construction of roller-compacted concrete dams, and renovation of heavy rail systems have been optimized using a computer-based simulation system called PROSIDYC (Project Simulation Dragados y Construcciones). PROSIDYC was developed in conjunction with the Division of Construction Engineering and Management at Purdue University. This system is particularly effective in analyzing repetitive work processes and does not require the construction of a visual or 4D model. PROSIDYC uses the CYCLONE (Cyclic Operations Network) modeling paradigm, which breaks work processes into active (work-producing) and delay (waiting) states. Construction of the large caissons used in the development of the port of Valencia, Spain, is an example of a project that used PROSIDYC simulation (**Fig. 2**).

On the Valencia project, the upper caisson pouring operation was investigated using PROSYDIC. It was the most important process for study, since this work represented 85% of the concrete poured and an estimated 74% of the fabrication time. A flow unit of 1 m³ was established for production modeling purposes. Since this process cycle depended on both concrete mixing and pumping, cycle capacities were checked to make sure that they did not represent a bottleneck.

Two alternatives for production improvement were studied. Alternative 1 doubled the capacity of the pump hoppers so that they could receive two batches at a time. Alternative 2 extended Alternative 1 by reducing the number of times the concrete delivery pipe needed to be moved/repositioned by modifying the shape of the working deck structure (see **table**).

Alternative 1 yielded an increase in productivity of 24% over the original setup. Alternative 2 resulted in a 38% increase in productivity over the process as originally configured. According to Dragados experts, simulation analysis typically results in savings of $200,000 per 100 hours of analysis using PROSIDYC, a return of $2000 for each engineer-hour invested.

Material and equipment technologies. Some technologies relate to physical systems rather than data manipulation. Many improvements in materials and their performance have been achieved in the past decade. New construction methods and techniques for equipment control have also been developed.

Researchers at the U.S. Department of Energy's National Laboratory at Los Alamos have discovered that small, bone-shaped fibers mixed into concrete increased its strength and toughness, and resisted crack propagation by bridging the crack. Composites, which use a polymer (plastic) matrix with embedded fibers, have very high strength characteristics, are lightweight (easy to handle), and corrosion-resistant. They are, however, expensive. Fiber-reinforced polymers (FRP) are used in a number of applications such as the construction of marine facilities and subsurface waste conduits where resistance to salt-water corrosion or the corrosive effect of chemicals is critical and outweighs cost considerations. Tests are in progress to evaluate FRP deck sections for use in vehicular bridges that are subjected to heavy salt use for melting ice and snow.

Automated equipment. Equipment automation is the focus of research and systems development at the Construction Automation and Robotics Laboratory at North Carolina State University. One such development is a trench excavation and pipe installation system that provides protection for the workers by automating the process.

The main components of the system, called Pipeman, are the human-machine interface consisting of a laser beam and feedback system (**Fig. 3**). The human-machine interface is used to keep the operator in a safe area. It allows the operator to guide the equipment intelligently and to accurately position the pipe while remaining outside the trench excavation. The human-machine interface includes manipulation and visualization functions.

The Pipeman system has a 3D spatial positioning system (SPS) that is interfaced with an excavator to provide location. A pipe manipulator, which is capable of handling pipes of various sizes, is attached to the bucket of the excavator. A laser beam is used to help the operator align pipes. Integration of SPS with a CAD system updates the excavator position in real-time and provides an as-built drawing of the pipe location. Development of this system is sponsored by the National Institute for Occupational Safety and Health (NIOSH), and it is currently being tested for use with different soils.

For background information *see* COMPOSITE MATERIAL; CONSTRUCTION ENGINEERING; CONSTRUCTION EQUIPMENT; CONSTRUCTION METHODS;

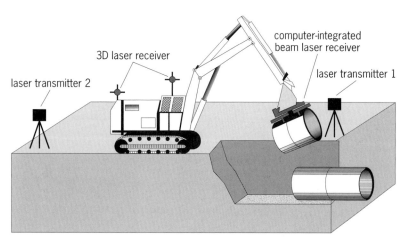

laser transmitter 2

3D laser receiver

computer-integrated beam laser receiver

laser transmitter 1

Fig. 3. Pipeman robotic trenching and pipe installation system. (*Huang & Bernold, 1993*)

ENGINEERING DESIGN; INTERNET; VIRTUAL REALITY; WORLD WIDE WEB in the McGraw-Hill Encyclopedia of Science & Technology.   Daniel W. Halpin

Bibliography. D. W. Halpin and L. S. Riggs, *Planning and Analysis of Construction Operations*, Wiley, New York, 1992; X. Huang and L. E. Bernold, Experimental work on robotics excavation and obstacle recognition, *Proceedings of the 5th Topical Meeting on Robotics and Remote Systems*, Knoxville, TN, pp. 83–88, April 25–30, 1993.

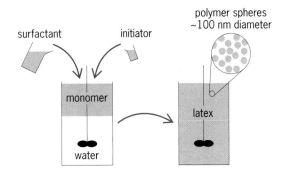

Fig. 1. Process of emulsion polymerization.

Emulsion polymerization

Emulsion polymerization is a variety of free-radical polymerization, a method in which individual molecules of high-molecular-weight polymer are rapidly generated by a free-radical chain reaction. In emulsion polymerization, a monomer with low water solubility is dispersed in small droplets (~10 micrometers in diameter) in water, and stabilized with surfactant to prevent separation of these two phases (**Fig. 1**). A dilute "seed" latex of minute polymer particles in water may also be added. The reactants are agitated and heated, and a free-radical initiator is added, leading to the formation of a latex—a dispersion of polymer particles of colloidal dimensions in water.

The term "emulsion polymerization" is a misnomer, as polymerization does not take place to any significant extent in the emulsified monomer droplets. The misnomer arose because the process was originally designed to polymerize in emulsion droplets, and the pioneers in the field did not realize that a much more complex process actually occurred. The first patents describe attempts to reproduce products similar to the natural polyisoprene latex produced by the rubber tree (*Hevea brasiliensis*), by dispersing monomers in water. Emulsion polymerization became an industrial process by the 1930s, with production of neoprene and of butadiene-acrylonitrile rubbers in Germany and the United States. Development and understanding of the process dramatically accelerated during World War II, when a research effort second only to the Manhattan Project was made in the United States for the industrial preparation of synthetic rubber.

Mechanism. A typical initiator of emulsion polymerization is an aqueous-phase thermal initiator (for example, ammonium persulfate, $NH_4S_2O_8$), although a redox couple (for example, potassium persulfate/sodium metabisulfite, $K_2S_2O_8/Na_2S_2O_5$) may also be used. Decomposition of the initiator in the aqueous phase generates hydrophilic radicals, which rapidly add to the small but not negligible amount of monomer present in the aqueous phase [reaction (1)]. The

$$SO_4^{\bullet-} + \text{methyl methacrylate (M)} \longrightarrow {}^-OSO_3M^\bullet \quad (1)$$

monomeric radical formed may then add a second

monomer molecule [reaction (2)], and so on. Two

$$^-OSO_3M^\bullet + M \longrightarrow {}^-OSO_3MM^\bullet \quad (2)$$

of these radical species may combine to form an inert product (bimolecular termination) in the water phase. As more monomer units are added, the radical is increasingly hydrophobic. At a critical degree of polymerization z, it becomes surface-active and rapidly adsorbs to the nearest hydrophobic interface. In the early stages of a typical emulsion polymerization, this interface is a micelle—an aggregate of surfactant molecules with a typical diameter of a few nanometers—or a preformed polymer seed particle. Entry to emulsion droplets is highly unlikely, because although emulsion droplets are large they are much less numerous than micelles. In the environment of a micelle or small particle, the reactive center immediately encounters a concentration of monomer much greater than in the aqueous phase and propagates rapidly to form polymer [reactions (3) and (4)]

$$^-OSO_3(M)_{z-1}^\bullet + M \longrightarrow {}^-OSO_3(M)_z^\bullet \quad (3)$$

$$^-OSO_3(M)_z^\bullet + n(M) \longrightarrow$$
$$\text{polymeric radical (P}^\bullet) \text{ in the particle phase} \quad (4)$$

(**Fig. 2**). The growing polymer chain will terminate either by bimolecular termination [reaction (5)] with

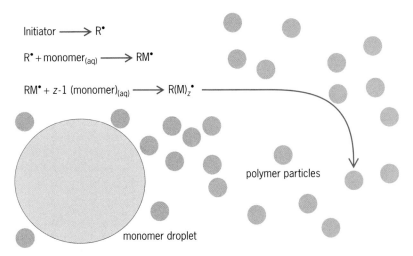

Fig. 2. Entry in emulsion polymerization.

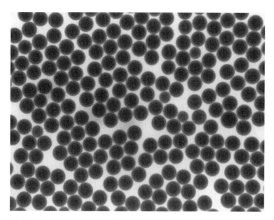

Fig. 3. Spontaneous packing of low-polydispersity poly(styrene) particles about 200 nm in diameter.

$$P^\bullet + P^\bullet \longrightarrow \text{polymer molecule} \qquad (5)$$

an entering radical, or by abstraction of hydrogen from another species (typically monomer) to generate a radical capable of exiting to the aqueous phase [reaction (6)].

$$P^\bullet + M \longrightarrow \text{polymer molecule} + M^\bullet \qquad (6)$$

As polymerization proceeds, the polymer particles will grow until no free surfactant remains to stabilize additional polymer/water interfaces. The monomer droplets merely serve as reservoirs of monomer, which will diffuse from the droplets to the growing polymer particles. Once the micelles have disappeared, new z-meric radicals in the water phase will all enter latex particles. (While this micellar nucleation mechanism is the most common way that particles are formed, there are other mechanisms,

for example, homogeneous nucleation, involving precipitation of polymer chains of degree of polymerization $j_{crit} > z$, which occurs in surfactant-free systems.) Eventually, all the monomer droplets will disappear, and the polymerization will consume the monomer present in the particles. It is common in industry to use a controlled feed, whereby ingredients, such as monomer, are added during the polymerization process. Thus, it is possible to maintain an approximately constant monomer concentration during much of the polymerization process, or to add a different monomer after the original monomer has been consumed.

Polymer solid contents of 50% by weight are routinely obtained in industry, with negligible amounts of coagulation and typical particle diameters of 100 nanometers. Each particle contains a large number of polymer chains, but the size of a particle is not related to the molecular weight of the component polymer.

Benefits. Applications of emulsion polymerization include latexes for use as paints, adhesives, paper coatings, carpet backings, barrier products (for example, rubber gloves), and leather coatings. These latexes are frequently based on styrene, butadiene, acrylates, methacrylates, vinyl acetate, and acrylonitrile, with small amounts of water-soluble comonomers, such as acrylic acid, playing a crucial role in ensuring colloidal stability of the resulting latex. In addition to products used as a latex, there are many products where the polymer is isolated from the latex by coagulation: for example, styrene-butadiene rubber (SBR), used for tires, polytetrafluoroethylene (Teflon), and neoprene [poly(2-chlorobutadiene)].

Many advantages of emulsion polymerization arise from the compartmentalization of the growing chains, which reduces termination far below typical levels in bulk or solution polymerization to enable higher-molecular-weight products and faster reaction rates. Dispersion of the polymer in water keeps the viscosity low and makes the end product easy to handle, as opposed to the highly viscous and intractable materials produced by bulk or solution polymerization. As the process avoids use of flammable and toxic organic solvents, it has environmental and workplace safety benefits.

Recent developments. The mechanistic understanding of emulsion polymerization is now at a level where the size, composition, and topology of the colloidal particles can be controlled to achieve desired end-use properties.

The properties of a film cast from a latex are optimal when the particles can pack optimally, as is the case when they are all the same size (monodisperse) [**Fig. 3**]. While particle nucleation is now well understood, the process is sensitive to small disturbances (for example, changing amounts of oxygen present in the starting materials). Hence, the early stages of an emulsion polymerization in the absence of preformed latex may give an uncontrolled number of particles with a broad distribution of sizes. After nucleation, each particle grows at approximately the

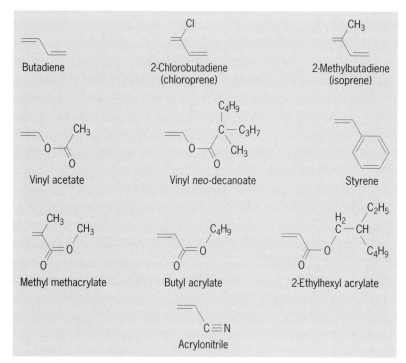

Fig. 4. Monomers used in emulsion polymerization.

same rate in terms of volume, so the size distribution in terms of diameter narrows as the reaction proceeds. For this reason, industrial emulsion polymerizations are often seeded with a preformed latex of small diameter. Control of size and size distribution requires accurate measurement or prediction of particle-size distribution during polymerization, and this has been achieved by a number of groups in recent years by combining calorimetry or online spectroscopy with occasional sampling and novel process-control algorithms.

Most novel emulsion polymer products are likely to be made by combining, or more efficiently controlling, already common monomers, rather than by introducing new commodity monomers (**Fig. 4**). Possibilities to be considered are copolymers, polymer blends, and novel topologies. Indeed, the production of nanostructured particles has been routine in the industry for many years.

In copolymers, different strategies must be used, depending on whether the two monomers should be distributed homogeneously within the particles or concentrated in separate domains. This can also be achieved by the online monitoring of the particle composition to optimize a mathematical model of the process that can be used for system control.

In some cases, it is desired to form two populations of particles with different copolymer compositions. For example, blocking of paints (the adhesion between two painted surfaces that can make a painted window frame difficult to open) can be minimized by generating a paint latex containing a population of small, relatively hard particles in addition to the larger, softer particles required for film formation. This may be done by varying the monomer feed composition in a single, continuous feed process. Water-soluble monomers are frequently incorporated in latexes to generate surfactant, during the course of the reaction, which will be physically grafted to the surface of the polymer particles. Unlike conventional surfactants, these will not separate from the particle surfaces on freezing or shearing to give particle coagulation, nor will they diffuse out of the formed films into the surrounding environment.

Core-shell particles, in which a softer outer shell [for example, poly(butyl acrylate)] may give good film formation, while a harder core [for example, poly(methyl methacrylate)] provides strength, can be prepared by a two-stage controlled monomer feed.

The most rapid advances in emulsion polymerization in the near future are likely to arise from application of the methods of controlled radical polymerization (CRP). Reversible addition fragmentation chain transfer (RAFT), atom transfer radical polymerization (ATRP), and nitroxide-mediated radical polymerization have been used recently on the laboratory scale to give control of molecular weight and allow formation of novel block copolymers in emulsion polymerization systems. These methods use a reversible capping of the growing polymer radical to greatly reduce termination processes. Rather than

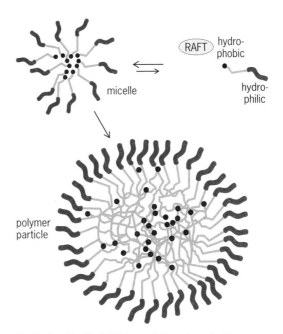

Fig. 5. Amphiphilic RAFT in emulsion polymerization.

the few growing chains of very short lifetime of conventional free-radical polymerization, CRP gives many (slowly) growing chains of very long lifetime. This leads to a much narrower distribution of molecular weights and the ability to create a range of polymer such as blocks and combs. Recent work suggests the practicality of eliminating all surfactant from emulsion polymerization through use of an aqueous-phase RAFT agent, and being able to make polymer chains of any desired architecture (**Fig. 5**). In the first stage, a "living" (termination-free) aqueous-phase polymer is generated, which reacts with the hydrophobic monomer added in the second phase to form polymeric surfactant molecules. These form rigid micelles which grow into polymer particles by continued feeding of the hydrophobic monomer. It has been suggested that because the individual polymer molecules in such a system are irreversibly attached to the micelles, unlike conventional surfactants which undergo rapid exchange, the number of micelles originally present can be correlated directly to the number of polymer particles. Thus any desired polymer microstructure can be built up. These methods also provide a pathway for incorporating specific functionalities on the surface of a polymer particle, for instance, to attach antibodies for use in latex agglutination tests in biomedicine.

For background information *see* COLLOID; COPOLYMER; EMULSION; FREE RADICAL; INTERFACE OF PHASES; MICELLE; POLYMER; POLYMERIZATION; SURFACTANT in the McGraw-Hill Encyclopedia of Science & Technology.　　　　Robert G. Gilbert;
Christopher M. Fellows

Bibliography. C. Burguiere, C. Chassenieux, and B. Charleux, Characterization of aqueous micellar solutions of amphiphilic block copolymers of poly(acrylic acid) and polystyrene prepared via ATRP: Toward the control of the number of particles in emulsion polymerization, *Polymer*, 44:509–518, 2002; C. J. Ferguson et al., Effective ab initio emulsion polymerization under RAFT control, *Macromolecules*, 35:9243–9245, 2002; R. G. Gilbert, *Emulsion Polymerization: A Mechanistic Approach*, Academic Press, London, 1995; K. Matyjaszewski, New (co)polymers by atom transfer radical polymerization, *Macromol. Symp.*, 143:257–268, 1999; M. Vicente et al., Control of microstructural properties in emulsion polymerization systems, *Macromol. Symp.*, 182:291–303, 2002.

Environmental endocrine disruptors

Endocrine disruptors are chemical compounds that interfere with the proper function of endocrine systems in humans and other organisms. There has been much scientific and medical research on pharmaceuticals as endocrine disruptors. However, endocrine disruptors are increasingly being discovered in the environment, and scientists and engineers are taking steps to reduce the risks posed by them.

Endocrine system. An organism's glands produce chemical compounds, known as hormones, and secrete them into the bloodstream, which delivers them to cells throughout the body. Cells contain specific sites, known as receptors, made up of molecules that bind to specific hormones (**Fig. 1**). When binding occurs, the cell has accepted the signal sent from the gland, and responds in a specific way. These chemical signals transmitted by endocrine system are needed for an organism's reproduction, development, and homeostasis (the regulation of temperature, fluid exchange, and other internal processes). Although sex hormones (for example, estrogen and testosterone) have received the most attention, endocrine disruption occurs in all glands, notably the thyroid, adrenal, pituitary, and pineal glands.

So vital is the endocrine system for an organism's normal growth, development, and reproduction that even small disturbances in endocrine function at the wrong time may lead to long-lasting and irreversible effects. Organisms are particularly vulnerable to endocrine disruption during highly sensitive times of development, such as prenatal and pubescent periods, when small endocrine disturbances may have delayed consequences that may not appear until adult life or future generations. A well-know case of multigenerational endocrine disruption concerns the synthetic hormone diethylstilbestrol (DES), which from 1940 to 1971 was prescribed to prevent miscarriages. Since then, DES has been classified as a known carcinogen. The major concern was not with the women treated, but that in-utero DES exposure led to a high incidence of cervical cancers in their daughters.

Types of disruptors. Chemicals that act as natural hormones by binding to a cell's receptor are called agonists. Chemicals that inhibit binding receptor are called antagonists, and have an effect opposite that of the natural hormone (for example, an anti-estrogen masculinizes an organism). Environmental endocrine disruptors can be of either type (see **table**). In addition, indirect disruptors affect the endocrine system because of their effects on another system. For example mercury, a strong neurotoxin, does not bind to a hormone receptor site, but its effect on the brain can interfere with glandular and therefore hormonal functioning. An organism's nervous, immune, and endocrine systems are interconnected chemical messaging systems, such that a change in one system may lead to changes to others.

Environmental endocrine disruption was observed throughout much of the twentieth century. Abnormal mating patterns in bald eagles on the east coast of North America were observed in the 1940s. In her 1962 book, *Silent Spring*, Rachel Carson observed that predatory birds, such as eagles, accumulated DDT (dichlorodiphenyltrichloroethane) and other chlorinated hydrocarbon pesticides in their fatty tissues. The eggshells produced by DDT-contaminated birds were abnormally thin, and the chicks were less successful in hatching than those with lower concentrations of the pesticide were. During the last 20 years, scientists have found associations between exposure to chemical compounds and changes affecting endocrine systems in humans and animals. These compounds were first called "environmental estrogens," but since other hormonal effects were found, the term "environmental hormone" is now preferred. "Hormone mimicker" was also sometimes used to designate compounds that could elicit a response similar to a natural hormone. For example, certain pesticides bind very easily to estrogen receptors, resulting in increased feminization of the organism. The terms "hormonally active

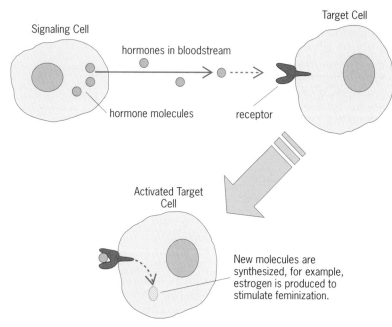

Fig. 1. Process for endocrine signaling between cells. A signaling cell releases into the bloodstream hormones which reach the receptor of a target cell. When a hormone binds to the receptor, new molecules are synthesized at the activated target cell.

Some compounds found in the environment that have been associated with endocrine disruption[†]

Compound[‡]	Endocrine effect	Potential source
2,2′,3′,4′,5,5′-Hexachloro-4-biphenylol and other chlorinated biphenylols	Antiestrogenic	Degradation of PCBs released into the environment
4′,7-Dihydroxy daidzein and other isoflavones, flavones, and flavonals	Estrogenic	Natural flora
Aldrin[*]	Estrogenic	Insecticide
Alkylphenols	Estrogenic	Industrial uses, surfactants
Bisphenol A and phenolics	Estrogenic	Plastics manufacturing
DDE (1,1-dichoro-2,2-bis(p-chlorophenyl)ethylene)	Antiandrogenic	DDT metabolite
DDT and metabolites	Estrogenic	Insecticide
Dicofol	Estrogenic or antiandrogenic in top predator wildlife	Insecticide
Dieldrin	Estrogenic	Insecticide
Diethylstilbestrol (DES)	Estrogenic	Pharmaceutical
Endosulfan	Estrogenic	Insecticide
Hydroxy-PCB congeners	Antiestrogenic (competitive binding at estrogen receptor)	Dielectric fluids
Kepone (Chlorodecone)	Estrogenic	Insecticide
Lindane (γ-hexachlorocyclohexane) and other HCH isomers	Estrogenic and thyroid agonistic	Miticide, insecticide
Lutolin, quercetin, and naringen	Antiestrogenic (as in uterine hyperplasia)	Natural dietary compounds
Malathion[*]	Thyroid antagonist	Insecticide
Methoxychlor	Estrogenic	Insecticide
Octachlorostyrene[*]	Thryroid agonist	Electrolyte production
Pentachloronitrobenzene[*]	Thyroid antagonist	Fungicide, herbicide
Pentachlorophenol	Antiestrogenic (competitive binding at estrogen receptor)	Preservative
Phthalates and their ester compounds	Estrogenic	Plasticizers, emulsifiers
Polychlorinated biphenyls (PCBs)	Estrogenic	Dielectric fluid
Polybrominated diphenyl ethers (PDBEs)[*]	Estrogenic	Fire retardants, including in-utero exposures
Polycyclic aromatic hydrocarbons (PAHs)	Antiandrogenic (aryl hydrocarbon-receptor agonist)	Combustion by-products
Tetrachlorodibenzo-*para*-dioxin and other halogenated dioxins and furans[*]	Antiandrogenic (aryl hydrocarbon-receptor agonist)	Combustion and manufacturing (such as halogenation) by-product
Toxaphene	Estrogenic	Animal pesticide dip
Tributyl tin and tin organometallic compounds[*]	Sexual development of gastropods and other aquatic species	Paints and coatings
Vinclozolin and metabolites	Antiandrogenic	Fungicide
Zineb[*]	Thyroid antagonist	Fungicide, insecticide
Ziram[*]	Thyroid antagonist	Fungicide, insecticide

[*] The source for asterisked compounds is T. Colborn et al., http://www.ourstolenfuture.org/Basics/chemlist.htm.

[†] For full list, study references, study types, and cellular mechanisms of action, see Chapter 2 of National Research Council, *Hormonally Active Agents in the Environment*, National Academy Press, Washington, DC, 2000.

[‡] Not every isomer or congener included in a listed chemical group (for example, PAHs, PCBs, phenolics, phthlates, and flavonoids) has been shown to have endocrine effects. However, since more than one compound has been associated with hormonal activity, the entire chemical group is listed here.

agent" and "endocrine disruptor" are general classifications for any chemical that causes hormonal dysfunction.

Like other environmental contaminants, endocrine disruptors vary in physical and chemical forms. These different forms possess numerous and distinct ways of breaking down in the environment. Resistance to breakdown is known as persistence. Chemicals also vary in their toxicity, as well as in their potential to build up (bioaccumulate) in the food chain. Persistent bioaccumulating toxics (PBT), also called PBT endocrine disruptors, are those that have long half-lives (the amount of time for half of the mass of the compound to break down in the environment). A substance's half-life will vary by where it is found in the environment. For example, chemicals may break down more readily in water (half-life of hours or days) than in soil or sediment (half-life of months or years). Bioaccumulating compounds easily build up in the food chain and lead to toxic effects, including hormonal dysfunction. The PBT endocrine disruptors can be organic compounds (known as persistent organic pollutants, or POPs), inorganic (certain metals and their salts), or organometallic (such as the butylated or phenylated forms of tin, which have been associated with endocrine effects in aquatic animals).

Prevention, engineering, and control. Engineers and scientists can reduce or eliminate the risks from endocrine disruptors. Actions can be taken at various stages, from the synthesis of compounds, to the release of chemicals into the environment, to removal and treatment after release. The understanding of endocrine disruption has been enhanced by scientific research in three major areas: laboratory studies (including animal testing and cellular studies such as receptor binding studies), epidemiology (the study of the incidence and distribution of diseases in human populations and ecosystems), and natural experiments (unplanned or uncontrolled events that can be compared before and after). One natural experiment involved a large spill of dicofol [a mixture of the pesticide DDT and its by-product 1,1-dichloro-2,2-bis(*p*-chlorophenyl)ethylene (DDE), both suspected endocrine disruptors] into Lake Apopka, Florida, in 1980. Subsequent studies indicated that wildlife, notably male alligators, showed marked reductions in gonad size following the spill.

Chemical screening. Although many pesticides and some industrial chemicals have undergone toxicological testing (that is, testing the potential for adverse health effects), these tests are in many ways inadequate to determine how a substance will interact with the endocrine system. Scientific knowledge related to endocrine disruptors is evolving. But there is general scientific agreement that better endocrine screening and testing of existing and new chemicals are needed. To this end, the U.S. Environmental Protection Agency has established the Endocrine Disruptor Screening Program (EDSP) to evaluate the potential hormonal effects of hundreds of chemicals (**Fig. 2**).

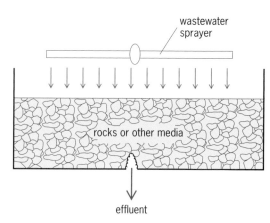

Fig. 3. Trickling filter wastewater treatment. This is known as a mixed-treatment system because it involves both aerobic and anaerobic processes. Wastewater is sprayed onto the filter media. The combination of microbes, water, and wastes forms a film on the media. The film near the top provides an environment (food, oxygen, and water) for the growth of microbes which break down complex wastes into simpler compounds, ultimately to carbon dioxide and water. At lower levels of the filter, different microbes thrive in an anoxic (oxygen-depleted) environment, where anaerobic metabolism breaks down organic compounds, ultimately to methane. Trickling filters work well for diluted wastes; for concentrated wastes, dedicated anaerobic treatment units perform better. (*Adapted from D. A. Vallero, Engineering the Risks of Hazardous Wastes, Butterworth-Heinemann, Boston, 2003*)

Chemical modification. Manufacturers of hormonally active substances must find ways of eliminating them. One means is to change their chemical structure so that they do not bind or block receptor sites on cells (known as green chemistry). The addition or deletion of a single atom or the arrangement of the same set of atoms (that is, an isomer) can significantly reduce potential toxic effects.

Waste treatment. Endocrine disruptors enter the environment in many ways. Wastes from households and medical facilities may contain hormones that reach landfills and wastewater treatment plants, where they enter waterways after passing through untreated or incompletely treated. Fish downstream from treatment plants have shown symptoms of endocrine disruption. Engineers designing treatment facilities must consider the possibility that wastes will contain hormonally active chemicals, and must find ways to treat them.

Municipalities may also be able to treat an endocrine disruptor discharged into the waste stream. For example, chemicals that are broken down relatively easily by microbes may be treated by common techniques, such as the trickling filter commonly used by cities in the United States (**Fig. 3**). However, compounds that are more persistent will require intensive treatment, such as increased oxidation using tapered aeration systems, which enhance the degradation of endocrine disruptors either chemically or microbiologically with bacterial genera such as *Pseudomonas, Rhodococcus,* and *Mycobacterium* (**Fig. 4**). Concentrated wastes may require anaerobic treatment (where molecular oxygen is absent) or facultative treatment (where bacteria

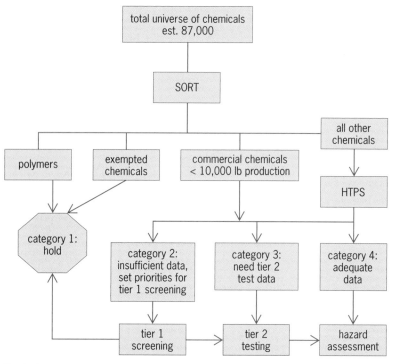

Fig. 2. Schematic of the endocrine disruptor screening program of the U.S. government (led by the U.S. Environmental Protection Agency). The chemicals are first sorted to identify those with potential endocrine disruptor mechanisms for the estrogen, androgen, and thyroid hormone systems. This is followed by testing to determine whether an endocrine-active substance causes adverse effects, to identify the adverse effects, and to quantify dose and effect. (*U.S. Environmental Protection Agency, Endocrine Disruptor Screening Program, Report to Congress, 2000*)

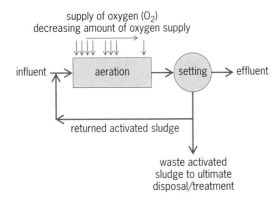

supply of oxygen (O$_2$)
decreasing amount of oxygen supply

influent → aeration → setting → effluent

returned activated sludge

waste activated
sludge to ultimate
disposal/treatment

Fig. 4. Aeration waste treatment system, an aerobic treatment approach for breaking down toxic substances, including endocrine disruptors, from household or manufacturing sources. The waste is combined with recycled biomass and aerated to maintain a target dissolved oxygen (DO) content. Organisms use the organic components of waste as food, decreasing the organic levels in the wastewater. Oxygen concentrations must be controlled to maintain optimal treatment efficiencies. One means of achieving optimal DO content is tapered aeration, as shown in the diagram. The tapered system provides high concentrations of oxygen near the influent to accommodate the large oxygen demand from microbes as waste is introduced to the aeration tank. (*Adapted from D. A. Vallero, Engineering the Risks of Hazardous Wastes, Butterworth-Heinemann, Boston, 2003; photo courtesy of D. J. Vallero*)

grow with or without molecular oxygen) to degrade complex, hormonally active compounds into simpler compounds.

Soil and ground-water treatment. Contaminated soil and sediment often must be removed and then treated off-site. Contaminated soil may be excavated and transported to kilns, where it is burned with combustible material. Contaminated soil may also be spread over an impermeable surface, allowing the more volatile compounds to evaporate. In a faster process, called thermal desorption, the soil is heated to evaporate the contaminants, which are then burned.

Generally, ground water is treated by first drilling recovery wells and pumping contaminated ground water to the surface. The water is then exposed to treatment approaches such as air stripping, filtering with granulated activated carbon (GAC), and air sparging. Air stripping processes increase a contaminant's vapor pressure, thereby transferring volatile compounds from the water to the air, where they are trapped and treated above ground. Ground water is allowed to drip downward in a tower filled with a permeable material through which a stream of air flows upward. Another method bubbles pressurized air through contaminated water in a tank. Filtering ground water with GAC entails pumping the water through the GAC to trap the contaminants. In air sparging, air is pumped into the ground water to aerate the water. Most often, a soil venting system is combined with an air sparging system for vapor extraction.

Bioremediation. Endocrine disruptors can also be treated where they are found. This is known as in-situ treatment. Bioremediation is the use of living microorganisms to break down toxic chemicals or to render them less hazardous. This is often done by extracting microbes (bacteria, and in some instances algae and fungi) from the soil, sediment, or water and exposing them to increasing amounts of a compound, so that the organisms will use the compound as an energy (food) source. This process is known as acclimation. The acclimated microbes are then applied to the waste in a treatment facility or in situ (in the field).

The most passive form of bioremediation is natural attenuation, where no engineering intervention is used and the contaminants are degraded by resident microbes over time. In this case, engineers monitor the soil and ground water to measure the rate at which the chemicals are degrading. Natural attenuation can work well for compounds that are found in the laboratory to break down under the conditions found at the site. For example, if a compound degrades at low pH conditions in the laboratory, it may also degrade readily in soils with these same conditions (for example, in deeper soil layers where bacteria are naturally adapted to these conditions).

Phytoremediation. Plants seeded in contaminated areas may also be used to remove hazardous chemicals. As the plants grow, their roots extract the chemicals from the soil. The harvested plants are either treated on site (for example, by composting) or transferred to a treatment facility. Poplar trees can help to treat areas contaminated with agricultural chemicals. Plants have even been used to treat the very persistent polychlorinated biphenyls (PCBs) and wood preservatives. Petroleum products have been broken down by planting grasses and field crops. Plants have also been used to extract heavy metals and radioactive substances from contaminated soil. In reality, both microbial and plant processes occur simultaneously.

Various methods are available for treating substances after they have been released into the environment. Eliminating the wastes before they are released, however, is the best means of reducing hormonal risks to humans and other organisms.

For background information *see* ENDOCRINE SYSTEM (VERTEBRATE); ENVIRONMENTAL ENGINEERING; ENVIRONMENTAL TOXICOLOGY; GLAND; HORMONE; INDUSTRIAL WASTEWATER TREATMENT; ORGANIC CHEMISTRY; RISK ASSESSMENT AND MANAGEMENT in the McGraw-Hill Encyclopedia of Science & Technology.   Daniel Vallero

Bibliography. R. Carson, *Silent Spring*, Houghton Mifflin, Boston, 1962; T. Colborn, D. Dumanoski, and J. P. Myers, *Our Stolen Future*, Penguin, New York, 1997; *Endocrine Disruptor Screening and Testing Advisory Committee (EDSTAC) Final Report*, 1998; S. Goodbred et al., *Reconnaissance of 17β-Estradiol, 11-Ketotestosterone, Vitellogenin, and Gonad Histopathy in Common Carp of United States Streams: Potential for Contaminant-Induced Endocrine Disruption*, USGS Open-File Rep. 96-627, Sacramento, CA, 1997; National Academy of Sciences, *Hormonally Active Agents in the Environment*, National Academy Press, Washington, DC, 1999; United Nations Environment Programme (Chemicals), *Regionally Based Assessment of Persistent Toxic Substance*, North American Regional Report, 2002; U.S. Environmental Protection Agency, *Special Report on Environmental Endocrine Disruption: An Effects Assessment and Analysis*, 1997.

Equine abortion epidemic

During late April and May of 2001 and 2002, the bluegrass area in the central part of Kentucky experienced an epidemic of equine abortions. Early fetal loss (days 35–100 of gestation) and late-term abortion occurred in recently bred mares and mares nearing foaling, respectively. These losses have been termed mare reproductive loss syndrome (MRLS). For the period, estimates placed the economic loss due to MRLS at over $330 million.

Clinical findings. During the first few weeks of April 2001, the incidence of equine abortions in central Kentucky was similar to prior years. However, in late April and during the first few weeks of May, veterinarians performing ultrasonography on pregnant mares reported fetal abnormalities in early gestation. Most of these mares subsequently aborted. Concurrently, mares in late gestation (from several weeks prior to due date to full term) began aborting in increased numbers. In general, no other illness was noted in mares that aborted. Various breeds of horses over a wide area of central Kentucky were affected simultaneously. Estimates placed the losses at 25–30% of the 2002 foal crop due to early fetal loss in 2001, with the 2002 losses being approximately one-third of the previous year. Over 500 late-term abortions occurred in 2001, and over 300 in 2002. These numbers represent a five- and threefold increase, respectively, over the 5-year mean for the same time period for late-term abortion of various causes.

Ultrasonography of early-gestation fetuses affected by MRLS typically revealed either a dead fetus or one showing decreased viability, as evidenced by decreased heart rate and size abnormalities. The fetal fluids exhibited increased echogenicity (tendency to absorb or reflect sound waves) due to increased particulate material. Losses typically affected fetuses

older than 35 days of gestation; fetuses of less than 30 days of gestation were unaffected. Mares that experienced early fetal loss were unresponsive to treatment measures to resume normal reproductive cyclicity and often could not be rebred until the following year. This was due to persistence of the endometrial cups following abortion. These normal endocrine structures develop in the uterus of mares and produce equine chorionic gonadotrophin (a hormone that has both follicle-stimulating and luteinizing properties) between approximately days 35–100 of gestation and then regress.

In many late-term abortion mares, the abortion was sudden and intense. Often the allantochorion (vascular fetal membrane) appeared prior to presentation of the fetus, with concurrent delivery of the entire placenta and fetus (red bag delivery). However, most mares with late-term abortion had no residual problems and rebred in a normal fashion.

Pathologic findings. Necropsy (autopsy) findings of aborted fetuses and placentas showed characteristic features. Early fetal loss fetuses usually showed moderate autolysis (self-induced enzymatic degradation of tissues and cells), but typically lacked diagnostically useful lesions. Late-term abortion fetuses were characterized by normal gestational development and weight, and there were often superficial hemorrhages on the surface of organs and tissues. The lungs were typically nonaerated (indicating that the foal did not breathe) and of normal consistency. In some cases, however, the lungs had increased firmness, suggesting pneumonia. The allantochorion and amnion (membrane around the fetus) sometimes were edematous (filled with excess fluid) and contained hemorrhages. The umbilical cord, typically the intra-amniotic portion, was thickened, edematous, roughened, and hemorrhagic.

Microscopic examination of the fetal organs and placenta revealed pneumonia and inflammation of the umbilical cord (funisitis) and amnion (amnionitis). Bacteria were often present on the surface of the umbilical cord and within the lung. In most cases, routine culture produced bacteria from the tissues and placental membranes, with non-beta-hemolytic streptococci or actinobacilli most commonly isolated. All other routine laboratory testing was typically normal or negative. Initial investigation into the cause of MRLS failed to incriminate microorganisms routinely associated with equine abortion, including viruses and the causative agents of fescue toxicosis (disease resulting from the ingestion of tall fescue grass infected with the fungus *Neotyphodium coenophialum*, formerly *Acremonium coenophialum*), bacterial placentitis, and leptospirosis (an acute febrile disease caused by bacteria of the genus *Leptospira*).

Case definition. After the 2001 outbreak had subsided, the cases submitted for pathologic work-up during the spring were reviewed, and a case definition was developed to identify cases of MRLS abortion to allow differentiation from other causes of abortion and stillbirth. Based on the medical history, pathologic findings, and microbiologic and toxico-logic testing, fetuses were given a global assessment score of being likely or unlikely associated with MRLS. A fetus was considered a probable MRLS abortion if its global assessment score indicated that it was likely associated with MRLS, the gestational age was greater than 269 days, and no other cause of abortion was found. A fetus was considered a definite MRLS abortion when the latter criteria for probable association were met and at least two of the following were present: non-beta-hemolytic streptococci or actinobacilli cultured from the tissues, premature placental separation at delivery, placental edema, placentitis, funisitis, pneumonia, or lack of lesions in the fetus.

Epidemiologic investigation. Using a case-control approach [an observational study involving comparison of subjects with ("case") and without ("control") the condition of interest], studies were undertaken to identify risk factors associated with both early fetal loss and late-term abortion. Forty-three farms for early fetal loss and 62 farms for late-term abortion were studied. The results showed that feeding hay in pasture, increased amounts of white clover in pastures, and increased numbers of eastern tent caterpillars (*Malacosoma americanum*) in the pastures were associated with increased risk for early fetal loss. Likewise, increased time on pasture, consumption of concentrate feed on the ground, increased grazing of pasture, and drinking of water from a trough were associated with heightened risk for late-term abortion.

In both studies, it was found that factors or characteristics associated with pasture predisposed mares to MRLS-related abortions, but the exact cause was not identified. Since the features were not suggestive of a transmissible infectious agent, it was concluded that the epidemic was indicative of a point-source exposure (simultaneous exposure to a common source or factor).

Search for causative agent. During and following the occurrence of the abortion epidemic, many potential causes of MRLS were considered. Samples were collected and analyzed for the presence of ergot alkaloids (poisonous substances produced by the fungus *Claviceps purpurea*) in tall fescue grass, mycotoxic (due to a fungal poison) or increased estrogenic (estrogen-like) activity in pasture material, and cyanide in pasture plants, but testing failed to incriminate these agents in the epidemic. However, researchers noted that the 2001 abortion epidemic occurred shortly after an especially large hatch of eastern tent caterpillars. The caterpillars defoliated the food source trees and prematurely abandoned the trees in search of other foodstuff. Pastures, fences, waterers, and barns were reported to be inundated with eastern tent caterpillars. In addition, several epidemiologic surveys found that the occurrence of increased abortions on farms correlated with the presence of eastern tent caterpillars and cherry trees. The association of high numbers of eastern tent caterpillars and the abortion epidemic led to the hypothesis

Eastern tent caterpillars emerging from their nest within a tree.

that eastern tent caterpillars could in some way be responsible for the abortions.

Eastern tent caterpillar association. Eastern tent caterpillars are widespread over most of the eastern United States. They hatch in early spring and feed on the leaves of several species of trees, notably black cherry trees (*Prunus serotina*; see **illus.**). After feeding and development, the larvae leave the trees, form a cocoon, and emerge a short time later as a moth. The moth deposits egg masses on the limbs of food source trees and then dies. The eggs hatch the following spring around the time of bud break. In general, eastern tent caterpillars are considered innocuous to vertebrates.

Cyanide transport. It was suggested that the eastern tent caterpillars could be transporting cyanide to horses, since the leaves of black cherry trees are known to contain cyanide and livestock poisonings have occurred following ingestion of cherry tree leaves. Subsequent studies demonstrated the presence of cyanide in eastern tent caterpillars, but at a level lower than that found in leaf material. Eastern tent caterpillars are known to metabolize the cyanide-containing compounds, and it was shown that there was progressively less cyanide detectable from the caterpillar's foregut to its droppings (frass). Further, direct administration of potassium cyanide to pregnant mares was not found to induce abortion. Together these findings suggested that eastern tent caterpillars were not causing abortion by delivering cyanide to horses.

Abortigenic potential. Subsequent studies were designed to determine if the eastern tent caterpillar was indeed associated with equine abortions by exposing horses to the caterpillars. In one study, natural exposure of pregnant mares to the caterpillars was mimicked by placing mares on pasture plots. The plots were pens specially designed to retain eastern tent caterpillars. Mares exposed to eastern tent cater-pillars and frass experienced significant abortions. A second study using the same plot system evaluated the effect of fasted eastern tent caterpillars (little frass) or frass only. Several abortions occurred in the horses exposed to eastern tent caterpillars, and none in the frass-exposed mares. Other experiments confirmed the abortigenic potential of eastern tent caterpillars by placing macerated caterpillars into the stomachs of pregnant mares via nasogastric instillation. The majority of mares receiving eastern tent caterpillars aborted; however, there were no abortions in the frass-only and control groups.

Although multiple studies have confirmed the ability of eastern tent caterpillars to induce abortion in pregnant mares, the pathogenesis of the abortions associated with eastern tent caterpillars remains to be elucidated.

For background information *see* AGRICULTURAL SCIENCE (ANIMAL); CATERPILLAR; EPIDEMIC; HORSE PRODUCTION; REPRODUCTIVE SYSTEM in the McGraw-Hill Encyclopedia of Science & Technology.

Neil M. Williams

Bibliography. N. D. Cohen et al., Case-control study of late-term abortions associated with mare reproductive loss syndrome in central Kentucky, *J.A.V.M.A.*, 222(2):199–209, 2003; T. D. Fitzgerald, *The Tent Caterpillars*, Cornell University Press, Ithaca, NY, 1995; D. G. Powell, Mare reproductive loss syndrome (MRLS), *Equine Dis. Quart.*, 9:4–6, 2001; R. Thalheimer and R. G. Lawrence, The economic loss to the Kentucky equine breeding industry from mare reproductive loss syndrome (MRLS) of 2001, Department of Equine Business, College of Business and Public Administration, University of Louisville; N. M. Williams et al., Gross and histopathological correlates of MRLS, *First Workshop on Mare Reproductive Loss Syndrome (MRLS)*, p. 29, 2002.

Estrogen actions and cardiovascular risk

Coronary heart disease is the major cause of death and disability among women in the developed world. Several observational clinical studies have reported that postmenopausal women who take hormone replacement therapy have lower incidence of coronary heart disease; therefore, hormone replacement therapy has been recommended for secondary prevention of heart disease (which aims to decrease risk of disease recurrence) in postmenopausal women. Initially, estrogen alone was prescribed as hormone replacement therapy. However, progestins (natural or synthetic forms of progesterone) are now added to estrogen for women with an intact uterus in order to reduce, or eliminate, the excess risk for endometrial cancer observed with estrogen alone (unopposed estrogen); this treatment modality is now commonly referred to as combined hormone replacement therapy.

The clinical findings from observational studies show favorable changes in lipid levels, vessel walls,

fibrinogen (clot-forming plasma protein) and antithrombin III (clot-dissolving plasma protein) concentrations, antioxidant action, and insulin secretion in women on hormone replacement therapy. Positive effects on the vessel walls include a rapid increase in nitric oxide production (which decreases blood pressure). In addition, vascular smooth muscle proliferation, which reduces or prevents blood flow, is inhibited by estrogen. However, the possibility that the observed association between hormone replacement therapy and reduced coronary heart disease risk might be attributable to several biases (selection bias, compliance bias, and diagnostic detection and follow-up bias) has been pointed out. In order to provide reliable unbiased information on the incidence of coronary heart disease, randomized prevention trials have been conducted. The first randomized primary prevention trial, the Women's Health Initiative (WHI), however, was terminated early due to increased risk of breast cancer, stroke, and pulmonary embolism in study participants receiving estrogen plus progestin treatment.

Randomized secondary prevention trials. The Heart and Estrogen/progestin Replacement Study (HERS) is the first large randomized secondary prevention trial of hormone replacement therapy in postmenopausal women with established coronary heart disease. The primary outcome was nonfatal myocardial infarction or death caused by coronary heart disease. Other studied cardiovascular outcomes included coronary revascularization surgery, unstable angina (incomplete obstruction of coronary artery), congestive heart failure (inability of heart to adequately pump blood throughout the body), resuscitated cardiac arrest, stroke or transient ischemic attack ("ministroke"), and peripheral arterial disease (narrowing of arteries, usually in the legs). HERS compared 2763 women with coronary heart disease, randomized to receive either conjugated equine estrogen (derived from the urine of pregnant mares) plus a progestin (medroxyprogesterone acetate, MPA) or placebo. Surprisingly, HERS found no overall difference in the primary outcome or in any of the secondary cardiovascular outcomes between intervention groups, although significant decreases (11%) in low-density lipoprotein ("bad") cholesterol (LDL-C) levels and significant increases (10%) in high-density lipoprotein ("good") cholesterol (HDL-C) levels were observed in the HRT group compared with the placebo group. A significantly increased risk of cardiac events in the first year of inclusion in the trial was noted in women given hormone replacement therapy, followed by a reduced risk in years 3–5. However, this trend was not sustained with further followup. After 6.8 years, hormone replacement therapy did not reduce the risk of cardiovascular events in women with coronary heart disease.

The Estrogen in the Prevention of Reinfarction Trial (ESPRIT), also a randomized secondary prevention trial, included postmenopausal women who had survived a first myocardial infarction. The purpose of this study was to assess whether the effect of unopposed estrogen reduced the risk of further cardiac events in 1017 postmenopausal women. There was no overall difference in the frequency of reinfarction or cardiac death between the hormone replacement therapy and placebo groups. The results of ESPRIT indicate that hormone replacement therapy with unopposed estrogen does not reduce the overall risk for further cardiac events in postmenopausal women who have survived a myocardial infarction.

The Estrogen Replacement and Atherosclerosis Trial examined the effects of hormone replacement therapy on the progression of coronary atherosclerosis (hardening of the coronary arteries) in women. A group of 309 postmenopausal women who after coronary angiography (roentgenographic visualization of coronary arteries after injection of a radiopaque substance) were confirmed to have coronary heart disease at base line, were randomly assigned to receive unopposed estrogen, estrogen plus MPA, or placebo. The results showed that estrogen and estrogen plus MPA induced significant reductions (9.4% and 16.5%, respectively) in LDL-C levels and significant increases (18.8% and 14.2%, respectively) in HDL-C levels. However, hormone replacement therapy (which includes unopposed estrogen treatment) did not alter the progression of coronary atherosclerosis as analyzed by angiography (average 3.2 years). Thus, the results of these studies taken collectively do not support the use of hormone replacement therapy for secondary prevention of coronary heart disease in women.

Randomized primary prevention trial. The WHI was the first randomized trial designed to directly address the effect of hormone replacement therapy on the incidence of coronary heart disease in predominantly healthy women, including an assessment of overall risks and benefits. The WHI had two treatment arms: estrogen alone in 10,739 women who had undergone hysterectomy and estrogen plus progestin in 16,608 postmenopausal women. However, the estrogen plus progestin treatment component was terminated early because of an unacceptable risk profile. This included increased risks for breast cancer, stroke, and pulmonary embolism. The risk for invasive breast cancer exceeded the stop boundary for this adverse effect, and the global index summarizing the balance of risks and benefits supported the conclusion that the risks exceeded the benefits. (The addition of progestins to estrogen-based hormone replacement therapy may increase the risk of breast cancer to a level above that observed with estrogen alone. This is in consonance with the fact that mitotic activity in the breast during a normal menstrual cycle is highest when progesterone levels are high.) Moreover, the risk of cardiovascular disease was increased by 22% in the estrogen plus progestin group, although this group had lower LDL-C (−12.7%) and higher HDL-C (7.3%) and triglycerides (6.9%) compared with the placebo group, consistent with results of the randomized secondary prevention trials. Study of the other treatment arm, estrogen alone, is

continuing, with the planned end of this trial being March 2005.

HRT failure to prevent CHD. Although the HERS, WHI, and Estrogen Replacement and Atherosclerosis trials show that hormone replacement therapy significantly reduces LDL-C and also significantly increases HDL-C concentrations, results from these randomized primary and secondary prevention trials do not support the use of hormone replacement therapy, especially not an estrogen plus progestin regimen, for primary and secondary prevention of coronary heart disease. Why does hormone replacement therapy fail to prevent coronary heart disease despite established beneficial effects of estrogen on lipid metabolism, endothelial function, and other factors involved in the pathogenesis and progression of atherosclerosis?

One possible explanation is that progestins down-regulate (suppress) the activity of estrogen receptors. They may also have direct effects that oppose the beneficial effects of estrogen. In the Postmenopausal Estrogen-Progestin Interventions Trial, the inclusion of MPA in addition to estrogen resulted in a 75–80% reduced increase in HDL-C compared with values observed with estrogen alone. In monkeys, MPA decreases the coronary artery dilation due to estrogen and prevents the protective effect of estrogen on coronary artery atherosclerosis.

Another possible explanation is that estrogen has pro-inflammatory actions that offset its beneficial effects. Inflammation plays an important role in mediating all stages of atherosclerosis from initiation through progression. The earliest type of atherosclerotic lesion, the so-called fatty streak, is an inflammatory lesion that includes macrophages and T lymphocytes. It has been reported that women taking unopposed estrogen or estrogen plus progestin have significantly higher levels of C-reactive protein, a marker of systemic inflammation. C-reactive protein levels are clearly associated with risks of cardiovascular events and are stronger than LDL-C levels as predictors of these events.

Role of estrogen receptors. Experimental studies using estrogen receptor gene knockout mice have provided useful information about the roles of the two types of estrogen receptors, ERα and ERβ, in the cardiovascular system. In wild-type (natural) mice, carotid artery injury results in increases in vascular intimal and medial area and thickness. Estrogen treatment suppresses these vascular injury responses. ERα knockout and ERβ knockout mice differ in their responses to carotid artery injury. In ERα knockout mice, estrogen is no longer able to suppress the vascular injury response, demonstrating a beneficial and critical role for ERα at the level of vascular wall inflammation. However, this lack of response to estrogen is not observed in ERβ knockout mice. On the other hand, ERβ knockout mice develop sustained systolic and diastolic hypertension and demonstrate multiple abnormalities of vascular function. These results highlight the different roles of the two estrogen receptors in the cardiovascular system.

Studies in knockout animals have also suggested that estrogen signaling plays a role in cardiovascular physiology and possibly in cardiovascular disease in humans. However, studies of improved ERα and ERβ selective agonists in various animal models of cardiovascular disease will be needed to explore the therapeutic opportunities suggested by these findings.

Therapeutic opportunities. Today HRT does not provide a validated therapeutic opportunity for either primary or secondary prevention of coronary heart disease. However, other hormone-related therapies may prove more successful.

Selective estrogen-receptor modulators. Selective estrogen-receptor modulators, which have selective agonist or antagonist effects in various estrogen target tissues, might provide new therapeutic opportunities. Ideally, such compounds will be found to retain most of the beneficial effects of estrogen while avoiding most adverse effects. For example, whereas estrogen or estrogen/progestin change C-reactive protein levels, the selective estrogen-receptor modulator raloxifene does not. A large randomized trial, Raloxifene Use in the Heart (RUTH), to investigate whether raloxifene is cardioprotective for postmenopausal women at risk for coronary heart disease, may provide an answer when it is completed in 2005.

Estrogen receptor polymorphisms. The therapeutic potential of estrogen receptor polymorphisms (variations in the gene encoding the estrogen receptor) should also be explored. For example, it has been reported that women with the ERα IVS1-401 C/C genotype, which is a common ERα polymorphism, showed twice the increase in HDL-C levels in response to hormone replacement therapy, compared with women with other genotypes. This report suggested the possibility of using genomic screening to tailor hormone replacement therapy according to individual genotypes. However, the relationship between polymorphisms and cardiovascular events remains to be established.

For background information *see* ARTERIOSCLEROSIS; ESTROGEN; GENE; HEART DISORDERS; MENOPAUSE; PROGESTERONE in the McGraw-Hill Encyclopedia of Science & Technology.

Michio Otsuki; Karin Dahlman-Wright; Jan-Åke Gustafsson

Bibliography. N. Cherry et al., Oestrogen therapy for prevention of reinfarction in postmenopausal women: A randomised placebo controlled trial, *Lancet*, 360(9350):2001–2008, December 21–28, 2002; D. Grady et al., Cardiovascular disease outcomes during 6.8 years of hormone therapy: Heart and Estrogen/progestin Replacement Study follow-up (HERS II), *JAMA*, 288(1):49–57, July 3, 2002; D. M. Herrington et al., Effects of estrogen replacement on the progression of coronary-artery atherosclerosis, *NEJM*, 343(8):522–529, August 24, 2000; S. Hulley et al., Randomized trial of estrogen plus progestin for secondary prevention of coronary heart disease in postmenopausal women, Heart and Estrogen/progestin Replacement Study (HERS) Research Group, *JAMA*, 280(7):605–613, August 19, 1998; B. L. Riggs and L. C. Hartmann, Selective estrogen-receptor modulators—mechanisms of action and application to clinical practice, *NEJM*, 348(7):618–629, February 13, 2003; J. E. Rossouw et al., Risks and benefits of estrogen plus progestin in healthy postmenopausal women: Principal results from the Women's Health Initiative randomized controlled trial, *JAMA*, 288(3):321–333, July 17, 2002.

Extreme-ultraviolet lithography

Lithography is a photographic process that is used to transfer the geometric patterns describing an integrated circuit to the silicon wafer. After a lithography step, chemical, etching, impurity deposition, or oxidation process steps are performed on the silicon wafer to further delineate the circuit structure. These sequential lithography and processing steps precisely form the three-dimensional electrical circuit on the silicon surface, thus producing the integrated circuits used in all types of electronic communications, displays, and signal processing equipment.

The lithography tool is an extremely high precision manufacturing machine. Typically the tool transfers the geometric circuit pattern from a mask, which is normally four times the actual circuit size, through a photoreduction process using a series of lenses. The smallest feature size printed by a lithography tool is equal to feature size $= k_1 \lambda/\text{NA}$, where λ is the wavelength of the illumination, NA is the numerical aperture of the camera, and k_1 is a proportionality constant which depends on the photoresist, specific system illumination characteristics, mask geometries, and manufacturing processes. For conventional optical lithography, the characteristic wavelength has decreased from 365 nanometers to 157 mm. The 157-nm lithography system is expected to print 70-nm features and perhaps 45-nm features with optical tricks such as the use of phase-shift masks and optical proximity correction (OPC).

Extreme-ultraviolet lithography (EUVL) extends optical lithography by using much shorter wavelengths in the range of 11 to 14 nm to allow much larger k_1 values for a specified feature size. The system that will be described uses an extreme-ultraviolet wavelength of 13.4 nm. Extreme-ultraviolet lithography can theoretically print features smaller than 30 nm using a 0.25-NA camera without the use of optical proximity correction.

Although extreme-ultraviolet lithography is similar to optical lithography, a major difference is caused by the fact the very short 13.4-nm light used in extreme-ultraviolet lithography is absorbed by all materials and gases. For an extreme-ultraviolet lithography system, reflective optics and masks coated with distributed quarter-wave Bragg reflectors and vacuum system operation are required, since conventional optics and masks would absorb the extreme-ultraviolet radiation used in the photoreduction process, and air or other gases in a conventional system would also absorb this radiation.

System operation. The 13.4-nm radiation for an extreme-ultraviolet lithography system (**Fig. 1**) is produced by focusing a high-power neodymium: yttrium-aluminum-garnet (Nd:YAG) laser beam on a xenon gas, liquid, or solid target to produce a 25- to 75-eV plasma that emits visible and extreme-ultraviolet illumination. A condenser consisting of multilayer-coated collector and grazing-incidence mirrors collects and shapes the extreme-ultraviolet beam into an arc field 6 mm (0.24 in.) wide by 104 mm (4.1 in.) long to illuminate the reflective mask or reticle. This reflective reticle has a very low coefficient of thermal expansion (less than $2 \times 10^{-8}/^{\circ}\text{C}$) and is clamped to a scanning reticle stage to move the mask across the illumination beam. Reflective reduction optics containing aspheric mirrors are used to demagnify the mask image by a factor of 4. Finally, a scanning wafer stage, containing a wafer coated with extreme-ultraviolet-sensitive photoresist, scans the wafer across the extreme-ultraviolet beam in perfect synchronism with, and at one-fourth of the speed of, the scanning reticle stage.

To prevent the buildup of carbon on the reflective surfaces in the presence of extreme-ultraviolet radiation, the partial pressures of hydrocarbon-containing gases in the vacuum must be controlled. Since the reflection of extreme-ultraviolet radiation from the reflective surfaces is approximately 70%, the deposited extreme-ultraviolet flux causes localized heating of the reticle and optical surfaces, and this heating requires thermal management of critical surfaces. In addition, the velocity and position of the scanning reticle and scanning wafer stages, which are magnetically levitated, must be controlled with nanometer precision.

System integration and imaging results. All components and systems to carry out extreme-ultraviolet lithography have been demonstrated in an alpha tool (first-generation research and development tool) called the Engineering Test Stand (**Fig. 2**), which uses a scanning process to print complete circuit test patterns. The completed system consists of two major environmental enclosures, a source chamber or illuminator and the main exposure chamber containing the reticle and wafer stages and projection optics. The illuminator is isolated from the main chamber by a membrane-type spectral-purity filter to remove out-of-band radiation and to provide an environmental barrier between the two systems.

The Engineering Test Stand is fully instrumented and contains approximately 100 temperature, vibration, and extreme-ultraviolet flux sensors to collect data as a function of various system operating conditions. Small samples of materials called witness plates are placed at specific locations to sample the environment and contamination occurring during operation. These samples are removed from the monitoring locations after specific periods of time and chemically analyzed to determine the types of con-

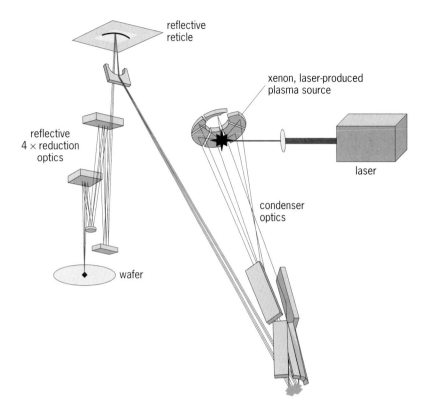

Fig. 1. Extreme-ultraviolet lithography system with fourfold (4×) reduction optics and the illuminator system consisting of the condenser optics and a xenon, laser-produced plasma source. (*After C. Gwyn, EUV lithography update, oe magazine, 2(6):22, June 2002, SPIE—The International Society for Optical Engineering*)

tamination present in the tool. Although witness plate data indicate that the Engineering Test Stand environment is benign in the absence of extreme-ultraviolet radiation, this radiation can induce oxidation of the molybdenum/silicon multilayers in the presence of water vapor. Unique gas mixtures have been developed to prevent contamination and provide in-situ cleaning of the optics. Early tests of the Engineering Test Stand indicate that the partial pressure of gases with atomic mass numbers greater than 44 is below 10^{-11} torr (10^{-9} pascal).

Scanned images of circuit test patterns with 100-nm and 80-nm line widths were printed in a step-and-scan mode (**Fig. 3**). This mode of operation is a way of dealing with the physical limitations in the size of optics and required resolutions that prevent the entire circuit from being printed at one time and allow only a narrow rectangular region (the width of the circuit in one direction and a few millimeters in the other direction) to be printed instantaneously. To print the entire circuit, the rectangular slit is scanned in the direction perpendicular to the wide dimension of the rectangle to print overlapping regions until the entire circuit is printed. The wafer is then "stepped" to the next circuit location on the wafer and the process is repeated.

The first images of complete circuits were printed using an experimental resist that had been developed for 248-nm lithography (termed deep ultraviolet or DUV) and was modified slightly in chemical composition to make it sensitive to extreme-ultraviolet

(a)

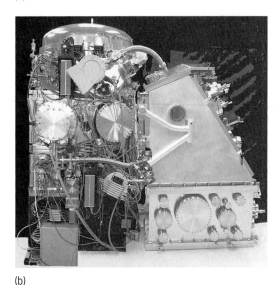

(b)

Fig. 2. Engineering Test Stand. (*a*) Front view, showing wafer handling area. (*b*) Side view of main exposure chamber. (*Sandia National Laboratories, Livermore, CA*)

light (13.4-nm wavelength). To reduce flare and print small features, the system uses a dark-field mask, which is a negative of the image to be printed, so that areas of the integrated circuit in which there are no lines to be printed are dark. The system has been used to print static and scanned images for lines with 90° bends (elbows) with 80-nm widths and 70-nm spacing, and 70-nm-square geometries.

Challenges. Although substantial progress has been made and essentially all aspects of the technology have been demonstrated, several challenges remain that must be addressed to assure the successful development of production equipment to support manufacturing of integrated circuits with 50-nm features. These include (1) providing incremental technology extensions to improve the production tool throughput and reduce the tool and operational costs (2) accelerating fabrication of beta tools (next-generation tools which improve upon alpha tools but lack some features of production tools) and production tools, (3) and developing the infrastructure of the companies and suppliers that manufacture and provide components and measuring sys-

tems to support the prime supplier of the lithography tools.

Technology extensions. Incremental technology extensions are needed to improve mirror reflectivity and lifetime; system environmental control; source efficiency, power, and lifetime; and spectral purity filter efficiency. The target improvements include 70%-reflectivity multilayer coatings with good thermal stability, environmental control to prevent carbon deposition or in-situ cleaning to support mirror lifetimes of 5 years, extreme-ultraviolet source power to support a tool throughput of one-hundred, 300-mm wafers per hour, and a filter that attenuates radiation outside the extreme-ultraviolet range to provide spectrally pure extreme-ultraviolet flux for illuminating the reticle.

Tool commercialization. When the major technology development began in 1997, the goal was to have beta tools in 2003 and production tools in 2005 to support the printing of 100-nm features. Since then, 248-nm, 193-nm, and 157-nm technologies have been extended to print smaller features. This development has shifted the targeted process introduction for extreme-ultraviolet lithography to 50 nm. In addition, the following issues have contributed to the time delays in producing beta extreme-ultraviolet lithography tools: (1) the technology is not a simple extension of optical lithography and additional resources are required, (2) the development of 157-nm lithography at the same time as extreme-ultraviolet lithography dilutes and competes for the available resources, (3) 157-nm materials issues and a possible extension to smaller feature sizes introduce timing uncertainty, (4) the downturn in the worldwide economy has reduced investments in research and development, (5) the lack of competition in developing the first extreme-ultraviolet lithography tools has reduced the competitive pressure, and (6) the readiness of the infrastructure to support the beta and production lithography tools is uncertain.

Infrastructure. The third major challenge consists of establishing the resist and mask infrastructure to support extreme-ultraviolet lithography in production. Imaging experiments using the Engineering Test Stand and a 10× microstepper (a lithography tool that prints only a small region of a mask, with the mask having a magnification of 10 times the actual printed area produced on the integrated circuit) have demonstrated the extendibility of modified deep-ultraviolet resists to extreme-ultraviolet lithography. Although photoresist optimization will be required to reduce line edge roughness, photoresists for extreme-ultraviolet lithography are not considered to be a problem.

The reflective extreme-ultraviolet lithography mask requires a number of additional manufacturing steps over conventional binary masks, which give rise to commercialization challenges. Mask blanks are constructed using a low-defect, low-thermal-expansion material and a square substrate that is highly polished with a flatness of 50 nm. The substrate is coated with a smoothing multilayer to cover

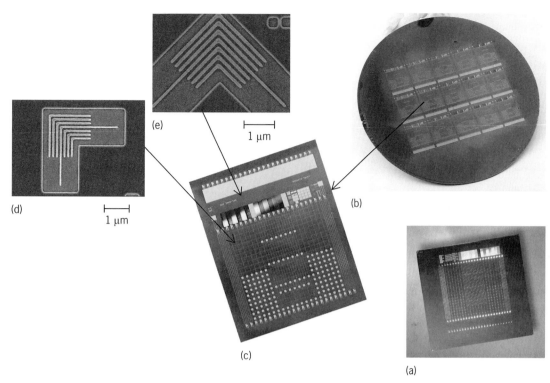

Fig. 3. Extreme-ultraviolet reflective mask and wafer with printed die and enlarged images. (*a*) 150-mm (6-in.) reflective mask. (*b*) 200-mm (8-in.) wafer with 15 printed dies. (*c*) One of the printed dies with a complete array of test patterns, measuring 24 mm (0.945 in.) wide by 32.5 mm (1.28 in.) in the scan direction. (*d*) Enlarged images of portions of the die with elbows (bent lines) with 80-nm width and 80-nm spacing, and (*e*) 100-nm width and 100-nm spacing. (*Part a from Intel Corp.; parts b–e from Sandia National Laboratories, Livermore, CA*)

all defects smaller than 50 nm, followed by the reflecting multilayer stack of 40 to 60 alternating layers of molybdenum and silicon. A capping layer is deposited to protect the multilayer stack, and an absorber of chromium or tantalum nitride (TaN) is deposited as the final layer on the mask blank. Circuit patterns are produced on the reflective mask blank by depositing a layer of photoresist and then using a conventional electron-beam patterning tool to write the pattern on the blank. Following the patterning process, the mask is inspected and repaired using a focused ion beam either to deposit or to remove absorber material. When using the mask in the exposure tool, extreme-ultraviolet radiation is reflected from regions where the absorber has been removed and absorbed by the remaining absorber pattern. The reflected pattern passes through the reflective optics and is reproduced on the resist-coated silicon wafer.

The mask patterning, inspection, and repair processes use extensions of conventional mask patterning equipment. Because of the small feature sizes, extra care is required for the handling, inspection, cleaning, and repair steps. In addition, a conventional optical pellicle to protect the mask from contamination will not work for extreme-ultraviolet masks, and an alternate solution must be found such as thermophoretic protection for the mask inside the tool and use of a removable cover for storage and transportation.

Advantages. Extreme-ultraviolet lithography has many advantages over other lithographies proposed for printing features smaller than 70 nm. For example, it is an optical lithography that builds upon the learning and supplier infrastructure established for conventional lithography. Extreme-ultraviolet lithography provides good depth of focus and linearity for both dense and isolated lines without optical proximity correction, using low-numerical-aperture systems. It uses robust 4× masks that are patterned using standard mask writing and repair tools. Although the masks are reflective, similar inspection methods can be used as for conventional optical masks.

The masks used in electron-beam lithography rely on blocking the passage of electrons from the areas of the mask that are to be printed, and are therefore stencil-like. The stencil image is supported by a very thin membrane of silicon or other material that allows passage of the electrons through the material where the stencil absorbing material is not present. This thin membrane is very fragile and is easily broken. Extreme-ultraviolet masks do not suffer from this fragility, and the low-thermal-expansion substrates provide good critical dimension control and image placement.

For background information *see* INTEGRATED CIRCUITS; OPTICAL MATERIALS; ULTRAVIOLET RADIATION in the McGraw-Hill Encyclopedia of Science & Technology.   Charles W. Gwyn

Bibliography. C. Gwyn, EUV lithography update, *oe magazine*, 2(6):22, June 2002, SPIE—The International Society for Optical Engineering; C. W. Gwyn et al., Extreme ultraviolet scanning lithography supports extension of Moore's law, *Future Fab Int.*, no. 11, pp. 196–203, July 2001; N. Harned et al., Engineers take the EUV lithography challenge, *oe magazine*, 3(2):18, February 2003; SPIE—The International Society for Optical Engineering; H. J. Levinson, *Principles of Lithography*, vol. PM97, SPIE Press, 2001; D. Sweeney, Extreme ultraviolet lithography: Imaging the future, *Sci. Technol. Rev.*, November 1999.

Finite element analysis (paleontology)

Biologists have long commented on how the vertebrate skeleton appears to be designed in accordance with engineering principles. Bones and their associated soft tissues are adapted to transmit, resist, and take advantage of the many forces the skeleton experiences during normal function. How bones respond to forces, that is, how they are stressed and strained, is therefore linked to their function. A great deal of functional information can be gained from examining how a skeleton responds to stresses and strains; yet paleontologists are limited because fossil bones are petrified (and thus have different properties than biomaterials in living animals), and skulls in particular are complex shapes not applicable to straightforward mathematical analysis. Recently, however, biologists and paleontologists have begun to borrow from the sophisticated design toolkit of engineers in order to examine skeletal construction.

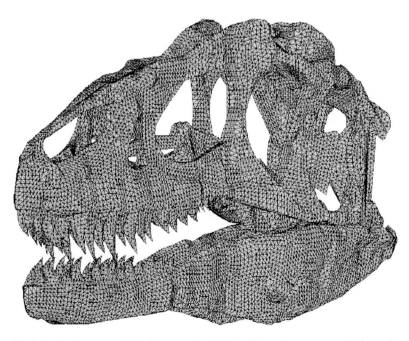

Fig. 1. Finite element model of *Allosaurus fragilis* skull. Elements are represented by each individual triangle. Geometric equations can be used to calculate strain and stress in simple structures. FEA permits the application of these geometric equations to complex, nongeometric shapes by calculating stress/strain in each individual element. (*Reprinted with permission,* © *Emily Rayfield*)

Finite element analysis (FEA) is one such tool. FEA is a means by which stress, strain, and displacement in a two- or three-dimensional structure may be deduced using mathematical principles. FEA can be performed using predesigned computer software or specific user-written code. The exponential increase in computing power over the past 10–20 years has increased the accessibility of FEA to researchers who traditionally lie outside the engineering sphere. By creating digital models of skeletons, FEA can potentially offer a new avenue of inquiry to vertebrate paleontologists who are interested in the function of extinct animal skeletons and the mechanical principles and evolutionary pressures underlying their construction.

How FEA works. In the first stage of FEA, a digital two- or three-dimensional model of the structure to be tested is divided into a finite number of simple geometric shapes called elements. Elements are joined to each other at discrete points called nodes, and the element and node construct is known as a mesh (**Fig. 1**). Specific material properties that approximate those of the actual structure, such as stiffness and density, are applied to the element mesh. Boundary conditions are then applied to the mesh: constraints are applied to prevent the structure from moving in a particular direction; and loadings are applied as forces, pressures, or accelerations, which mimic loads the structure would experience during life or use. This process of model creation is known as preprocessing.

The next stage is analysis, which involves the calculation of force vectors and displacements at each individual node, taking into account the material properties of the structure. Stress and strain at each nodal point are subsequently calculated to provide a composite picture of the mechanical behavior of the structure. Mathematical solvers integrated into FE software perform this stage of the analysis.

Finally, during postprocessing, results are visualized and interpreted (**Fig. 2**). Emphasis is placed upon the checking of errors and refinement of the original mesh and boundary conditions to ensure the model represents the original structure as accurately as possible.

FEA in zoology and paleontology. The basic principles of FEA were originally derived by engineers in the late 1950s and early 1960s. Since the early 1970s, this method has been used widely in orthopedic medicine and bioengineering; however, there have been only a handful of studies utilizing FEA in zoology and paleontology. In 2003, the application of this technique to these fields was still in its infancy. Interestingly, an offshoot of FEA known as finite element scaling analysis (FESA), which is concerned with the calculation of displacements only, and not stress and strain, has been used since the mid-1970s to quantitatively examine shape change (such as comparing the prominence of facial features in early hominoids).

A 200-element FE model of the bill of a shoebill (a type of stork), published in the mid-1980s, appears to represent the first application of FEA to zoology or paleontology. Stress plots and displacements were displayed for two bill-loading regimes; however, actual loads and material properties were not specified. Not until the late 1990s did further zoological studies appear, mainly focused on primate lower jaws and teeth and horses' hooves.

Ammonites. The first published application of FEA in paleontology was an investigation into the shell strength of ammonites (an extinct group of mollusks) in the late 1990s. FEA showed that increasingly complex septal (internal shell wall) construction weakened ammonite shells (with weakening indicated by higher stress magnitudes). FEA models used in this analysis consisted of around 10,000 elements subject to hydrostatic pressure. A more recent analysis of the problem involved the creation of more morphologically accurate septal models, created from 20,250 realistic eight-node curved elements with the strength and stiffness of *Nautilus* nacre (the shell lining of a genus of mollusks), rather than the four-noded flat plate elements used in the initial study. In contrast to the previous analysis, the new study discovered complex septal morphology was in fact stronger than simple septal morphology. These results highlight accuracy problems in model geometry and element choice, a situation all users of FEA must face.

Evolutionary questions. General questions in cranial morphology have begun to be addressed using FEA. From the late 1990s to the present day, a few simple FEA models of three-dimensional beam and structural models and flat two-dimensional planes containing holes have been used to investigate the adaptive and mechanical significance of fenestra (openings) in the skulls of amniotes and the effect of increased nasal and braincase size in mammals. FEA has also been used to examine the ossification of limb bones during growth and to investigate why separate ossification centers are found at the ends of limb bones in some vertebrates but are absent from the limbs of others. Since the analysis was placed within an evolutionary framework, the results have implications for extinct and living animals.

Extinct animal models. Currently, the use of FEA in vertebrate paleontology has generally focused upon the mechanical behavior and function of skulls.

Snouts: basal synapsids. In the late 1990s, two three-dimensional FE models of the snout of a gorgonopsid and a therocephalian synapsid (mammal-like reptiles) were created. Models were geometric approximations of snout morphology, but for the first time an attempt was made to apply realistic bite forces to a FE vertebrate model in numerous directions, representing bilateral, unilateral, or shear biting. Actual material properties were not estimated; however, a close association of model stress distribution to actual cranial buttressing and mobile joint articulation was discovered. Predictions on the role of such predatory animals in Permian ecosystems have been made based partly on these results.

Snouts: archosaurs. Other snout models have investigated stress within the anterior skull of the theropod dinosaur *Megalosaurus* and the archosauriform *Pro-*

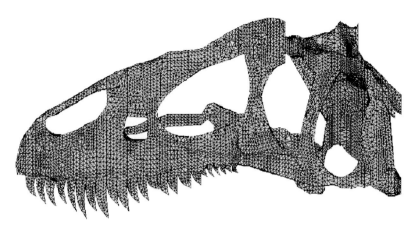

Fig. 2. FEA-generated stress plot of skull in Fig. 1. A color-coordinated map of stress distribution within the structure may be produced from a FEA. Strain maps, vector plots, and displacement values are produced in a similar fashion. Here color and white areas of the mesh represent increased compressional stress. (*Reprinted with permission, © Emily Rayfield*)

terosuchus. The *Megalosaurus* model utilized appropriate material properties and investigated the effect on stress distribution of introducing kinetic (mobile) joints into the skull.

Whole-skull models: Allosaurus. Snout models are useful indicators of cranial stress within a particular region of the skull. However, errors may occur where the connection of the snout to the back of the skull must be estimated. It is, therefore, advantageous to model complete crania, and indeed this has been achieved on two occasions. In 2001, a 200,000-element model of the skull and lower jaw of the theropod dinosaur *Allosaurus fragilis* was completed (Figs. 1 and 2). This is currently the most complete and complex FE vertebrate model. Material properties, jaw muscular forces, bite force, and condylar (jaw joint) forces were accurately estimated. Stiffness and density values of structurally similar bovine bone were taken as an estimate of allosaur bone material properties. Size and force production of jaw adductors was estimated and used to calculate bite and jaw joint force, and all forces were applied to the model in the correct anatomical position. The model was constrained at insertion points of neck musculature and vertebral elements on the posterior surface of the skull. Numerous biting regimes were modeled, including bilateral, unilateral, and tearing bites. FEA revealed that the skull is extremely strong, and appears designed to accommodate high-magnitude stresses produced during prey capture and killing.

Whole-skull models: pterosaurs. Recently, a FE model of a pterosaur skull was created. The model is a reasonably accurate geometric representation of skull form incorporating bony material properties. The pterosaur model and a FE tooth model with properties of dentin (bonelike tissue) and enamel are currently part of an investigation into element formation and material property determination. Bite force analysis is being undertaken; yet the results are currently not available in the literature.

Foot: Gorgosaurus libratus. Finally, a recent FEA of the long bones in the foot of *Gorgosaurus libratus* elucidated the dynamic strengthening function of foot bones and associated ligaments. Researchers examined strain energy in the long bones of the foot of *G. libratus* during locomotion. Their results supported the idea that associated ligaments helped transfer footfall energy along the long axis of the splintlike middle foot bone and prevented damage from bending.

Problems. The above examples provide a review of the current status of FEA in vertebrate paleontology. Use of the technique will surely increase, as FEA has the potential to address both specific and wide-ranging questions concerning the functional morphology and evolution of fossil animals. However, a number of technical and theoretical problems face future FEA users.

First, experimental evidence from living animals has shown that not all structures are adapted to the functions they undertake or that functional signatures are muddled in bones that undertake numerous functional tasks. With structure decoupled from function, the elucidation of skeletal stress patterns may not yield satisfactory hypotheses of function. Nevertheless, FEA still bears the potential to test predetermined hypotheses of function and adaptation, and the strengths of the technique lie in this particular area.

Second, creation of model geometry is difficult, and problems are faced deciding how abstract a model should be when created. Material properties of bone and other tissues must be estimated from analogs in living animals. Boundary conditions (constraints and loading forces) must be estimated if not absolutely, then relatively. Moreover, there are technical issues involving element choice, mesh size, and position of constraints that could potentially influence the output of a FEA.

Conclusion. Taking these cautionary notes into account, the importance of FEA as a tool to investigate the mechanical behavior and function of extinct animal skeletons is evident. Paleobiologists will be able to utilize FEA, a technique relatively new to paleontology, in order to address old, fundamental questions such as why the skeletons of extinct animals were shaped the way they were.

For background information *see* DINOSAUR; FINITE ELEMENT METHOD; FOSSIL; PALEONTOLOGY; SKELETAL SYSTEM; STRESS AND STRAIN; SYNAPSIDA in the McGraw-Hill Encyclopedia of Science & Technology.

Emily Rayfield

Bibliography. R. M. Alexander, *Bones: The Unity of Form and Function*, Macmillan, New York, 1994; M. Fastnacht et al., Finite element analysis in vertebrate palaeontology, *Senckenbergiana lethaea*, 82(1):195–206, 2002; C. McGowan, *A Practical Guide to Vertebrate Mechanics*, Cambridge University Press, 1999; E. J. Rayfield et al., Cranial design and function in a large theropod dinosaur, *Nature*, 409:1033–1037, 2001; E. F. Weibel, C. R. Taylor, and L. Bolis, *Principles of Animal Design*, Cambridge University Press, 1998.

Flower diversity and plant mating strategies

The diversification in form and function of flowers, the reproductive structures of angiosperms (flowering plants), provides some of the most compelling examples of adaptation by natural selection. Flowers vary enormously in size and display greater structural variation than the equivalent structures in any other group of organisms. This diversity provides outstanding opportunities for investigating the functional association between floral traits and plant mating. The recent integration of theoretical, comparative, and experimental approaches in evolutionary biology is giving rise to new insights into the relations between floral diversity and plant mating strategies.

Mating in plants. Mating is of profound evolutionary significance because it directly influences the amount and organization of genetic variation in populations and their responses to natural selection. Floral traits that influence mating are of particular importance because they govern not only their own transmission but also the transmission of all the other genes within the genome. Because most plants are hermaphroditic and produce multiple reproductive structures (flowers and inflorescences, segregated flower clusters), mating patterns can vary considerably among individuals, populations, and species. This variation can involve different rates of self- and cross-fertilization, levels of biparental inbreeding, and mate number resulting from multiple paternity. The complexity of plant mating can be revealed through the use of neutral co-dominant genetic markers, particularly allozymes (different forms of the same enzyme) and microsatellites (short repeated sequences of deoxyribonucleic acid). Such marker gene studies of mating patterns are a crucial prerequisite for determining the selective forces directing floral evolution and the diverse mating strategies that characterize flowering plants.

Floral diversity. Analysis of the relation between pollination and mating patterns is crucial for understanding the function and evolution of floral characters. This is because the selective forces determining reproductive diversification in angiosperms largely exert their influence through pollination and its effect on mating and fertility. Floral diversity in most angiosperm groups has been shaped by the evolution of adaptations associated with pollen vector type, the avoidance of inbreeding, and effective pollen dispersal.

Pollen vector divergence. Because plants are sessile, most floral diversity is associated with the particular vectors employed to achieve successful cross-fertilization. Pollen dispersal may occur via biotic (animals) or abiotic (wind and water) agents, resulting in contrasting suites of floral traits. Due to wide variations in morphology and physiology among animals, it is not surprising that adaptive radiations associated with pollen vector divergence are a prominent feature of many animal-pollinated families (for example, Orchidaceae, Polemoniaceae, Scrophulariaceae). Recent work has focused on attempts to understand the ecological mechanisms responsible for evolutionary shifts from one pollinator group to another (for example, the evolution of bird pollination from bee pollination).

Flowers adapted for wind and water pollination exhibit less striking floral variation because of the absence of showy, attractive structures. However, even in abiotically pollinated groups, diverse structural mechanisms promoting successful cross-pollination are evident, although less is known about their biophysical characteristics and functional significance. Although wind pollination is known to have evolved from animal pollination in many angiosperm families, we are still largely ignorant of the microevolutionary forces causing this shift in pollination system.

Inbreeding avoidance. Not all floral diversity is directly linked to the particular agents of pollen dispersal. For example, considerable variation in the spatial and temporal arrangement of sexual organs is associated with mechanisms that reduce the harmful effects of self-fertilization (selfing) on plant fitness. Experimental studies of outcrossing species commonly demonstrate that offspring resulting from selfing exhibit reduced viability and fertility compared with those arising from cross-fertilization. The intensity of this fitness difference is particularly evident under field conditions, when plants are exposed to a full range of biotic and abiotic stresses. This phenomenon, known as inbreeding depression, largely results from the exposure of deleterious recessive alleles that are normally hidden from selection in the heterozygous condition in outcrossing populations. Inbreeding depression features in most theoretical models for the evolution of mating systems, and is widely recognized as a major selective force influencing reproductive traits.

Pollen dispersal. Because of the harmful genetic consequences of selfing, many plant species are protected from inbreeding depression through physiological self-incompatibility (in which pollen is unable to fertilize ovules of the same plant). Since self-incompatibility systems are essentially passive in nature, and therefore cannot influence pollen dispersal, most outcrossing species also possess floral mechanisms that actively promote pollen dispersal among plants. For example, in species with the sexual polymorphisms heterostyly and enantiostyly, the reciprocal placement of stigmas (pollen receivers) and anthers (pollen bearers) in the floral morphs promotes pollinator-mediated crosspollination between the morphs (**Fig. 1**). The positioning of sexual organs also reduces interference between female and male function in the same plant, resulting in less pollen wastage from self-pollination and more economical use of pollen. These sexual polymorphisms, therefore, reduce the sexual conflict that hermaphroditic plants encounter by achieving effective cross-pollen dispersal while avoiding sexual interference between female and male function.

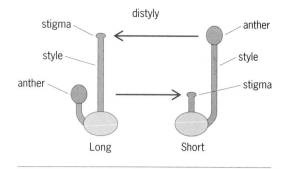

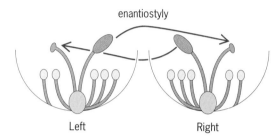

Fig. 1. Floral sexual polymorphisms distyly (a form of heterostyly) and enantiostyly. In distylous populations, two floral morphs occur: long (L)- and short (S)-styled. They differ reciprocally in stigma and anther height. Only cross-pollinations between the floral morphs result in seed set (arrows). In enantiostylous populations, two floral morphs occur, with styles deflected either to the left or right side (L and R morph, respectively) of the flower.

In plants with large floral displays, such as mass-flowering trees and large clonal herbs, a considerable amount of selfing can potentially occur when pollinators transfer pollen grains among flowers on a single plant (geitonogamy). Geitonogamy removes pollen from the pool of male gametes available for cross-pollination, thus reducing fitness through male reproductive function (pollen discounting). It has recently been proposed that many features of floral design and display serve as "antidiscounting mechanisms," reducing male gamete wastage and promoting fitness gain through outcrossed siring success. This hypothesis is challenging to investigate experimentally because of the difficulties in measuring pollen dispersal and male-outcrossed siring success in plants. Nevertheless, several recent experimental studies using genetic markers have provided convincing evidence that floral traits such as dichogamy (temporal segregation of female and male sex function) can function as antidiscounting mechanisms.

Evolutionary transitions. Spatial variation in ecological conditions and its influence on the local pollination environment elicit most evolutionary transitions in the reproductive biology of plants. Closely related species of angiosperms often possess different pollination and mating systems, indicating that plant reproductive traits can be evolutionarily labile (liable to change). Indeed, intraspecific variation in mating patterns is not uncommon, particularly in wide-ranging herbaceous species occupying diverse environments (for example, continents as well as islands). Two contrasting evolutionary changes to

Fig. 2. Intraspecific variation in plant mating strategies. (*a*) Outcrossing and (*b*) selfing populations of the neotropical annual aquatic *Eichhornia paniculata* (Pontederiaceae) differing in flower size and showiness. (*c*) Hermaphroditic, (*d*) female, and (*e*) male plants of *Sagittaria latifolia* (Alismataceae), a North American clonal aquatic. Populations of this species are either monoecious [hermaphroditic plants with separate female (bottom of inflorescence) and male (top of inflorescence) flowers] or dioecious [separate female and male plants].

tor visits, resulting in pollen limitation of seed production and low fertility. If these conditions persist, genetic variants capable of autonomous self-pollination will be favored, leading to the evolution of autogamy as a mating strategy. In contrast, inferior pollinator service results when sufficient pollination occurs to overcome pollen limitation, but the patterns of pollen dispersal result in offspring with low fitness because of increased selfing or biparental inbreeding. Models of the evolution of gender dimorphism indicate that if the product of the selfing rate (s) and inbreeding depression (δ) is >0.5, unisexual mutants will spread, leading to the evolution of separate sexes (dioecy). Autogamy and dioecy are represented in many plant families, and each mating strategy is associated with a distinctive group of floral characters that arises by convergent evolution (that is, has multiple independent origins).

Evolution of selfing. The evolutionary pathway from predominant outcrossing to habitual selfing is the most well-known example of mating system evolution in plants. Selfing populations can be distinguished from their outcrossing progenitors by a suite of traits, including smaller flowers with reduced allocation to attractive structures, lower pollen production, and sex organs in close proximity, which promotes autonomous self-pollination. Selfers often occur in ecologically or geographically marginal sites with uncertain pollinator service, implicating reproductive assurance as the principal selection pressure causing this transition. Phylogenetic methods have enabled reconstruction of the evolutionary history of selfing, and in some genera this transition has originated on multiple occasions (for example, *Amsinckia*, *Eichhornia*, *Linanthus*). To understand whether selfing phenotypes will spread in outcrossing populations, it is necessary to have information on the pollination ecology of populations; the mode of selfing and its genetic basis; and the relations between selfing rates, inbreeding depression, and pollen discounting.

Evolution of dioecy. Although dioecy is the dominant sexual system in many animal groups, it is relatively rare in flowering plants. Approximately 7% of angiosperm species have separate female and male plants, yet this form of gender dimorphism is represented in nearly half of all angiosperm families. This distribution implies multiple independent origins of unisexuality. Comparative studies of trait correlations among angiosperm families indicate that dioecy is commonly associated with a suite of traits, including small, inconspicuous, white or green flowers; pollination by wind, water, or generalist pollinators; fleshy fruits; and woody growth forms. Most dioecious species exhibit conspicuous sexual dimorphism, with male plants often producing larger flowers in greater numbers than female plants. Population sex ratios commonly deviate from unity, with male-biased sex ratios most often reported as a result of earlier flowering in males or greater mortality in females because of higher reproductive costs.

mating biology are particularly evident in flowering plants (**Fig. 2**). These transitions are the evolution of predominant self-fertilization (autogamy) from outcrossing and the evolution of dioecy (separate female and male plants) from cosexuality (hermaphroditism). Although these transitions involve strikingly different functional endpoints (for example, uniparental versus biparental reproduction), there is evidence that in animal-pollinated groups both can occur if populations encounter unsatisfactory pollinator service. This may at first seem paradoxical; however, it can be reconciled by recognizing that a plant's pollination environment can be unsatisfactory in at least two distinct ways.

Changes in pollinator service. Insufficient pollinator service occurs when plants receive too few pollina-

Models for the evolution of dioecy highlight the importance of a few key parameters, of which the most important are the selfing rates and inbreeding depression of ancestral cosexual populations, the genetics of sex determination, and the reallocation of resources to female and male function. Future empirical work on the ecological factors resulting in changes in pollinator service and the increased selfing required to drive the evolution of dioecy is required.

For background information *see* FERTILIZATION; FLOWER; PHYSIOLOGICAL ECOLOGY (PLANT); PLANT EVOLUTION; POLLEN; POLLINATION; REPRODUCTION (PLANT) in the McGraw-Hill Encyclopedia of Science & Technology.   Spencer C. H. Barrett

Bibliography. S. C. H. Barrett, The evolution of mating strategies in flowering plants, *Trends Plant Sci.*, 3:335–341, 1998; S. C. H. Barrett, The evolution of plant sexual diversity, *Nat. Rev. Genet.*, 3:274–284, 2002; S. C. H. Barrett (ed.), *Evolution and Function of Heterostyly*, Springer, Berlin, 1992; S. C. H. Barrett and L. D. Harder, Ecology and evolution of plant mating, *Trends Ecol. Evol.*, 11:73–78, 1996; M. A. Geber, T. E. Dawson, and L. F. Delph (eds.), *Gender and Sexual Dimorphism in Flowering Plants*, Springer, Berlin, 1999; S. D. Johnson and K. E. Steiner, Generalization versus specialization in plant pollination systems, *Trends Ecol. Evol.*, 15:140–143, 2000; D. G. Lloyd and S. C. H. Barrett (eds.), *Floral Biology: Studies on Floral Evolution in Animal-Pollinated Plants*, Chapman and Hall, New York, 1996; J. Lovett Doust and L. Lovett Doust (eds.), *Plant Reproductive Ecology: Patterns and Strategies*, Oxford University Press, 1988; M. T. Morgan and D. J. Schoen, The role of theory in an emerging new plant reproductive biology, *Trends Ecol. Evol.*, 12:231–234, 1997; R. Wyatt (ed.), *Ecology and Evolution of Plant Reproduction: New Approaches*, Chapman and Hall, New York, 1992.

Focused ion beam machining

Focused ion beam machining (FIBM), also termed focused ion beam milling, provides a tool for processing materials on the nanometer scale. It enables selective removal of minute quantities of matter with high spatial resolution in a precisely controlled manner which was hitherto impossible. Focused ion beam machining instruments have been developed in university and industrial research laboratories over the last 25 years and have been commercially available since the mid-1980s. Functionally, the instruments are similar to a conventional scanning electron microscope (SEM), but use an intense source of ions in place of an electron emitter. Imaging, as in a scanning electron microscope, is done by sweeping the beam pixel-by-pixel across the surface of a specimen in a raster pattern. However, in focused ion beam machining both secondary electrons and secondary ions can be collected to form an image. This provides additional image contrast information when the in-

strument is operating as a scanning ion microscope (SIM). Deposition of conducting or insulating material on selected areas with nanometer resolution can also be done in the same instrument. Combining focused ion beam maching with focused ion beam deposition (FIBD) provides a means for ultraprecision fabrication, modification, and patterning of materials. Instruments have also been developed which provide dual electron and ion capabilities, using an electron beam for imaging and a focused ion beam for machining and deposition.

Technology. Although other alternatives exist, most modern high-resolution focused ion beam systems use a liquid-metal ion source (LMIS). A low-melting-point metal, held in a reservoir, is heated just above the melting point and allowed to flow along the specially prepared surface of a sharpened tungsten needle. Normally, gallium (melting point 29.8°C or 85.6°F) would be used as a source of Ga^+ ions. To prevent contamination and ensure stable operation, the liquid-metal ion source must operate under high vacuum of around 10^{-9} torr (10^{-7} pascal). The tip of the tungsten needle is held close to the aperture of an extraction electrode, and a potential difference of 10–100 kV is maintained between the two. The electric field pulls the molten metal into a conical shape having tip radius of order 1 nm. The intense electric field at the tip of this cone (the Taylor cone) is sufficiently high to cause the metal atoms to be ionized. Metal ions are extracted from the liquid-metal ion source by field ionization, providing an extremely intense (but also divergent) source of ions. The ion beam is accelerated along the axis of a column having electrostatic lenses which focus the beam into a spot with an approximately gaussian intensity distribution. The column also has electrodes to blank (cut off) the beam and to deflect and position the beam with nanometer precision. Advanced low-aberration ion optical systems are capable of producing spot sizes below 6 nm, maximum beam currents of 30 nA, and current densities greater than 25 A cm^{-2}.

Micromachining process. Positive gallium (Ga^+) ions carry much more momentum than electrons and are capable of removing material from a target by physical sputtering. The ions impart sufficient momentum in colliding with target atoms to eject them from the surface. Typical material removal rates are around 1 μm^3 nA^{-1} s^{-1}. Rates can be increased significantly by gas-assisted etching: Halogen gases are introduced into the differentially pumped processing chamber and increase the sputtering rate by forming volatile compounds with the target material. As focused ion beam machining is carried out under high vacuum, most of the sputtered material is removed by the vacuum pumps. However, material redeposited in the vicinity of the point of impact of the ion beam can be a problem, especially if high-aspect-ratio features (having high depth-to-width ratio) are being machined. Three-dimensional profiles can be formed by controlling the dwell time of the beam at each pixel position. The amount of material removed is related to the dwell time, the beam parameters (beam energy, current, and spot size), and the sputter yield of the material, that is, the number of atoms removed per incident ion. The capabilities of the instrument as a scanning ion microscope allow specific areas for milling to be located and, while milling is in progress, allow the region to be inspected in real time. To minimize unwanted specimen damage, imaging is carried out at low (picoampere) beam currents and milling carried out at higher (nanoampere) beam currents.

Applications. Early applications of focused ion beam technology arose from the needs of the semiconductor processing industry: photolithography mask repair, integrated circuit failure analysis, and prototype application specific integrated circuit (ASIC) reworking and repair. However, applications have now extended beyond these areas and include specimen preparation for microanalysis, microfabrication and process characterization for micro-electro-mechanical systems (MEMS), and numerous niche applications.

Lithography mask repair. Chrome-on-glass photomasks are susceptible to wear and damage in use. The thin, patterned, layer of chrome has geometric features that must be faithfully transferred to semiconductor substrates, but may have opaque or clear defects, often smaller than a micrometer in size, rendering it useless. Opaque defects result from the presence of unwanted absorbing regions, and clear defects occur where chrome regions have been inadvertently removed. Repair of opaque or clear defects is achieved by focused ion beam machining or focused ion beam deposition respectively, and is an economically viable alternative to fabricating a new mask. These techniques were later successfully extended to the more challenging problems of phase-shift and x-ray lithography mask repair.

Integrated circuit failure analysis. Process-related problems can lead to catastrophic device failure during integrated circuit manufacture. Each oxide, nitride, and metallization layer and each region of dopant diffusion through the cross section may need checking for integrity. Previously, the only possibility was to cleave the device by crude mechanical means, in itself damaging and contaminating the functional layers, with little control over the precise area examined. Focused ion beam machining provides a unique diagnostic tool by allowing ultraprecise sectioning at site-specific areas on the chip and ensuring side-wall quality and smoothness that would be impossible by any other means (**Fig. 1**).

ASIC rework and repair. The ability to modify a prototype device at the die or wafer level can significantly reduce the time for design change evaluation and eliminate costly iterations around the design-fabricate-test loop. Simple design errors can also be corrected at the prototype stage, allowing the verified, functional design to go into production much more quickly. Focused ion beam machining has been used with considerable success to achieve circuit modifications by selectively milling through interconnects and, combined with selectively depositing

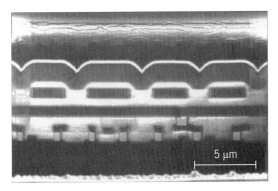

Fig. 1. Electron image of an integrated circuit chip sectioned by focused ion beam machining. (*From P. Krueger, Dual-column (FIB-SEM) wafer applications, Micron, 30:221–226, 1999*)

metal tracks of tungsten, aluminum, or copper by focused ion beam deposition, in changing the electrical topology of the circuit.

Specimen preparation for microanalysis. Focused ion beam machining techniques developed for the semiconductor arena have subsequently been refined for use in numerous other fields, including metallurgy, materials science, earth sciences, polymer chemistry, and the biosciences. A key development is the ability to prepare electron-transparent sections at site-specific locations on a specimen. Two approaches are used: samples are cut and polished prior to finishing by focused ion beam machining, or prepared in situ by milling to expose a section which can be removed by a lift-out technique. In both cases section wall thicknesses of about 100 nm are achievable.

Finely detailed microstructure in alloys and polymers can be revealed following precision sectioning and polishing by focused ion beam machining (**Fig. 2**). Cross sections can then be examined by transmission electron microscopy (TEM) or even

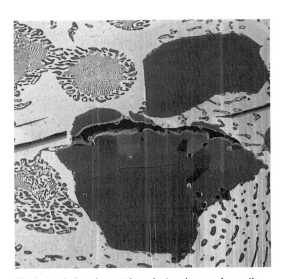

Fig. 2. Ion-induced secondary electron image of a section of a zinc-aluminum alloy with titanium diboride (TiB_2) inclusions that has been polished by focused ion beam machining. (*From M. W. Phaneuf, Applications of focused ion beam microscopy to materials science specimens, Micron, 30:277–288, 1999*)

Fig. 3. Ion-induced secondary electron image of a 100-nm section of carbonaceous chondrite that has been prepared by focused ion beam maching and is ready for lift-out by a precision micromanipulator. (*From P. J. Heaney et al., Focused ion beam milling: A method of site-specific sample extraction for microanalysis of Earth and planetary materials, Amer. Mineralogist, 86:1094–1099, 2001*)

electron energy loss spectroscopy. Often the resolution and image contrast provided by scanning ion microscopy alone can provide an abundance of information.

By allowing researchers to work with extremely small volumes of material, focused ion beam machining has become an attractive proposition for specimen preparation in rare geological samples, precious gemstones, or extraterrestrial materials (**Fig. 3**). In addition to transmission electron microscopy, electron diffraction or energy dispersive x-ray spectroscopy of selected areas can be used to analyze ultrathin specimens prepared in this way.

Applications in MEMS. MEMS technology is based on batch microfabrication methods used by the semiconductor industry, but with additional surface micromachining requirements. Focused ion beam machining is beginning to play a role in the characterization of production processes and also as a direct-write tool. The structural layers of micromotors, which are among the most complex MEMS devices that have been produced, are revealed in cross section by focused ion beam machining (**Fig. 4**).

Extreme accuracy is required in fabricating the magnetic pole gap for present-day magnetic storage unit read/write heads. Magnetoresistive write heads with storage densities of 100 Gbits/in.^{2} (15 Gbits/cm^2) demand a gap of 60 nm. Focused ion beam machining is used commercially as a maskless process for magnetoresistive head trimming to produce ultraprecision, narrow-tip aligned structures.

Integration of optical elements with MEMS (micro-opto-electro-mechanical systems, or MOEMS) will challenge microfabrication technology yet further. Among the integrated optical elements that have been fabricated by focused ion beam machining is a spiral Fresnel zone plate whose function is to measure position to an accuracy of better than 1 nm (**Fig. 5**).

Emerging applications. As frontiers in microfabrication and nanofabrication technology are pushed back, focused ion beam machining is finding an increasing stream of niche applications. Examples include the shaping of microcutting tools that can be used for milling or turning in mesoscale manufacturing

Fig. 4. Scanning electron microscope image of a MEMS micromotor fabricated by a five-level polysilicon process. The structural polysilicon layers, revealed in cross section by focused ion beam machining, are about 2 μm thick. (*From J. M. Bustillo, R. T. Howe, and R. S. Muller, Surface micromachining for microelectromechanical systems, Proc. IEEE, 86:1552–1574, 1998*)

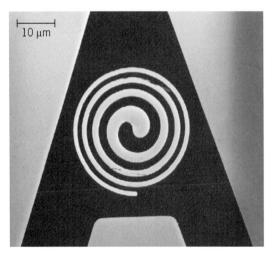

Fig. 5. Gold-coated silicon nitride cantilever with an integrated spiral Fresnel zone plate designed to measure position to an accuracy of better than 1 nm. (*From L. R. Senesac et al., Fabrication of integrated diffractive microoptics for MEMS applications, Optical Manufacturing and Testing IV, SPIE 4451, pp. 295–305, 2001*)

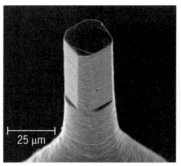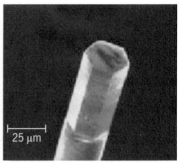

Fig. 6. Three micromilling tools formed by focused ion beam machining. (*From G. L. Benavides, D. P. Adams, and P. Yang, Meso-Machining Capabilities, Sandia Nat. Lab. Rep., SAND2001-1708, June 2001*)

(**Fig. 6**), trimming of silicon nitride cantilevers (used in atomic-force microscopy) to increase their resonant frequency, structuring of magnetic nanoelements for magnetoelectronics, fabrication of superconducting quantum interference devices (SQUIDs), and production of carbon nanotube cantilevers for scanning probe metrology. Many future applications will exploit the ultrahigh resolution and maskless processing capabilities of focused ion beam machining.

For background information *see* ELECTRON MICROSCOPE; INTEGRATED CIRCUITS; MACHINING; METALLOGRAPHY; MICRO-ELECTRO-MECHANICAL SYSTEMS (MEMS); MICRO-OPTO-ELECTRO-MECHANICAL SYSTEMS (MOEMS); SCANNING ELECTRON MICROSCOPE; SECONDARY ION MASS SPECTROMETRY (SIMS); SPUTTERING in the McGraw-Hill Encyclopedia of Science & Technology.   Sam T. Davies

Bibliography. G. L. Benavides, D. P. Adams, and P. Yang, Meso-machining capabilities, *Sandia Nat. Lab. Rep.*, SAND2001-1708, June 2001; P. J. Heaney et al., Focused ion beam milling: A method of site-specific sample extraction for microanalysis of Earth and planetary materials, *Amer. Mineralogist*, 86:1094–1099, 2001; R. Krueger, Dual-column (FIB-SEM) wafer applications, *Micron*, 30:221–226, 1999; J. Orloff, High-resolution focused ion beams, *Rev. Sci. Instrum.*, 64:1105–1130, 1993; M. W. Phaneuf, Applications of focused ion beam microscopy to materials science specimens, *Micron*, 30:277–288, 1999; P. D. Prewett and G. L. R. Mair, *Focused Ion Beams from Liquid Metal Ion Sources*, Research Studies Press, Taunton, England, 1991.

Fuel cells

Like traditional power generation systems, fuel cells extract the chemical energy bound in fuel and, in combination with air as an oxidant, transform it into electricity. Traditional systems, however, use combustion and mechanical devices to achieve the energy transformation. Combustion releases the fuel chemical energy as thermal energy (**Fig. 1**). The hot gases produced move pistons (as in automobiles) or turbine wheels (as in stationary gas turbines) which rotate the shafts of electrical generators. The mechanical pistons and turbines give rise to friction, which affects the overall conversion efficiency of fuel chemical energy to electricity. Efficiencies of 18% (automobiles) to 35% (electrical production) are common. The majority of the fuel chemical energy (82% for automobiles, 65% for electrical production) is dissipated as heat. Combustion also gives rise to the formation and emission of criteria pollutants, species that have been shown to degrade air quality, such as carbon monoxide (CO), partially oxidized hydrocarbons (HC), and oxides of nitrogen (NOx).

Fuel cells use electrochemistry to transform the fuel and oxidant to electricity in one step. Neither combustion nor mechanical components are required. In a typical fuel cell, hydrogen is transformed to a hydrogen proton at the anode (**Fig. 2a**). The electrons released provide the "electricity" and drive an electrical load, while the protons diffuse through an electrolyte to the cathode. At the cathode, the electrons ionize oxygen, and the resultant charged oxygen combines with the hydrogen protons to form the exhaust product, water.

Fuel cells vary by the material used for the electrolyte, the species that diffuse through the electrolyte, the operating temperature, and the fuel-to-electrical conversion efficiency (see **table**). High-temperature fuel cells (for example, solid oxide and molten carbonate) are capable of 50% fuel-to-electrical efficiency. When the waste heat is recovered and used, the combined heat and power (CHP) efficiencies can approach 80%. Potential uses of the waste heat include heating of hot water, space heating, and air conditioning using technologies such as absorption cooling (in which energy from the waste heat is used to produce a cold water stream).

To apply fuel cell technology, a balance-of-plant (supporting components for operating of the the fuel cell) is required (Fig. 2b). A fuel processor is required to extract hydrogen from a hydrogen source such as a fossil fuel (for example, natural gas) or water. Since the fuel cell produces direct current (dc), an inverter is needed to convert the dc to alternating current (ac). Where dc can supply the loads directly, the inefficiency and complexity of an inverter can be avoided.

Galileo application areas and performance requirements

	Application area			
Performance area	Aviation	Land transport	Leisure	Geodesy and time
Coverage	Local to global	Regional to global	Global	Local
Accuracy*	1–200 m	10 m and less (rail)	10 m	To centimeter range; 10 ns
Integrity	High	Moderate	Not guaranteed	Yes
Availability	Very high	High	Not guaranteed	High
Mask angle	Low	High	Not critical	Low to high, depending on application

*1 m = 3.3 ft.

Launch vehicles. Various launch vehicles are available for placing the Galileo spacecraft into orbit. The Ariane V can carry up to eight spacecraft at a time (**Fig. 3***a*), the Proton can take up to six (Fig. 3*b*), and the Soyuz ST-2003 booster can launch two. The Delta IV launch vehicle family offers an alternative for placing two to eight spacecraft in orbit at a time. *See* SATELLITE LAUNCH VEHICLES.

Control segment. The control segment for Galileo will consist of both monitoring sites that receive the signals from the satellites and uplink sites that transmit to the spacecraft the clock corrections, updated orbital parameters, and other data (**Fig. 4**). The spacecraft will then continuously transmit these data to the users.

Implementation schedule. The planned implementation schedule for Galileo involves (1) program definition (which has been accomplished), (2) a development and validation phase (which is in progress), (3) an in-orbit validation phase (beginning with the first spacecraft launch, planned for about 2005, and (4) full deployment of the spacecraft constellation and the control segment. User receiver system and application development will be a continuous process during these activities. Full implementation and operational status for the system is planned for 2008.

Applications. The application areas for Galileo and the principal performance capabilities associated with these applications are shown in the **table**. Augmentation services are frequently required to assure users of the integrity of the signals and to provide various types of corrections for obtaining greater accuracy or other performance improvements.

Analyses relating to the need for Galileo and its prospective usage predict that by 2015 the greatest European usage of the GNSS will be in cars (41%); followed by integrated personal communications (33%); police, fire, and ambulances (9%); trucks and buses (8%); personal outdoor recreation (4%); geographical information systems (1%); and aviation (about 0.7%), leaving about 3% divided among all other uses.

For background information *see* ATOMIC CLOCK; SATELLITE NAVIGATION SYSTEMS in the McGraw-Hill Encyclopedia of Science and Technology.

Keith D. McDonald

Bibliography. G. Hein et al., The Galileo frequency structure and signal design, *Proceedings of ION GPS 2001*, pp. 1273-1282, Institute of Navigation, 2001; D. Iron, Galileo's challenge—Procuring the public/private partnership, *Galileo's World*, 4(2):20-23, Advanstar Communications, 2002; G. Lachapelle et al., How will Galileo improve positioning performance?, *GPS World*, 13(9):38-48, Advanstar Communications, 2002; K. D. McDonald, The European Galileo system: An American perspective, *Proceedings of the 58th Annual Meeting & CIGTF Test Symposium*, pp. 30-50, Institute of Navigation, 2002; O. Onidi, Launching Galileo, *Galileo's World*, 4(2):16-19, Advanstar Communications, 2002.

Geoarcheology

Geoarcheology entails the use of geologic concepts, methods, and knowledge for solving archeological problems. Geology and archeology are historical sciences based largely on the study of a complex stratigraphy that contains mineral, fossil, and cultural remains in a spatial and implicitly chronological context, and that is used to reconstruct the succession of events that produced the sedimentary record. Of all the natural sciences now incorporated into archeology, geology has the longest association with it. A union of the young science of geology with the even younger discipline of archeology occurred in the midnineteenth century, growing out of the same intellectual ferment that gave birth to evolutionary biology and much of modern geology. The importance of geology in the solution of archeological problems is now well understood by archeologists. Colin Renfrew, a leading British archeologist, wrote in 1976 that "since archaeology, or a least prehistoric archaeology, recovers almost all of its basic data by excavation, every archaeological problem starts as a problem in geoarchaeology." The term "archeological geology" is essentially synonymous with geoarcheology.

In the last half of the twentieth century, geoarcheology became a recognized discipline with its own journals, scientific organizations, and graduate programs. Although geoarcheology is not theory-driven—it adopts theories as needed from both

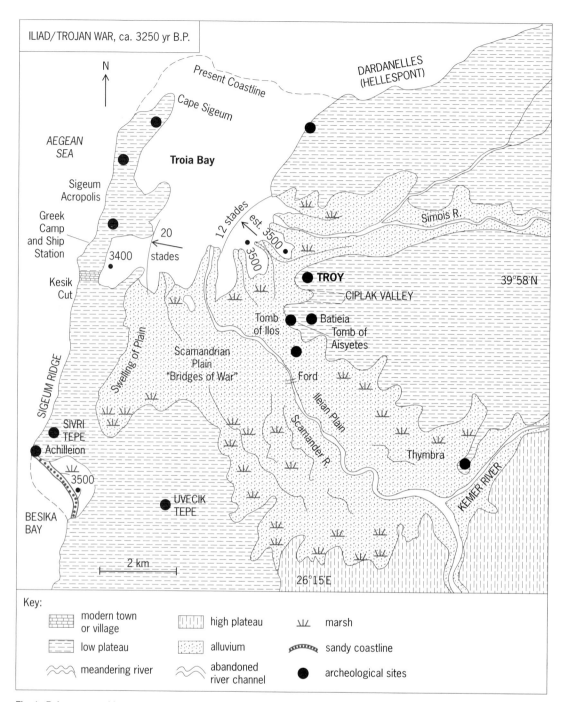

Fig. 1. Paleogeomorphic reconstruction of the Trojan landscape at the time of the *Iliad*, based on core drilling at numerous sites, shows that Homer was accurate in his descriptions.

geological sciences and archeology—its progress often depends on new instrumental techniques that have been borrowed or adapted from physics and chemistry. Examples are ground-penetrating radar and instrumental neutron activation analysis.

Prospecting techniques. Locating buried archeological sites presents special problems for geoarcheologists. Many prospecting techniques are available, including geophysical, geochemical, core drilling, and aerial satellite remote sensing. Geophysical methods such as ground-penetrating radar can de-

tail the location, extent, and character of modified terrain. Surface geophysical surveys can be combined with soil magnetic analysis and geochemical prospecting to give a fair picture of the extent of anthropogenic remains of a buried site or feature.

The first known geophysical survey for an archeological application in America was carried out in 1938 at the historic site of Colonial Williamsburg, Virginia. Now geophysical surveys are used regularly to locate buried archeological sites and to delineate features at sites in a cost-effective and nondestructive manner. For example, in 2001 R. E. Chavez and coworkers used ground-penetrating radar on the eastern flank of the Pyramid of the Sun at Teotihuacan, Mexico, to locate the continuation of a tunnel discovered beneath its western main entrance.

Core drilling can provide an accurate picture not only of the sedimentary record but also of ancient landscapes. For at least 2000 years, scholars have debated the location of Troy and the events and geographic features described in Homer's *Iliad*. Recently, J. C. Kraft and coworkers used geological evidence to show what the Trojan plain looked like at the time of the Trojan War, over 3000 years ago. The correlations among the written word, the archeological record, and the sedimentological and paleontological data from core drilling show clearly the reality of Homer's description of place, event, and topography, although the ancient landscape has since been radically altered (**Fig. 1**). Coring has also revealed two important deeply buried ancient cities in China.

Provenance and age analysis. The provenance of an artifact material is its origin or source area (for example, locality, site, or mine). In geoarcheological terms, this means the geographic/geological source of the raw material from which the artifact was made, that is, a specific geological deposit—normally a quarry, mine, geological formation, outcrop, or other geological feature. A large number of chemical, physical, and geological techniques have been used to source artifactual materials, including trace-element concentrations, isotopic composition, diagnostic minerals or assemblages, microfossils, and geophysical parameters. Throughout the Neolithic of the Old World and the prehistoric of the New World, the main material traded widely over long distances was obsidian, a jet-black volcanic glass. One of the earliest and perhaps the most successful sourcing based on chemical characterization was accomplished for this material in 1964, and many new, nondestructive analytical techniques are now under development. Obsidian is not a common rock type in most regions of the world, and therefore the number of possible sources is limited.

Chemical fingerprinting of rocks as a guide to raw material sources now extends far beyond obsidian. J. Greenough and coworkers used major- and trace-element concentrations of Egyptian basalts to source Egyptian basalt artifacts. The chemical data showed that First Dynasty basalt vessels (Abydos), Fourth Dynasty basalt paving stones (Khufu's funery temple, Giza), and Fifth Dynasty paving stones (Sahure's complex, Abu Sir) came from the Haddadin lava flow in northern Egypt. Other geoarcheologists are using isotope ratios, electron spin resonance, and petrography (using a polarizing microscope) to source the marble used in ancient monuments throughout the Mediterranean. Petrographic studies of tempers have been an excellent tool to determine geological sources and distribution of ancient ceramics. (Tem-pers are coarse materials such as sand, shells, and small pottery fragments added to clay before forming to reduce shrinkage in firing or to improve workability in shaping a pottery vessel.) Others have used mineralogy and texture, combined with trace-element concentrations, to present a detailed picture of the Medieval trade in iron ore.

In North America north of the Rio Grande River, utilitarian copper artifacts appeared initially about 5500 years ago in the archeological record. Use of native copper flourished in the western Great Lakes area because of the wealth of available copper in the form of nuggets and lode deposits outcropping at the surface. In North America, in contrast with most of the rest of the world, the indigenous peoples did not melt or smelt copper. Hence the concentrations of chemical elements were not altered from the original raw materials—an essential criterion for sourcing by chemical characterization. G. Rapp and coworkers analyzed native copper deposits throughout the United States and Canada using instrumental neutron activation analysis (INNA) as a basis for sourcing (**Fig 2**). In INNA, samples are irradiated to produce unstable radioactive nuclides that omit gamma rays of characteristic energies, which can be compared to those of standard reference materials to determine the presence and concentration of particular elements in the sample.

Geoarcheologists also use a number of physical methods in the absolute dating of archeological materials (that is, in estimating their actual age in years). These methods include fission-track dating of the damage in a mineral caused by fission of uranium-238; archeomagnetic dating based on the fact that the Earth's magnetic poles migrate and reverse polarity over time; and electron spin resonance and thermoluminescence dating, both based on the accumulation of trapped electrons in minerals. Fission-track dating was used to resolve the date of the earliest example of the genus *Homo* outside Africa.

Fig. 2. A piece of native copper flattened (but not made into a tool) by Native Americans over 3000 years ago. The small hole was drilled for INAA analysis; also shown are the two pieces drilled out.

Sediment and soil analysis. Most archeological data, as well as environmental and climatic data, are recovered from sedimentary deposits or associated soils. Consequently, much of geoarcheology is involved with detailed field and laboratory studies of site or regional sediments and soils. The term "archeological sediment" is used to distinguish deposits in the sedimentary record that result directly from past human activities. A major interpretational problem for geoarcheologists can be determining whether artifacts were part of the initial deposit or were introduced later by human activities or by natural mixing processes.

An understanding of soils is vital to archeology. Both sediments and soils are horizontally layered but have very different origins. Unfortunately, semantic confusion has been common in archeology because the word "soil" is used for two very different things. "Soil" has been used incorrectly, and loosely, for surface and near-surface sediments of many descriptions. More correctly, soil is that portion of the Earth's surface materials that supports plant life and is altered by continuous chemical and biotic activity and weathering. It may be useful to think of sediments as biologically dead. In contrast, soils develop from the weathering of a variety of rock material at the surface of the Earth and are very much chemically and biologically alive. Because they indicate the presence of stabilized landscape surfaces, soils can contain evidence of possible human occupation and artifact accumulation.

Geoarcheologists are turning more and more to a broad range of chemical analyses of archeological sediments and soils to recover indications of ancient lifeways. The amount and chemical nature of the phosphorus can distinguish burial grounds and agricultural and habitation areas, as well as indicate a variety of industrial remains. Organic carbon is perhaps the most important element for animal and plant life. When found in soils and sediments, it retains many characteristics of the role it played in ancient human landscapes. For example, many of the carbon compounds in oil, wine, meat, and plants are different and recognizable if not degraded.

Most chemical nutrients required for all life on land are supplied from the soil. The chemical elements nitrogen (N), phosphorus (P), potassium (K), calcium (Ca), magnesium (Mg), and sulfur (S) are macronutrients. Because all plants use these elements, the removal of vegetation by human activities depletes their concentration in soils and underlying sediments. Eight chemical elements are considered important micronutrients (required in much smaller amounts than the macronutrients): iron (Fe), manganese (Mn), zinc (Zn), copper (Cu), boron (B), molybdenum (Mo), cobalt (Co), and chlorine (Cl). One of the ways human activities alter the soil environment is by addition or subtraction of these elements, depleting some nutrients to the point of deficiency and augmenting others to the point of toxicity. Toxicities are often recorded in human paleopathologies. Examples can be seen in the raw ma-

terial processing and use of lead and arsenic. Most of these chemical activities leave clear records in archeological sediments, as trace elements released from anthropogenic sources become part of normal biogeochemical processes.

For background information *see* ARCHEOLOGICAL CHEMISTRY; ARCHEOLOGICAL CHRONOLOGY; ARCHEOLOGY; GEOCHEMICAL PROSPECTING; GEOLOGY; PROSPECTING; PROVENANCE (GEOLOGY); SOIL; SOIL CHEMISTRY; STRATIGRAPHY in the McGraw-Hill Encyclopedia of Science & Technology.

George (Rip) Rapp

Bibliography. R. E. Chavez et al., Site characterization by geophysical methods in the archaeological zone of Teotihuacan, Mexico, *J. Archaeol. Sci.*, 28: 1265–1276, 2001; J. Greenough, M. Gorton, and L. Mallory-Greenough, The major- and trace-element whole-rock fingerprints of Egyptian basalts and the provenance of Egyptian artefacts, *Geoarchaeol. Int. J.*, 16:763–784, 2001; J. C. Kraft et al., Harbor areas at ancient Troy: Sedimentology and geomorphology complement Homer's *Iliad*, *Geology*, 31:163–166, 2003; G. Rapp et al., *Determining Geologic Sources of Artifact Copper: Source Characterization Using Trace Element Patterns*, University Press of America, Lanham, MD, 2000.

Geologic mapping

Geologic mapping is a highly interpretive, scientific process that can produce a range of map products for many different uses, including assessing groundwater quality and contamination risks; predicting earthquake, volcano, and landslide hazards; characterizing energy and mineral resources and their extraction costs; waste repository siting; land management and land-use planning; and general education. The value of geologic map information in public and private decision-making (such as for the siting of landfills and highways) has repeatedly been described anecdotally, and has been demonstrated in benefit-cost analyses to reduce uncertainty and, by extension, potential costs.

The geologic mapper strives to understand the composition and structure of geologic materials at the Earth's surface and at depth, and to depict observations and interpretations on maps using symbols and colors (**Fig. 1**). Within the past 10 to 20 years, geographic information system (GIS) technology has begun to change some aspects of geologic mapping by providing software tools that permit the geometry and characteristics of rock bodies and other geologic features (such as faults) to be electronically stored, displayed, queried, and analyzed in conjunction with a seemingly infinite variety of other data types. For example, GIS can be used to spatially compare possible pollutant sources (such as oil wells) with nearby streams and geologic units that serve as ground-water supplies. In addition, GIS can be used to compare the position of a proposed road with the surrounding geology to identify areas of high excavation costs or unstable slopes. These comparisons have always been possible, but GIS greatly facilitates the analysis and, as a result, offers geologists the opportunity to provide information in map form that is easily interpreted and used by the nongeologist.

The public has come to expect near-instantaneous delivery of relevant, understandable information via the Internet, which in turn has begun to affect the methods used in geologic mapping, as well as the nature of the product. Geologists are rapidly incorporating GIS and information technology (IT) techniques into the production and dissemination of geologic maps, as described below.

Field methods. Geologists traditionally record in field notebooks their observations, including sketches, measurements (for example, the angle of tilted strata), and narratives. The validity of these observations remains; however, digital photographs now frequently supplement sketches, and instrumentation enhances measurement accuracy (for example, more precise locations are possible with a global positioning system instrument than with simple reference to position in relation to topographic and cultural features). Increasingly, field narratives are written and organized on a notebook computer or a personal digital assistant (PDA).

The development of field systems for recording and managing information has accelerated in the past 10 years. It is expected that the most useful systems will be widely adopted, thereby helping to standardize the techniques and formats for field observations. Increasingly, geologists not only are recording field observations but also are making field interpretations of the position of features such as geologic contacts and faults (that is, "drawing lines") using rudimentary GIS software on PDAs or notebook computers.

Geologic descriptors. When recording observations, geologists use descriptive terms and rock names that are in common use or unique to an area. These terms are then synthesized and rewritten into formal map unit descriptions that are published with the map. With the advent of GIS and the ease with which digital maps can be obtained and queried, geologists are recognizing the importance of a well-defined, standard terminology in order to help users, at a desktop computer or on the Internet, query simultaneously two maps made by different geologists. National and international efforts are now underway to define standard classifications for geologic information such as rock composition and texture. This standard language then can be used in the field or in the office to organize and interpret field observations. It is recognized that because of the many geologic terrains and geologic mapping agencies, and because of long historical usage of certain terms, multiple standard classifications will be necessary to accommodate regional variations in terminology. However, this system should function well, provided each classification is well defined and can be correlated with other classifications to ensure ease of translation from one format to another.

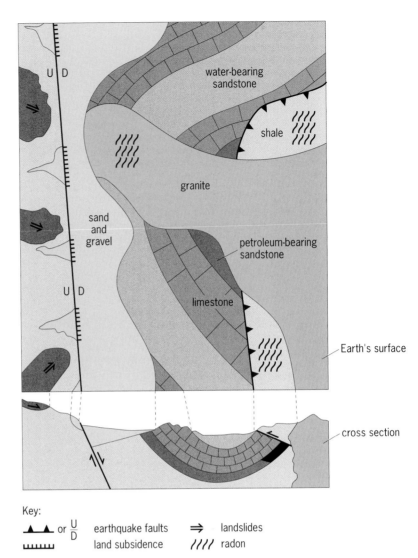

Key:

▲▲ or U/D   earthquake faults   ⇒   landslides
⊥⊥⊥⊥⊥   land subsidence   //// radon

Fig. 1. Graphic representation of some of the typical information contained in a general-purpose geologic map and some of the many applications of that information. A geologic map can be used to identify geologic hazards, locate natural resources, and facilitate land use planning. (*After R. L. Bernknopf et al., 1993*)

Geologic map databases. The main goal in many geological surveys no longer is to create a single geologic map but to create a database from which many types of geologic and engineering geology maps can be derived. This requires a database design or "data model" that is sufficiently robust to manage complex geologic concepts such as three-dimensional (spatial) and temporal relations among map units, faults, and other features (**Fig. 2**). This is especially challenging because the software that manages these databases is not static but continues to evolve, thereby requiring an adaptable database design. Also, to permit the exchange of databases among agencies and users, either a common database design or a common interchange format is required. These design efforts are underway in North America and other parts of the world. As a consequence, geologists now are reevaluating how information is managed in the field, what kind of information is gathered and for what purpose, and the extent to

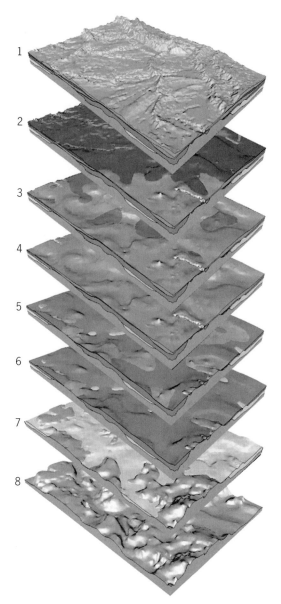

Fig. 2. Three-dimensional stack of glacial geologic layers in east-central Illinois. Layer 1 is land surface; layer 8 is the underlying bedrock. The light-colored unit in layer 7 is a sand and gravel aquifer filling a bedrock valley, and buried by low-permeability glacial till. (*After D. R. Soller et al., Three-Dimensional Geologic Maps of Quaternary Sediments in East-Central Illinois: USGS Geol. Investig. Ser. Map I-2669, scale 1:500,000, 1999*)

which map information, once it has been created, can be updated to include new observations and interpretations. Traditionally, once the geologist delivered a manuscript map to the cartographer, the job was finished. Map databases are meant to be maintained, and so a map can now be more accurately considered a progress report that can be updated. A critical aspect of the map database is the ability to manage the information and to preserve its integrity, for example, by migrating it to a new data format or structure or to a new standard terminology.

Cartography. GIS and graphic design software have radically changed the techniques by which map information is published. Traditional film (peelcoat)

technologies have been supplanted by digitally prepared negatives for offset printing of paper maps and by plot-on-demand approaches, which generate one map at a time upon request by users. Plot-on-demand is useful for producing customized, single-attribute maps (for example, showing only the shear strength of the geologic materials or the susceptibility of a groundwater supply to contamination). Digital cartographic techniques are evolving and becoming capable of producing sophisticated map layouts and products. It is anticipated that digital cartographers will be continually challenged to develop new techniques as software evolves and as geologists and users demand more complex and informative products.

Dissemination of maps and databases. In addition to the conventional venues for obtaining maps (such as bookstores and sales offices), GIS and the Internet have made it possible to reach and educate new and potential users of geologic maps. Maps and databases now are available on the Internet in a variety of formats. Some formats (for example, Portable Document Format, or PDF) are designed for the visual display of a map and do not require specialized software, whereas other formats (for example, ArcInfo™) are intended for use by professionals who have access to the appropriate software for the detailed analysis of a map database. To provide the public with access to such analyses without requiring them to purchase the software, numerous agencies are experimenting with software that permits users to view maps and to submit queries and view results within a Web browser. This technology holds great promise and should provide useful information after certain conditions are met. These include (1) standardized database structure, terminology, and interchange format; (2) increased availability of engineering geology information; (3) increased Internet bandwidth; and (4) evolutionary advances in the software that serves map data.

New advances in mapping and preparation of map products, made possible by advances in GIS and IT, have not altered the basic science of geology but offer new techniques for organizing, maintaining, and analyzing map data, and, potentially, for increasing its use by the public and by scientists.

For background information *see* CARTOGRAPHY; DATABASE MANAGEMENT SYSTEM; ENGINEERING GEOLOGY; GEOGRAPHIC INFORMATION SYSTEMS; GEOLOGY; INFORMATION TECHNOLOGY; MAP REPRODUCTION in the McGraw-Hill Encyclopedia of Science & Technology.   David R. Soller

Bibliography. R. L. Bernknopf et al., *Societal Value of Geologic Maps*, USGS Circ. 1111, 1993; S. B. Bhagwat and V. C. Ipe, *Economic Benefits of Detailed Geologic Mapping to Kentucky*, Illinois State Geol. Surv. Spec. Rep. 3, 2000; R. R. Compton, *Geology in the Field*, Wiley, New York, 1985; D. R. Soller, *Digital Mapping Techniques '02—Workshop Proceedings*, USGS Open-file Rep. 02-370; D. J. Varnes, *The Logic of Geological Maps, with Reference to Their Interpretation and Use for Engineering Purposes*, USGS Prof. Pap. 837, 1974.

of an exhaustable one (petroleum). Variation in the chemical and physical properties of the resulting plastics is achieved by controlling the separation and chemical modification steps used on the corn feedstock. The process steps have been designed to be as green as possible, including incentives for the corn suppliers to use sustainable agricultural practices.

Education. Education plays a key role in green chemistry. How people learn chemistry affects how they will solve real-world problems. If chemical education fails to incorporate green principles, the next generation of chemists will continue with waste-generating practices. **Figure 2** compares a conventional chemical laboratory experiment with one designed to incorporate green principles. The new, greener procedure preserves all of the pedagogical elements of the older laboratory, but also teaches the powerful lesson that green chemistry can create knowledge and compounds without harming the natural world or creating toxic waste.

For background information *see* CATALYSIS; CHLORINE; CONSERVATION OF RESOURCES; HYDROGEN PEROXIDE; LIGNIN; ORGANIC SYNTHESIS; PAPER; SOLVENT; SUPERCRITICAL FLUID; SUPRAMOLECULAR CHEMISTRY in the McGraw-Hill Encyclopedia of Science & Technology.   S. W. Gordon-Wylie

Bibliography. P. T. Anastas and J. C. Warner, *Green Chemistry: Theory and Practice*, Oxford University Press, 1998; M. C. Cann and M. E. Connelly, *Real World Cases in Green Chemistry*, American Chemical Society, 2000; S. S. Gupta et al., Rapid total destruction of chlorophenol pollutants by activated hydrogen peroxide, *Science*, 296:326–328, 2002; K. Kamata et al., Efficient epoxidation of olefins with ≥99% selectivity and use of hydrogen peroxide, *Science*, 300:964–966, 2003; R. J. Lempert et al., *Next Generation Environmental Technologies: Benefits and Barriers*, RAND Science and Technology Policy Institute, 2003; B. M. Trost, The atom economy: A search for synthetic efficiency, *Science*, 254:1471–1477, 1991.

Heat (biology)

Animals and plants release heat in order to stay alive. This heat, or enthalpy, is due to exothermic chemical reactions such as Eq. (1) between a sugar and

$$C_6H_{12}O_6 + 6O_2 = 6CO_2 + 6H_2O$$
$$\Delta_r H^0 = -2810 \text{ kJ mol}^{-1} \quad (1)$$

oxygen ($\Delta_r H^0$ is the standard enthalpy of reaction, where ΔH^0 is the change in enthalpy under standard conditions and the subscript r is the symbol for chemical reaction). The combustion of other organic molecules such as amino acids, fatty acids, and carbohydrates produces exothermic amounts of heat and is involved in maintaining the body temperatures of living organisms. Because of these combustion processes, the heat produced by organisms can often be directly related to their oxygen uptake.

Typical power P (heat production rate) per gram of some organisms, animals, plants, and social systems

Living species	Temperature, °C	$P(W \cdot g^{-1})$
Organism		
Bacterium (*Escherichia coli*)	37	0.52
Yeast (*Saccharomyces cerevisiae*)	30	0.21
Insect		
Bee	30	$180 \cdot 10^{-3}$
Animal		
Lizard	25	$2 \cdot 10^{-3}$
Mouse	38	$8.3 \cdot 10^{-3}$
Dog	39	$1.8 \cdot 10^{-3}$
Human	37	$1.4 \cdot 10^{-3}$
Cow	38	$0.9 \cdot 10^{-3}$
Plant		
Germinating corn	24	$0.7 \cdot 10^{-3}$
Voodoo lily (*Sauromatum guttatum*)	25	$100 \cdot 10^{-3}$
Social system		
Bumblebees (*Bombus lapidarius*)	25	$300 \cdot 10^{-3}$
Anthill	20	$0.23 \cdot 10^{-3}$
Hornet colony	32	$23 \cdot 10^{-3}$

The rate at which heat is produced in living organisms varies considerably according to the type of animal or plant. This rate is a function not only of the type of chemical reactions taking place but also of the rate of the reactions and the metabolism involved. Heat production in living organisms is an index of biological activity. Typical heat production rates of some living systems are listed in the **table**.

Temperature effect. The effect of temperature on living organisms is a reflection of the chemical processes involved in maintaining life. As with many chemical reactions, the rates of biological processes are usually governed by the Arrhenius law [Eq. (2)],

$$\ln k = \ln A - E_a/RT \quad (2)$$

where k is the rate constant for the reaction, A is a constant, E_a is the activation energy (the energy added to make the reaction go), R is the universal gas constant, and T is the Kelvin temperature. For many biological processes, the rate of a reaction decreases by twofold for every 10°C (18°F) drop in temperature; this is why refrigeration inhibits decomposition processes (biochemical reactions involving bacteria) in foodstuffs.

An interesting pair of experiments that illustrates the application of Arrhenius' law to biological processes involves the chirping of cicada beetles and the flashing of fireflies. If the log of the frequency of the chirping of a cicada and the log of the frequency of the flashes of a firefly's light are plotted against the reciprocal temperature (K), the activation energy can be calculated [Eq. (2)]. Both plots yield an activation energy of 51 kJ mol^{-1}, which is equivalent to a twofold rise in the rate for a 10°C (18°F) rise in temperature. These experiments were done in an environment which changed in temperature from 16°C (60.8°F) to 30°C (86°F).

Measurements of heat flow. The heat change q of a mass of material m of specific heat s undergoing a temperature change ΔT is given by Eq. (3). Today,

$$q = ms\Delta T \qquad (3)$$

temperature changes of less than $10^{-5}\,°\text{C}$ can readily be detected with thermopiles (arrays of thermocouples), with thermistors, or by platinum resistance thermometers. Heat changes are measured in calorimeters, and heat dissipation rates of less than $10^{-6}\,\text{W}$ are today easily measurable, especially in flow calorimeters. The latter type involves a flow of two reactant solutions, meeting in the heating chamber that contains an electrical heater and a thermopile. *See* MICROCALORIMETRY; TEMPERATURE MEASUREMENT.

The heats involved in living systems are complex and involve consecutive and parallel reactions of proteins, DNA, and enzymes, as well as oxidative and anaerobic processes. In spite of this complexity, the speed, precision, and reproducibility of calorimetry make it an ideal tool for monitoring biological and cell processes. One example is the monitoring of cultures of animal cells used in the production of target proteins. The optimum conditions can be found by counting the viable cells as a function of types of nutrients and rates of feeding. However, a more convenient way of doing this is to replace the counting process with the monitoring of the heat released as a function of time (that is, the heat flux) using a flow microcalorimeter. Calorimetry of biological processes is a nonspecific physicochemical method that integrates over many simultaneously produced heats and is best at monitoring changes in a steady-state situation of living systems. This form of calorimetry has been used to investigate many forms of living material, including microorganisms, plants, individual insects, social collections of insects, and humans.

Microorganisms. Bacteria and fungi can be grown in aqueous solutions, with generation times as short as 30 min. They can easily be investigated using sensitive flow microcalorimeters, since reproduction involves the production of heat, which can easily be measured using these instruments. The brewing, baking, fermentation, and compost industries, as well as the fields of environmental pollution and ecophysiology, have benefited greatly from experiments used to determine the optimum conditions for nutrients, temperature, and pressure of carbon dioxide (CO_2) and oxygen (O_2). Recently, the optimum conditions for the production of antibiotics and enzymes have been investigated.

Plants. The temperature of a plant is usually the same as the ambient temperature and is a reflection of its large surface area–to–volume ratio. The metabolic rate, reflected by the heat production rate, typically is much lower in plants than in animals. However, at certain stages of a plant's life cycle, such as the germination stage, there are large bursts of energy (see table). For some flowering plants, the flowers themselves produce bursts of energy in order to volatilize odorous compounds as a way of attracting potential pollinators. The voodoo lily is one such thermogenic plant and, at a certain stage, produces heat at a rate of 0.1 W—more than the heat produced by many animals.

Experiments have shown that the metabolic rate of tissue material of some trees is a good indicator of their long-term growth rate. The tissue is placed in a photomicrocalorimeter (sample exposed to a light source) and fed with nutrients, and its metabolic rate (that is, heat production rate) is monitored. As a result, screening tissue-cultured material can save time and effort in the commercial cultivation of trees.

Insects. Insects have a high heat dissipation rate (see table) and, as a result, can be easily monitored by calorimetry. Insect colonies (bees, hornets, and ants) have also been monitored for heat dissipation. Application of the alarm pheromone, 2-methyl-3-buten-2-ol, to a hornets nest was shown to produce a twofold increase in heat dissipation.

The change of heat production due to human-made pesticides and other pollutants on insect colonies has shed light on the physiology of such interactions and has been useful in determining optimum levels of chemical control. Much the same idea has been used to understand and monitor the influence of heavy metals on crustaceans. For example, 1 microgram of cadmium chloride has been shown to reduce the heat production rate of fresh-water snails to irreversibly low levels in just a few minutes.

Animals. Heat flow data can be considered as just another parameter in the study of life processes. It has the advantage of being easy to measure. Animal and human whole-body calorimeter experiments have been used to monitor heat flow under stress conditions of temperature, pressure, oxygen levels, and food deprivation. The term "whole-body calorimeter" often refers to instrumentation used to measure oxygen uptake or carbon dioxide release. These measurements are then used to calculate heat production from known enthalpy data of biochemical reactions. It is an indirect method. Direct calorimetry on human subjects is possible and has been investigated. The indirect method has been used to determine the optimum environmental temperature for newborn babies and the optimum feeding formula for babies. Surgical treatment of liver tumors involves cutting off the blood supply, which can lead to permanent damage. But how long can the liver withstand such treatment? Calorimetric experiments on pieces of rat liver tissue injured by ischemia (decreased blood flow) showed that the limit was 90 min, after which permanent damage occurred.

Experiments on human cells are ideally suited to microcalorimetry. The effect on cell metabolism by trace elements (which are considered hazardous) and by drugs and antibiotics has been tested on blood cells, including erythrocytes, lymphocytes, and granulocytes. For instance, results show that the antibiotic gentamicin increased erythrocyte and decreased granulocyte heat production rates.

Outlook. The influence of temperature on the rates of biochemical processes gives some insight into their mechanisms. Measuring the heat production rates of biological systems is useful for understanding and monitoring complex life processes, particularly in assessing the optimal conditions and the effect of stress on these processes. Among other things, such measurements may prove to be of great importance in predicting growth patterns under global warming conditions.

For background information *see* BIOCALORIMETRY; BIOLOGICAL OXIDATION; BIOPHYSICS; CALORIMETRY; CHEMICAL THERMODYNAMICS; ENTHALPY; HEAT CAPACITY; METABOLISM; TEMPERATURE MEASUREMENT; THERMISTOR; THERMOCOUPLE; THERMODYNAMIC PRINCIPLES; THERMOMETER; THERMOREGULATION in the McGraw-Hill Encyclopedia of Science & Technology.   Trevor M. Letcher

Bibliography. M. J. Dauncey, Whole body calorimetry in man and animals, *Thermochimica Acta*, 193:1–40, 1991; R. B. Kemp et al. (eds.), *Chemical Thermodynamics for the 21st Century*, Blackwell Science, Oxford, 1999; K. J. Laidler, Unconventional applications of the Arrhenius law, *J. Chem. Educ.*, 49:343–345, 1972; M. Monti, Calorimetric studies of lymphocytes, *Thermochimica Acta*, 193:281–285, 1991; I. Wadso, Trends in isothermal microcalorimetry, *Chem. Soc. Rev.*, 26:79–104, 1997.

Higgs boson

Over the past decade, experiments in laboratories around the world have elevated the electroweak theory to a law of nature that holds over a remarkable range of distances, from the subnuclear (10^{-19} m) to the galactic (10^{20} m). The electroweak theory offers a new conception of two of nature's fundamental interactions, ascribing them to a common underlying symmetry principle. It joins electromagnetism with the weak interactions—which govern radioactivity and the energy output of the Sun—in a single quantum field theory. Dozens of measurements have tested and confirmed the agreement between theory and experiment at the level of one part in a thousand.

Weak interactions and electromagnetism. Weak interactions and electromagnetism are linked through symmetry, but their manifestations in the everyday world are very different. The influence of electromagnetism extends to unlimited distances, while the influence of weak interactions is confined to dimensions smaller than an atomic nucleus, less than about 10^{-17} m. The range of an interaction in quantum theory is inversely proportional to the mass of the force particle, or messenger. Accordingly, the photon, the force carrier of electromagnetism, is massless, whereas the W and Z particles that carry the weak forces are heavyweights, with masses nearly a hundred times that of the proton, the nucleus of hydrogen.

The many successes of the electroweak theory and its expansive range of applicability are most impressive, but the present understanding of the theory is incomplete. We have not yet learned what differentiates electromagnetism from the weak interactions—what endows the W and Z with great masses while leaving the photon massless. It is commonplace in physics to find that the symmetries observed in the laws of nature are not manifest in the consequences of those laws. A liquid is a disordered collection of atoms or molecules held together by electromagnetism that looks the same from every vantage point, reflecting the fact that the laws of electromagnetism are indifferent to direction. A crystal is an ordered collection of the same atoms or molecules, held together by the same electromagnetism, but it does not look the same from every vantage point. Instead, a crystal displays ranks, files, and columns that single out preferred directions. The rotational symmetry of electromagnetism is hidden in the regular structure of a crystal.

Higgs condensation. The central challenge in particle physics today is to understand what hides the symmetry between the weak and electromagnetic interactions. The simplest guess goes back to theoretical work by Peter Higgs and others in the 1960s. According to this picture, the diversity of the everyday world is the consequence of a vacuum state that prefers not a particular direction in space but a particular particle composition. The vacuum of quantum field theory—and of the world—is a chaotic confusion of many kinds of virtual particles winking in and out of existence. The laws of physics do not change if we interchange the identities of different particles. At high temperatures, as in the disordered liquid, the symmetry of the laws of physics is manifest in the egalitarian throng of particles. But at low temperatures, as in the ordered crystal, one kind of particle is preferred, and condenses out in great numbers, hiding the symmetry.

The condensate of Higgs particles can be pictured as a viscous medium that selectively resists the motion of other particles through it. In the electroweak theory, the drag on the W and Z particles caused by their interactions with the Higgs condensate gives masses to the weak-force particles. Interactions with the condensate could also give rise to the masses of the constituent particles—quarks and leptons—that compose ordinary matter. Today's version of the electroweak theory shows how this could come about, but does not predict what the mass of the electron, the top quark, or other particles should be.

Experimental searches. If it were possible to heat up the vacuum enough, we could see the symmetry restored: all particles would become massless and interchangeable. For the present, it is beyond human means to heat even a small volume of space to the energy of 1 TeV (10^{12} electronvolts)—the temperature of 10^{16} kelvins. However, there is hope of succeeding on a smaller scale: it is possible to excite the Higgs condensate and see how it responds. The minimal quantum-world response is the Higgs boson: an electrically neutral particle with zero spin.

Although the Higgs boson has not been observed, experiments have begun to offer some evidence for its presence, beyond the shadowy role as giver of mass for which it was conceived. Experiments carried out over the past decade, using the Large Electron-Positron Collider (LEP) at CERN (European Laboratory for Particle Physics) and other instruments, are sensitive not just to the structure of the electroweak theory but also to quantum corrections. Long before the top quark was discovered at Fermilab in 1995, precision measurements detected its virtual quantum effects and anticipated its extraordinarily great mass. The influence of the Higgs particle is subtle, but increasingly incisive experiments strongly hint that a Higgs boson with mass less than about 200 GeV (1 GeV = 10^9 electronvolts) is required by precision measurements.

The electroweak theory predicts that the Higgs boson will have a mass, but it does not predict what that mass should be. (Consistency arguments require that it weigh less than 1 TeV.) Nevertheless, enough is known about the Higgs boson's properties—about how it could be produced and how it would transform into lighter particles or decay (as do all elementary particles except stable particles like the electron)—to guide the search.

LEP searches. The most telling searches have been carried out in experiments investigating electron-positron annihilations at energies approaching 210 GeV at LEP. The quarry of the LEP experimenters was a Higgs boson produced in association with the *Z*. A Higgs boson accessible at LEP would decay into a *b* (bottom) quark and a *b* antiquark, and have a total width (inverse lifetime) smaller than 10 MeV. The **illustration** shows how the decay pattern of a standard-model Higgs boson depends on its mass.

In the final year of LEP experimentation (before the machine was shut down in November 2000 to make way for construction of the Large Hadron [LHC]), accelerator scientists and experimenters made heroic efforts to stretch their discovery reach to the highest Higgs mass possible. At the highest energies explored, a few tantalizing four-jet events showed the earmarks of Higgs + *Z* production. A statistical analysis shows a slight preference for a Higgs-boson mass near 116 GeV, but not enough to establish a discovery. Time will tell whether the LEP events were Higgs bosons or merely background events. The LEP observations do place an experimental lower bound on the Higgs-boson mass: the standard-model Higgs particle must weigh more than about 114 GeV.

Prospects. In a few years, experiments at Fermilab's Tevatron, where 1-TeV protons collide with 1-TeV antiprotons, may be able to extend the search, looking for Higgs + *W* or Higgs + *Z* production. The best hope for the discovery of the Higgs boson lies in experiments at the LHC at CERN, where 7-TeV beams of protons will collide head-on. When the LHC is commissioned around the year 2007, it will make possible the study of collisions among quarks at energies approaching 1 TeV. A thorough exploration of the 1-TeV energy scale should determine the mechanism by which the electroweak symmetry is hidden and reveal what makes the *W* and *Z* particles massive.

Further questions. Once a particle that seems to be the Higgs boson is discovered, a host of new questions will come into play. Is there one Higgs boson or several? Is the Higgs boson the giver of mass not only to the weak *W* and *Z* bosons but also to the quarks and leptons? How does the Higgs boson interact with itself? What determines the mass of the Higgs boson? To explore the new land of the Higgs boson in more ways than the LHC can do alone, physicists are planning a TeV linear electron-positron collider.

Origins of mass. While the Higgs boson may explain the masses of the *W* and *Z*, and perhaps the masses of the quarks and leptons, it is not accurate to say that the Higgs boson is responsible for all mass. Visible matter is made up largely of protons and neutrons, and their masses reflect the energy stored up in the strong force that binds three quarks together in a small space. The origins of mass suffuse everything in the world around us, for mass determines the range of forces and sets the scale of all the structures we see in nature.

Extensions of electroweak theory. The present inability to predict the mass of the Higgs boson is one of the reasons many physicists believe that the standard electroweak theory needs to be extended. The search for the Higgs boson is also a search for extensions that make the electroweak theory more coherent and more predictive. Supersymmetry, which entails several Higgs bosons, associates new particles with all the known quarks, leptons, and force particles. Dynamical symmetry breaking interprets the Higgs boson as a composite particle whose properties we may hope to compute once its constituents and their interactions are understood. These ideas and more will be put to the test as experiments explore the 1-TeV scale.

For background information *see* ELECTROWEAK INTERACTION; HIGGS BOSON; INTERMEDIATE VECTOR BOSON; LEPTON; PARTICLE ACCELERATOR; QUANTUM FIELD THEORY; QUARKS; STANDARD MODEL; SUPERSYMMETRY; SYMMETRY BREAKING; SYMMETRY LAWS (PHYSICS); WEAK NUCLEAR INTERACTIONS in the McGraw-Hill Encyclopedia of Science & Technology.

Chris Quigg

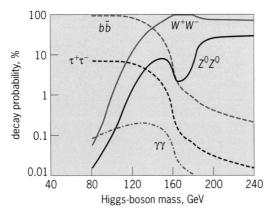

Dependence of the decay probabilities of a standard-model Higgs boson on its mass.

Bibliography. F. Close, *Lucifer's Legacy: The Meaning of Asymmetry*, Oxford University Press, 2001; M. Kado and C. Tully, The searches for Higgs bosons at LEP, *Annu. Rev. Nucl. Part. Sci.*, 52:65–113, 2002; G. 't Hooft, Nobel Lecture: A confrontation with infinity, *Rev. Mod. Phys.*, 72:333–339, 2000; M. J. G. Veltman, Nobel Lecture: From weak interactions to gravitation, *Rev. Mod. Phys.*, 72:341–349, 2000; F. Wilczek, Masses and molasses, *New Scientist*, 162(2181):32–37, April 10, 1999.

Human evolution in Eurasia

Humans evolved in Africa and were confined to that continent for much of their early history. Just what triggered the first dispersals from Africa is poorly understood, but it is clear that people began to colonize Eurasia shortly after 2 million years ago (Ma). These migrants were probably representatives of *Homo erectus* (sometimes called *H. ergaster*). This extinct species was the first to attain a body size (stature) equal to that of modern humans, although the brain remained low in volume and thus smaller relative to body size than our own. Archeological traces left by these people have been recovered from the Jordan Valley in Israel and the Georgian Caucasus. Such early occupations were likely transitory and did not result in permanent settlements. However, groups of *H. erectus* were able to travel across southern Asia to the Far East, where they established themselves in both Java and China by about 1.6 Ma.

The first penetration westward into Europe apparently came much later. There are indications that humans were moving into the Mediterranean region after 1.0 Ma, but the initial populating of Europe north of the major mountain barriers is documented only after 500 thousand years ago (Ka). The biological identity of the first Europeans is unclear, but it is agreed that these people differ from *H. erectus*. Many of the ancient fossils are presently assigned to the species *H. heidelbergensis* (named originally from a mandible found at Mauer, near Heidelberg in Germany). This species seems to have been more advanced behaviorally than its predecessor, and there is evidence that *H. heidelbergensis* was able to make relatively sophisticated stone tools and wooden spears for hunting.

'Ubeidiya and Dmanisi. Some of the oldest signs of occupation in Eurasia come from 'Ubeidiya in the central Jordan Valley, where there are large collections of stone artifacts. As determined from the accompanying animal fossils, the lower levels in the site are about 1.4 Ma. There are few human remains from 'Ubeidiya, but the tools attest to the presence of incoming bands of *H. erectus*. This species lived also at Dmanisi in the Caucasus. Here excavations have uncovered stone chopping tools (pebbles from which several flakes have been removed) and fossils of extinct animals deposited above a lava flow that was erupted about 1.8 Ma. A human mandible was discovered at Dmanisi in 1991. Paleoanthropologists agree that this specimen has the characteristics of *H. erectus*. More recently, two crania, another lower jaw and a complete skull, have been recovered from the same sedimentary levels. These finds show that the Dmanisi population included relatively small individuals, with brains at the low end of the size range expected for *H. erectus*. The cranial vault and especially the facial skeleton exhibit some resemblances to earlier *H. habilis* from East Africa, but the fossils are best grouped with *H. erectus* as known from Africa and the Far East. The evidence from Dmanisi demonstrates that, contrary to earlier belief, hominins (the taxonomic group to which modern humans and all extinct members of the human lineage belong) with small brains and equipped with the simplest of tool kits were able to move out of Africa relatively quickly after the emergence of the genus *Homo*.

Homo erectus in the Far East. *Homo erectus* was first discovered in Java late in the nineteenth century. The Dutch paleontologist Eugene Dubois was very lucky to find a skullcap and later a femur at Trinil in 1891–1892. In the Sangiran area closeby, there are volcanic deposits from which radiometric dates can be obtained, and results suggest the presence of *H. erectus* by 1.6 Ma. At Sangiran, the fossils include several partial crania, mandibles, and teeth. With their thickened brows, low braincases with midline keeling, and heavy chinless jaws, the Java hominins are much like the people that inhabited China. *Homo erectus* was unearthed at Zhoukoudian in China in the 1920s, and discoveries of bones and stone artifacts at other sites now show that the species was living in this region before 1.5 Ma. The upper layers in the cave at Zhoukoudian are (only) about 300 Ka in age, suggesting that *H. erectus* may have survived in the Far East until late in the Middle Pleistocene. This pattern contrasts with that in the West, where the species seems to disappear at a relatively early date.

Species that followed Homo erectus. In western Eurasia and in Africa, it is likely that the *H. erectus* lineage split to produce at least one new form of hominin. This speciation event probably occurred at or before the beginning of the Middle Pleistocene (780 Ka). In Africa, fossils from this time period are clearly different from *H. erectus* in brain size, width of the frontal bone, proportions of the occipital region, and anatomy of the underside of the skull. Where it is preserved, the face is still heavily constructed, but the brows, nasal profile, and bony palate more closely resemble the condition seen in later humans. In many instances, the fossils are found with stone tools that are more sophisticated than the choppers and relatively crude handaxes associated with *H. erectus*. From Bodo in Ethiopia to Elandsfontein in South Africa, a shift toward the manufacture of thinner, more finely flaked handaxes is documented in the Middle Pleistocene, and this change in behavior may be linked to speciation.

The African skulls from Bodo, Broken Hill (now Kabwe) in Zambia, Elandsfontein, and other localities are remarkably similar to specimens found in Europe. The cranium from Petralona in Greece resembles the African individuals in the massive construction of its upper face and cheek, facial projection, configuration of the brows, and many aspects of vault shape. Much the same conclusion applies to the less complete cranium from Arago Cave in France, dated to about 450 Ka. The skulls from the Sima de los Huesos in Spain (ca. 400 Ka) also share many features with their African contemporaries, although they possess a few traits that suggest evolutionary ties to later Neandertals. If the anatomical similarities of the African and European populations are emphasized, all of the hominins may be referred to a single species. One name that is available for this group is *H. heidelbergensis*. Like the Africans, the people at several of the European localities made finely flaked hand axes as well as other tools. The level of technical skill that is evident (and perhaps also the cooperation of individuals that is implied) surpasses that of earlier *H. erectus*.

TD6 assemblage from Spain. Additional information bearing on the first peopling of Europe is accumulating from the site of Gran Dolina in northern Spain. A layer deep in these deposits designated by the excavators as TD6 has produced stone core-choppers and flakes, animal bones, and human remains dated slightly in excess of 780 Ka. The specimens include a juvenile face, an adult cheek bone, part of a subadult frontal including some of the brow, and a piece of the cranial base on which most of the joint for the mandible is preserved. There is also a broken lower jaw with teeth, along with vertebrae, ribs, and bones of the hand and foot.

Juan Luis Arsuaga and colleagues argue that the TD6 people are not *H. erectus*. Morphology of the hollowed cheek region, orientation of the nasal aperture, features of the hard palate, a wide frontal, and the apparently modern mandibular joint all suggest that the Gran Dolina fossils are different from *H. erectus* and more like later humans. Also, there can be little doubt that this population is distinct from the Neandertals. The hollowed cheek (bearing a canine fossa) points toward this conclusion, and neither in the juvenile nor in the adult faces is there much sign of the specialized Neandertal condition (in which the surface of the cheek is inflated and angled forward toward the nose). The partial mandible is generalized in its morphology, whereas the teeth resemble those of European and African Middle Pleistocene hominins.

Given this complex of traits, the Gran Dolina material may represent a new species. The name *H. antecessor* was proposed by José Maria Bermúdez de Castro and coworkers in 1997. However, the number of fossils is still quite small, and several of the craniodental remains are fragmentary and/or subadult. A fair question is whether there is presently enough evidence to separate the TD6 assemblage from other penecontemporary (from about the same time) fossils already on record from about the same time. In particular, it must be asked whether the Gran Dolina bones and teeth differ from those of other early Europeans such as Mauer and Arago. Much attention has been focused on the development of a canine fossa in the midface, but this feature is variable in its expression in other populations, and its significance is unclear. In the mandible, teeth, and postcranial bones there are few traits that differentiate the Gran Dolina hominins from Europeans of the Middle Pleistocene.

Conclusion. People differing from *H. erectus* appeared in southern Europe before 780 Ka and in Africa at about the same time. One reading of the record suggests that these groups represent a single lineage that can be called *H. heidelbergensis* (see **illus.**). Later in the Middle Pleistocene, some populations dispersed northward within Europe, where they were subject to long episodes of extreme cold. During glacial advances occurring over several hundred thousand years, they adapted to the harsh conditions and evolved the specialized craniofacial characters and body build of the Neandertals. In this same interval, other representatives of *H. heidelbergensis* in Africa were becoming more like modern humans. Fossils from Irhoud in Morocco, the Omo region of Ethiopia, and Laetoli in Tanzania document this evolutionary progression. Also, there is a wealth of information from molecular comparisons to confirm that *H. sapiens* evolved in Africa.

Alternatively, it can be argued that *H. antecessor* is the ancestor to all later humans. Soon after its

Alternative evolutionary trees showing the relationships among *Homo erectus*, Middle Pleistocene hominins, and modern humans. Bars depict the time range (in millions of years) estimated for each species. Broken lines indicate likely links of ancestors with descendants. (*a*) Scenario showing *H. heidelbergensis* to be descended from *H. erectus*. After spreading widely across Africa and western Eurasia at the beginning of the Middle Pleistocene, *H. heidelbergensis* was the antecedent to both Neandertals in Europe and recent humans. (*b*) Alternative interpretation in which *H. antecessor* is recognized as the descendant of *H. erectus*. In turn, *H. antecessor* evolved into European *H. heidelbergensis*, and this species gave rise (only) to the Neandertals. African *H. rhodesiensis* was ancestral to *H. sapiens*.

first appearance in Spain, *H. antecessor* must have given rise to *H. heidelbergensis*. In this scenario, the *heidelbergensis* lineage was confined exclusively to Europe, where its members gradually acquired the morphology of the Neandertals. Also, *H. antecessor* is presumed to have evolved an African offshoot, represented at localities such as Bodo, Kabwe, and Elandsfontein. Although these Middle Pleistocene hominins are acknowledged as similar to their European contemporaries, they are not assigned to *H. heidelbergensis*. Instead, the African fossils are lumped in a separate species called *H. rhodesiensis*. Whether this taxonomic view can be accepted will depend largely on the outcome of a search for more fossils in the TD6 level at Gran Dolina. In any case, there is little doubt that mid to later Pleistocene populations from Africa should be recognized as ancestral to *H. sapiens*.

For background information *see* EARLY MODERN HUMANS; FOSSIL HUMANS; NEANDERTALS; PALEOLITHIC; PHYSICAL ANTHROPOLOGY; PREHISTORIC TECHNOLOGY in the McGraw-Hill Encyclopedia of Science & Technology.   G. Philip Rightmire

Bibliography. M. Balter, In search of the first Europeans, *Science*, 291:1722–1725, 2001; E. Delson et al., *Encyclopedia of Human Evolution and Prehistory*, 2d ed., Garland Publishing, New York, 2000; A. Gibbons, A new face for human ancestors, *Science*, 276:1331–1333, 1997; R. G. Klein, *The Human Career: Human Biological and Cultural Origins*, University of Chicago Press, 1999.

Human Genome Project

On April 14, 2003, the International Human Genome Sequencing Consortium, led in the United States by the National Human Genome Research Institute and the Department of Energy, announced the successful completion of the Human Genome Project (HGP) more than 2 years ahead of schedule. The DNA sequence produced by the HGP covered about 99% of the human genome's gene-containing regions and was sequenced to an accuracy of greater than 99.99%. In addition, to help researchers better understand the meaning of the human genetic instruction book, the project had taken on a wide range of other goals, from sequencing the genomes of model organisms to developing new technologies to study whole genomes. All of those goals were met or surpassed (see **table**).

Scientific strategy. The scientific strategies that were employed evolved over the years, but the basic concept for the project that was initially proposed by the National Research Council (NRC) of the U.S. National Academy of Sciences in the mid-1980s turned out to be effective. Because the human genome is so big (human DNA consists of about 3 billion nucleotides connected end to end in a linear array; **Fig. 1**), it was necessary to break the task down into manageable chunks. The first step was to create a genetic map of the whole genome. Such a map is generated by analyzing how frequently markers are inherited together in families. Markers that are in proximity are inherited together more frequently than those that are far apart. The resulting map shows a series of signposts along the DNA (**Fig. 2**).

The second step was to create a physical map of the DNA. The DNA was broken into small pieces that could be cloned and replicated in bacteria to generate enough material for study. The pieces were then fitted together by studying how they overlapped with each other. The genetic markers, as well as other types of landmarks, were used for this purpose. Eventually the whole genome was covered with overlapping clones.

The final step was to sequence the clones and piece together the entire genome sequence. Before

Human genome project goals			
Area	Goal	Achieved	Date
Genetic map	2- to 5-centimorgan-resolution map (600–1500 markers)	1-centimorgan-resolution map (3000 markers)	September 1994
Functional analysis	Develop genomic-scale technologies	High-throughput oligonucleotide synthesis DNA microarrays Whole-genome knockouts (yeast) Scale-up of two-hybrid system for protein-protein interaction	1994 1996 1999 2002
Physical map	30,000 sequence-tagged sites	52,000 sequence-tagged sites	October 1998
Capacity and cost of finished sequence	Sequence 500 megabases/year at < $0.25 per finished base	Sequence >1400 megabases/year at < $0.09 per finished base	November 2002
Human sequence variation	100,000 mapped human single-nucleotide polymorphisms	3.7 million mapped human single-nucleotide polymorphisms	February 2003
Gene identification	Full-length human cDNAs	15,000 full-length human cDNAs	March 2003
DNA sequence	95% of gene-containing part of human sequence finished to 99.99% accuracy	99% of gene-containing part of human sequence finished to >99.99% accuracy	April 2003
Model organisms	Complete genome sequences of *Escherichia coli, Saccharomyces cerevisiae, Caenorhabditis elegans, Drosophila melanogaster*	Finished genome sequences of *Escherichia coli, Saccharomyces cerevisiae, Caenorhabditis elegans, Drosophila melanogaster,* plus whole-genome drafts of several others, including mouse, rat, *Caenorhabditis briggsae, Drosophila pseudoobscura*	April 1996–April 2003

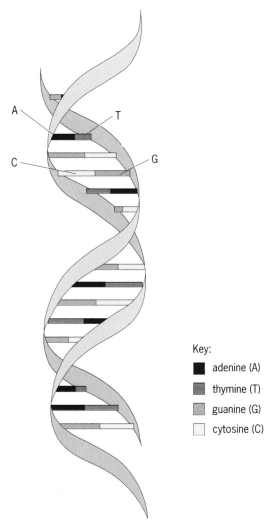

Key:

■ adenine (A)

■ thymine (T)

■ guanine (G)

□ cytosine (C)

Fig. 1. Structure of DNA. DNA is a long thin molecule made up of chemical units called nucleotide bases. The four bases are adenine, thymine, guanine, and cytosine (A, T, G, and C). They are arranged in two parallel strands that are complementary to each other. An A always appears opposite a T, and a G always appears opposite a C. Sequencing the DNA involves determining the order of the bases in the molecule.

this could be done, much research and development was needed to make sequencing more efficient and less costly. With a major investment in research, the technology did improve substantially and the costs came down dramatically. The improvements included streamlined methods, the use of robots to handle large numbers of samples simultaneously, more accurate and faster sequencing machines, and better computer programs for tracking and assembling the data.

Model organisms. An important element of the overall strategy was to include the study of model organisms in the HGP. There were two reasons for this: (1) Simpler organisms provide good practice material. (2) Comparisons between model organisms and humans yield very valuable scientific information. All life forms have much in common, including some of their DNA sequences. Without the information that

can be gained from the study of model organisms, it would be very difficult to know what the human DNA sequence means.

The HGP initially adopted five model organisms to have their DNA sequenced: the bacterium *Escherichia coli*, the yeast *Saccharomyces cerevisiae*, the roundworm *Caenorhabditis elegans*, the fruitfly *Drosophila melanogaster*, and the laboratory mouse *Mus musculus*. Subsequently, the HGP supported the sequencing of additional organisms, including the rat *Rattus norvegicus*, the roundworm *Caenorhabditis briggsae*, and the fruitfly *Drosophila pseudoobscura*.

In December 2002, a study comparing the human genome sequence with the mouse genome sequence found that 99% of human genes have a counterpart in mice, even though the two mammals diverged from a common ancestor more than 75 million years ago. In addition to sharing most gene functions with humans, the mouse is a valuable model for biomedical research because it can be selectively bred and other experiments can be conducted that are not possible on humans.

Project evaluation. Originally, the proponents of the HGP suggested that a 15-year time frame was appropriate for obtaining the human genome sequence. In order to make sure that this schedule was followed, 5-year plans were developed to set intermediate milestones. Remarkably, all the milestones were met, many times well before the planned deadlines. The sequence of yeast was released in April 1996, the sequence of *E. coli* was published in September 1997, *C. elegans* was completed in December 1998, the *Drosophila* sequence was released in early 2000, a first draft of the rat sequence was released in November 2002, and the mouse sequence was published in December 2002.

A first draft of the entire human sequence was released in June 2000 with publications describing it following in February 2001. A high-quality, comprehensive version of the human genome sequence was completed in April 2003, more than 2 years quicker than the 15-year period and coinciding with the 50th anniversary of the discovery of the double-helical structure of DNA. From the outset, sequence information from the HGP was immediately and freely distributed to scientists around the world, with no restrictions on its use or redistribution. In parallel with the publicly funded effort, a private United States company called Celera also generated a draft of the human sequence, which was completed about the same time as the HGP's first draft, though not made freely accessible. The Celera strategy used a different approach. Instead of mapping the cloned DNA pieces first and sequencing them later, random pieces were sequenced directly and subsequently correlated with maps. Some aspects of this approach were incorporated into the strategy for sequencing additional organisms, such as the mouse and the rat.

Findings. How many genes are there is probably the most common question regarding the human genome. Although a definitive count of human genes must await further experimental and computational analysis, the HGP's initial analysis of the human genome in 2001 led to an estimate that humans have only about 30,000 genes. That was quite a surprise because previous estimates were 80,000 to 100,000 genes. The lower gene count does not necessarily mean that the human genome is less complex, because many genes can produce more than one protein by alternate splicing of their exons (protein-encoding regions of the gene) during translation into the constituents of proteins.

Another fascinating feature of the human genome sequence, as well as the genome sequences of other mammals, is the distribution of genes on the chromosomes. It turns out that mammalian chromosomes have areas with many genes in proximity to one another, but these regions are interspersed with vast expanses of DNA devoid of protein-coding genes. This uneven distribution of genes stands in marked contrast to the more uniform distribution of genes throughout the genomes of many other organisms, such as the roundworm and the fruitfly.

In their analysis, HGP scientists also found that about 50% of the DNA in the human genome is made of repetitive sequences, which is much greater than the percentage of repetitive DNA found in the roundworm (7%), the fruitfly (3%), and a variety of other nonmammalian species. These repeated sequence elements provide a rich record of clues to the evolutionary past of humans. It is possible to date various types of these repeats to the times when they appeared in the evolutionary process and to follow their fates in different regions of the genome and in different species. Because repeated sequences are more likely to misalign when DNA is being copied or repaired, areas with a high concentration of repeats are more prone to mutation than other regions of the genome. Consequently, the various types of repeats have helped to reshape the genome in a multitude of ways by rearranging and modifying existing genes, as well as by creating entirely new genes.

Such repeats and the relatively predictable patterns of DNA shrinkage and growth associated with them are also being used as "DNA dating" tools to explore evolution in a rapidly emerging area of research known as phylogenetics. The pace of evolutionary processes, sometimes referred to as a molecular clock, appears to vary among different types of repeats. However, researchers in some instances have been able to select a specific type of repeat, examine its prevalence and length among the genomes of a wide range of species, and then use that genetic information to construct a family, or phylogenetic, tree of various species showing when in evolution that repetitive element was "born" and how it "moved" among species during the course of evolution.

For example, two types of repeats, called DNA transposons and long terminal repeat (LTR) transposons, present in the genome of the mouse are

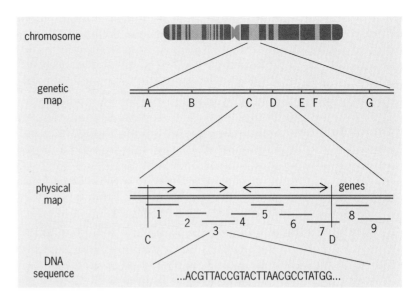

Fig. 2. Steps in analyzing a genome. (1) Markers are placed on the chromosomes by genetic mapping, that is, observing how the markers are inherited in families. (2) A physical map is created from overlapping cloned pieces of the DNA. (3) The sequence of each piece is determined, and the sequences are lined up by computer until a continuous sequence along the whole chromosome is obtained. Steps 2 and 3 can be reversed or done in parallel. As the pieces are sequenced, the sequences at the overlapping ends can be used to help order the pieces. If the sequencing is done before the pieces are mapped, the process is called whole-genome shotgun sequencing.

much rarer in the human genome. Consequently, a phylogenetic tree built using data on DNA transposons and LTR transposons would branch in a manner that reflects how the prevalence of these repeat elements has changed since rodents and primates diverged from a common ancestor an estimated 75 million years ago.

Future research. The availability of large amounts of DNA sequence information as well as other genomic resources has profoundly affected biomedical research and thinking. Information about whole genomes that was previously gained one gene at a time has now been obtained. With the complete sets of genes of organisms available, how genes are turned on and off and how genes interact with each other can be studied. The possibilities are endlessly exciting, and the demand for more sequence is increasing all the time. Many more organisms will be sequenced. But having the sequence is not enough. What the different genes do and how they affect human health must also be learned.

Among the most important large-scale opportunities for scientists in the genome era are the development of innovative technologies and the creation of informational frameworks to make the results of genomic research applicable to individual health. Although all humans are 99.9% identical in their genetic makeup, the 0.1% variation in DNA sequences is thought to hold key clues to individual differences in susceptibility to disease.

The International HapMap Project was launched in October 2002 with the goal of building a catalog of human genetic variations and determining how these variations are organized into neighborhoods, or haplotypes, along the human chromosomes. The HapMap will serve as a tool of researchers trying to discover the common genetic variations associated with complex diseases, as well as variations responsible for differences in drug response.

In March 2003, scientists also set out to develop efficient ways of identifying and precisely locating all of the functional elements contained in the human DNA sequence. The ultimate goal of the Encyclopedia of DNA Elements (ENCODE) project is to create a reference work that will help scientists mine and fully utilize the human sequence, gain a deeper understanding of human biology, and develop new strategies for the prevention and treatment of disease. In the first phase of ENCODE, researchers will work cooperatively to develop high-throughput methods for rigorously analyzing a defined set of DNA target regions constituting approximately 1% of the human genome. It is hoped this pilot project will pave the way for scaling up this effort to characterize efficiently and effectively all of the protein-coding genes, nonprotein coding genes, and other sequence-based functional elements contained in human DNA. As has been the case with the public effort to sequence the human genome, data from the ENCODE project will be collected and stored in a database that will be freely available to the entire scientific community.

Other frontiers include the establishment of publicly available libraries of chemical compounds for use by basic scientists in their efforts to chart biological pathways, as well as genomic initiatives to understand the life processes of single-cell organisms, or microbes, with the ultimate aim of using the capabilities of these organisms to address needs relating to health, energy, and the environment.

Many challenges lie ahead. Getting the full sequence of human DNA is a dramatic achievement. But what was viewed as an end point has turned out to be just a beginning: a new era of genomic biology lies ahead.

For background information *see* DEOXYRIBONU-CLEIC ACID (DNA); GENE; GENETIC CODE; GENETIC ENGINEERING; HUMAN GENETICS; MOLECULAR BIOLOGY; NUCLEIC ACID in the McGraw-Hill Encyclopedia of Science & Technology.   F. Collins; E. Jordan

Bibliography. M. D. Adams et al., The genome sequence of *Drosophila melanogaster, Science*, 287: 2185–2195, 2000 (also related articles about the completion of the DNA sequence, *Science*, 287: 2181–2184 and 2196–2224, 2000); *C. elegans* Sequencing Consortium, Genome sequence of the nematode *C. elegans*: A platform for investigating biology, *Science*, 282:2012–2018, 1998 (also related articles, *Science*, 282:2018–2046, 1998); F. S. Collins, Medical and societal consequences of the Human Genome Project, *N. Engl. J. Med.*, 341:28–37, 1999; F. S. Collins et al., A vision for the future of genomics research: A blueprint for the genomic era, *Nature*, 422:835–847, 2003; F. S. Collins, M. Morgan, and A. Patrinos, The Human Genome Project: Lessons from large-scale biology, *Science*, 300:286–290, 2003; A. Goffeau, R. Aert, and M. L. Agostini-Carbone, The Yeast Genome Directory, *Nature*, supplement to vol. 387, May 29, 1997; E. D. Green, The Human Genome Project, in *Metabolic and Molecular Bases of Inherited Disease*, 8th ed., 2001; International Human Genome Sequencing Consortium, Initial sequencing and analysis of the human genome, *Nature*, 409:860–921, 2001; Mouse Genome Sequencing Consortium, Initial sequencing and comparative analysis of the mouse genome, *Nature*, 420:520–562, 2002; National Research Council, *Mapping and Sequencing the Human Genome*, 1988; J. C. Venter et al., The sequence of the human genome, *Science*, 291:1304–1351, 2001.

Human/parasite coevolution

Human evolution is fundamentally a story of the spread of people and their habitat-altering mode of life throughout the world. A consistent cost of this evolution—specifically of first hunting, then agriculture and domestication—has been the accumulation of a large number of parasitic diseases.

Origins of parasite associations. Some human parasites are the direct result of domestication. Numerous species of parasites infect humans and at least one other species related to humans only by domestication and civilization. For example, the sheep liver fluke (*Fasciola hepatica*) infects humans when they eat aquatic vegetation contaminated with the metacercariae (the infective juvenile stage of flukes). *Fasciola hepatica* does not infect any other primates, so we believe that the association between humans and *F. hepatica* originated with the domestication of sheep, bringing humans into more frequent contact with the same aquatic vegetation as sheep.

Other parasites, acquired by humans long before the advent of agriculture and domestication, are the result of humans competing with other species to take control of the surroundings. For example, humans are known to host three species of tapeworms, *Taenia solium*, *T. saginata*, and *T. asiatica*, acquired by eating raw or poorly cooked pork (*T. solium*) or beef (*T. saginata* and *T. asiatica*). Studies comparing phylogenetic relationships among *Taenia* spp. with those of their hosts indicate that humans acquired *Taenia* on two separate occasions, correlating with the shift from scavenging to predation (hunting) in humans more than a million years ago in Africa. The closest relative, or sister species, of *T. solium* is *T. hyaenae*, occurring in hyaenas and African hunting dogs. *Taenia saginata* and *T. asiatica*, themselves sister species, are most closely related to *T. simbae*, which inhabits lions. This suggests that humans acquired *Taenia* as a by-product of competition with carnivores, the original hosts for *Taenia* spp. (**Fig. 1**). As humans secured their prey farther from competitors through domestication, they isolated strains of *Taenia* that eventually became distinct species.

Yet other parasites inhabiting humans are legacies of our primate origins. **Figures 2** and **3** show the phylogenetic trees of two groups of roundworm parasites, pinworms of the genus *Enterobius* and their closest relatives and hookworms of the genus *Oesophagostomum*, subgenus *Conoweberia*. Comparison of the two trees reveals host relationships that are similar, but not identical, to each other and to the phylogeny of the great apes and their relatives (**Fig. 4**).

Association complexity. Parasites and their hosts can become associated with each other in two fundamental ways, by descent or by host switching. In cases of association by descent, sister species of parasites inhabit the same host species, the same species of parasite inhabits hosts that are sister species, or sister species of parasites inhabit sister species of hosts. Each of these evolutionary outcomes produces a distinctive phylogenetic signature (**Fig. 5**). In cases of association by host switching, the same species of parasite inhabits hosts that are not sister species (and may not even be closely related), or sister species of parasites occur in hosts that are not sister species. Each of these evolutionary outcomes also produces a distinctive phylogenetic signature (**Fig. 6**). Some hosts do not harbor parasites that are harbored by other close relatives. Is that because they were never associated with the parasite group, or because the member of the parasite group with which they were associated is now extinct? Each of these possibilities also produces a distinctive phylogenetic signature (**Fig. 7**).

Brooks parsimony analysis (BPA) is a method for assessing complex histories of coevolution. All members of a parasite group, and their ancestral relationships, can be denoted numerically (as in Figs. 2 and 3); each parasite species has a distinctive code, so each host for each parasite also has a distinctive code. All codes for all parasites and their hosts can be combined into a single host cladogram (a branching diagram, such as those in the illustrations), which is the most parsimonious depiction of the evolutionary histories of the included parasite groups—that is, which assumes the simplest evolutionary explanations for all observed data—in the context of their hosts. Different evolutionary options, as depicted in Figs. 5–7, are inferred from the analysis a posteriori. Each case of association by descent among the parasites is consistent with one set of host relationships, producing a general pattern. Each case of association by host switching produces a unique host relationship. The general pattern of host relationships among the parasites should generally be consistent with the phylogeny of the hosts.

Roundworm and primate phylogeny. Figure 8 is the host cladogram produced by Brooks parsimony analysis of the pinworms and hookworms

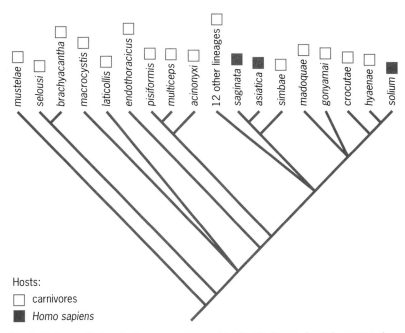

Fig. 1. Phylogenetic tree for tapeworms in the genus *Taenia*. Boxes above the names of species refer to host groups, either carnivores or humans. Humans appear to have acquired *Taenia* species twice, *T. solium* and the ancestor of *T. saginata* + *T. asiatica*, which appears to have evolved into different species in humans. (*Modified from E. P. Hoberg et al., Out of Africa: Origins of the Taenia tapeworms in humans, Proc. Roy. Soc. London, Ser. B, 268:781–787, 2000*)

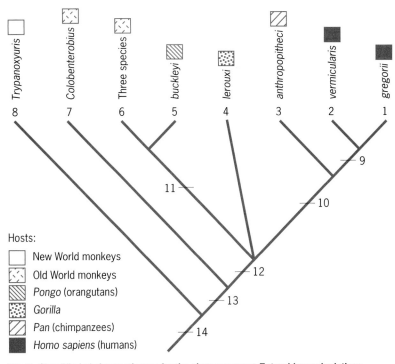

Fig. 2. Simplified phylogenetic tree for the pinworm genus *Enterobius* and relatives, coded for coevolutionary analysis. Each branch of the phylogenetic tree has its own number; therefore, each parasite species has a unique numerical code embodying its evolutionary history that can be used in assessing coevolutionary associations with its host. 6 = the monophyletic group *E. brevicauda* + *E. bipapillatus* + *E. macaci*. (*Modified from D. R. Brooks and D. A. McLennan, Extending phylogenetic studies of coevolution: Secondary Brooks parsimony analysis, parasites, and the great apes, Cladistics, 19:104–119, 2003*)

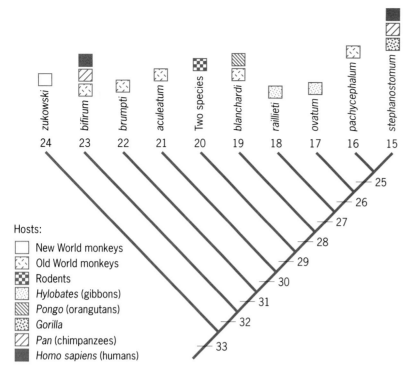

Fig. 3. Phylogenetic tree for the hookworm genus *Oesophagostomum (Conoweberia)*. The numerical codes on this phylogenetic tree differ from those on the *Enterobius* tree (Fig. 2), so the information from both parasite groups can be integrated into a single analysis. 20 = *O. xeri* and *O. susannae*. (Modified from D. R. Brooks and D. A. McLennan, Extending phylogenetic studies of coevolution: Secondary Brooks parsimony analysis, parasites, and the great apes, Cladistics, 19:104–119, 2003)

(Figs. 2 and 3). The analysis suggests that both groups of roundworm parasites have been jointly associated at least since the common ancestor of the anthropoid primates (ancestors 14 and 33 in Fig. 8). There is a

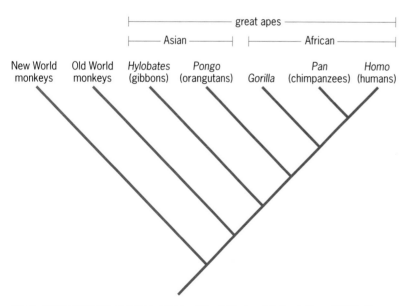

Fig. 4. Phylogenetic tree for the great apes (Hominidae) and their relatives. Old World monkeys include the Cercopithecinae (baboons, macaques) and the Colobinae (colobines, langurs). New World monkeys include howler monkeys, spider monkeys, and tamarins. (Modified from M. Goodman et al., Toward a phylogenetic classification of primates based on DNA evidence complemented by fossil evidence, Mol. Phylogen. Evol., 9:585–598, 1998)

strong historical backbone of cospeciation among these organisms; bold lines in Fig. 8 indicate the general pattern of association by descent among the parasite groups, which is also consistent with the phylogeny of great apes (Fig. 4). In terms of *Enterobius*, there was a cospeciation event occurring with (1) the *Pan* + *Homo* bifurcation (ancestor 10 giving rise to species 3 and ancestor 9), (2) the *Gorilla* + sister-group split (ancestor 12 giving rise to species 4 and ancestor 10), and (3) at the *Pongo* + sister-group split (presence of species 5 in *Pongo*). (Note that "ancestors" are inferred from phylogenetic studies but, unlike "species," are not specifically characterized and named.) There are three apparent cases of *Oesophagostomum* lineage duplication—the simultaneous evolution of more than one parasite species in a single host—in the ancestor of the Old World monkeys + great apes (denoted by the joint occurrence of ancestors 30, 31, and 32). No lineage duplications are postulated for *Enterobius*.

There have been six episodes of host switching in *Oesophagostomum*. The most obvious of these is the occurrence of the hookworms *O. xeri* + *O. susannae* (both given species number, 20; Fig. 2) in South African rodents, which represents an episode of speciation by host switching from apes to rodents. This type of exchange is not unique; pinworms in the genus *Trypanoxyuris* include species that evolved as a result of switching from New World monkeys to squirrels. The remaining host switches, *O. pachycephalum* (species 16) in cercopithecines, and *O. ovatum* (species 17) and *O. raillieti* (species 18) in *Hylobates*, represent episodes of speciation by host switching among primates, specifically from one or more species ancestral to the African great apes. The only pinworm to switch primate hosts was species 6 (which actually refers to the ancestor of a monophyletic group of three species inhabiting Old World monkeys; Fig. 2). Ancestor 6 appears to have colonized Old World monkeys from the *Pongo* lineage, where it subsequently diversified in the new host group.

The occurrence of *O. bifurcum* (species 23) in *Pan* and *Homo* represents postspeciation dispersal by the parasite from its original Old World monkey hosts. The host ranges of *O. stephanostomum* (species 15) and *O. bifurcum* were not gained by repeated host switching but by the parasite species not speciating when its host did. *Homo* and *Pan* are sister groups, so it is most parsimonious to postulate that *O. bifurcum* colonized the common ancestor of *Pan* + *Homo*, a single host switch. Once in that ancestor, *O. bifurcum* did not speciate, even though its new host did. Likewise, the presence of *O. stephanostomum* in *Gorilla*, *Pan*, and *Homo* is best explained as the result of parasite speciation in the common ancestor of the African great apes, without subsequent speciation. Finally, *O. brumpti* (species 22) and *O. aculeatum* (species 21), inhabiting various Old World monkeys, have apparently not participated in the exchange between cercopithecids

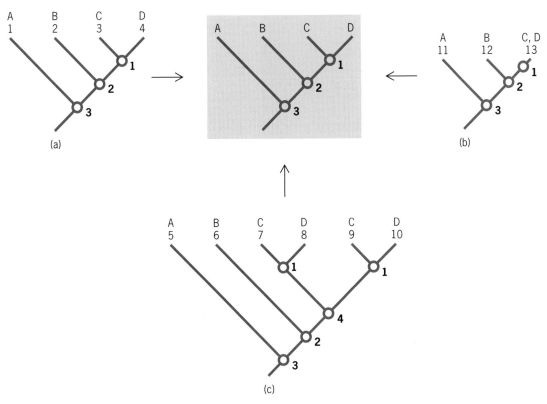

Fig. 5. Association by descent. The phylogenetic tree inside the box represents the phylogeny of the host (letter codes); open circles represent speciation events. Nodes numbered 3 refer to the speciation event giving rise to host A + the ancestor of B + C + D; those numbered 2 refer to the speciation event giving rise to host B + the ancestor of C + D; and those numbered 1 refer to the speciation event giving rise to hosts C + D. (a) Parasites (number codes) may speciate each time the host group speciates, (b) may speciate less often than the host group, or (c) more often than the host group. The node numbered 4 in c refers to a parasite speciation event in the ancestor of hosts C + D, without speciation in the hosts.

(baboons and macaques, OWM₁) and anthropoids (chimpanzees, Pan₂, and humans, Homo₂).

Migration impacts. Ancient movements of primates throughout the tropical regions of the Old World account for many host switches. *Oesophagostomum bifurcum* and *O. brumpti* occur in both Africa and Asia, while *O. aculeatum* is endemic to Asia, possibly suggesting dispersal from Africa to Asia (although *O. xeri* + *O. susannae* are endemic to South Africa). *Oesophagostomum blanchardi* (species 19), *O. raillieti* (species 18), and *O. ovatum* (species 17) are Asian endemics, while *O. pachycephalum* (species 16) and *O. stephanostomum* (species 15) are African endemics, suggesting dispersal from Asia to Africa by the common ancestor(s) of, not surprisingly, the African great apes (*Gorilla*, *Pan*, and *Homo*). Thus, by the time early hominids moved out of the African forest into the African savannah, *O. bifurcum* already existed, so our ancestors' movements into the same habitat as Old World monkey hosts of *O. bifurcum* resulted, in part, in the addition of *O. bifurcum* to the repertoire of human parasites. Phylogenetic data indicate that *Homo* moved to Asia, leaving *O. stephanostomum* behind in Africa (humans in Asia do not have this hookworm). The occurrence of *O. bifurcum* in humans in Africa and Asia is not surprising; the host cladogram suggests that *O. bifurcum*

occurred in Old World monkeys on both continents long before *Homo* arrived in Asia. This period may also have seen the differentiation of the pinworms *E. vermicularis* (species 2) and *E. gregorii* (species 1). If true, this suggests that African and Asian humans were isolated sufficiently for their pinworms to differentiate into sister species, but not sufficiently for humans themselves to speciate. As a result, subsequent migrations of humans throughout the world have reinforced the identity of *H. sapiens* as a single species, and have moved *E. enterobius* and *E. gregorii* about to such an extent that they coexist today in most parts of the world. It is possible that this enforced commingling of sister species of very recent origin may result in the fusion of *E. vermicularis* and *E. gregorii*; already, there is disagreement among taxonomists as to whether or not there are consistently recognizable differences between the species. Similarly, *T. saginata* and *T. asiatica*, sister species occurring in Africa and Asia, respectively, diverged from each other between 750,000 and 1,700,000 years ago, perhaps at the same time *E. vermicularis* and *E. gregorii* were differentiating. Subsequent movements of humans have dispersed *T. saginata* widely, but not *T. asiatica*; thus, these two species are better differentiated than the two species of *Enterobius*.

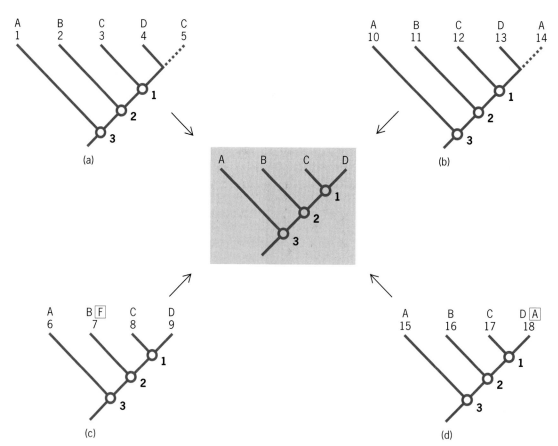

Fig. 6. Association by host switching. The phylogenetic tree inside the box represents the host phylogeny; open circles represent speciation events; dashed branches represent potential parasite speciations. (a) Parasites may switch to a host outside the host group inhabited by other members of its own group and speciate, (b) may switch to a host inside the host group inhabited by other members of its own group and speciate, (c) may switch to a host outside the host group inhabited by other members of its own group and not speciate, or (d) may switch to a host inside the host group inhabited by other members of its own group and not speciate.

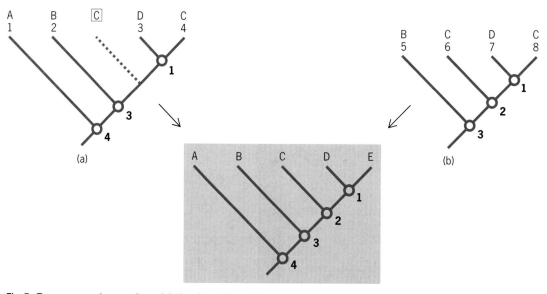

Fig. 7. Two reasons why parasites might be absent from a host species. The phylogenetic tree inside the box represents the host phylogeny; open circles represent speciation events. (a) The parasite clade (species 1–4) became associated with the host clade (species A–E) before the evolution of host C; therefore, the lack of a member of the parasite group in C is most parsimoniously explained as the result of secondary loss (extinction) of the parasite group from that member of the host group. (b) The parasite clade (species 5–8) became associated with the host clade (species A–E) after the evolution of A; therefore, the lack of a member of the parasite group in A is most parsimoniously explained as a case in which the parasite group never inhabited that member of the host group.

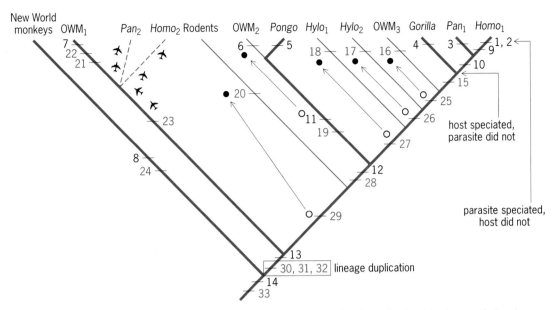

Fig. 8. Brooks parsimony analysis–generated host cladogram for pinworms (black numbers) and hookworms (colored numbers) of great apes. Bold lines = association by descent between the primates and their pinworms and hookworms. Dashed lines and airplanes = host switching not leading to parasite speciation; open circle + solid circle = host switching leading to speciation; numbers accompanying slash marks = species codes (from Figs. 2 and 3). Box = lineage duplication; OWM = Old World monkeys; *Hylo* = *Hylobates*. Each subscript represents a different origin of an association between the host group and at least one of the parasite groups. (*Modified from D. R. Brooks and D. A. McLennan, Extending phylogenetic studies of coevolution: Secondary Brooks parsimony analysis, parasites, and the great apes, Cladistics, 19:104–119, 2003*)

For background information *see* MEDICAL PARASITOLOGY; PHYLOGENY; PRIMATES in the McGraw-Hill Encyclopedia of Science & Technology.

Daniel R. Brooks

Bibliography. D. R. Brooks and D. A. McLennan, Extending phylogenetic studies of coevolution: Secondary Brooks parsimony analysis, parasites, and the great apes, *Cladistics*, 2003; D. R. Brooks and D. A. McLennan, *The Nature of the Organism: An Evolutionary Voyage of Discovery*, University of Chicago Press, 2002; D. R. Brooks and D. A. McLennan, *Parascript: Parasites and the Language of Evolution*, Smithsonian Institution Press, 1993; E. P. Hoberg et al., Out of Africa: Origins of the *Taenia* tapeworms in humans, *Proc. Roy. Soc. London, Ser. B*, 2000.

Humanoid robots

In the foreseeable future, robots will be used in everyday life. Some robots are already commercially available for vacuuming, for mowing, and as "pets." Currently these robots do not have a humanoid form, but eventually they will.

In science fiction, robots take on a humanoid form, with a head, a torso, two arms, and two legs. In industrial applications, however, robots are designed to execute one specific task, and their bodies rarely, if at all, resemble humans. Car assembly robots are typically fixed to the floor, with a simple arm and gripper rather than a hand with four fingers and a thumb. The robots sent to Mars were wheeled rovers with cameras rather than two-legged robots with heads.

The main reason that these robots do not have a humanoid form is that the human body and mind are extremely complex and difficult to replicate. The human hand, for instance, can move in more than 22 independent ways (degrees of freedom), providing the ability to execute many different tasks. These degrees of freedom are controlled by over 35 muscles in the palm and the forearm. The hand is also remarkably strong. In a robot, the motors that would allow the same degrees of freedom and strength would not fit into the same space as a hand. Moreover, a battery to drive all of these motors would be too large and heavy to be included as a part of the humanoid body.

Even with these challenges, it is important to build robots with humanoid features. A single humanoid robot will be able to accomplish multiple human tasks currently performed by multiple task-specific robots. Humans will also be able to teach humanoid robots daily tasks intuitively. If a physically disabled person wants to use the robot as a teleoperator, a humanoid robot would be able to mimic the movements of the human operator. In addition, many believe that humanoid robots should interact with humans in a natural way, requiring the robots to be able to speak, gesture, and express and respond to emotion.

Developments. Within the last decade, huge strides have been made in humanoid robot development. In 1996, the Honda Motor Co., Ltd., announced P2, the first humanoid robot capable of autonomously walking and climbing stairs. P2 was 71.7 in. (182 cm) tall and 463 lb (210 kg). Asimo (**Fig. 1**), the most recent humanoid robot from Honda, has the same capabilities, yet is only 47.3 in. (120 cm) tall and weighs

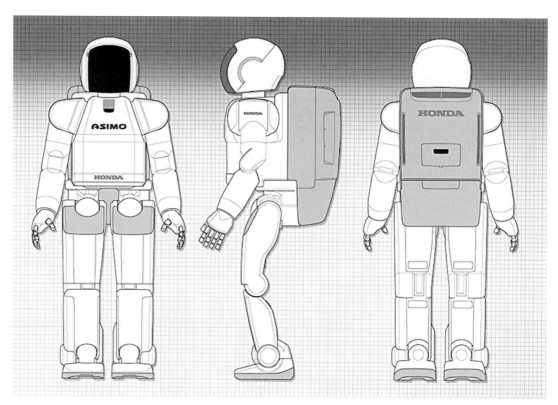

Fig. 1. Honda's Asimo is intended for assisting people. (*Honda Motors Co., Ltd.*)

95 lb (43 kg). Even though its degrees of freedom are limited (the hand has only one joint and the whole body has a total of 26 degrees of freedom) and the battery lasts for only 30 min, it has an undeniable charm. Honda's stated goal with Asimo is to assist people, for example, the elderly.

More recently, Sony has developed a much smaller humanoid robot, SDR-4X. It is under 23 in. (58 cm) tall and weighs only 14 lb (6.3 kg). SDR-4X can even recognize a few different people using face recognition and their emotion from facial expressions. Sony aims to use this robot for entertainment just like its predecessor, Aibo, the robotic dog.

NASA has constructed Robonaut to perform tasks in the harsh environments of space, which requires human dexterity, force, and size. To achieve such dexterity, Robonaut's hand has a total of 12 degrees of freedom. The robot is designed to perform a number of tasks, ranging from assisting an astronaut during extravehicular activity to performing geological surveys autonomously on distant planets.

Anatomical hand. In one of the most extreme cases of building a humanoid robot (or part of one), researchers at Carnegie Mellon University have constructed an anatomically based robotic hand using materials and structures very similar to those found in human hands (**Fig. 2**). Cadaver bones were used to construct a three-dimensional model, and the bones were machined with accurate size and shape. The joints in the human hand were mimicked by encapsulating viscous materials with a ligamentlike structure. Muscles were constructed with a combination

of elastic materials and mechanical actuators. Using this hand, the biomechanical functions of complicated weblike tendon structures have been discovered, new sets of muscle control strategies have been investigated, and new hand surgical procedures have been proposed.

While this robotic hand has shown how crucial it is to construct an anatomically correct body part for a future humanoid robot, it has also highlighted some difficulties in doing so. In particular, the anatomical structures in the robotic hand suffer from considerable wear and tear on the ligament and tendon materials. Every day the human body repairs such damage by replacing damaged cells with new ones.

Fig. 2. Anatomically based robotic hand constructed at Carnegie Mellon University that will become a humanoid hand one day. (© *2002 Yoky Matsuoka*)

Future humanoid robots must include self-repairing capabilities, but until this technology becomes available, researchers at Carnegie Mellon University are constructing humanoid body parts that preserve the most important functional features.

Cognition and adaptation. While robots may take on a physically humanoid form and move like humans, cognitively they are not very human. A humanoid robot must have the ability to adaptively learn to interact with humans and the world. At the Massachusetts Institute of Technology (MIT), a humanoid robot, Cog, is being developed to investigate human adaptation ability and cognition. It is hypothesized that the exploratory behavior of infants is a vehicle of cognitive development. If we could understand more about how infants learn, we might be able to understand human cognition better. In 1995, Cog was used in the first attempt to investigate the human manipulation-learning strategy using a physical system. Learning manipulation requires a nonlinear controller to respond in a nonlinear system that contains a significant amount of sensory input and noise. Starting from hardwired reflex behavior that babies are born with (curling the fingers when the palm is touched and opening the hand when the fingers are pinched), the humanoid's hand was exposed to a variety of objects. Using three neural networks that were intimately connected, Cog has adaptively learned to grasp different objects with different grasping strategies. Its manipulation ability was similar to that of infants. Originally the humanoid grasped the objects almost as an accident when its reflexes were activated. With trial and error, the humanoid has learned to grasp objects without dropping them, and its grasping strategies have become more sophisticated and tuned to object shape, size, and hardness. Researchers at MIT are currently investigating how these adaptation experiences may help develop cognition.

Outlook. Although the human body and mind are complex and replicating them is extremely difficult, rapid developments are being made in humanoid robots in terms of their physical bodies and adaptive cognitive abilities. Humanoid robots will become more and more commonplace in the research world and soon will begin to appear as commercial products. In all likelihood, the first applications will be for entertainment and as toys. Later, improved functionality will lead to other applications, such as use as tour guides, and eventually to use as household helpers and companions.

For background information *see* ADAPTIVE CONTROL; ARTIFICIAL INTELLIGENCE; COGNITION; DEGREE OF FREEDOM (MECHANICS); INTELLIGENT MACHINE; NEURAL NETWORK; ROBOTICS in the McGraw-Hill Encyclopedia of Science & Technology.

Yoky Matsuoka

Bibliography. R. O. Ambrose et al., Robonaut: NASA's space humanoid, *IEEE Intellig. Sys.*, 15(4): 57–63, July/August 2000; Y. Matsuoka, Primitive manipulation learning with connectionism, *Advances in Neural Information Processing Systems*, 8:889–895, MIT Press, 1996; D. D. Wilkinson, M. Vande Weghe, and Y. Matsuoka, An extensor mechanism for an anatomical robotic hand, *Proceedings of the 2003 IEEE International Conference on Robotics & Automation*, 2003.

Hyperspectral remote sensing (mapping)

Hyperspectral imagery offers new and unique opportunities for landscape mapping. It is an offshoot of multispectral imagery, which measures reflected light in 1 to approximately 10 bands and has been an important tool in landscape mapping since deployment of *Landsat* in the 1970s. Both methods measure bandwidths in the infrared, visible, and ultraviolet regions of light. Compared to multispectral sensors used for most remote mapping prior to the late-1990s, however, hyperspectral sensors record narrower bandwidths and more bandwidths of reflected visible and shortwave infrared light. For example, the AVIRIS (Airborne Visible/Infra-Red Imaging Spectrometer) hyperspectral sensor records 224 different bandwidths of light between 400 and 2500 nanometers. In contrast, the *Landsat* multispectral Thematic Mapper records reflected light in only six bands. The narrower bandwidths of hyperspectral sensors make possible the detection of specific features that have strong reflectance in a narrow band of light (such as certain minerals). The larger number of bands enables detection of subtle differences between similar features, because the cumulative differences become large when summed over many bands.

Landscape mapping with hyperspectral sensors requires five steps, which include acquisition of imagery; image preprocessing to remove distortion and separate signal from noise; collection of ground data that are used to "train" the image and evaluate mapping accuracy; application of algorithms that identify different types of features on the image; and postmapping accuracy assessment. These same general steps are used for multispectral mapping, but hyperspectral mapping requires special considerations and offers unique capabilities compared to multispectral mapping. The following sections outline unique components of hyperspectral mapping and provides working examples from stream habitat mapping projects.

Image acquisition. All hyperspectral imagery prior to 2000 was acquired from aircraft. Although the *Hyperion* sensor launched aboard the *Earth Observing 1* (EO-1) satellite in late 2000 now provides a satellite-based alternative, it is likely that much hyperspectral acquisition will continue to be from aircraft because of the need for higher spatial resolution and better signal-to-noise ratios than *Hyperion* can provide. For example, mapping of small but critical stream habitat features like riffles and pools requires the use of hyperspectral sensors mounted on aircraft to achieve 1–2-m (3.3–6.6-ft) spatial resolution. The spatial resolution obtained from aircraft is an order of magnitude finer than even the most ambitious hyperspectral satellite sensors are planned to provide.

The relatively small number of hyperspectral sensors that are readily available through commercial and government sources worldwide (approximately 20) and the reliance on aircraft create problems not encountered with satellite-mounted multispectral sensors, which can provide repeat imagery at a site over periods ranging from approximately 1 hour to 1 month. Most hyperspectral researchers only want imagery that is obtained under specific weather conditions (such as clear skies without cloud shadow) and times (such as summertime, to capture imagery with vegetation). This creates a logjam of mapping missions, with instrument operators scheduling only several hours to several days at any one site. If the weather conditions are poor, researchers may not get hyperspectral imagery for that year. In addition, the cost of image acquisition with hyperspectral sensors can be $25,000 or more per day.

Image processing. After acquiring imagery, the remote sensing expert must process the data to remove errors and separate the signal from the noise. Corrections for geometric distortions (such as uneven scale across the image) and sun angle effects follow identical procedures to those used for multispectral imagery. However, hyperspectral imagery enables corrections for atmospheric distortions that multispectral imagery cannot do at all or as well. The narrow bandwidths of hyperspectral imagery enable "atmospheric soundings" that measure the adsorption of light by water vapor. Mathematical removal of the water vapor adsorption reveals the true reflectance of an object on the ground, regardless of humidity or distance from the sensor to the feature. Because atmospheric adsorption can obscure a feature or change its reflectivity over time or between places, the atmospherically corrected reflectance allows better identification of features, better detection of change over time, and better comparisons between sites.

Hyperspectral imagery also enables better separation of signal from noise, because multivariate statistics such as principal components work more effectively with many bandwidths. For example, principal component images created from a 128-band, 1-m (3.3-ft) image from the airborne Probe-1 hyperspectral remote sensing system isolate the spectral signal into lower-number bands and separate the noise into later bands. Principal component images of the Lamar River in Yellowstone National Park, WY, show how hyperspectral imagery enables segregation of signal from noise. Principal component band 1 captures the major brightness variations across the scene (**Fig. 1a**). Principal component band 5 shows more subtle features (Fig. 1b). Principal component band 40 is largely noise (Fig. 1c). Out of the 128 total bands originally available, most of the signal has been compressed into the first 25 bands. Isolating the signal in this manner reduces the size of the data set because the noise bands can be discarded. It also improves the mapping accuracy because the classification procedures are not confused by the noise.

Ground truth data. Most remote sensing mapping projects require collection of field data by ground teams. These "ground truth" data are used both to train the image—that is, the computer system is trained to associate specific spectral response patterns with specific landscape classes—and to evaluate the accuracy of maps produced by the image.

Because hyperspectral images collected from aircraft often have fine spatial resolution, location errors of only several meters can lead to major mapping errors. For example, based on field coordinates, a

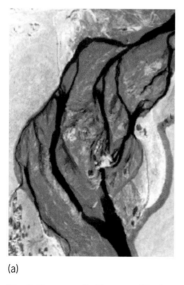 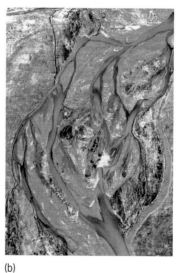 

(a)   (b)   (c)

Fig. 1. Hyperspectral images of the Lamar River in Yellowstone National Park, WY. (*a*) Principal component band 1 clearly highlights the river (black), gravel bars (medium gray), vegetation (light gray), and woody debris (white); (*b*) band 5 shows variations in stream turbulence, shown by shades of gray within the river; (*c*) band 40 is largely salt-and-pepper noise.

researcher might identify a bushy feature on an image as a willow, when in fact it is a sage brush located 3 m (10 ft) from the willow. When the researcher uses this feature to train the image, the computer algorithms will identify all the sage brush as willows. Such errors are not normally a problem with satellite-based multispectral imagery, because location errors of several meters will still fall within the same pixel on the image. (Each pixel represents a ground area of 30 by 30 m, or 98 by 98 ft, for much of the satellite-based *Landsat* data.)

Solutions to high-precision "co-registration" of field maps and imagery can be as simple as drawing field maps onto printouts of the imagery, but this assumes the imagery can be downloaded during the field season, which rarely happens. Alternatively, complex automated procedures can be used that require digital elevation models, complex three-dimensional ray tracing software, and a precision Global Positioning System (GPS) and inertial devices on the aircraft that monitor location and pitch, yaw, and roll. These data and algorithms document the position and orientation of the sensor, the distance and direction to the ground surface, and the resultant position of the pixels relative to the sensor with tremendous precision and accuracy.

If the mapping goal is to detect a specific feature (such as a type of mineral) as opposed to a general cover type (such as all bare rock), field crews will often collect ground-based measurements of that feature with a handheld spectrometer. These measurements are used to identify the spectral signature of the feature. This signature can then be used to search the hyperspectral image and find locations with similar spectral signals.

Classification and mapping. The many bands that make up a hyperspectral image enable many mapping approaches. These mapping algorithms fall into two broad categories: feature detection and general land cover mapping.

In general terms, algorithms for feature detection mathematically compare the spectral signature of a feature to the spectra in each pixel on the image. When a close match is found, the feature is mapped to that location. But hyperspectral imagery can go further and detect a feature that makes up only a portion of a pixel. Conceptually, this is like trying to "unmix" a shade of gray. The sensor detects the relative darkness of the gray. The unmixing algorithm then determines what proportions of black and white are required to create that shade of gray.

The maps that result from feature detection algorithms can therefore show relative abundance of a feature in each pixel. For example, woody debris along streams is crucial to aquatic habitat, flow hydraulics, and sediment storage. Mapping wood is difficult, however, because logs are long linear features that often make up only a portion of a pixel and therefore cannot be readily detected on multispectral imagery. The ability of hyperspectral imagery to unmix pixels enables mapping of these key habitat features at the subpixel scale (**Fig. 2**).

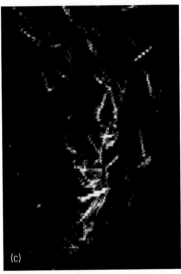

Fig. 2. Hyperspectral mapping of wood. (*a*) Large logs and smaller woody debris on the banks of the Lamar River, Yellowstone National Park, WY, August 1999. (*b*) Hypothetical 1-m pixel covering the wood. (*c*) Hyperspectral wood map of the area shown in *a*. White indicates that the wood fills the entire pixel. Increasingly darker shades of gray indicate less wood per pixel. Black indicates no wood.

The ability of hyperspectral imagery to clearly separate signal from noise also enables cover mapping of features that are spectrally similar. In this situation, the spectral signal is not sufficiently unique to clearly separate its spectra from all others on the image. The cumulative effect of subtle differences over many bandwidths, however, allows the cover types to be separated. For example, in-stream habitats, such as riffles, pools, glides, and eddy drop zones, are all water features with similar spectral signatures. Hyperspectral imagery can detect these subtle variations where a multispectral sensor cannot.

Accuracy assessment. Once a hyperspectral map has been produced, it is important to evaluate its accuracy using field data. Hyperspectral maps, however, challenge the ability to evaluate accuracy, because they can be better than the field maps that—in theory—are supposed to provide the ground truth for measuring accuracy.

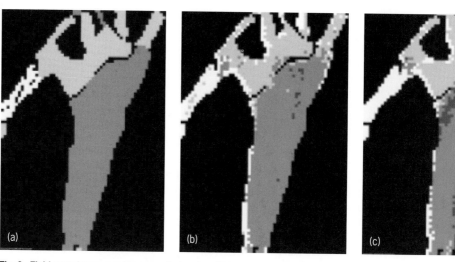

Fig. 3. Field mapping versus hyperspectral and multispectral mapping of the Lamar River. (*a*) Map created by the ground survey team. (*b*) Hyperspectral map created with 1-m (3.3-ft) resolution and 128-band imagery. (*c*) Map created using simulated 6-band multispectral imagery. Black represents exposed islands in the channel, boundaries between units, or areas outside the channel. White represents eddy drop zones (areas where fine sediments are deposited), light grays are riffles, medium grays are glides, and dark grays are pools.

Stream habitat comparison maps created at the Lamar River illustrate how accurate a hyperspectral map is when compared with a field map and a multispectral map (**Fig. 3**). The survey team mapped stream habitats as homogenous polygons because every tiny variation could not be mapped in the time available. The hyperspectral map picked up detail at a finer level than the ground crews could map, including fine-sediment deposits in low-velocity zones along the channel edges. Using the survey team's field map to assess accuracy resulted in the small local variations captured by the hyperspectral map being classified as mistakes when in fact they were probably real. (The multispectral map picked up some of the fine detail but also misclassified large portions of the stream as pool.) Since there is no easy way to survey in-stream habitats with the same level of detail as a hyperspectral image, it can be argued that in this case a hyperspectral map is actually a more accurate mapping tool than a ground-based survey. On occasions like this, the accuracy of hyperspectral mapping challenges fundamental procedures for error analysis and model validation in remote sensing.

For background information *see* IMAGE PROCESSING; REMOTE SENSING; TOPOGRAPHIC SURVEYING AND MAPPING; VEGETATION AND ECOSYSTEM MAPPING in the McGraw-Hill Encyclopedia of Science & Technology.   W. Andrew Marcus

Bibliography. J. B. Campbell, *Introduction to Remote Sensing*, Guilford Press, 2002; C.-I. Chang, *Hyperspectral Imaging: Techniques for Spectral Detection and Classification*, 2004; R. G. Congalton and K. Green, *Assessing the Accuracy of Remotely Sense Data: Principles and Practices*, Lewis Publishers, New York, 1999; R. J. Frouin, *Hyperspectral Remote Sensing of the Ocean*, 2001; W. L. Smith and Y. Yasuoka (eds.), *Hyperspectral Remote Sensing of the Land and Atmosphere*, 2001; Q. Tong (ed.) et al., *Multispectral and Hyperspectral Image Acquisition and Processing*, 2001.

Impulsive aggression (forensic psychiatry)

Forensic psychiatry has been defined by the American Academy of Psychiatry and the Law as the "subspecialty of psychiatry in which scientific and clinical experience is applied to legal issues in legal contexts embracing civil, criminal, correctional or legislative matters." In essence, forensic psychiatry can be understood as psychiatric consultation to lawyers and courts in order to assist in the resolution of legal issues wherein a person's mental state is of legal significance. In criminal law, examples include competency to stand trial and mental responsibility at the time of the offense (whether the defendant satisfies criteria for acquittal based on insanity). In civil law, a person's mental state can come into question when considering tort (civil wrong) claims based on emotional injury and various civil competencies such as competency to make out a will (testamentary competency).

Behavioral dyscontrol and the law. Various legal tests of responsibility and competence involve criteria of cognitive ability; that is, whether the person is able to think adequately in order to handle the particular task. The specific type of thinking required varies depending on the nature of the task as well as the relevant law. For example, in many states to satisfy an affirmative test of insanity the defendant must prove that she or he did not know the nature and quality of what she or he was doing; or if the defendant did know what he or she was doing when committing the crime, that it was not known this was wrong. In civil law, too, cognitive criteria conform to the substantive law, and from a clinical perspective cognitive abilities may be determined with adequate clarity and confidence.

From the psychiatric standpoint, however, the noncognitive ability to control one's actions is just as relevant to states of mind with legal significance. A person could be unable to cooperate with an attorney in preparing a defense, unable to conform his conduct to the requirements of the criminal law, or unable to hold assets, let alone bequeath them because of profound deficiencies in the ability to self-control, even where thinking is otherwise generally intact. Though sometimes considered to be a legitimate element of the insanity defense or a component of competence to stand trial, controlling processes generally receive much less attention and understanding in forensic consultations than cognitive processes. How does one determine, for example, whether a criminal defendant was unable to control an impulse to commit a criminal act or simply unwilling to exercise self-control? Attempts to make this distinction have been considered more speculative than scientific and therefore more prejudicial than probative. But this concern does not refute the very powerful role that the ability to self-control plays in regulating human behavior.

Classification of impulsive aggression. For many years classifications of mental disorders have recognized pathological disturbances in which poor control over aggressive impulses is exhibited. The purest example of this type of disturbance is termed intermittent explosive disorder (IED) in the official diagnostic manual of the American Psychiatric Association (*DSM-IV-TR*, 2000). This disorder is manifested by aggressive outbursts that can be violent or destructive and that are entirely out of proportion to any possible provocation or stimulus. A substantial amount of recent research has been concerned with the phenomenon of impulsive aggression, a condition that shows similarities with IED. Both impulsive aggression and IED involve explosive aggression, but in the former the aggressive acts are manifestations of impulsiveness rather than isolated events. Both impulsive aggression and IED can account for some criminal acts and civil wrongs. Improved diagnosis and treatment could lead to prevention, benefiting potential victims, the potential actor (person committing such an act), and society.

Impulsive aggression is a sudden, minimally if at all provoked eruption of agitation and destructive or assaultive behavior. Impulsive aggression has been distinguished from premeditated aggression, the latter being planned acts that are intended to benefit the actor socially or materially. Impulsive aggression is also distinguished from secondary or medically related aggression resulting from a mental or neurological disorder. Secondary aggression should subside with effective treatment of the primary disorder.

Though a complete understanding of the mechanisms of impulsive aggression remains elusive, substantial advancement has been made in recent years through brain imaging and electrophysiological examination of cerebral cortical phenomena, as well as research on neurotransmitters, molecular genetics, and relevant psychosocial factors.

Brain imaging. Imaging studies have demonstrated an association between impulsive aggression and structural changes in the frontal or temporal cerebral cortex, or occurring diffusely throughout the cortex. Studies using single-photon-emission computer tomography (SPECT) showed reduced blood flow in the prefrontal region. Hypometabolism in prefrontal cortex has been found in aggressive subjects, using positron emission tomography (PET) to measure alterations in glucose metabolism. Particularly in murderers with benign (nonabusive) psychosocial backgrounds, significantly diminished glucose metabolism has been demonstrated in the prefrontal cortex. Prefrontal dysfunction can result in impaired inhibition of subcortical structures resulting in poor judgment, impulsivity, emotional dysregulation, and aggressive behavior. Mechanisms for the control and release of impulsive aggression are thought to involve a number of brain structures (prefrontal cortex, frontostriatal tracts, anterior cingulate cortex, insular cortex, amygdala, hippocampus, and ventral striatum).

Electrophysiology. An event-related potential (ERP) is the measured average of all electroencephalographic (brain wave) segments that are related in time to an event such as an auditory or visual stimulus. An ERP is specified by the electrical polarity (P for positive or N for negative) and the latency in milliseconds. An ERP of P300, the most studied of the late ERP components, is thus an event-related potential of positive polarity with a latency of 300 ms. The P300 occurs only when the event is processed during consciousness. A so-called oddball paradigm is a P300 experiment in which the subject selectively identifies an unusual target stimulus from among many irrelevant stimuli (for example, distinguishing a visual or auditory target from a distracting background). During oddball tasks, subjects with impulsive aggression have prolonged latencies and diminished P300 amplitudes in the temporal and parietal lobes. Most intriguing, it has been demonstrated that phenytoin, an anticonvulsant, both reduces impulsively aggressive outbursts and normalizes the ERPs in the same subjects.

Neurotransmitters and molecular genetics. Three neurotransmitters have been implicated in impulsive aggression: serotonin (5-HT), norepinephrine (NE), and dopamine (DA). Of these, 5-HT has been the most studied. Decreased serotonin activity or availability in the central nervous system is consistently associated with impulsive aggression in humans.

A recent study provides epidemiological evidence that a specific molecular genetic impairment combined with an abusive upbringing increases the probability of boys showing aggressive and antisocial behavior. A low-activity monoamine oxidase A

(MAOA) genotype and maltreatment were associated with a disproportionate level of convictions for violent offenses. There are two types of MAO: MAOA and MAOB. Constituting about 80% of the MAO in the brain, MAOA catabolizes 5-HT and NE by deaminating these neurotransmitters into their corresponding aldehydes. DA is oxidized by both MAOA and MAOB. Thus all three neurotransmitters (5-HT, NE, and DA) are inactivated by MAOA. Because genes for both MAOA and MAOB are located on the X chromosome, deletion of either gene results in complete deficiency of the gene in males. A syndrome was described in which males who lacked a functional MAOA gene manifested borderline mental retardation and engaged in acts such as arson and inappropriate sexual behaviors as well as impulsive aggression. Other investigators found that mice with a mutated MAOA gene similarly displayed increased aggressive behaviors as adults.

A specific neurotransmitter mechanism for reduced MAOA activity and aggression remains to be established. One hypothesis maintains that with diminished level or activity of "impulse controlling" neurotransmitter(s), such as 5-HT, an individual's ability to self-control is reduced. Alternatively, rather than the specific level of 5-HT alone, the ratio of impulse controlling versus "impulse driving" neurotransmitter(s), such as DA, may be more determinative. Low MAOA activity should result in reduced breakdown of 5-HT, resulting in increased 5-HT with greater self-control rather than less. From this perspective the findings of the MAOA genetic study are a surprising paradox. Perhaps the absence of the MAOA gene during development results in a compensatory down regulation of serotonin receptors with a "low serotonicity" phenotype (that is, a decrease in the number of 5-HT receptors as a result of increased exposure to 5-HT). In other words, abnormally high 5-HT activity in early development leads to low serotonicity in adolescence or adulthood with resultant impulsive aggression. Any attempt to reconcile the MAOA-serotonin paradox, however, remains speculative.

Overview. A pattern of recurrent and extreme impulsive aggression is likely a result of a pathological imbalance between aggressive impulses and opposing controlling mechanisms. The association between abusive early backgrounds and impulsive aggression suggests the importance of psychological processes such as social learning, conditioning, and sensitizing. The combination of specific neurophysiological abnormalities and an unfavorable psychosocial background is likely to have a stronger association with impulsive aggression than either alone. In an extremely aggressive individual with a benign psychosocial background, marked presence of neurophysiological dysfunction is especially likely. The hypometabolism in the prefrontal cortex suggested by imaging studies is consistent with the neurotransmitter hypothesis of impulsive aggression. Through psychological mechanisms such as judgment, restraint, and planning, the prefrontal cortex serves adaptive and self-controlling functions mediated through 5-HT.

Controlling and exciting mechanisms can be altered through other structures in the temporal lobes and the Papez circuit (anatomical structures within the brain believed to mediate emotion) as well. Implication of the parietal lobes through cortical ERP studies raises the possibility that difficulty in processing information mentally may not simply represent an association but may also represent a pathogenetic factor in impulsive aggression. Without the ability to process afferent and subjective information efficiently and to adapt accordingly, immediate impulses will be given automatic expression. In any event the stabilization of nerve membranes brought about by specific anticonvulsants appears to benefit both improved cognition and improved control of impulsive aggression where adequately defined.

Future directions. Concepts such as free will, consciousness, understanding, and rationality are important presumptions behind a variety of forensic psychiatric issues. Not only cognitive but also volitional components contribute to a person's ability to choose and to act or to refrain from acting. Important to the advancement of research on impulsive aggression will be the application of consistent definitions of the phenomenon. As the psychology, underlying biology, psychosocial determinants, and epidemiology of impulsive aggression become better understood, such knowledge will serve forensic psychiatrists whose consultations involve subjects with impulsive aggression. Whether this will inform public policies on psycholegal issues will depend upon the politics of competing social interests. Future research that applies a consistent definition of impulsive aggression and attempts to integrate multiple areas of relevant scientific inquiry can be expected to produce valuable results.

For background information *see* AGGRESSION; BEHAVIOR GENETICS; BRAIN; ELECTROENCEPHALOGRAPHY; FORENSIC EVIDENCE; SYNAPTIC TRANSMISSION in the McGraw-Hill Encyclopedia of Science & Technology.   Alan R. Felthous

Bibliography. E. F. Coccaro (ed.), *Aggression: Psychiatric Assessment and Treatment*, Marcel Dekker, New York, 2003; *Diagnostic and Statistical Manual of Mental Disorders, Fourth Edition, Text Revision (DSM-IV-TR)*, American Psychiatric Association, Washington, DC, 2000; A. Raine, *The Psychopathology of Crime: Criminal Behavior as a Clinical Disorder*, Academic Press, San Diego, 1993; R. Rosner (ed.), *Principles and Practice of Forensic Psychiatry*, 2d ed., Arnold, London, 2003.

Indoor fungi

Fungal (or mold) spores are commonly dispersed in air, and on some days 10,000 or more fungal spores can be found in a cubic meter of air. With every breath humans take, there is a high potential that fungal spores will be inhaled. Usually this is not a problem, but if the number of spores inhaled is high enough, susceptible and sensitized individuals may develop adverse effects, such as an allergic response. Recently it has been suggested that the spores of some fungi may contain toxins and that inhalation leads to deleterious effects when the toxins are released in the lungs. However, this conclusion is controversial.

Fungal spores, so common in outside air, can easily be transported indoors via air currents through openings such as windows, doors, and ventilation systems. Thus, one might expect to find many of the same molds indoors as well as outdoors. However, the number and kinds of fungi often differ substantially between the inside and the outside of buildings. This may indicate that amplification of the quantity and types of fungi is due to growth of molds inside the home. In this situation, areas for mold amplification inside buildings should be located, and the conditions that favor amplification should be corrected.

Common indoor molds. A recently published study describes many of the types of fungi found in indoor and outdoor air across the United States, as well as the ranges of airborne concentrations at which they were detected. Described below are the more common airborne fungi.

Cladosporium. In nature, *Cladosporium* grows in abundance on plant debris and dead or senescent grasses (including cereals such as corn and wheat). It sporulates prolifically on these materials and is the most common spore found in outdoor air (**Fig. 1**). *Cladosporium* is also the most commonly found fungus indoors in North America. It grows especially in bathrooms; the brown stains often noticeable around the edge of the bath or shower stall or sometimes brownish spots found on the bathroom walls or ceilings are caused by colonies of *Cladosporium*. The water droplets left around the edge of the bath and the high humidity in the bathroom, preventing rapid drying, provide the free water that *Cladosporium* needs to thrive. *Cladosporium* can also cause brown stains of growth on window frames, where water condenses on the panes and runs down to soak the frame. However, *Cladosporium* cannot cause serious wood rot by itself.

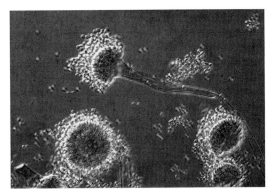

Fig. 2. Photomicrograph of *Aspergillus fumigatus.* (*Courtesy of B. G. Shelton*)

Fig. 3. Photo of rotting orange with *Penicillium* growth. (*Courtesy of G. L. Barron*)

Aspergillus. *Aspergillus* is also very common in indoor and outdoor air (**Fig. 2**). It could be called the "basement" mold because that is where it is most likely found in homes. This remarkable fungus is a xerophyte; that is, it is adapted to life in areas where the water supply is limited. In fact, some species of *Aspergillus* do not require free water for growth; they can grow when relative humidities exceed 60%, resulting in increased moisture content of growth substrates. Basements in many parts of North America are commonly above 60% relative humidity during summer months, conditions under which the growth of *Aspergillus* is relatively slow. Yet, over the course of one or more summer seasons, there is often time for the fungus to exploit anything stored in the basement, particularly leather goods such as shoes and belts.

Penicillium. *Penicillium* is another ubiquitous fungus in both indoor and outdoor air (**Fig. 3**). Some species of *Penicillium* enjoy cool temperatures, and one of its favorite locations in buildings for growth and sporulation is inside the refrigerator. *Penicillium* can grow on almost any food, especially fruit, cheese, pizza, and bread. In fact, *Penicillium* is even used to produce some foods. For example, Roquefort cheese

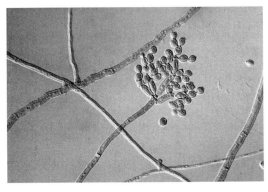

Fig. 1. Photomicrograph of *Cladosporium.* (*Courtesy of G. L. Barron*)

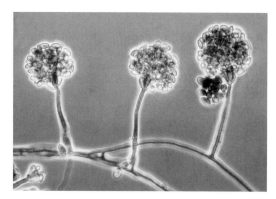

Fig. 4. Photomicrograph of *Stachybotrys chartarum*. (*Courtesy of B. G. Shelton*)

is produced using *P. roqueforti. Penicillium* can also survive at room temperature and is commonly found growing in homes or buildings on exterior walls with moisture resulting from water condensation. In addition, a few *Penicillium* species can grow under the same xerophytic conditions as *Aspergillus*.

Penicillium is sometimes called the blue mold because of the pigment in its spores, which imparts a blue color to the heaping masses of spores that form over the surface of colonies. As with most fungal genera, the individual spores of *Penicillium* are so tiny that they are invisible to the eye, only a few micrometers in diameter. Disturbing heavily moldy food or other moldy substance potentially releases millions of spores in an invisible airborne cloud.

Stachybotrys. Stachybotrys is not as common in indoor or outdoor air as *Cladosporium, Aspergillus,* and *Penicillium,* but on careful inspection it can probably be found in most buildings where cellulose building materials have become wet (**Fig. 4**). The effect of *Stachybotrys* on human health have caused concern. Some claim that the spores of *Stachybotrys* contain potent toxins that when inhaled can cause serious or even fatal effects in humans. However, this conclusion is highly controversial, and to date has not been proven. Nevertheless, these allegations have caused concern for the occupants of houses or buildings where *Stachybotrys* is found.

In nature, *Stachybotrys* utilizes cellulose as an energy source. It is not a fast grower or spore producer. For growth and spore production, it requires wet conditions for longer periods of time than other common indoor molds. In buildings, therefore, it is found where leakage or flooding has soaked cellulose materials such as paper and cardboard or other cellulose-based products, particularly paper covering on wallboard or ceiling tiles. It often grows on the inside surface of walls following flooding, as the inside space between walls stays wetter for longer and is difficult to dry out. *Stachybotrys* may also grow on the outside surface of interior walls exposed to prolonged flooding and consequent high humidity in the surrounding air.

Effects on human health. Potential health effects from exposure to airborne fungi have gained considerable attention recently. This concern can be traced

back to a reported outbreak of pulmonary hemosiderosis (a disorder causing bleeding of the lungs in infants and childern) that occurred in 1993 in Cleveland. Investigators alleged that fungal toxins produced by *Stachybotrys* were associated with this outbreak. Media attention given to this reported outbreak was significant, and subsequently many legal cases occurred, alleging toxic exposures to indoor fungi. However, this conclusion was not supported by the Centers for Disease Control and Prevention in Atlanta, which ultimately published an official statement that the reported association should be considered "not proven." A closer look at other published reports alleging toxic outcomes from indoor airborne fungal exposure also suggests that the investigations contained significant flaws and limitations. Therefore, at this time, evidence for any toxic outcome from exposure to fungi in indoor air is not officially accepted.

Nevertheless, molds in the home can be a serious problem for many people. Possible health effects from airborne fungal exposure in indoor environments include allergic reactions, onset of an attack in certain asthmatics, and rarely infections. Such reactions are usually limited to susceptible and sensitized individuals.

Allergic reactions. Allergic reactions are similar to those experienced by individuals susceptible to allergens from sources such as pollen, animal dander, and dust mites.

Asthmatic reactions. In addition to allergic-type reactions, fungi may sometimes trigger asthmatic reactions in certain individuals. However, this is considered rare, as asthmatic reactions are more commonly induced by triggers such as cat allergens, exercise, and cold-air shock. Even more rarely, *Alternaria alternata* (a common plant pathogen that is occasionally recorded inside homes) may be associated with severe asthmatic reactions, including death, in a small portion of asthmatics.

Infections. Infections are another possible, but rare, outcome of exposure to fungi in indoor air. For normal, healthy individuals this is not a problem, as infections from indoor molds are usually limited to people with weakened immune systems and usually occur in immunocompromised individuals in hospital settings.

Since fungal spores in outdoor air can reach very high levels, it is often difficult to attribute adverse health effects to indoor fungi to the exclusion of outdoor fungi. Furthermore, since most people are exposed to many fungal types on a daily basis it is difficult to attribute adverse health effects to a particular species.

Removal and control of indoor molds. Removal and control of molds in homes and other buildings is, for the most part, a matter of common sense. Prevention is the key, and the focus should be directed at controlling water, moisture, and high humidity. If there is a leak, flood, or water infiltration, the area should be dried and cleaned before any fungi have time to grow and sporulate.

Other strategies to help minimize indoor mold growth are as follows:

1. Ventilate the bathroom to the exterior after showers and baths, and wipe the free water from the tiles and edge of bath or shower stall with a cloth or sponge.

2. Keep a dehumidifier running in the basement in the summer to maintain the relative humidity level below 60% to control xerophytic fungi.

3. If a home or building is flooded, drain and dry out the area as quickly as possible. Fungi will start growing almost immediately, but there are a few days of lag time before the colonies start to sporulate. It is also important to quickly drain and dry the space between the walls.

4. Carefully remove moldy food in the refrigerator, so as not to disturb the colonies, and discard outside.

5. Ventilate cooking stoves and other high-moisture producers to the exterior. Cooking is a major source of organic volatiles that float around the house and eventually condense on the walls to form an excellent nutrient source for fungal growth, should the conditions become appropriate.

6. Wipe down areas where molds are already established with a dilute solution of household bleach, which will act as an effective fungicide. Since this will kill the fungus at the site but will not remove the conditions that allowed the fungus to develop in the first place, the growth may well return as affected areas are recolonized by new airborne spores.

For background information *see* ALLERGY; ASTHMA; FUNGI; MEDICAL MYCOLOGY; MYCOLOGY; MYCO-TOXIN in the McGraw-Hill Encyclopedia of Science & Technology. Brian G. Shelton

Bibliography. Centers for Disease Control and Prevention, *Morbidity and Mortality Weekly Report*, 49(9):180–184, March 10, 2000; Environmental Protection Agency, Office of Air and Radiation, Indoor Enviornments Division, *A Brief Guide to Mold, Moisture, and Your Home*, EPA Publ. 402-K-02-003; B. G. Shelton et al., Profiles of airborne fungi in buildings and outdoor environments in the United States, *Appl. Environ. Microbiol.*, 68(4):1743–1753, 2002.

Inhalation drug therapy

Historically, drug delivery technologies have been used to extend the duration of effect, or to speed the onset of action, of drugs or to improve the convenience of their administration. Today the role of drug delivery technology has been expanded to include improving the efficiency and specificity with which drugs can be delivered to their target organs or systems. Inhalational drug therapy, in which drugs are delivered to the lungs in the form of an aerosol, is one example. Because inhaled drugs are absorbed directly into the bloodstream, the development of pulmonary methods of drug delivery to replace oral methods avoids exposure of the drug to the harsh gastrointestinal environment, minimizing gastroin-testinal problems such as low solubility, low permeability, drug irritability, unwanted metabolites, and dosing uncertainty caused by meals of varied composition. Because they are delivered directly to diseased lungs, certain inhalable antibiotics may be more effective than oral or injectable antibiotics for treating lung infections.

Advances in pulmonary drug delivery have also opened the door to noninvasive administration of a wide variety of molecules, large and small. In the case of large molecules, such as peptides and proteins, which because of their size are not orally bioavailable (that is, they do not get absorbed from the gastrointestinal tract), pharmaceutical companies can now consider developing pulmonary delivery systems rather than being forced to develop small-molecule substitutes with similar properties (mimetics) or to abandon promising molecules altogether. This article focuses on the promise of new inhalational therapies for improved treatment of systemic diseases via drug delivery to the lungs, as well as the treatment of pulmonary diseases by targeting drugs directly to the lungs that in the past were delivered orally or by injection.

Systemic drug delivery through lungs. About 84 protein and peptide drugs, also referred to as macromolecules, biomolecules, or biotherapeutics, are currently being marketed in the United States, and about 350 more are being tested in clinical trials. Oral biomolecule drugs are difficult to use in humans because they are quickly digested by gastrointestinal enzymes before they can reach target tissues. Therefore, most of these drugs are given by injection, leading to less than ideal compliance due to aversion to needles by some patients. In contrast, pulmonary delivery of such drugs eliminates the need for injections while providing rapid onset of action comparable to intravenous therapy. This makes inhalable drugs ideal for treating such conditions as the severe acute "breakthrough" pain usually caused by malignant tissue infiltration, migraine headache, acute anxiety (panic attacks), and nausea.

Researchers have discovered that in order for drugs to be absorbed via the lungs, the aerosol particles must range in size from 1 to 3 micrometers in aerodynamic diameter (a unit of measurement that takes account of particle shape and density, and thus is a measure of particle behavior in air). Such small particles readily pass through the mouth and throat, travel down the airways, reach the deep lung, dissolve in airway or alveolar lining fluid, and are absorbed into the bloodstream. The surface area of the lungs is enormous, equivalent to the size of a tennis court (80 m^2), which facilitates absorption of drugs. The lungs also have much lower levels of protein and peptide-destroying enzymes and greater permeability than the gastointestinal tract. Moreover, pulmonary delivery avoids the problem of slow and variable (meal-related) absorption in the gastrointestinal tract and avoids first-pass hepatic metabolism (that is, the liver's removal of a drug from the blood before it reaches the systemic circulation), which may greatly reduce bioavailability and/or lead to injury of liver cells.

Aerosol generation and delivery systems. During the past decade, there has been an evolution of three main types of therapeutic aerosol generators: (1) small-volume nebulizers, which generate aerosols from aqueous solutions; (2) metered-dose inhalers, which generate aerosols using propellant liquids; and (3) dry powder inhalers, which deliver powder drug formulations.

Powder versus liquid aerosols. Formulation is the key to successful aerosolization with all delivery systems. To engineer particles small enough to target the deep lung—where they must be deposited, dissolved, and absorbed into the bloodstream—requires formulating proteins, peptides, or small molecules into dispersible powders or liquid aerosols.

Powder aerosols are usually superior to liquids because they deliver much higher doses with each breath, thus usually allowing aerosol therapy to be administered rapidly. Furthermore powders can be stored at room temperature, and do not support bacterial growth if kept extremely dry during storage. Liquid aerosol drug concentration—and thus the volume required to achieve the prescribed dose and the time required for aerosol therapy—is limited by the drug's solubility. This concentration rarely exceeds 20% (and is often much lower) because very concentrated drug solutions may precipitate and often have a high viscosity, making them very difficult to aerosolize using small-volume nebulizers. Metered-dose inhalers are generally limited in their ability to deliver more than 500–1000 micrograms per puff and thus cannot be used to deliver drugs such as antibiotics, which require delivery to the lungs of doses ranging from 50 to 100 mg. In contrast, dry powder aerosols commonly support a drug load of 50–90%. This translates into a much more patient-friendly delivery system; since a dry powder inhaler may deliver up to 35 mg of medication in each inhalation, treatment takes no more than 2 min, whereas current liquid aerosol inhalers require 15–30 min to deliver a comparable mass of drug. Powder formulations of proteins and peptides are remarkably stable and do not require refrigeration, making them more convenient to use and less costly to distribute and store.

Formulating powders. There are several ways to develop drug powder formulations. One method is glass stabilization, in which small-molecule glass formers (oxides that can readily form a glass) such as simple sugars are mixed in a solution with a drug macromolecule. As water is removed from the solution, the drug molecules at first pack randomly into an amorphous solid, or rubbery state, which is unstable. Then, as more water is removed, the rubbery form turns into an amorphous glass state, resembling hard rock candy. This glass stabilization process protects the fragile, active protein drug as long as a minute amount of water is present to keep it stable and clinically effective. Each amorphous glass has a specific glass transition temperature, similar to a melting point, at which it reverts to its undesirable rubbery form. Consequently, dry powder drugs must be manufactured to have a glass transition temperature well above any temperatures at which they will be stored, shipped, or used. The drug powder (such as insulin) can be stored for a year or more by means of double-foil vacuum-sealed blister packs that block moisture and deliver a unit dose of drug. The unit packaging accommodates flexible dosing, since patients require individualized and varying doses over the course of the day.

Engineered particles. For the past 60 years, particles used for inhalation were generally created by micronization, a variation on grinding with a mortar and pestle. Such particles have high interparticulate adhesive forces and require considerable dispersive energy to provide inhalable therapeutic aerosols.

A method has recently been perfected to engineer powder, liquid, or metered-dose inhaler formulations of drugs using a lipid-based emulsion process that produces hollow, porous, spongelike light drug particles in the aerodynamic range of 1 to 5 μm, which is optimal for pulmonary delivery. They are created by a spray-drying procedure that makes use of escaping volatile perflubron to produce very low density particles of water-soluble, or lipophilic, large- and small-molecule drugs that can be efficiently delivered to the airways or deep lung. Since particle-particle interactions are extremely low, simple and inexpensive devices such as passive dry powder inhalers (which require the patient's vigorous inhalation to actuate them) and metered-dose inhalers can readily be used to disperse the aerosol and make it suitable for inhalation even at relatively low inspiratory flow velocities in the range of 15–30 L/min. By contrast, micronized particles generally require inspiratory flow velocities above 30 L/min; for optimal dispersion, depending on the device and the drug formulation, velocities of 60–100 L/min may be required.

Treating the lung directly. When systemically administered orally or via injection, only about 1 or 2% of the dose of most medications used to treat lung disease are delivered to the lungs. For the past 50 years, increasing numbers of drugs have been delivered directly into the lungs by means of metered-dose inhalers, dry powder inhalers, and small-volume nebulizers to effectively treat asthma, chronic obstructive pulmonary disease, cystic fibrosis, and other lung diseases while minimizing systemic adverse effects.

Compliance with therapeutic regimens utilizing metered-dose inhalers has been inadequate primarily due to problems coordinating aerosol discharge with inhalation. This has led to the development of breath-activated metered-dose inhalers and inexpensive add-ons in the form of valved holding chambers. The addition of valved holding chambers to metered-dose inhalers is generally superior to metered-dose inhalers alone because the add-ons (1) ensure aerosol delivery by obviating the need to precisely coordinate aerosol discharge and inhalation; (2) enable the selection of small particles (less than approximately 2–3 μm), thus optimizing pulmonary delivery; (3) usually increase the small-particle dose; and

(4) reduce the upper respiratory tract dose by approximately 90% and total body dose by approximately 75%, further minimizing the oropharyngeal and systemic effects associated with drugs such as inhaled corticosteroids. Other inexpensive add-on devices greatly improve the versatility of metered-dose inhalers; by the addition of appropriately sized masks and the use of various accessory devices, metered-dose inhalers are now available to treat infants, the elderly, patients in emergency departments, people on ventilators, and even horses with heaves. For treating asthma, this approach to aerosol therapy is increasingly replacing wet nebulization, thus reducing the cost of treatment considerably.

For background information *see* DRUG DELIVERY SYSTEMS; LUNG; RESPIRATION; RESPIRATORY SYSTEM; RESPIRATORY SYSTEM DISORDERS in the McGraw-Hill Encyclopedia of Science & Technology.

Michael T. Newhouse

Bibliography. A. L. Adjei and R. K. Gupta (eds.), *Inhalation Delivery of Therapeutic Peptides and Proteins*, Marcel Dekker, New York, 1997; R. W. Niven, Delivery of biotherapeutics by inhalation aerosol, *Crit. Rev. Therap. Drug Carrier Sys.*, 12:151–231, 1995; J. S. Patton, Mechanisms of macromolecule absorption by the lungs, *Adv. Drug. Delivery Rev.*, 19:3–36, 1996; J. S. Patton, J. Bukar, and S. Nagarajan, Inhaled insulin, *Adv. Drug Delivery Rev.*, 35:235–247, 1999; J. S. Skyler et al., for the Inhaled Insulin Phase II Study Group; Efficacy of inhaled human insulin in type 1 diabetes mellitus: A randomized proof-of-concept study, *Lancet*, 357(9253):331–335, Feb. 3, 2001; J. S. Skyler et al., for the Inhaled Insulin Phase II Study Group, Inhaled human insulin treatment in patients with type 2 diabetes mellitus, *Ann. Intern. Med.*, 134(3):203–207, Feb. 6, 2001.

Inkjet printing

Inkjet printing is a method of forming a hardcopy image (text or graphics) by generating very small drops of ink and "jetting" them onto a substrate. This concept was introduced commercially in the mid-1980s to replace dot-matrix (low cost but slow and noisy) and laser-based (high speed and high resolution but also high hardware cost) printing systems for home and office applications. As inkjet color inks and photographic-quality paper development progressed, a new goal of supplanting traditional silver halide photographs was established. Today, the home photographer can produce inkjet prints that exceed the quality and light-fade stability of color photographic prints.

Inkjet print quality has improved continuously by using a systems approach of providing smaller drop volumes along with better media (paper) and inks. The colorants used in the inks may be dyes (water/solvent-soluble) or pigments (insoluble). Dye-based inks historically provided more vivid color and were somewhat easier to manufacture and keep in suspension, but pigment-based inks are closing the gaps in these areas. Pigmented inks, which do not penetrate the media as readily as dye-based inks, have been developed with the equivalent quality of laser printers or typewriters for black text printed on plain paper. Compared to dye-based inks, pigmented inks provide water-fastness, light-fade resistance, and better smear resistance. Recently, color pigmented inks have found use in large format inkjet printers for outdoor signs. Photo imaging and color adoption in the office applications have been instrumental in increasing use of inkjet products.

The human eye can detect dots on paper 25 micrometers (0.001 in.) or smaller if the color is dark (high optical density). A dot of this size on photo paper requires an ink drop of 2 picoliters or less, assuming typical dot spreading on the media. Therefore, drop volume reductions beyond 1 pL are probably not necessary for grain-free photos. Color printing speed and quality also have been improved by adding more colors, some with reduced dye or pigment loads of the black, yellow, cyan, and magenta colorants, but this adds product cost. In the future, important product features will be speed and ease of use.

Printhead technologies. There are two popular techniques used to jet inks with adequate drop ejection velocity, size, and directional control. They are continuous and drop-on-demand (DOD), as seen in the **table**. The earliest inkjet (continuous) used a piezoelectric material that changes shape when a voltage is applied. This shape change is used to increase or decrease the volume of ink confined in the ink chamber. A column of ink exits the firing chamber through a small hole or nozzle and then breaks into a droplet. The ink flow is continuous as a periodic voltage is supplied to the piezoelectric material. Unwanted drops are deflected into an ink-recycling channel. Piezoelectric materials are also used in DOD printheads. Thermal inkjet (TIJ) is another DOD technology and has gained over 75% of the market. Here the ink is superheated, and the resulting large volumetric expansion is used to propel the droplet. Typical drop volumes range from 2 to 50 pL, with velocities of 10 m/s, and a head-to-media spacing of 1 mm.

Continuous inkjet. This technology is used primarily for high-volume applications such as magazine labels, can-dating codes, bar-code labels, and carton marking. Inkjetting frequencies of 100 kHz or more can be attained since the head is operated in a stable mode. The ability to switch between ink systems (such as dye and pigmented inks), known as ink flexibility, and the short distance from the head to the substrate are advantages. However, the nozzle packing density (number of nozzles per printhead area) and the high printhead cost have prevented this technology from penetrating consumer printing applications. New applications for continuous inkjet are emerging as new inks with improved durability and specific attributes that allow direct printing on glass, ceramics, fabrics, and other substrates are

Comparison of inkjet printhead technologies

| | Continuous (piezo) | Drop-on-demand | |
		Thermal (TIJ)	Piezo (PIJ)
Advantages	High volume High frequency Ink flexibility Long throw distance Many substrates	High efficiency Low cost Integrated chip processes Integrated drive High frequency Nozzle density	High efficiency Ink flexibility Long throw distance Many substrates Multiple drop volumes Long-life head
Disadvantages	Nozzle density High head cost	Ink flexibility Unwanted heat Head reliability	Nozzle density High head cost Electrical integration

developed. Systems using continuous inkjet technology typically cost from $50,000 to nearly $1 million.

Drop-on-demand inkjet. In digital printing, the most efficient use of energy is to place ink only when and where needed, or DOD. The volume, velocity, and direction of each ejected drop are critical for maintaining consistent quality and operating conditions. A computer generates the desired image and sends the dot firing sequence to the printer, which either energizes a piezo crystal or fires a thermal resistor in a printhead with hundreds or thousands of small nozzles lined up in several rows. The printhead scans over the media, printing a swath of one or more colors. This process is repeated until the entire document is printed.

Thermal DOD inkjet. Since the mid-1980s, thermal inkjet has shown continuous improvement in quality, speed, and cost. A major breakthrough in the technology was the development of the replaceable printhead. This allowed both the ink and printhead to be placed in one inexpensive cartridge. There are two primary technologies used in thermal inkjet: edge and top shooter (see **illus.**). Both techniques use the physical principle of converting the electrical energy delivered to the firing resistor into heat in order to superheat a small volume of ink. Most thermal inkjet now use the top shooter design since it is more thermally efficient. Thermal inkjet primarily uses established integrated circuit manufacturing processes and equipment, so costs are generally lower than piezoelectric inkjet (PIJ). Its silicon-based substrate allows the integration of electronic drive circuitry, which reduces the overall printing system costs and improves performance. The latest thermal inkjet technology uses silicon etching for ink feed slots, thermal inkjet resistor formation, and ink channel/nozzle formation. Drop ejection frequencies of up to 36 kHz have been achieved. The upper limits for firing frequency are determined by the overheating and fluid refill characteristics of the chamber.

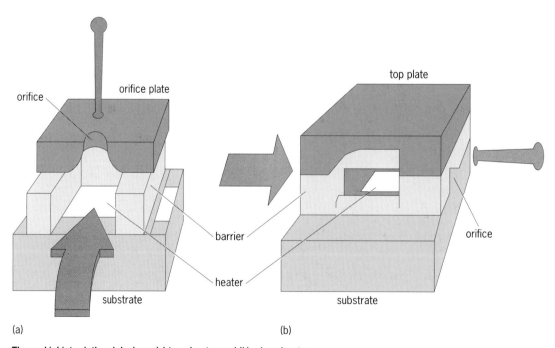

Thermal inkjet printhead designs: (*a*) top shooter and (*b*) edge shooter.

Firing chamber densities are currently at 600 dots per inch (dpi). For most applications, acceptable text and graphics quality can be achieved at 600 dpi, so a single pass of the printhead is sufficient. Higher-quality printing is achieved with several passes of the printhead at a sacrifice in printing speed. Thermal inkjet is lower-cost, and allows a higher degree of circuit integration and higher nozzle packing density than piezoelectric inkjet.

Piezoelectric DOD inkjet. Piezoelectric inkjet did not find wide acceptance until the mid-1990s, when costs, image quality, and ease of use began to match thermal inkjet. Early DOD piezoelectric inkjet piezoceramic printheads required very costly machining operations. The development of low-cost thick-film piezo elements and high-volume manufacturing contributed to this technology. Materials used in piezoelectric inkjet printhead construction are more robust than thermal inkjet, which has enabled them to jet a wider range of inks. In contrast to thermal inkjet, the piezoelectric inkjet printhead is permanent and only the ink cartridge is changed. One popular use is a phase change ink, or wax-based ink. The printhead is operated above the melting temperature of the wax, so the jetted drops "freeze" on the media. This allows a wide choice of substrates, but the printhead and maintenance costs are high. These products are used in the office and low-volume commercial markets. Since the ink is "squeezed" from the chamber, more control of the fluid dynamics is possible. As many as three different drop volumes ranging in an order of magnitude are possible from a single firing chamber, reducing thermal inkjet's advantage of higher nozzle packing density. Drop dynamics are similar to thermal inkjet with fewer unwanted satellite drops formed, but slightly lower drop velocity and operating frequency.

Outlook. It is rare to find one technology that can satisfy user needs from a personal photo printer to an outdoor highway sign printer. The print quality for most home and office printing by inkjet has now achieved a level where it is comparable to all competing technologies. Ease of use and speed will continue to improve with advances in printhead design for longer print swath, higher operating frequencies, and greater drop dynamics control. The versatility of inkjet printing will continue to move "upscale" to rival even offset presses as page-wide array printheads become commercially available.

For background information *see* DYE; INK; PIEZO-ELECTRICITY; PIGMENT (MATERIAL); PRINTING in the McGraw-Hill Encyclopedia of Science & Technology.
                                                Rob Beeson

Bibliography. R. Beeson, Is print quality an issue anymore—Or is it performance, performance?, *GIGA 25th Global Ink Jet and Thermal Printing Conference*, 2002; H. Le, Fabrication materials and process on ink jet printheads, *IS&T Non-impact Printing Conference Tutorial*, 2000; R. Mills, Ink jet in a competitive market place, *IMI 10th Annual European Ink Jet Printing Conference*, 2002.

James Webb Space Telescope (JWST)

The National Aeronautics and Space Administration (NASA) plans to launch the James Webb Space Telescope, which in many ways is the scientific and technological successor of the Hubble Space Telescope, in 2011. In 2002, the company now known as Northrop Grumman Space Technology (which was then TRW Space and Electronics) and its partners, Ball Aerospace and Eastman Kodak, were awarded the prime contract to build the observatory, formerly known as the Next Generation Space Telescope.

Observing early stars and galaxies. Equipped with a large, 6.6-m-class (260-in.) deployable mirror (**Fig. 1**) and a suite of revolutionary, infrared-sensing cameras and spectrometers, JWST will make it possible to see even farther into space than is currently possible with Hubble (**Fig. 2**) and will help to analyze faint sources of light that Hubble cannot even detect. Light from these nascent stars and galaxies, emitted early in the history of the universe, has traveled so far by the time it reaches us that it has stretched into the infrared wavelength band and is invisible to the human eye.

Consequently, no one up to now had the tools to observe this cosmic "dark zone," but this "first light machine" will finally reveal what the universe looked like when it was a fraction of its current age and size, and the first stars and galaxies were beginning to take form. In addition, JWST will demonstrate new technologies needed for future missions. For this reason, the National Academy of Science has ranked JWST as one of NASA's top science goals.

Additional scientific goals. In addition to observing these young galaxies, JWST will tackle four other major objectives over its 5–10-year lifetime.

Fig. 1. Conception of the James Webb Space Telescope (JWST), with a 6.6-m-class (260-in.) mirror and large sunshield. (*NASA Goddard Space Flight Center*)

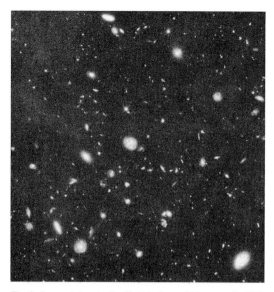

Fig. 2. Image taken by the Hubble Space Telescope in 1998 of a small area of the southern sky with galaxies whose light was emitted 12 billion years ago, early in the history of the universe, and has thus traveled 12 billion light-years to reach us. The James Webb Space Telescope will be able to observe even farther into space, to a time when stars and galaxies were beginning to take form. (*Space Telescope Science Institute; NASA*)

JWST will help determine the geometry of the universe, its age, and its ultimate fate. In 2000, two teams of astronomers found evidence that the expansion of the universe is accelerating rather than slowing down. Their observations seemed to confirm the existence of a new form of energy that causes the expansion of the universe to accelerate. JWST is capable of studying this phenomenon.

Although mission planners designed the spacecraft primarily to observe the farthest reaches of the universe, it also can look closer to home. With JWST, scientists can study the history of the Milky Way and its nearby neighbors by studying the old stars and star remnants that formed over the galaxy's lifetime.

Astronomers also will use JWST to study star birth and formation. Its infrared sensors can pierce the dust and gas that surround stellar nurseries and reveal the processes that dictate the mass and composition of stars, as well as the production of heavy elements.

Finally, NASA designed JWST to study the origin and evolution of planetary systems like our own. JWST may be able to directly detect large, Jupiter-sized planets around nearby stars. Although smaller planets cannot be imaged directly, JWST's high resolution will make it possible to see how they behave as a planetary system, especially when they are in the process of formation, which will provide a larger picture of their evolution.

Reduction of size and cost. A thousand times more sensitive in the infrared than Hubble, JWST will accomplish far more than what current ground- and space-based observatories can do. Yet, JWST will achieve this at a fraction of Hubble's size and overall cost. JWST will weigh about 5000 kg (11,000 lb), compared with Hubble's 11,000 kg (24,000 lb), and a medium-sized rocket, such as the Atlas 5, will likely launch the spacecraft. The European Space Agency has agreed to provide an Ariane 5 launcher. *See* SATELLITE LAUNCH VEHICLES.

NASA could not have considered a mission of this magnitude just a few years ago. Since NASA began studying the mission in 1995, the agency has made significant progress in advancing technologies and management approaches that would allow it to pack a lot of scientific capability into a relatively small package.

Orbital considerations. JWST's orbit (**Fig. 3**)—at the second Lagrangian point (L2) of the Sun-Earth system, located 1.5 million kilometers (940,000 mi) from the Earth in the anti-Sun direction—allows NASA to perform this mission. (The point L2 is one of the five Lagrangian points in the orbital plane of the Earth and Sun at which a third object of negligible mass can remain in equilibrium.) The L2 orbit offers a thermally stable environment. At the L2 point, JWST will be in orbit around the Sun rather than the Earth, as with the Hubble Space Telescope. This arrangement will allow JWST to reside in the shadow of a lightweight sunshield, the size of a tennis court, which will deploy in orbit (Fig. 1). In this shadow, JWST can passively cool to about 35 K (about −400°F). Although passive cooling is an old concept, NASA has never flown a mission before that uses this method to reach these extreme temperatures.

Cryogenic cooling. To observe the farthest reaches of the universe, temperature is an essential consideration. Observations in the near- and mid-infrared wavelength bands (0.6–0.9 to 28 μm) cannot be conducted at temperatures above 35 K. Anything warmer would create too much of its own infrared

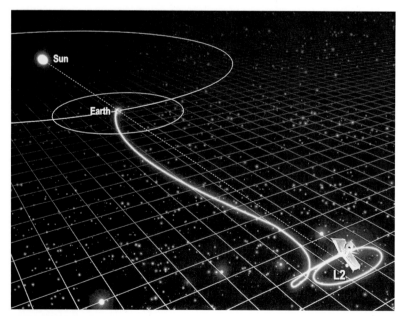

Fig. 3. Trajectory of the James Webb Space Telescope from Earth to the neighborhood of the second Lagrangian point (L2) of the Sun-Earth system, and its orbit near L2. (*NASA Goddard Space Flight Center*)

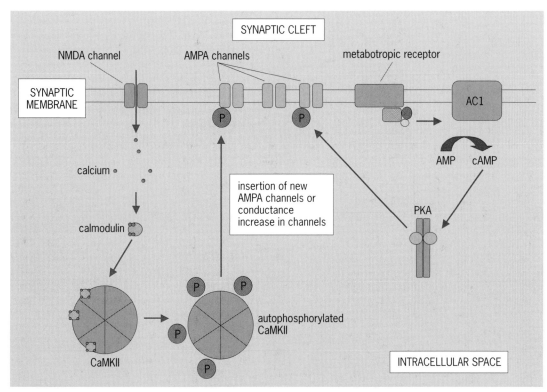

Fig. 2. Signaling pathways controlling synaptic plasticity. The calcium-calmodulin-type II kinase (CaMKII) pathway is shown on the left. Calcium entering through NMDA receptors or voltage-gated ion channels (not shown) binds to calmodulin, which binds to CaMKII. CaMKII is a multimeric protein composed of two six-member rings (only one ring shown here). By autophosphorylating (circles) at the T286A site (see text), the enzyme becomes active in the absence of calcium/calmodulin and can phosphorylate substrates, including AMPA channels. The protein kinase A (PKA) pathway is shown on the right. Activation of a metabotropic receptor (which is not an ion channel and thus affects neurotransmission indirectly, for example, a glutamate receptor or perhaps a catecholamine receptor) causes adenyl cyclase activity to increase (ACI). The result is an increase in cyclic AMP (cAMP) and hence PKA activity. PKA is a four-member kinase composed of two regulatory domains (circles) and two catalytic domains (bars). Once activated, the kinase can phosphorylate substrates such as the AMPA channel.

cannot remember where it is from one day to the next. Furthermore, it is sufficient to alter a single amino acid at the autophosphorylation site on CaMKII (T286A) to decrease CAMKII function and impair spatial memory (Fig. 2). Other forms of memory are also affected by CaMKII mutations, including fear conditioning in which the animal has to remember the context in which an aversive foot shock was received.

Plasticity. Knockout mice have also been used to show that CaMKII mutations affect neocortical plasticity (the ability of the brain to modify itself in response to changing conditions). The body is represented in a map in the somatosensory cortex; that is, adjacent parts of the body are mapped onto adjacent parts of the somatosensory cortex. In the mouse, the map is largely occupied by whisker representation, much the same way as the human map is largely occupied by hand representation. The functional map of the whiskers can change depending on whisker use. If all but a single whisker is removed, the representation of that spared whisker expands into new territory within the cortex. This shift in the map occurs in normal mice but is absent in CaMKII knockouts

and in T286A point mutants, indicating that CaMKII plays a fundamental role in many forms of plasticity in the brain.

Development. Occasionally, natural knockouts occur that have a clear phenotype that leads to identification of the knocked-out gene and its role in a particular process. One example is the "barrelless" mouse. The whisker representation in the mouse somatosensory cortex has a very distinct pattern that can easily be seen in horizontal sections of cortex processed with anatomical strains. Each whisker corresponds to a discrete "barrel" in the cortex; therefore, the same pattern made by the whiskers can be seen in the pattern of barrels in the cortex. The distinguishing feature in the barrelless mouse is that the barrels do not form. Otherwise, the animal looks relatively normal. Analysis of the genome revealed that the animal lacked the gene for adenyl-cylase I (ACI) which is an important enzyme in the brain for turning a linear molecule (AMP) into a ring form, cyclic AMP (cAMP), which in turn leads to activation of a molecular signaling cascade involving protein kinase A (PKA) (Fig. 2). Following this discovery, it was found that the ACI/cAMP/PKA pathway plays an important role in long-term potentiation at the thalamocortical synapse, which is the synapse bringing information from the whiskers to the barrels within the cortex. So the study of a serendipitous knockout has led to a logical examination of the signaling pathway leading to and from ACI in an attempt to understand a basic developmental event within the cortex.

Genetic models of neurological disease. Animal models of disease are extremely useful for studying the basis of the disorder and developing treatments that work in the animal before transferring them to the clinic. Some diseases are due to a missing or inactivated gene, but generally they are best modeled by overexpressing a gene; for example, Alzheimer's disease has been modeled by overexpressing the gene encoding beta-amyloid, and Huntington's disease has been modeled by overexpressing a gene with extra CAG (cytosine, adenine, guanine) repeat sequences in the code.

Alzheimer's disease. Alzheimer's disease is usually associated with abnormalities in genes encoding APP (amyloid precursor protein) and PS1 and PS2 (presenilins 1 and 2). Genetic mutations causing over- or underexpression of APP and/or PS1 have not generated animal models that mimic the Alzheimer's disease state. A more successful method has been to overexpress a mutated form of beta-amyloid gene known as the Swedish mutation (APPswe). Animals with this mutation show memory impairment and the high levels of beta-amyloid deposition and plaques that are characteristic of Alzheimer's disease. These animals, therefore, appear to provide a useful test bed for developing therapies for Alzheimer's disease.

Huntington's disease. Huntington's disease is a movement disorder associated with a variable number of CAG repeats in a gene (known as Huntingtin) carried only by those at risk of succumbing to the disorder. The number of CAG repeats tends to increase from one generation to the next, and the more repeats present, the higher the risk of showing symptoms at earlier ages. Transgenic mice that have been generated to have 115–150 CAG repeats show some motor dysfunction (diskinesias and tremors), suggesting that the disease can be modeled this way. The natural progression for studies of neurodegenerative diseases in mutant models is to treat the disease process in the animal models, cure it, and transfer that treatment to the clinic. It is likely that many studies of this type will evolve over the next few years.

For background information *see* ALZHEIMER'S DISEASE; GENE; GENE ACTION; GENETIC ENGINEERING; HUNTINGTON'S DISEASE; MEMORY; NEUROBIOLOGY; SYNAPTIC TRANSMISSION in the McGraw-Hill Encyclopedia of Science & Technology.   Kevin Fox

Bibliography. M. R. Capecchi, The new mouse genetics: Altering the genome by gene targeting, *Trends Genet.*, 1989; J. Z. Tsien et al., Subregion- and cell type-restricted gene knockout in mouse brain, *Cell*, 87:1317–1326, 1996; P. Popko (ed.), Mouse models in the study of genetic neurological disorders, *Advances in Neurochemistry*, vol. 9, Kluwer Accademic, New York.

Leptospirosis

Leptospirosis is a bacterial disease that is now recognized as one of the new, emerging infectious diseases. The Centers for Disease Control and Prevention (CDC) has defined such diseases as "new, reemerging, or drug-resistant infections whose incidence in humans has increased within the past two decades or whose incidence threatens to increase in the near future." Leptospirosis, presumed to be the most widespread zoonosis (a disease of animals that may be transmitted to humans) in the world, is caused by pathogenic *Leptospira* species and is characterized by a broad spectrum of clinical manifestations, varying from a subclinical infection (showing no symptoms) to a fatal disease.

Etiologic agents. Leptospires are spirochetes (flexible, spiral-shaped bacteria with flagella) that are coiled, thin, and highly motile; have hooked ends; and contain two flagella, which enable the bacteria to move in a watery environment and burrow into and through tissue. Traditionally, the genus *Leptospira* contains two species: the pathogenic *L. interrogans* and the free-living *L. biflexa*. Leptospires are phylogenetically related to other spirochetes that have a genome approximately 5000 kilobases in size. Recently, a number of leptospiral genes have been cloned and analyzed.

Epidemiology. Leptospirosis is primarily an animal disease, with a worldwide distribution that affects at least 160 mammalian species. The incidence is significantly higher in warmer climates than in temperate regions, mainly due to longer survival of leptospires under warm, humid conditions.

Animals can be divided into maintenance hosts and accidental hosts.

Maintenance hosts. Maintenance hosts are a species in which infection is endemic and is usually transferred from animal to animal by direct contact. Infection is usually acquired at an early age, and the prevalence of chronic excretion of the leptospires in the urine increases with the age of the animal. The disease is maintained in nature by chronic infection of the renal tubules of maintenance hosts. Rats, mice, opossums, skunks, raccoons, and foxes are the most important environmental maintenance hosts, with farm animals and dogs being domestic maintenance hosts. All of these animals shed the bacteria in their urine.

Accidental hosts. Humans are not commonly infected with leptospires and are considered an accidental (incidental) host. Approximately 100 human cases of leptospirosis are reported to the CDC each year, but this is probably an underestimation of the total number of cases.

Transmission. Transmission of leptospires may follow direct contact with urine, blood, or tissue from an infected animal or exposure to a contaminated environment. Human-to-human transmission is rare. Since leptospires are excreted in the urine, they can survive in water for many months. Leptospires can enter the host through abrasions in the skin or though mucous membranes such as the conjunctiva. Drinking contaminated water may introduce the bacteria through the mouth, throat, or esophagus. The incubation period is usually 1–2 weeks.

Certain occupational groups are at high risk for acquiring leptospires. These include veterinarians, sewage workers, agricultural workers, livestock farmers, slaughterhouse workers, meat inspectors, rodent-control workers, and fish workers. These individuals acquire the bacteria by direct exposure or contact with contaminated soil or water. In western countries, there is a significant risk of acquiring a leptospirosis infection from recreational exposure (canoeing, windsurfing, swimming, water skiing, white-water rafting, fresh-water fishing) and domestic animal contact.

Pathogenesis. The mechanisms by which leptospires cause disease are not well understood. A number of putative factors have been suggested, but their role in pathology remains unclear. It is known that after the leptospires enter the host they spread to all organs of the body. The bacteria reproduce in the blood and tissues. Although any organ can be infected, the kidneys and liver are most commonly involved. In these organs, the leptospires attach to and destroy the capillary walls. In the liver this leads to jaundice, and in the kidneys it causes inflammation (nephritis) and tubular cell death (necrosis), resulting in renal failure. After antibodies are formed, the leptospires are eliminated from all organ sites except the eye and kidneys, where they may persist for weeks or months; the reason for this is unknown.

Clinical manifestations. The clinical manifestations of leptospirosis can be divided into two forms: anicteric (not jaundiced) and icteric (jaundiced).

Anicteric. The majority of infections caused by the leptospires are anicteric (either subclinical or of very mild severity), and patients usually do not seek medical attention. In this mild form, leptospirosis may be similar to an influenzalike illness with headache, fever, chills, nausea, vomiting, and muscle pain. Fever is the most common finding during a physical examination. Most patients become asymptomatic within 1 week.

Icteric. Between 5 and 15% of all patients with leptospirosis have the icteric (severe) form of the disease, which is characterized by jaundice, renal and liver dysfunction, hemorrhaging, and multiorgan involvement and can result in death. This severe form is also referred to as Weil's syndrome (named in 1886 after Adolf Weil, a German physician). The mortality rate for this form of the disease is 5–15%. The jaundice occurring in leptospirosis is not associated with

hepatocellular necrosis, and liver function returns to normal after recovery. Acute infections in pregnancy have also been reported to cause abortion and fetal death.

The clinical presentation of leptospirosis is biphasic, with the acute or septicemic (characterized by bacteria in the blood) phase lasting about a week. This is followed by the immune phase, characterized by antibody production and excretion of the leptospires in the urine. Most of the complications of leptospirosis are associated with localization of the leptospires within the tissues during the immune phase and thus occur during the second week of the infection.

Laboratory findings. Since the kidneys are invariably involved in letospirosis, related findings include presence of white blood cells, red blood cells, and granular casts (proteinaceous products of the kidney in the urine); mild proteinuria (protein in the urine); and in severe leptospirosis, renal failure and azotemia (an increase in nitrogenous substances in the blood). The most common x-ray finding in the kidney is a patchy pattern that corresponds to the many hemorrhages that occur due to the destruction of blood vessels.

Diagnosis. A definitive diagnosis of leptospirosis is based either on the isolation of the leptospires from the patient or on immunological tests. Leptospires can be isolated from the blood and/or cerebral spinal fluid during the first 10 days of the illness, and from the urine for several weeks beginning about the first week of the infection. They can be stained using carbol fuchsin, counterstained, and cultured in a simple medium enriched with vitamins.

Treatment. Treatment of leptospirosis differs depending on the severity and duration of symptoms at the time of presentation. Patients with only flulike symptoms require only symptomatic treatment but should be cautioned to seek further medical help if they develop jaundice. The effect of antimicrobial therapy for the mild form of leptospirosis is controversial, but such treatment is definitely indicated for the severe form of the disease. Treatment should begin as soon as a severe diagnosis is made. In mild cases, oral treatment with tetracycline, doxycycline, ampicillin, or amoxicillin should be considered. For more severe cases of leptospirosis, intravenous administration of penicillin G, amoxicillin, ampicillin, or erythromycin is recommended. Most patients with leptospirosis recover. Mortality is highest in the elderly and those who have Weil's syndrome. Leptospirosis during pregnancy is associated with high fetal mortality. Long-term follow-up of patients with renal failure and hepatic dysfunction has documented good recovery of both renal and liver function.

Prevention. Individuals who may be exposed to leptospires through either their occupation or involvement in recreational water activities should be informed about the risks. Some measures for controlling leptospirosis include avoiding exposure to urine and tissues from infected animals, vaccination of domestic animals, and rodent control.

For background information *see* BACTERIA; LEPTOSPIROSIS; MEDICAL BACTERIOLOGY; SPIROCHETE; ZOONOSES in the McGraw-Hill Encyclopedia of Science & Technology.   John P. Harley

Bibliography. R. W. Farrar, Leptospirosis, *Crit. Rev. Clin. Lab. Sci.*, 21(1), 1995; R. W. H. Gillespie, Epidemiology of leptospirosis, *Amer. J. Pub. Health*, 53, 1963; P. N. Levett, Leptospirosis, *Clin. Microbiol. Rev.*, 14(2), 2001; R. van Crevel, Leptospirosis in travelers, *Clin. Infect. Dis.*, 19:132, 2001.

Lysosome-related organelles

Lysosome-related organelles are a diverse group of specialized intracellular membrane-enclosed compartments that exhibit certain characteristics of lysosomes (major membrane-bound digestive structures involved in intracellular degradation of biological materials in all eukaryotic cells). Lysosomes contain a unique collection of digestive enzymes and membrane proteins, maintain an acidic luminal pH, and receive ingested as well as newly synthesized materials. Although most of these features are shared to a certain extent by all lysosome-related organelles, each of these organelles also contains cell-type–specific components responsible for its specialized functions.

Examples. Lysosome-related organelles include the following, each found in a different type of cell.

Melanosomes. Melanosomes are specialized membrane-bound organelles produced in melanocytes in the skin and retinal pigmented epithelial cells in the eye for synthesis and storage of the pigment melanin. Melanosomes share several features with lysosomes and appear to coexist with conventional lysosomes in melanocytes. In mammalian skin, melanosomes transport melanins from melanocytes to neighboring keratinocytes (epidermal cells that produce keratin, the principal protein of the skin, hair, and nails) to generate the color of skin and hair and form a barrier against the harmful effects of ultraviolet light. In the eye, in addition to being the primary determinant of eye color, melanosomes absorb stray light that enters the eye and may function as scavengers of toxic metabolic intermediates.

Platelet dense granules. Platelets, circulating cellular fragments derived from precursor cells known as bone marrow megakaryocytes, are crucial for repair of blood vessels after injury. They come equipped with three major types of secretory granules: α-granules, dense granules, and lysosomes. Platelet dense granules store high concentrations of small molecules, such as adenosine diphosphate, adenosine triphosphate, serotonin, and calcium, which are important in mediating platelet aggregation leading to blood clot formation. Their membranes are enriched in proteins typically found in lysosomes.

Lamellar bodies. Lamellar bodies of lung type II epithelial cells are lysosome-related organelles specialized for the storage and secretion of pulmonary surfactant, a phospholipid-rich luminal material arranged in the form of tightly packed membrane sheets, or lamellae. Once secreted, surfactant forms a phospholipid film at the air-liquid interface that reduces surface tension and prevents pulmonary alveoli (air sacs) from collapsing. In addition to surfactant, lamellar bodies contain lysosomal membrane proteins and soluble digestive enzymes.

Lytic granules. Cytotoxic T lymphocytes of the mammalian immune system provide protection against intracellular pathogens, such as viruses and some bacteria and parasites that multiply in the host-cell cytoplasm. Once activated, a cytotoxic T-lymphocyte recognizes the infected cell and discharges on it a lethal dose of cell-killing agents that have been stored in its specialized secretory organelles known as lytic granules. In addition to their storage and secretory function, lytic granules serve as lysosomes in cytotoxic T-lymphocytes.

Vesicular trafficking and biogenesis. Lysosome-related organelles are part of the eukaryotic endomembrane system that exports newly synthesized proteins via the secretory pathway (in which proteins are synthesized in the endoplasmic reticulum and then sent to the Golgi apparatus, where they are sorted into transport vehicles destined for lysosomes) and imports materials via the endocytic pathway (in which endosomes transport newly ingested material to lysosomes). Transport within this complex system of organelles is mediated by membrane-enclosed carriers that bud from a donor compartment, move through the cytoplasm, and then fuse with an acceptor compartment—a process known as vesicular trafficking. Each step of this process involves a set of vesicular trafficking "machinery" that ensures specificity and efficiency. For example, transport carrier formation and selection of its cargo (the material it will transport) often involve specific cytoplasmic proteins that form a coat on the membrane of the donor compartment. One major class of coating contains the protein clathrin, which can be assembled into a cagelike structure, and one of several related adaptor protein (AP) complexes consisting of four different polypeptide subunits. By connecting cargo receptors in the donor membrane with scaffold proteins (such as clathrin) in the cytoplasm, adaptor proteins recruit specific cargo molecules from inside the donor compartment and deliver them to the budding transport carriers (see **illus.**). Of the four known adaptor protein complexes (AP-1, AP-2, AP-3, and AP-4), AP-2 functions in clathrin-mediated endocytosis at the plasma membrane, whereas the other AP complexes apparently function to produce transport carriers from the trans-Golgi network and/or endosomes.

Biogenesis, or formation of organelles, within the endomembrane system is also thought to involve vesicular trafficking. For example, delivery of newly synthesized lysosomal enzymes to lysosomes requires selecting and packaging these enzymes into transport carriers at the trans-Golgi network, a process that involves clathrin, AP-1, and the proteins associated with them.

Current research. The features that lysosome-related organelles share with lysosomes reflect some shared cellular processes and pathways involved in their biogenesis. This notion is supported by the observation that a single mutation can affect the structure and function of several of these organelles. For example, the human genetic disease Hermansky-Pudlak syndrome is characterized at the cellular level by defects in multiple cytoplasmic organelles, including melanosomes, platelet dense granules, lamellar bodies of lung type II epithelial cells, and lysosomes. Consequently, patients with this disease exhibit reduced pigmentation of skin and eyes, prolonged bleeding, and progressive accumulation of partially degraded materials in lysosomes. Our understanding of lysosome-related organelle formation comes largely from the recent identification of mutated genes associated with disorders of lysosome-related organelles in animal models and humans.

The first clue of the existence of a specialized vesicular trafficking pathway for the biogenesis of lysosome-related organelles was provided by the discovery that a gene whose mutation alters the eye color of the fruit fly *Drosophila melanogaster* encodes a subunit of the AP-3 adaptor, and the mutant flies have abnormal pigment granules that, like mammalian melanosomes, are related to lysosomes. Since there are many types of fruit fly mutants with abnormal eye color, *Drosophila* has emerged as a model system for the identification of genes involved in the biogenesis of lysosome-related organelles. Recent analyses of the nine eye-color genes of the fruit fly suggest that at least two vesicular trafficking pathways are involved in pigment granule biogenesis, one of which requires the AP-3 adaptor protein complex.

Studies on mouse coat color mutants indicate that at least 16 distinct mutations produce phenotypes related to Hermansky-Pudlak syndrome in mice; that is, abnormalities in melanosomes, platelet dense granules, and lysosomes. So far, 12 mouse genes related to Hermansky-Pudlak syndrome have been identified, and 6 of them are mouse counterparts of the genes mutated in patients with Hermansky-Pudlak syndrome. Several of them, including genes encoding AP-3 subunits, are counterparts of the fruit fly eye-color genes, demonstrating that the molecular mechanisms involved in the biogenesis of lysosome-related organelles are conserved in evolution. Analyses of the mouse models of Hermansky-Pudlak syndrome further indicate that the products of most genes related to the syndrome are involved in the biogenesis of lysosome-related organelles by mechanisms independent of the AP-3 adaptor, consistent with the observations made in the fruit fly.

Human diseases. Genetic defects that affect multiple lysosome-related organelles are known to be associated with a group of autosomal recessive disorders such as Hermansky-Pudlak syndrome, Chediak-

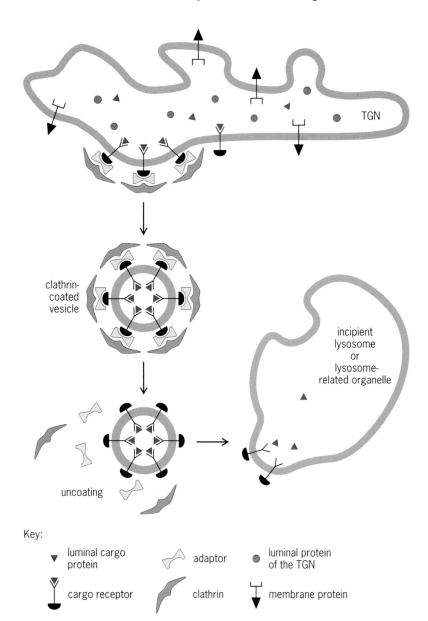

Key:

▼ luminal cargo protein	⟁ adaptor	● luminal protein of the TGN
⊻ cargo receptor	⌒ clathrin	⊥↓ membrane protein

Recruitment of cargo molecules into clathrin-coated vesicles at the trans-Golgi network (TGN) for delivery to incipient lysosomes or lysosome-related organelles. In the TGN, soluble lysosomal or lysosome-related organellar proteins (cargos) bind to the luminal domain of their receptors in the TGN membrane. The cytoplasmic domain of the cargo receptors contains sorting signals that are recognized by the cytoplasmic adaptors, which are heterotetrameric protein complexes. The adaptors, in turn, bind to clathrin, which assembles into a cagelike coat, thus trapping specific cargo molecules in the budding transport vesicles. Other proteins responsible for the targeted delivery of the vesicle to the incipient lysosome or lysosome-related organelle are presumed to also be segregated into the same vesicle.

Higashi syndrome, Griscelli syndrome, and Elejalde syndrome.

Hermansky-Pudlak syndrome. Patients with Hermansky-Pudlak syndrome exhibit clinical symptoms that include hypopigmentation (reduced pigmentation) of skin and eyes, prolonged bleeding episodes, progressive pulmonary fibrosis, and accumulation of partially degraded materials in lysosomes. These features result from defective melanosomes, platelet dense granules, lamellar bodies of lung type II epithelial cells, and lysosomes, respectively. Mutations in at least six genes can cause Hermansky-Pudlak syndrome. The most common type of the disease results from mutations in the *HPS1* gene. Currently, the only gene with known biochemical function is *HSP2*, which encodes a subunit of the AP-3 adaptor.

Chediak-Higashi syndrome. Chediak-Higashi syndrome is characterized by hypopigmentation of skin and eyes, bleeding tendency, progressive neurological impairment, and severe immunodeficiency. At the cellular level, a diagnostic feature of Chediak-Higashi syndrome is the presence of giant lysosomes, melanosomes, platelet dense granules, and lytic granules of cytotoxic T-lymphocytes, probably resulting from uncontrolled organelle fusion. The gene defective in Chediak-Higashi syndrome encodes a large protein whose biochemical function has yet to be identified.

Griscelli syndrome. Griscelli syndrome shows hypopigmentation of the skin and hair, due to lack of melanosome transfer from melanocytes to keratinocytes, and immune system abnormalities that result in part from reduced exocytosis of lytic granules in cytotoxic T-lymphocytes. The disease results from mutations in the gene encoding Rab27A, a small guanosine triphosphate–binding protein involved in vesicle trafficking.

Elejalde syndrome. Elejalde syndrome patients have a hypopigmentation phenotype indistinguishable from that of Griscelli syndrome, but no apparent immune abnormalities. Instead, they show a severe neurological impairment early in life, and their cells accumulate abnormal inclusion bodies that appear to represent abnormal lysosomes. The gene defective in Elejalde syndrome encodes Myosin-5A, an actin-based molecular motor. The identical hypopigmentary phenotype of Griscelli syndrome and Elejalde syndrome patients can be explained by recent findings showing that in melanocytes Rab27A and Myosin-5A act together to keep melanosomes in the peripheral dendrites or extensions of the melanocyte, where transfer of melanosomes to surrounding keratinocytes takes place. Thus, defects in either Rab27A (as in Griscelli syndrome) or Myosin-5A (as in Elejalde syndrome) will result in lack of melanosome transfer from melanocytes to keratinocytes and hypopigmentation of the skin.

In addition to multiorganelle disorders, many human genetic diseases are associated with defects that affect a single lysosome-related organelle function. For example, loss-of-function mutations in the gene encoding tyrosinase, the melanosomal enzyme that catalyzes the rate-limiting step of melanin biosynthesis, result in reduced or even absent pigmentation without affecting other organelles.

For background information *see* CELL (BIOLOGY); ENDOCYTOSIS; ENDOPLASMIC RETICULUM; ENZYME; GENE ACTION; GOLGI APPARATUS; SECRETION in the McGraw-Hill Encyclopedia of Science & Technology.

Mindong Ren

Bibliography. E. C. Dell'Angelica et al., Lysosome-related organelles, *FASEB J.*, 14:1265–1278, 2000; V. Lloyd, M. Ramaswami, and H. Kramer, Not just pretty eyes: Drosophila eye-colour mutations and lysosomal delivery, *Trends Cell Biol.*, 8:257–259, 1998; R. A. Spritz et al., Human and mouse disorders of pigmentation, *Curr. Opin. Genet. Dev.*, 13:284–289, 2003.

Mantophasmatodea (insect systematics)

With almost a million species known, the insects include more than two-thirds of all animal species in the world. Moreover, several thousand insect species new to science are described every year. Many of these are freshly collected in the wild, but many others are traced among the millions of insect specimens stored in each natural history museum. While this work is rarely noted by the public, the description of two closely related insect species from Africa—*Mantophasma zephyra* from Namibia and *M. subsolana* from Tanzania—in April 2002 was reported by media around the world. These wingless creatures are inconspicuous, about 1–2.5 cm (0.4–1 in.) long, and superficially resemble a mixture between a praying mantis (Mantodea) and a stick insect (Phasmatodea). However, the fact that they constitute a previously unknown order of insects—now called the Mantophasmatodea (heel-walkers, gladiators)—makes their discovery one of the greatest zoological events of recent years.

Systematic hierarchy. Biologists sort all organisms into a hierarchical system that ideally reflects their evolutionary relationships (that is, the phylogenetic tree): closely related species are classified as being in the same genus; related genera constitute a family; related families are included in the same order; and related orders together form a class, such as the Insecta. Each group in this system is characterized by particular modifications (called apomorphies) that have evolved in the stem lineage of this group and have been inherited by the descendants. Such apomorphies can be changes in the body structure, cell structure, deoxyribonucleic acid (DNA) sequence, behavior, metabolism, or any other pattern exhibited by organisms. Nonetheless, because apomorphies can become lost in some descendants and the same apomorphies can evolve independently in different lineages (homoplasy), reconstructing the tree of life is full of conflicts; much of the tree has so far remained unresolved.

Insect orders. Phylogenetic orders in insects essentially correspond to what are considered by nonzoologists to be the "basic" types of insects. They total approximately 35 and include the Coleoptera (beetles), Lepidoptera (butterflies and moths), Blattaria (cockroaches), Odonata (damselflies and dragonflies), and Phasmatodea (stick insects). (The number of orders varies slightly, because in different classifications some groups are ranked as several closely related orders, or as several suborders of a single order.) Each order is characterized by its own particular set of apomorphies. The transformation of the forewings into hard elytra (protective coverings of hindwings) in the beetles, or of body hairs into scales in the Lepidoptera, are examples of easily recognized apomorphies; others are more subtle, such as changes in the location of muscle insertions or in the supply route of a nerve.

Each insect species discovered during the past decades displayed the set of apomorphies characterizing some particular insect order and, thus, could be assigned to one of the known orders. However, Mantophasmatodea do not display the characteristic modifications of any of the previously recognized insect orders and, therefore, had to be assigned to a new order. The significance of their discovery is best reflected by the fact that the last two discoveries of new insect orders date back almost a century. These are the tropical Zoraptera (presently 30 species known) and the Notoptera, or ice-crawlers, of northeastern Asia and northern America (presently 26 species known), discovered in 1913 and 1914, respectively. In addition, the silverfish-like *Tricholepidion gertschi* from California, first described in 1961, constitutes the only case in which ordinal status has remained conjectural.

Discovery of Mantophasmatodea. In 2001, the discovery of Mantophasmatodea was initiated when a specialist of stick insects (O. Zompro) was shown a strange-looking, unidentified male insect (collected in Tanzania in 1950) at the British Museum in London, and somewhat later found a similar but female specimen (collected in Namibia in 1909) in the insect collection of the Museum für Naturkunde in Berlin. Subsequent studies conducted by a specialist of insect anatomy and phylogenetics (K.-D. Klass) yielded the scientific basis for ranking these insects as a new order. The female and male specimens were eventually described as *M. zephyra* and *M. subsolana*, respectively, and on these the new order Mantophasmatodea is based.

In addition to these extant species, an extinct member of Mantophasmatodea, *Raptophasma kerneggeri*, was identified in 45-million-year-old Baltic amber; nymphs (larvae) of these fossils had been known since 1997, while the first adults appeared in 2001. (However, because many of the characters relevant for the ordinal placement of an insect are not visible in fossils, only the anatomical studies in the two extant specimens could justify the creation of a new order, Mantophasmatodea, to which the fossils were assigned thereafter.) The amber-encased fossils show that in the early Tertiary the Mantophasmatodea also inhabited Europe, whereas all known extant species are restricted to Africa south of the Equator.

The first living Mantophasmatodea were traced in March 2002 on an expedition to the Brandberg, a mountain in northern Namibia. (Live populations in Tanzania still remain to be traced.) Findings included a new species, *Praedatophasma maraisi*, and specimens possibly belonging to *M. zephyra*. Based on these specimens, studies of sperm ultrastructure and DNA sequences were conducted. The DNA studies confirmed the ordinal status of Mantophasmatodea.

In May 2002, several additional species of Mantophasmatodea were found in a number of localities in western South Africa. It eventually became evident that Mantophasmatodea are quite common in southwestern Africa. The various species seem to occupy fairly restricted, largely exclusive geographic areas, thus displaying a high degree of endemism.

Life history. Mantophasmatodea inhabit several major biomes of southwestern Africa: the dry Nama Karoo and Succulent Karoo, and the more humid Fynbos. In the first, precipitation predominantly falls during summer, whereas the second and third are characterized by winter rainfall; all are essentially bare of trees.

Heel-walkers live singly, and there seems to be both diurnal and nocturnal activity. They are often found in tufts of grass or Cape reed, and they prey on other insects (up to their own size), which they grasp using their strong, spiny forelegs and midlegs. Movements are mostly slow but can be rapid when prey is caught or copulation is initiated. When walking, Mantophasmatodea usually have their basal tarsomere (upper foot segment) on the ground, while the fifth tarsomere (lowermost foot segment) and the terminal claws and arolium (a large lobe between the claws) are held in an elevated position. The males use the projection of their subgenital plate to knock on the ground (which may be a form of communication).

For copulation, the male jumps onto the female and bends his abdomen aside and downward to bring the copulatory organs into contact. The copulation, with the male sitting on the female's back, can last up to 3 days. The female then produces several sausage-shaped egg pods, each containing 10–20 eggs enclosed in a hard envelope made from gland secretions and sand. The nymphs, which resemble the adults, hatch in the early humid season, molt several times, and reach maturity near the end of the humid season. Accordingly, life cycles appear to differ diametrically in areas of winter versus summer rainfall.

Morphology. Except for their lack of wings (see **illus.**), Mantophasmatodea show a fairly generalized body structure, which includes the possession of an ovipositor (egg-laying device) on the posterior abdominal segments in the female. Yet, there are some apomorphies that distinguish Mantophasmatodea from other insects: (1) the location of the anterior tentorial pits (anterior invagination points of the endoskeleton of the head) far above the anterior articulation of the mandible; (2) the presence of a small triangular projection beyond the third segment of the tarsi (feet); and (3) the presence of a projection on the middle of the male's subgenital plate. The various species differ in details of the very complicated male and female genitalia and in the coloration of various body parts; however, coloration also varies strongly within species. *Praedatophasma maraisi* is characterized by its spiny thorax.

Phylogenetic placement. Anatomical details of the abdominal spiracles, the ventral muscles in the abdomen, and the ovipositor of the female show that despite their lacking wings the Mantophasmatodea belong to the Pterygota, which comprise all winged

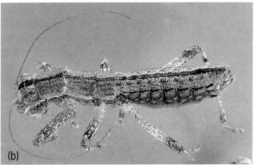

An undescribed species of Mantophasmatodea from western South Africa: (a) male, (b) female. Males and females can be distinguished by the structure of the hind part of the abdomen, and males are usually smaller and more gracile than females. (*Courtesy of Elke Klass*)

insects and many secondarily wingless ones. Furthermore, DNA studies suggest the wingless Notoptera to be the closest relatives of Mantophasmatodea. The morphological evidence has remained highly conjectural, though a close relationship to the Notoptera also appears possible based on the structure of the foregut.

Current research. A team of entomologists who cover a variety of research fields has been formed to continue the study of heel-walkers. This includes the taxonomic treatment of the eight species so far known from South Africa and the search for additional species. Further core issues are the study of the morphology, anatomy, embryology, nymphal development, chromosome sets, and life history, and the reconstruction of phylogenetic and phylogeographic relationships using morphological and molecular data.

For background information *see* ANIMAL SYSTEMATICS; FOSSIL; INSECTA; ORTHOPTERA; PHYLOGENY; TAXONOMY in the McGraw-Hill Encyclopedia of Science & Technology.   Klaus-Dieter Klass

Bibliography. A. Arillo, V. M. Ortuño, and A. Nel, Description of an enigmatic insect from Baltic amber, *Bulletin de la Société Entomologique de France*, 102:11–14, 1997; K.-D. Klass et al., Mantophasmatodea: A new insect order with extant members in the Afrotropics, *Science*, 296:1456–1459, 2002; M. D. Picker, J. F. Colville, and S. van Noort, Mantophasmatodea now in South Africa, *Science*, 297: 1475, 2002; O. Zompro, The Phasmatodea and *Raptophasma* n. gen., Orthoptera incertae sedis, in Baltic amber (Insecta: Orthoptera), *Mitteilungen des Geologisch-Paläontologischen Instituts Hamburg*, 85:229–261, 2001; O. Zompro, J. Adis, and W. Weitschat, A review of the order Mantophasmatodea (Insecta), *Zoologischer Anzeiger*, 241:269–279, 2002.

Marine habitat protection

One-fifth of the world's animal protein supply comes from the ocean, and activities related to extracting this food employ about 200 million people. The industry has changed considerably over the years as modest subsistence sailing vessels have evolved into industrialized fleets. Technology has substantially increased with the use of sonar and radar for navigation and tracking, and diesel-powered ships that can tow heavier gear and bring in increasingly higher-volume catches. While the food supply has remained steady, leveling off in recent years at 100 million tons of biomass removed annually, this does not reflect shifts in the actual makeup of the catch (from one type of fish to another). These shifts indicate dramatic fluctuations in localized fishery populations that some scientists say are caused by the impacts of fishing gear on sea-floor habitat. With evidence collected over the last decade, and supported by underwater video clips of ocean bottoms scraped bald, these scientists warn that the gear may be so damaging that current levels of extraction will be impossible to maintain. Activities that severely alter the sea floor are likely to destroy a great deal of oceanic biodiversity, as that is where 98% of marine species live, and recovery from this damage may take many decades.

Damage from trawling and dredging. The impact of fishing equipment on habitat varies, depending on gear design and the degree of contact it has with the sea floor. Trawls (large funnel-shaped nets) and dredges (scooping or suction devices) drag along the bottom, increasing in weight by hundreds or thousands of pounds as they accumulate catch, and are deemed responsible for the most severe habitat damage. Trawls are the dominant fishing method today, replacing other methods, such as longlining, which do not affect the floor at all. The large fishing vessels common today tow trawls wide enough to pass over a half acre (2000 m²) with one sweep. Additionally, equipment innovations such as the installation of large rollers to trawls allow the nets to drag over rough terrain such as rocky coral reefs without snagging, so that areas previously protected by sheer inaccessibility are now vulnerable.

The range of fishing activity has also expanded in recent years from the continental shelf to the often unexplored and previously undisturbed abyssal ocean plains. Recent discoveries of hundreds of new species associated with previously undocumented deep-sea features, such as hydrothermal vents, sea mounts off the coast of Tasmania, and extensive communities of cold water corals in the North Sea, highlight the amount of information we have yet to learn about the nature of the sea floor and the wildlife harbored there. Besides the myriad of species yet to be discovered, life cycles, sexual reproduction requirements, dispersal, and trophic interactions of known benthic (deep-sea) species are largely unknown.

Habitat structure. As dredges and trawls pass over the bottom, they scrape and smooth away surface sediments and most of the sponges, anemones, and corals in their path. Recovery from this destruction can be very slow. For example, annelid worms live in burrows in muddy bottom sediment. They build elaborate tube structures to access oxygen, and each of these tube structures can house 20–30 different species. If destroyed, these communities are not easily replaced, as many annelids live for 50 years or more but are incapable of rebuilding tubes once they age past a certain life stage.

Nutrient exchange. Trawling and dredging also contribute to constant high levels of clouding (turbidity) caused by the suspension of sediment. This can increase the rate of nutrient exchange between the water column and the sea floor, altering the biochemical environment, and can severely inhibit the population growth of some species. Coral reefs and sea grass beds harbor high levels of biodiversity and need relatively clear waters to reestablish after a disturbance.

Benthic community structure. Storms help shape an ocean floor into a patchwork of benthic communities in various stages of succession. Some argue that the effect of fishing gear is not much different from natural disturbances such as storms with intense wave action, so that long-term impacts will not be significant. Some species, especially those that live in shallower nearshore waters, are well adapted to frequent disturbances and rebound easily. However, others (including many commercially viable species) are not so well adapted. Besides disturbance, species composition along the floor depends on the type of substrate (mud, silt, sand, hard pavement, cobble, rocky, large boulders, or in combination) and topography. Topographic features range from steep slopes to gradual banks, shifting dunes, basins, cliffs, canyons, and pinnacles, each serving an important function for resident specialized communities. Fishing gear that cuts frequently and uniformly across all surfaces and depths will even cut the topography and shift complex communities into a more flat, uniform landscape, greatly decreasing the biodiversity.

Quantifying the damage. Negative impacts of fishing gear on ocean bottom habitat have been so well proven that most research is now focused on quantifying the degree of damage, where it occurs, and what to do about it. Studies have shown that it is not uncommon in the Gulf of Maine for the same site to be trawled four times in a year, while areas near New Zealand experience passes of up to 50 times per year. A single pass of a trawl results in, on average, a 68% decrease in anemone abundance and can break off thousands of pounds of coral reef. Preliminary estimates of damage along the Northeast-Atlantic shelf-break near Norway, Ireland, and Scotland are that 30–50% of coral reefs have been killed, some of which are hundreds of years old.

Marine fishery and ecosystem management. Several strategies are used to manage fishery populations and protect essential fish habitat from damage.

Stock assessments. Past efforts to manage fisheries have relied upon a species-by-species approach that focuses on estimating population sizes for exploited species and using models to calculate allowable catch given those "stock assessments." In the United States, the National Marine Fisheries Service reports that over 100 managed fish are currently overexploited, and another two-thirds of managed fish stocks have an unknown status. Global assessments estimate that half of the world's fisheries are fully exploited and another 22% overfished. Unfortunately, the situation could be much worse than stock assessments suggest, because these standard management approaches do not factor in fishing gear habitat destruction. In light of these findings, many experts are advocating a more precautionary approach aimed at keeping entire ecosystems intact. The issue has gained attention as United States regional fish management councils have been legally charged with protecting essential fish habitat (EFH) from damage.

Marine protected area. Nature reserves and parks are a common conservation tool in the terrestrial environment, with at least 5% of the world's surface protected, but only recently have such reserves been adopted in the marine environment. Marine protected areas (MPAs) are geographic regions temporarily or permanently designated with varying levels of restriction on extractive activities such as fishing and oil drilling. In the United States, less than 0.01% of the coastal marine environment has reserve status. There are currently 12 reserves operating in the United States, and preliminary studies indicate their potential to help increase ocean biomass and protect ecosystem functions. For example, Edmonds Underwater Park in the Puget Sound has been protected since 1970 and has been shown to have significantly larger and more numerous fish than surrounding unprotected areas. Such results are sometimes difficult to interpret, as scientists lack data for historic population levels, but further long-term monitoring efforts will help put preliminary findings in context.

Reserve advocates also hope to prove that marine protected areas not only will help repopulate the waters within the boundaries but also will supply fish and other marine organisms to surrounding waters beyond the protected borders. This hypothesis is based on the fact that even slight increases in fish body size can result in exponentially higher numbers of eggs produced. Thus, if reserves allow individuals to grow to their natural maximum size, their more numerous offspring will have greater potential to spillover, to the benefit of nearby fishing operations and ecosystems. A major concern associated with reserves is that they may increase fishing pressures on surrounding areas, and with little data to rely upon it is difficult for managers to know what the optimal size for a reserve should be.

Gear modification. Another management strategy is to advocate for policies that require gear modifications. Currently most gear is designed to catch the most fish at the lowest cost, though some gear changes have already been somewhat successful at reducing bycatch (the inadvertent harvesting of nontarget species). There are probably methods to lure or herd fish into nets that have not been explored that could lower impact on the sea floor.

For background information *see* BIODIVERSITY; FISHERIES ECOLOGY; MARINE CONSERVATION; MARINE ECOLOGY; MARINE FISHERIES; OCEANOGRAPHIC VESSELS in the McGraw-Hill Encyclopedia of Science & Technology.   Stacey Solie

Bibliography. L. W. Botsford et al., The management of fisheries and marine ecosystems, *Science*, 277(5325):509, 1997; M. Cryer et al., Modification of marine benthos by trawling: Toward a generalization for the deep ocean?, *Ecol. Appl.*, 12(6):1824–1839, 2002; E. Dorsey, and J. Pederson, *Effects of Fishing Gear on the Sea Floor of New England*, Conservation Law Foundation, Boston, 1998; M. A. Hixon et al., Oceans at risk: Research priorities in marine conservation biology, in *Conservation Biology: Research Priorities for the Next Decade*, pp. 125–154, Island Press, 2001; *Marine Protected Areas: Tools for Sustaining Ocean Ecosystems*, National Academy Press, 2001; National Academy of Sciences, *Effects of Trawling and Dredging on Seafloor Habitat*, 2001; National Research Council, *Marine Protected Areas: Tools for Sustaining Ocean Ecosystems*, 2001; Natural Resource Council, *Trawling and Dredging on Seafloor Habitat*, Academy Press, 2002; L. Watling and E. Norse, Disturbance of the seabed by mobile fishing gear: A comparison to forest clearcutting, *Conserv. Biol.*, 12:1180–1197, 1998; K. Wing, *Keeping Oceans Wild: How Marine Reserves Protect Our Living Seas*, Natural Resources Defense Council, 2001.

Mars Global Surveyor

The *Mars Global Surveyor* (*MGS*) was launched in November 1996 and attained Mars orbit in September 1997. In late 2003 it was still in operation, its mission having been extended twice since the nominal end date of January 31, 2001. *Mars Global Surveyor* was the first successful mission of the National Aeronautics and Space Administration (NASA) to be launched to Mars since *Viking* in 1976. The only other spacecraft to reach Mars in the intervening decades were the Soviet *Phobos 2* in 1989, which lasted only two months in Mars orbit before losing contact with Earth, and NASA's *Mars Pathfinder*, a lander-rover combination that was launched a month after *Mars Global Surveyor* but reached Mars 2 months ahead of it, and transmitted data from the planet's surface for 85 days. *Mars Global Surveyor* was conceived as a low-cost replacement for the *Mars Observer* spacecraft that was lost in 1993, and carries four of the original six *Mars Observer* science instruments.

The four scientific instruments carried by *Mars Global Surveyor* are the Magnetometer/Electron Reflectometer (MAG/ER), the Mars Orbiter Laser Altimeter (MOLA), the Thermal Emission Spectrometer (TES), and the Mars Orbiter Camera (MOC). Additional scientific measurements are made by tracking *Mars Global Surveyor*'s radio signal. The earlier Mars missions had shown that the atmosphere of Mars was cold and thin, $-81°F(-63°C)$ in average temperature and about 6 millibars (600 pascals) in surface pressure (0.6% of sea-level pressure on Earth). Water was understood to be very scarce outside the north polar ice cap. On the other hand, the Martian surface features revealed by these earlier orbiters appeared to tell the story of a Mars that, 4×10^9 years ago, had enough water, an atmosphere thick enough, and temperatures warm enough to support liquid water flowing on the surface. If early Mars was indeed warm and wet, conditions might have been just as favorable for the evolution of life as they were on Earth. The possibility of a warm and wet primordial Mars also begs the question of whether Mars might still conceal just enough water to support a few simple organisms. Thus, the Mars exploration program remains preoccupied, perhaps justifiably, with the search for water and life. All four of the *Mars Global Surveyor* science instruments have provided results that are highly relevant to this search.

Magnetometer/Electron Reflectometer. This instrument measures magnetic fields and electrons in the near-Mars space environment. This allows studies of the interaction of the solar wind with Mars, and allows measurements of the magnetic field of Mars itself. The Magnetometer/Electron Reflectometer has demonstrated conclusively that Mars, unlike the Earth, has no global intrinsic magnetic field at present. The core of Mars is therefore presumed to have cooled off to the point where it can no longer sustain the core convection necessary for magnetic field generation. However, the Magnetometer/Electron Reflectometer has detected strongly magnetized rock, occurring in bands of alternating polarity, in the oldest terrain of the Martian crust. This remanent magnetization was presumably induced by a global magnetic field at the time the crust formed. It implies that primordial Mars did have a magnetic field produced by a hot, convective core, and that the magnetic field probably turned off more than 4 billion years ago. Since a convective core is indicative of the heightened state of geological and volcanic activity necessary to replenish atmospheric gases, and since a global magnetic field acts to protect a planet's atmospheric gases from being stripped away by the solar winds, the disappearance of the Martian magnetic field was a harbinger of harsher times to come for any life forms that may have existed on Mars.

chains containing 16 carbon atoms. Some proteins contain both groups. These increase the association of proteins with membrane lipids. Lipid rafts support the assembly of many cytokine- and chemokine-dependent signaling complexes containing myristoylated and palmitoylated proteins. Caveolae typically support the assembly of protein complexes that are active in signal transduction initiated by growth factors and, in locomotory cells, those that are active in cell attachment mediated by integrins. In addition to acylation, caveolar signaling proteins are bound to caveolin by a motif (local pattern of amino acids) that is rich in aromatic amino acids. Palmitoylation also plays an important role both in maintaining caveolin itself at the cell surface and in stabilizing the association of the binding site with signaling proteins.

The high FC content of both lipid rafts and caveolae appears to be a major factor in targeting proteins to these microdomains. Depleting membrane FC by using extracellular lipid acceptors such as cyclodextrins (soluble synthetic FC acceptors) favors the translocation of bound signaling proteins to other parts of the membrane. Such movements have been detected both by fluorescence microscopy and by changes in the pattern of immunoprecipitated proteins.

Signal transduction. In addition to their initial activity in the assembly of signaling complexes, FC-rich microdomains can play a more complex role. Three examples follow.

Platelet-derived growth factor. In the case of the signaling cascade initiated by the extracellular signaling protein called platelet-derived growth factor (PDGF), which stimulates cell growth and division and is essential to wound healing, FC is required for the initial assembly of the PDGF receptor complex in caveolae, but the FC must be expelled for subsequent steps in the pathway to proceed effectively. Presumably these effects are mediated by FC-dependent changes in the tertiary (folded) structure of the complex during the activity of kinase (a phosphorylating enzyme that affects protein folding); however, crystallographic data are not yet available. Linkage between loss of FC and signal transduction is further demonstrated by the fact that addition of extracellular FC-binding proteins stimulates both FC efflux and kinase activities in response to PDGF. These data indicate that far from being inert, as FC has sometimes been described, FC in caveolae is dynamically regulated.

Endothelial nitric oxide synthase. Endothelial nitric oxide synthase (e-nos), an enzyme which regulates vascular nitric oxide production, is an important mediator of vascular tone. E-nos is localized within a signaling module (a complex of signaling molecules working together) in endothelial cell caveolae. Depletion of FC displaces e-nos from caveolae, rendering it inactive, an effect opposed by the delivery of FC to the caveola.

Signaling kinase Lyn. The signaling kinase Lyn is part of the B-cell antigen receptor (BCR) pathway which mediates the inflammatory response in lymphocytes, where it is mainly localized to lipid rafts. Lyn has a higher specific activity (kinase activity per protein molecule) and level of phosphorylation when purified from raft compared to nonraft membrane fragments. Raft FC mediates this effect dynamically by protecting Lyn protein phosphoamino acids within rafts from hydrolysis.

Caveolae, rafts, and cellular FC homeostasis. In addition to activities related to protein assembly and signal transduction, independent roles for caveolae and rafts in FC homeostasis have been proposed.

FC efflux. The accessibility of caveolar FC is borne out by its rapid transfer from the cell surface to extracellular lipid acceptors such as plasma lipoproteins, particularly high-density lipoprotein (HDL), and cyclodextrin. The caveolar FC pool has been shown to contribute preferentially to such lipid transfer from caveola-rich primary smooth muscle cells and fibroblasts. This effect was not found in virus-transformed or other continuously dividing cell lines, probably because of the much smaller numbers of caveolae present.

Owing to their short life, a possible contribution of lipid rafts to FC efflux has not been quantified, though FC from rafts, like that from caveolae, is easily transferred to cyclodextrin.

Intracellular FC. An additional role has been suggested for caveolae and caveolin as intracellular mediators of FC transfer between membrane-bound organelles and storage areas. Caveolin has been identified in the perinuclear trans-Golgi fraction, in lipid storage droplets, in a weakly acidic recycling endosome fraction, and in a chaperone complex proposed as a carrier for the newly synthesized FC originating from the endoplasmic reticulum. Many transformed cells transfected with caveolin complementary DNA (cDNA) retain the protein within such intracellular pools.

Some investigators have argued that redistribution of caveolin in cells pretreated with metabolic inhibitors implies that caveolin recycles to the cell surface. It has also been suggested that because caveolin redistributes from the surface to the cell interior in FC-depleted cells, it must have a role as a transporter of FC between membrane compartments. Neither of these conclusions necessarily follows on present evidence. The stability of caveolae at the cell surface argues against recycling, but does not exclude it. Caveolin is an FC-binding protein, and its distribution among membranes in the cell may simply reflect that of FC-rich membranes (including those of recycling endosomes, trans-Golgi membranes, and lipid droplets) rather than FC transport per se by caveolin.

Lipid rafts and viral infectivity. A recent development is the recognition that signaling proteins in lipid rafts are the binding sites by which many viruses gain entry to cells. The same structures can contribute coat lipids to newly synthesized virions. Lymphocyte raft signaling proteins CD4 and CCR5 are among the contributors to human immunodeficiency virus (HIV-1) binding. Depletion of FC in these microdomains via cyclodextrin significantly reduces infectivity in vitro. Other viruses shown to utilize lipid raft pathways for infection include Echovirus and Ebola virus. Details of the linkage between virus binding and internalization have not yet been worked out. Preliminary data suggest that caveolae may contribute to the internalization of other microorganisms.

Answers to the many questions remaining about caveolae, lipid rafts, and their functions may be gained from their tertiary structure, when this information becomes available. Present data suggest that structural differences might contribute significantly to heterogeneity within plasma membranes, generating the unique lipid and protein structures that play important roles in cellular transport and communication.

For background information *see* CELL MEMBRANES; CHOLESTEROL; LIPID; PROTEIN; SIGNAL TRANSDUCTION; SPHINGOLIPID in the McGraw-Hill Encyclopedia of Science & Technology.   Christopher J. Fielding

Bibliography. C. J. Fielding and P. E. Fielding, Relationship between cholesterol trafficking and signaling in rafts and caveolae, *Biochim. Biophys. Acta*, 1610:219–228, 2003; F. R. Maxfield, Plasma membrane microdomains, *Curr. Opin. Cell Biol.*, 14:483–487, 2002; R. G. Parton, Caveolae—from ultrastructure to molecular mechanisms, *Nat. Rev. Mol. Cell Biol.*, 4:162–167, 2003.

Microcalorimetry

Calorimeters measure all processes, chemical and physical, involving an exchange of heat energy with the surroundings. The calorimeter is nonspecific in its operation; that is, it will monitor and record all heats of reactions that occur. An isothermal microcalorimeter is operated at constant temperature and is used to measure heat flow in the microwatt range.

Isothermal microcalorimetry offers several advantages over other analytical techniques. Its sensitivity is such that materials can be studied at ambient temperatures (many conventional techniques require sample study at elevated temperatures and a response extrapolated to the temperature of interest). And materials can be measured in their normal physical form (solid, liquid, or gas), without any special preparation before study. As a result, isothermal microcalorimetry is a rapid, noninvasive (not influenced by the calorimeter), and nondestructive tool for the quantitative determination of a variety of thermodynamic and kinetic parameters.

For pharmaceutical systems, isothermal microcalorimetry is a widely accepted tool for studying active drugs and inactive ingredients used in drugs (excipients). Microcalorimetry is routinely used for compatibility studies (such as drug-excipient and excipient-excipient interactions), stability (shelf life), degradation, and for probing physical properties such as crystallinity.

Generally, the information sought is related to the kinetics of that system (for example, shelf life). It is essential that accurate quantitative information can be derived for parameters such as the rate and rate constant. Also important is the quantity of material undergoing change in the system, which can be explored for that reaction through the magnitude of the calorimetric signal and calculation of the enthalpy (heat transferred to a system during a specified change at constant pressure). Knowledge of these parameters then allows a host of other thermodynamic and kinetic information to be derived from the system under study.

Batch versus flow techniques. The majority of pharmaceutically relevant studies are performed using static/batch techniques. Batch-mode experiments require that the system under study has a half-life which permits observation of the reaction after the necessary preparation time (loading of ampules, equilibration period, and dissipation of frictional heat after lowering of the ampules into the measuring position) of about 1 hour after initiation of the reaction. An isothermal microcalorimeter operated in batch mode has sufficient sensitivity to monitor slow reactions with lifetimes of up to 10,000 years. It has been shown that only 50 hours of data is required to discriminate between a reaction that has a first-order rate constant of 1×10^{-11} s^{-1} and a reaction with a first-order rate constant of 2×10^{-11} s^{-1}.

Some systems are not particularly suited to batch-mode techniques, such as enzyme/substrate systems, activity of drugs and antibiotics on microbiological systems, and cellular systems. Such systems also are difficult to study using conventional techniques and usually require invasive sampling or study under nonrelevant conditions (for example, plating out of microorganisms). Therefore, flow-through calorimetric techniques offer several advantages (for example, noninvasive and nondestructive) over classical methods and can be used when batch techniques are not suitable.

Flow calorimeters have a shorter equilibration period (since a smaller volume of solution is examined), and hence reactions with short half-lives are amenable to study. Flow systems allow thorough mixing of reactants, allowing systems that are prone to sedimentation to be studied. Flow calorimetry also has been used to probe surface characteristics by passing gas or vapor over a solid surface.

Flow microcalorimetry. Depending on whether the process is endo- or exothermic, heat will flow from or to the heat sink in order to maintain isothermal conditions. This heat flow or thermal power is measured as a change in heat energy (joules) per unit time (seconds) by the calorimeter and is expressed as dq/dt (or Φ); it has units of J s^{-1} (watts). Since the calorimeter returns data as a function of time, it is essentially a rate meter, and therefore it is possible to derive thermodynamic and kinetic information from the calorimetric data.

Flow calorimeters operate on the principle that the inflow of a reacting solution at constant flow rate gives rise to a constant heat effect in the calorimetric cell that will eventually reach steady-state conditions where the heat generated in the cell, per unit time, will be equal to the heat transported from the cell by the thermopiles (array of thermocouples). Only under these conditions will the calorimetric signal be a true representation of the extent of reaction in the cell. Even under these ideal conditions, some heat will be lost. Heat loss is likely through the air gaps between parts of the cell not in contact with the thermopile, but more particularly with the effluent as it leaves the cell. This will result in inaccurate values for any derived parameters. Therefore, calorimetric output from flowing systems is dependent not only on the thermodynamic and kinetic parameters but also on the effective volume of the calorimetric cell (which in turn is governed by the flow rate of the solution through the calorimetric cell).

In order to elucidate the desired parameters from calorimetric data, it is necessary to know the equations which describe the calorimetric output in terms of the desired parameters. The calorimetric equation which describes a first-order solution phase reaction studied by flow-through calorimetry is given below. Here F is the flow rate of reacting solution

$$\Phi = -Fc\Delta H(1 - e^{-k\tau}) \cdot e^{-kt}$$

(dm^3 s^{-1}), c is the initial concentration of reactant (mol dm^{-3}), ΔH is the enthalpy (J mol^{-1}), k is the first-order rate constant (where $dc/dt = kc$), and τ is the residence time (s).

Until recently, it was not possible to determine accurately the thermal volume or, as a result, parameters such as enthalpy. This was overcome by the development of a chemical test and reference reaction for flow-through microcalorimetry. As well as providing traceability and validation of results from any experimental study, it allowed training of new personnel, troubleshooting for potential sources of instrumental error, and investigation of experimental design.

It is now possible to determine the desired thermo-kinetic parameters for simple (integral ordered) solution-phase reactions studied by both batch and flow techniques. However, for complex solution-phase reactions the individual constituent reactions must be integral-ordered and conform to the equations set out earlier. It is also possible to obtain thermo-kinetic parameters for solid-state reactions that are free from assumption and without imposing any model on the system.

Flow calorimetric experiment. One of the many applications of flow calorimetry is the study of enzyme/substrate systems to yield both thermodynamic and kinetic information from the system. The kinetic nature of enzyme systems was described by Leonor Michaelis and Maude L. Menten in 1913. In our discussion, the parameters sought are the enthalpy, rate constant, Michaelis constant (the substrate concentration at half the maximum rate), and the enzyme activity. The following example de-

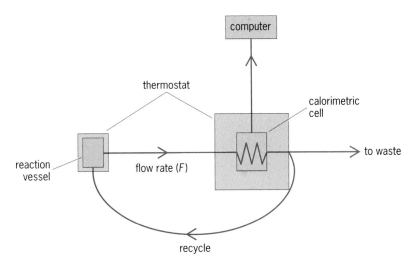

Fig. 1. Schematic for the flow-through calorimetric cell.

scribes a study on the well-known enzyme substrate system, urea/urease, to obtain such information.

Stock solutions of various concentrations of urea are prepared and buffered to pH 7.0. The concentrations of urea chosen are such that the complete kinetic profile of the reaction can be observed (that is, from first-order through mixed-order and finally to zero-order kinetics).

A 50-mL aliquot of the urea solution is preheated to the operating temperature of the calorimeter and is run in a continuous loop, at a known flow rate, until a stable baseline is achieved (**Fig. 1**). This solution is then inoculated with 4.55 mL of a standard, fixed-concentration, urease solution also buffered to pH 7.0, and the resulting calorimetric output is recorded as a function of time. This is repeated for all concentrations of urea. **Figure 2** shows a selection of typical calorimetric outputs for this enzyme system. Since the data are recorded as a function of time, the rate of reaction can be determined from the calorimetric output. **Figure 3** shows a plot of rate versus substrate concentration. This plot has a form that confirms that the system does conform to Michaelis-Menten kinetics.

Using the calorimetric data obtained from this study, it is possible to accurately calculate (using

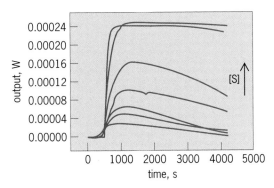

Fig. 2. Calorimetric outputs for first-order, mixed-order, and zero-order urea/urease enzyme reactions. [S] is the total substrate concentration (mol/dm^3).

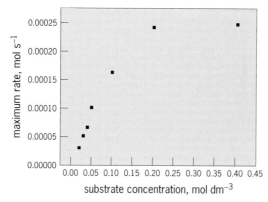

Fig. 3. Maximum rate versus substrate concentration for the urea/urease enzyme reaction.

a new calculational method, not described here) values for the parameters described earlier free from assumption. This new method also allows the effects of inhibitors of enzyme activity and their mechanism of action to be studied using flow calorimetry.

Flow-microcalorimetry applications. A number of pharmaceutically relevant systems have been explored through flow calorimetry. Much of the early work in this area of calorimetry centered on the study of microorganisms. It was shown that in principle it is possible to distinguish between different microbial species by the shape of the calorimetric curve produced during organism growth. Quantification of the number of organisms present in the system is possible by comparison of the areas under the curve for an uncharacterized system with that of a known standard.

Since calorimetric data are a measure of rate, it is possible to ally heat output with metabolic activity of human and animal cells. Both types of cell have been extensively studied by flow calorimetry; one example is the study and optimization of Chinese hamster ovary cells producing recombinant interferon γ.

Flow calorimetry has also been used to investigate dosage forms. Dissolution of tablets under various simulated conditions has been explored, for example. Here the tablet is presented with various solutions designed to mimic conditions in the gastrointestinal tract (pH 7 buffer) and stomach (pH 3 buffer for fasting and lipid solutions for fatty meals). The rate of dissolution can then be estimated for conditions in vivo.

For background information *see* CALORIMETRY; CHEMICAL THERMODYNAMICS; ENTHALPY; ENZYME; THERMOCHEMISTRY in the McGraw-Hill Encyclopedia of Science & Technology.   Michael A. O'Neill

Bibliography. A. E. Beezer et al., Thermodynamic and kinetic parameters from isothermal heat conduction microcalorimetry, *J. Phys. Chem.*, 105:1212–1215, 2001; P. K. Gallagher (ed.), *Handbook of Thermal Analysis and Calorimetry: Principles and Practice*, Elsevier, 1998; A. M. James (ed.), *Thermal and Energetic Studies of Cellular Biological Systems*, Wright, 1987; M. A. O'Neill et al., Practical and theoretical consideration of flow-through microcalorime-try: Determination of "thermal volume" and its flow rate dependence, *Thermochimica Acta*, 2003; M. A. A. O'Neill, *Calorimetric Studies of Complex Biological Systems: Theoretical Developments and Experimental Studies* (PhD Thesis), University of Greenwich, 2002.

Microfluidics

Microfluidics refers to technology that involves manipulating fluids in structures in which at least one linear dimension is less than a millimeter. Currently, the smallest dimension of common microfluidic elements is 10–500 micrometers. The most mature and commercially successful microfluidic systems are the print heads found in common inkjet printers. More recent efforts have been aimed at developing microfluidic systems for performing chemical processes. The principal target applications of these chemical systems are as portable analyzers of chemicals and biomolecules [proteins and deoxyribonucleic acid (DNA)] for biomedical applications and for defense against chemical and biological weapons.

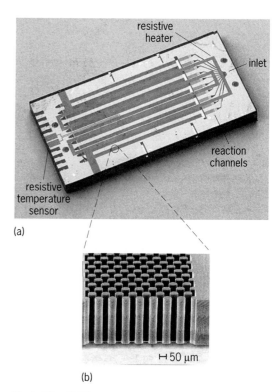

Fig. 1. Microfluidic chip for performing multiphase (liquid/gas) reactions with heterogeneous catalysis. (a) Photograph of chip. This system was used for the hydrogenation of alkenes to form alkanes. The channel structures are in silicon. The heater and temperature sensor are thin metal films (*photograph by Felice Frankel*). (b) Scanning electron micrograph showing some of the micropillars that are fabricated inside each reaction channel. These pillars provide a large surface area on which to present catalytic materials in the reactors (*from M. W. Losey et al., Design and fabrication of microfluidic devices for multiphase mixing and reaction, J. Microelectr. Sys., 11(6):709–717, 2002; © 2002 IEEE*)

Applications of microfluidic systems	
Application	Advantage of microscopic scale
Inkjet printing	High resolution, high operation speed
Analytical chemistry	Portability, high operation speed, small sample
Protein and DNA analysis for proteomics and genomics	volume (<1 μL)
High-throughput screening of chemicals (pharmaceuticals)	
Portable chemical detection	
Chemical synthesis	Small sample volume, high operation speed, fine
In-situ synthesis of toxic compounds	control of reaction parameters, rapid transfer of heat
Combinatorial chemistry	and mass
Analysis of chemical kinetics	
Fuel cell	Portability
Cooling for microelectronics	Portability, rapid heat transfer
Cell culture/cell sorting	Control of individual cells

Some chemical syntheses have also been performed in microfluidic devices (**Fig. 1**). Microfluidic technologies have only recently begun to be commercialized. The **table** lists applications of microfluidic systems. *See* MINIATURIZED ANALYSIS SYSTEMS.

Flow characteristics. Experimental and theoretical attention has been paid to the possibility that fluid behavior in microchannels deviates from that found in macroscopic flows. Except in unusual cases, such deviations are neither predicted nor observed; for flows of simple liquids, such as water and organic solvents, the same general equations (Navier-Stokes equations) that govern macroscopic flows, with the no-slip condition at solid boundaries (meaning that the fluid is stationary relative to the solid at the boundary; **Fig. 2a**), are valid on scales above tens of nanometers. One notable exception is the observation of slip of liquids over nonwetting surfaces, that is, surfaces on which the liquid beads up as water does on Teflon® (polytetrafluoroethylene). A proposed mechanism for this slip is that a layer of nanoscopic vapor bubbles covers the surface and acts as a lubricating layer of low viscosity between the liquid and the solid. The largest slip lengths that have been observed are about 1–2 μm (Fig. 2b). Thus, slip will lead to deviations from expected behavior only in flows of nonwetting liquids through channels with cross-sectional dimensions of less than 10 μm. In flows of gases (less commonly used than liquids), slip has long been predicted and has been observed in microchannels when the mean free path of molecules in the gas is comparable to the dimension of the channel.

While the basic governing equations are the same, there are several general features that distinguish flows in microstructures from flows in common macroscopic systems (such as water faucets and coffee cups). As the dimension of a flow decreases, the importance of forces that act on the volume of the fluid, such as inertia and gravity, diminishes relative to that of forces that act at the surfaces, such as viscous friction and surface tension. The ratio of inertial to viscous forces is expressed as the Reynolds number of the flow, Re $= \rho v w / \eta$, where ρ (measured in kg/m^3) is the density of the fluid, v (m/s) is the velocity, w (m) is the characteristic dimension of the flow ($\sim$volume/surface area of container; **Fig. 3a**), and η [kg/(m $\cdot$ s)] is the dynamic viscosity of the fluid. In microchannels, Re < 100, so flows are laminar rather than turbulent (flows in channels are typically laminar for Re < 2000). Coflowing streams in a laminar flow intermix only by diffusion because no spontaneous eddies carry mass and momentum between them. This characteristic of laminar flows allows spatial control of solute within the flow, but it hinders rapid mixing.

Fabrication. Most methods of fabricating microfluidic systems are adaptations of techniques for fabricating microelectronic structures; photolithography is at the core of these methods. These techniques lead to a flat, chiplike format in which microchannels are constrained in planar layers on a flat substrate (Fig. 1). This format facilitates integration of electronics and optics, but it restricts the geometries that are accessible for the design of fluidic elements. The channel structures are etched into a flat surface of silicon or glass, or molded into a soft or hard plastic sheet; the channels are closed by sealing the structured material to another flat surface.

Flow characterization. Simple characterization of flows in microfluidic devices can be achieved by measuring global variables such as the applied pressure difference (Pa) across a channel and the volumetric flow rate (m^3/s) through the channel. Electronic sensors of pressure and flow speed have been integrated into microchannels to provide local measurements of the flow. Fluorescence microscopy is commonly used to perform characterization of these

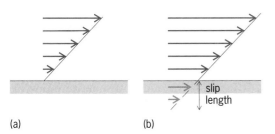

(a)   (b)

Fig. 2. Flow profiles at a solid boundary (a) with no slip and (b) with slip.

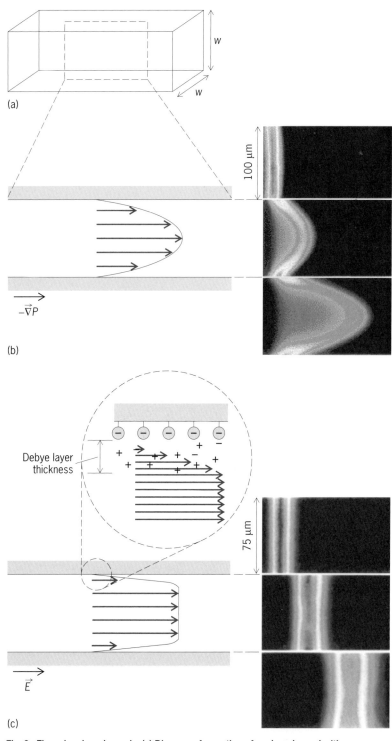

be followed through a microscope by coinjecting streams of fluorescent solution with streams of clear solution. In steady flows, the three-dimensional evolution of fluorescent streams can be imaged through a confocal microscope (**Fig. 4**).

Pumping fluids. The two most common methods for driving flows in microchannels are with externally applied pressure gradients and with externally applied electric fields. Capillary, acoustic, and magnetic forces have also been used to move fluids in microchannels.

Pressure gradients are often generated by macroscopic pumps that are connected via tubes to the inlets of the microfluidic chip. Integrated microdiaphragm and microperistaltic pumps have also been developed. In pressure-driven flow in channels, the flow speed is given roughly by (1), where ∇P is the

$$v \sim -w^2 \nabla P / \eta \qquad \text{(m/s)} \qquad (1)$$

applied pressure gradient (Pa/m). To maintain a given flow speed, the required pressure gradient grows rapidly as the channel shrinks. In pressure-driven flows, the flow speed varies from zero at the wall to its maximum value in the middle of the channel (Fig. 3b). This variation leads to dispersion (spreading) of solute along the direction of the flow. This dispersion is unfavorable for transporting narrow bands of solute such as in a chemical separation.

Electric fields are applied by placing electrodes in the inlet and outlet of a microchannel. In electrolyte solutions (such as salty water), flows are generated by the interaction of the applied electric field with ions that accumulate in a thin (less than 1 μm thick) layer of fluid, the Debye layer, adjacent to channel walls; these are called electroosmotic flows (Fig. 3c). Fluid in the Debye layer moves due to the electrical body force and entrains the remainder of the fluid. The flow speed in the bulk is given roughly by (2), where μ_{eo} (m^2/s × V) is the electroosmotic

$$v_{eo} \sim \mu_{eo}E \qquad (2)$$

mobility that depends on the surface charge density, the concentration of ions in the liquid, and the viscosity of the liquid, and E (V/m) is the magnitude of the electric field. The magnitude of the electroosmotic mobility is given roughly by (3). In uniformly

$$|\mu_{eo}| \sim 1\ \mu\text{m}/(\text{s} \cdot \text{V/m}) \qquad (3)$$

charged channels, the flow speed v_{eo} is constant over the cross section except in the Debye layer. This property allows bands of a single type of solute to be transported along the channel with little dispersion. Capillary electrophoresis is a useful tool for chemical analysis that exploits this behavior and is performed in microfluidic devices. The flow speed v_{eo} is also independent of the dimension of the channel down to the thickness of the Debye layer. Electroosmotic pumping is therefore appropriate in channels down to submicrometer dimensions. Disadvantages of electroosmotic pumping include the requirement of

Fig. 3. Flows in microchannels. (a) Diagram of a section of a microchannel with a square cross section. (b) Flow profile of a pressure-driven flow and (c) an electroosmotic flow in a channel. The two series of fluorescent micrographs show the evolution of a band of fluorescent dye in the respective flows. The blurring of the interface between fluorescent and nonfluorescent regions is due to molecular diffusion. (*After P. H. Paul, M. G. Garguilo, and D. J. Rakenstraw, Imaging of pressure and electrokinetically driven flows through open capillaries, Analyt. Chem., 70:2459–2467, 1998*)

flows with micrometer-scale resolution. Micro particle image velocimetry (μPIV) maps the streamlines in a flow by following the trajectories of submicrometer fluorescent tracer beads in sequences of micrographs. The evolution of streams in microflows can

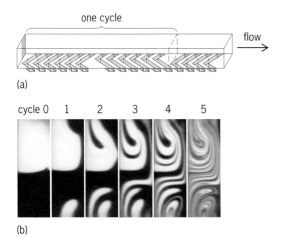

(a)

cycle 0   1   2   3   4   5

(b)

Fig. 4. Passive micromixer. (*a*) Section of a microchannel with grooves in the form of asymmetric herringbones on the bottom wall; the grooves lead to mixing in pressure-driven flows through the channel. One mixing cycle is made up of two regions of asymmetric herringbones; the direction of the asymmetry switches from one region to the next. (*b*) Confocal fluorescence micrographs of the vertical cross sections of flow through such a channel. The evolution of two coflowing streams (one of fluorescent liquid and one of clear liquid) in the mixer is seen: The number of folds approximately doubles with each cycle. The blurring of the interface between fluorescent and nonfluorescent regions is due to molecular diffusion. (*After A. D. Stroock et al., Chaotic mixer for microchannels, Science, 295:647–651, 2002*)

high voltages ($V \sim 1$ kV), sensitivity to the chemical characteristics of the walls, and the tendency for species to separate electrophoretically in the applied field. (This tendency is an advantage for analysis but a disadvantage for transporting general chemical mixtures.)

Mixing. In microchannels of common dimensions ($500 \, \mu$m $> w > 10 \, \mu$m), the motion of solute by diffusion across the laminar flow is often slow relative to the flow speed along the channel. In this situation, mixing is often the slow step in a chemical process, unless transverse flows that stir the fluid are purposely induced. (In turbulent flow, these transverse flows occur spontaneously.) An efficient stirring flow stretches and folds the fluid such that the interface between unmixed regions grows exponentially in time; this type of flow is called chaotic (Fig. 4). Active mixers create transverse motion in the principal flow with local, oscillatory forces generated with bubbles, applied electric or magnetic fields, or flows in cross-channels. These mixers have the potential to be very efficient but require sophisticated controls. Passive mixers use fixed geometrical features in the channel to induce transverse components in the principal flow (Fig. 4). These designs are simple to operate but may not achieve full mixing as quickly as active designs.

Challenges. With the exception of inkjet print heads and, perhaps, on-chip capillary electrophoresis, microfluidic technology is still in its infancy. No consensus has been reached on even the most basic procedures and designs. Optimal designs for pumps, valves, injectors, and so forth must still be invented

and characterized, and ideal materials and methods of microfabrication must be developed. Future applications of microfluidic devices must also be conceived of and explored. For example, microfluidic systems could act as implants to deliver drugs and monitor physiological parameters, or as active materials that control environmental parameters in clothing and buildings.

Nanofluidic systems. These systems, in which the characteristic size is less than a micrometer, are already being developed. In this regime, pressure may not be a useful driving force, whereas electroosmosis likely will. Diffusion will often transport solute molecules as rapidly as the flow does; mixing will not be a problem, but careful delivery of solute will. Whereas the design principles for microfluidics were largely borrowed from macroscopic chemical systems (pumps, channels, and so on), the best guides to designing nanofluidic systems may be biological systems such as cells: transport will be controlled on the molecular rather than the hydrodynamic level, with molecular motors carrying chemical building blocks along definite paths between distinct chemical environments.

For background information *see* BOUNDARY-LAYER FLOW; CONFOCAL MICROSCOPY; ELECTROPHORESIS; FLUID FLOW; FLUORESCENCE MICROSCOPE; MICRO-ELECTRO-MECHANICAL SYSTEMS (MEMS); NAVIER-STOKES EQUATIONS; REYNOLDS NUMBER; TURBULENT FLOW in the McGraw-Hill Encyclopedia of Science & Technology.   Abraham D. Stroock

Bibliography. C.-M. Ho and Y.-C. Tai, Micro-electro-mechanical-systems and fluid flows, *Annu. Rev. Fluid Mech.*, 30:579–612, 1998; M. W. Losey et al., Design and fabrication of microfluidic devices for multiphase mixing and reaction, *J. Microelectr. Sys.*, 11(6):709–717, 2002; P. H. Paul, M. G. Garguilo, and D. J. Rakestraw, Imaging of pressure and electrokinetically driven flows through open capillaries, *Analyt. Chem.*, 70:2459–2467, 1998; A. D. Stroock et al., Chaotic mixer for microchannels, *Science*, 295:647–651, 2002; D. C. Tretheway and C. D. Meinhart, Apparent fluid slip at hydrophobic channel walls, *Phys. Fluids*, 14:L9–L12, 2002; G. M. Whitesides and A. D. Stroock, Flexible methods for microfluidics, *Phys. Today*, 54(6):42–48, June 2001.

Mine waste acidity control

Acidic water drainage from both surface and underground mining waste is a source of pollution, since it often contains dissolved metals. Many new technologies for assessing and controlling mine water acidity have been tested in recent years.

Acid rock drainage. Acid rock drainage (ARD) is a naturally occurring process caused by the oxidation of certain sulfide minerals, especially iron sulfides such as pyrite (FeS_2). Acid rock drainage is known to occur in natural mineralized zones and is widespread in recently drained tidal sediments such as in the

Netherlands and southeast Asia, but is most often associated with mining areas.

Acidic and sulfate-enriched water with pH ranging from less than 2.0 to 4.5 may develop when rocks or sediments that contain pyrite are exposed to oxygen [reaction (1)].

$$FeS_2 + {}^{15}\!/_4 O_2 + {}^7\!/_2 H_2O \longrightarrow Fe(OH)_3 + 2SO_4^{-2} + 4H^+ \quad (1)$$

The acidity resulting from ARD can degrade water quality since many metals are more soluble at low pH than at neutral pH. Waters that have been in contact with sulfide-rich sediment or rock may also contain elevated concentrations of soluble iron, aluminum, and manganese. In some mining areas, rocks may also contain elevated amounts of copper, zinc, cadmium, nickel, lead, and other metals that are soluble at low pH.

Some pyritic rocks and sediments also contain an abundance of carbonates such as calcite ($CaCO_3$) or dolomite [$CaMg(CO_3)_2$]. When rocks containing both pyrite and carbonates break down chemically upon environmental exposure, the acid formed by sulfide oxidation is neutralized by carbonate dissolution. Most of the metals liberated by the sulfide reactions are insoluble at the resulting solution pH, ranging from 5.5 to 7.5 [reaction (2)].

$$FeS_2 + {}^{15}\!/_4 O_2 + {}^7\!/_2 H_2O + 2CaCO_3 \longrightarrow Fe(OH)_3 + 2Ca^{+2} + 2SO_4^{-2} + 2CO_2 \quad (2)$$

A few metals, such as manganese, zinc, or nickel, remain soluble at this pH range, so ARD need not have low pH values (less than 4.5) to degrade water quality. The solubility of some compounds, especially selenate or arsenate, actually increases at higher pH, making them problematic in certain mine waters. Mine water containing elevated sulfate concentration at neutral pH (∼7) is often called neutralized ARD.

Risk assessment. Laboratory tests have been devised to determine if rocks or sediments will become acidic upon exposure to oxygen. A. A. Sobek and coworkers first described "static tests" which identify the potential for forming acidic drainage by examining the balance of acid-generating and acid-neutralizing minerals. This technique, also known as acid-base accounting, measures the quantity of sulfide minerals (such as pyrite) that may form acid and the acid-neutralizing minerals, such as calcite. Static test results are reported in units of tons of $CaCO_3$ per 1000 tons of material, where the net neutralization potential (NNP) is equal to the acid neutralization potential (ANP) minus the acid generation potential (AGP).

The acid generation potential is based on the stoichiometry of reaction (1) and equals the pyritic sulfur (in weight percent) times 31.25. It assumes that all of the sulfide is present as pyrite and that all of the pyrite will oxidize and produce acid. The acid neutralization potential is determined by adding acid to a sample and determining the amount of acid consumed by carbonate, hydroxide, and certain alumino-silicate minerals.

The risk of ARD occurrence is determined by graphing the acid generation potential and acid neutralization potential (see **illus.**), where a line can be drawn designating a net neutralization potential of zero. Several criteria exist for the evaluation and interpretation of static test data. Generally, material with a net neutralization potential greater than +20 is considered non-acid-generating. Material with a net neutralization potential less than −20 is considered acid-generating.

Data from waste rock and tailings (waste material after extracting the ore) collected at three typical mines in the western United States are shown in the illustration. Most samples of waste rock from the mine in Montana had net neutralization potential values that were much less than −20 tons per 1000 tons, and interstitial water in contact with this material typically had a pH ranging from less than 2 to 3.5. The waste rock from the mine in Idaho had very low pyrite levels and contained abundant calcite, which prevented acid generation. Finally, tailings from the silver mine in Alaska had very low net neutralization

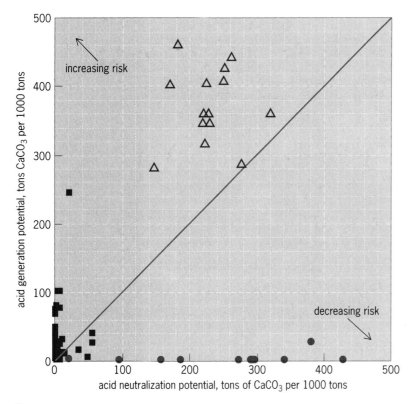

Key:

■ Montana gold mine

● Idaho gold mine

△ Alaska silver mine

Risk of acid rock drainage based on static test results for waste rock from mines in Montana and Idaho and tailings from a mine in Alaska.

potential values ranging from 0 to -250 tons per 1000 tons. Despite the low net neutralization potential values, the water in contact with the tailings had a pH (6.5 to 7.5) that was near-neutral. The lack of acid production is attributed to the abundance of calcite and dolomite, and the relatively slow rate of pyrite oxidation. Limiting the movement of oxygen into the tailings is generally thought to be the factor that limits the rate of pyrite oxidation.

Control. A recent global downturn in metal prices, as well as opposition from environmental groups, has caused a spate of mine closures. These trends have forced the mining industry and regulatory authorities to recognize the need for effective long-term ARD controls. Historic technologies have relied upon two primary approaches: (1) hydrologic isolation of potential acid-generating materials using surface water diversion and engineered covers, and (2) collection and treatment of acidic water. Recently, a number of newer ARD control techniques have been developed and tested. Many of these new technologies rely on reducing the supply of oxygen. If oxygen is prevented from contacting pyritic materials or if the rate of oxygen transport is reduced, ARD can be prevented or reduced in severity. The techniques are especially well suited to more mesic (moderately moist) areas typical of higher-elevation mine sites in the United States and Canada, and may have value in tropical or maritime climates in Latin America, on the Pacific rim, and in Alaska.

Three primary means of reducing oxygen supply to potentially acid-generating material have been developed, including water covers, dry covers, and oxygen-consuming covers.

Water covers. These have been used to control oxygen migration into tailings. Placement of mine waste under water is an effective method for reducing oxygen flux into potentially acid-generating materials because oxygen diffusion is approximately 10,000 times slower in water than in air. Mine waste can be submerged either through subaqueous disposal or through construction of water covers. Common examples of subaqueous disposal include lake or marine placement of tailings.

Typically, the depth of placement should be sufficient to prevent wind or wave action from disturbing the mine waste materials. The effect of subaqueously disposed materials on biota is an important design consideration. Potential environmental effects that need to be evaluated are the physical effects of the mine waste on the habitat, the potential toxicity of entrained pore water that may be released form the mine waste, and the leachability of metals in the submerged mine waste. Typically, submerged mine waste materials prove to be inert in most lake or marine environments.

Mine waste facilities can also be submerged if they are impounded within a water-retaining structure, such as a tailings dam, that has been designed for permanence. Water covers are most commonly used for tailings or for waste rock that is disposed of along with tailings in an impoundment (pond). The depth of water covers within a tailings pond is often much less than for subaqueously placed mine waste. For shallower water covers, the effects of convective movement of oxygen due to wave action and mixing must be considered. Models have been developed to predict oxygen diffusion through water covers.

A secondary benefit of water covers is that diffusion of oxygen within the saturated mine wastes is extremely slow because all of the pore space is water-filled. Additionally, production and seasonal decomposition of algae in the water cover will gradually create a carbon-rich sediment above the mine waste that will consume some or all of the oxygen before it reaches the mine waste.

M. Li and coworkers measured the oxygen flux into tailings reclaimed with a 1-m (3.3-ft) water cover at two mine sites (Falconbridge, Ontario, and Louvicourt, Quebec) in Canada. At Falconbridge, the oxygen flux was determined to be generally less than 2 mol of oxygen per square meter per year. The measured oxygen flux into the tailings prior to creation of the water cover was 234 mol/m^2/yr, so the water cover reduced oxygen flux by more than 99%.

At Louvicourt, the oxygen flux was less than 1 mol/m^2/yr, compared to 1150 mol/m^2/yr in exposed tailings. Oxygen penetrated less than 5 mm (0.2 in.) into the tailings layer. At the low rates of oxygen flux achieved by the water cover, sulfide oxidation was negligible.

Dry covers. A dry cover is a cap design that includes a compacted, fine-textured soil layer that retains water even in unsaturated conditions above the water table. Dry covers are composed of layers of engineered components (such as a flexible membrane liner or geosynthetic clay liner) or geologic materials (such as glacial tills, silts, or clays); they contain one or more layers with little air-filled porosity so that oxygen flux is slowed. The gradation and compaction of soil materials used to construct the oxygen-restricting layer in a cover system is critical to their performance. These designs work best in climates with a net surplus of precipitation, where potential desiccation of the layer due to transpiration of water by plants is avoided. Burying the oxygen-restricting layer beneath the zone of root activity can also minimize desiccation. Geomembranes can also be used to restrict airflow into the mine wastes, but this alternative is more costly.

Development rock (rock removed while excavating or mining the ore body) from the Greens Creek, Alaska, underground mine contains varying amounts of potentially acid-generating material that has been encapsulated and stored in the interior of the facility. A dry cover was designed, constructed, and monitored to eliminate advection (transport of oxygen by the bulk movement of air) and restrict the diffusion of oxygen into the waste rock. The principal of the cover design is that a compacted layer (for example, a finely textured and poorly sorted glacial till) at residual saturation contains very little air-filled porosity. The compacted till layer is maintained at residual saturation by placing beneath it a capillary break (which exploits the relative differences in porosity between soil types to inhibit water infiltration) and by burying the layer below the active root zone. The wet climate at Greens Creek decreases the risk of desiccation of the lower portion of the dry cover. Modeling and initial field monitoring results suggest that the dry cover will reduce the gaseous diffusion rate by as much as 99%.

Oxygen-consuming layers. These layers consume oxygen prior to its reaching the sulfide-enriched zones. Oxygen-consuming layers are generally carbon-rich or amended with carbon, because microbial decomposition of organic material consumes oxygen and may compete with the sulfide oxidation process for available oxygen. Placement of organically enriched layers overlying pyritic mine waste can be an effective means of removing oxygen from air prior to its reaching the buried mine waste zone. Simple carbon compounds may also support biological sulfate reduction, a microbial process which reduces sulfate to sulfide and produces alkalinity. These chemical byproducts may have an added benefit of reducing the levels of many dissolved metals, especially zinc and nickel.

Addition of organic carbon to tailings to support biological sulfate reduction has been evaluated at several mines. Three basic means of adding carbon to tailings are (1) a carbon-rich layer in a cover, (2) an amendment to tailings, and (3) a naturally accumulating component of sediment in tailings closed with a water cover.

A. Chtaini and others tested the use of papermill waste as covers or additives to acidified tailings as a means of abating the release of metals and acidity. Laboratory and field-scale tests were conducted on paper mill waste covers with and without partial incorporation of paper waste into the underlying tailings. In field test plots, the paper mill waste cover with 30 cm (12 in.) of incorporated paper waste increased the pH of pore water (in the amended layer) from 4.7 to 7.0, and decreased zinc levels from 162 mg/L to below the detectable level. Similar reductions in copper, cadmium, and nickel also occurred. The applied organic wastes did not remove all the oxygen from the system, but maintained neutral pH and low metal levels through dissolution of alkalinity in the waste, and sulfate reduction reactions.

N. Tassé and others measured increases in pH, carbon dioxide, and methane, as well as decreases in dissolved oxygen beneath a 2-m-thick (6.6-ft) wood waste cover placed over sulfide-enriched tailings. The test site was located at East Sullivan in northeastern Quebec. The primary advantage of the organic cover was the removal of oxygen prior to reaching the tailings. After placing the cover, the pH near the original tailings surface (which had acidified prior to cover placement) increased from 4.0 to 6.5 within 3 years. The ground water at the site is expected to continue to improve in water quality over a 10-year period.

For background information *see* ACID AND BASE; CALCITE; DOLOMITE; MINING; PH; PYRITE; WEATHERING PROCESSES in the McGraw-Hill Encyclopedia of Science & Technology.   William M. Schafer

Bibliography. British Columbia Acid Mine Drainage Task Force Report, *Draft Acid-Rock Drainage Technical Guide*, 2 vols., 1989; A. Chtaini et al., A study of acid mine drainage control by addition of an alkaline mill paper waste, pp. 1147–1161 in *Proceedings of the 4th International Conference on Acid Rock Drainage*, Vancouver, B.C., Canada, May 31–June 6, 1997; L. C. W. Elliott, L. Liu, and S. W. Stogran, Organic cover materials for tailings: Do they meet the requirements of an effective long term cover?, pp. 813–824 in *Proceedings of the 4th International Conference on Acid Rock Drainage*, Vancouver, B.C., Canada, May 31–June 6, 1997; M. Li, B. Aube, and L. St-Arnaud, Considerations in the use of shallow water covers for decommissioning reactive tailings, pp. 117–130 in *4th International Conference on the Abatement of Acidic Drainage*, Vancouver, B.C., 1997; M. Li, L. J. J. Catalan, and P. St-Germain, Rates of oxygen consumption by sulphidic tailings under shallow water covers—Field measurements and modelling, pp. 913–920 in *Proceedings from the 5th International Conference on Acid Rock Drainage*, Denver, CO, 2000; W. L. Lindsay, *Geochemical Equilibria in Soils*, Wiley, New York, 1979; D. D. Runnells, T. A. Shepherd, and E. E. Angino, Metals in water: Determining natural background concentrations in mineralized areas, *Environ. Sci. Technol.*, 26(12):2316–2323, 1992; A. A. Sobek et al., *Field and Laboratory Methods Applicable to Overburden and Mine Soils*, U.S. Environmental Protection Agency, EPA-600/Z-78-054, National Technical Information Service, Springfield, VA, 1978; N. Tassé et al., Organic-waste cover over the East Sullivan mine tailings: Beyond the oxygen barrier, pp. 1627–1642 in *Proceedings of the 4th International Conference on Acid Rock Drainage*, Vancouver, B.C., Canada, May 31–June 6, 1997; N. van Breeman, Soil forming processes in acid sulphate soils, in H. Dost (ed.), *Acid Sulphate Soils: Proceedings of the International Symposium*, Wageningen, ILRI Pub. no. 128, vol. 1, pp. 66–130, International Institute for Land Reclamation and Improvement, 1973.

Miniaturized analysis systems

The miniaturization of analytical instrumentation has been a driving force in analytical chemistry for several decades. Miniaturization brings about many advantages for chemical analysis, including decreased analysis time, sample consumption, and reagent consumption; increased sample throughput; and the ability to analyze very small samples. Miniaturized analytical systems are less expensive to manufacture, and therefore cost less. An additional benefit of miniaturization is the ability to include more functions on a single device, with the objective of integrating a chemical laboratory on a chip (lab-on-a-chip). One outcome of lab-on-a-chip technology will be portable devices for point-of-care medical diagnostics, environmental monitoring, and the detection of chemical and biological warfare agents in real time.

The development of lab-on-a-chip technology has become a very active area of research in analytical chemistry. Much lab-on-a-chip work focuses on the use of microfluidics devices for analysis. For our purpose, microfluidics may be defined as the movement and analysis of small fluid volumes (<10 microliters) in microfabricated devices. Microfluidics encompasses microchip separations, including electrophoresis and chromatography, and microfluidic flow injection analysis. The incorporation of additional functional elements, including enzymatic reactors and sample purification elements, will eventually lead to the development of complete lab-on-a-chip devices. *See* MICROFLUIDICS.

Microchip separations. For microchip separations, including microchip liquid chromatography and microchip capillary electrophoresis, the goal is to reduce analysis time and sample consumption, while increasing functionality. The earliest work in chemical separations focused on the separation of DNA for both sequencing and fragment analysis. Other areas of interest in chemical separations include protein identification and sequencing (proteomics), environmental monitoring, and clinical diagnostics. Chemical separations at the microchip scale are typically much faster than traditional chromatography and electrophoresis. **Figure 1** shows the separation of dopamine and hydroquinone using microchip capillary electrophoresis. The 30-s migration time is significantly reduced in comparison to the 20-min time required for this analysis using conventional instrumentation. The development of microchip liquid chromatography has lagged behind the development of microchip capillary electrophoresis because of difficulties associated with integrating mechanical pumps, injectors, and detectors. Only in the last few years have examples of microchip chromatography appeared in the literature.

Device fabrication. Microfabricated separation systems are constructed by etching or molding methods, depending on the substrate. Originally, all devices were made from either quartz or glass, as these two materials possess excellent optical properties and fabrication protocols from electrical engineering could be used. Glass-based devices are fabricated by first selectively etching one glass substrate to form channels and then thermally bonding it to a second piece of glass. This fabrication process is time consuming and costly.

Polymers are being explored as a replacement for glass because they are less expensive, and the

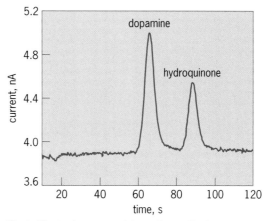

Fig. 1. Electropherogram of dopamine and hydroquinone using microchip capillary electrophoresis. Under the same conditions using conventional capillary electrophoresis, the separation time was almost 20 min.

device fabrication process is both simpler and less expensive. Polymer microchip systems are fabricated by micromolding. A micromold is fabricated by either of two methods, depending on the target polymer. In the simpler method, a thick layer of photoresist (a light-definable polymer) is patterned on a silicon wafer to produce the mold (**Fig. 2**). These molds are typically used with soft polymers such as poly(dimethylsiloxane). In the other method, a pattern is etched in a solid surface such as a silicon wafer or nickel plate. The resulting mold is used for hot-molding structures in poly(methylmethacylate) and other hard polymers.

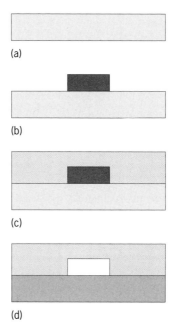

Fig. 2. Cross section of the micromolding procedure. (*a*) A silicon wafer (*b*) is coated with photoresist and the photoresist patterned to give a well-defined mold. (*c*) The pattern is then replicated by pressing the polymer against the mold. (*d*) After separation from the mold, the polymer is sealed to a second substrate to form a channel.

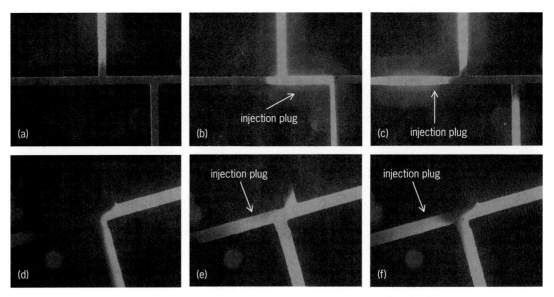

Fig. 3. Fluorescent photomicrograph of (*a*–*c*) cross-injection and (*d*–*f*) gated injection. (*a*) Preinjection state showing the approach of the sample solution to the channel intersection. (*b*) Injector being filled with sample solution. (*c*) Injection completed and sample plug beginning to be separated as it moves to the left. (*d*) Sample and buffer flowing through intersection prior to injection. The sample solution is fluorescent. (*e*) Injection has begun with sample flowing in two directions now. (*f*) After injection, the sample plug is moving to the left and the flow of buffer (clear) is reestablished.

To create a replicate, a polymer substrate is pressed against the mold. The replicate is sealed to a second flat surface (the fourth wall) to complete the channel. For soft polymers such as poly(dimethylsiloxane), the overall fabrication time (not including the time required to make the mold) is 1–2 h, while the fabrication time for hard polymers is 30 min to 1 h. Molds can be used to make hundreds of microchips, resulting in a decrease in the manufacturing costs.

Microchip operation. For microchip capillary electrophoresis, fluid flow is generated by applying a potential gradient across the capillary. The electrical field gives rise to a bulk solution transport called electroosmotic flow. The linear flow velocity of electroosmotic flow is dependent upon the applied field strength, the solution conditions, and the substrate material, and thus is not as reproducible as pressurized flow. However, electroosmotic flow is simple to actuate, can control the direction of flow without valves, and provides high-resolution separations, making it the most commonly used method of fluid transport at the microchip scale.

Another key to microchip capillary electrophoresis is the injection of the sample. Two methods exist for electrokinetic injection: cross-injection and gated injection (**Fig. 3**). In gated injection, the sample and buffer flow continuously through a channel intersection, with the buffer directed toward the separation channel and the sample directed toward the waste channel. Injection occurs when the sample is directed into the separation channel through precision application of voltage. Gated injection is the simpler of the two injection modes, but results in a biased injection with cationic species injected pref-

erentially over anionic species because they have a higher mobility in the initial stage of the injection process. Cross-injection uses electroosmotic flow to completely fill a channel intersection with sample solution before redirecting the solution down a separation channel. The volume of injected sample is controlled by the distance between the sample channel and the waste channel. In Fig. 3*c*, the total injected volume is 1.2 nanoliters. After injection, the components in the sample plug are separated based on their electrophoretic mobility in the separation channel. Detection occurs at some point down the separation channel.

Detection in microchip capillary electrophoresis has received significant attention due to its obvious importance. Initially, all detection was accomplished using laser-induced fluorescence. Laser-induced fluorescence is an excellent and sensitive detection technique for fluorescently labeled compounds; however, it is bulky and will be too expensive for the development of portable devices. Other detection modes have been pursued for specific applications, such as mass spectrometry. Mass spectrometry is attractive because it can identify compounds based on their mass and fragmentation pattern. Electrochemical techniques have also received significant attention for detection due to their low cost, high sensitivity, and ease of miniaturization. A variety of electrochemical detection modes are available, ranging from selective with amperometry to general with conductivity.

Microchip liquid chromatography. In addition to the established microchip techniques based on capillary electrophoresis, work has begun on the development of microchip liquid chromatography.

The potential advantages of microchip liquid chromatography are decreased analysis time, increased throughput, and reduced sample and reagent consumption. In one example of low-pressure affinity chromatography, a 500-fold reduction in sample and reagent consumption was noted, compared to traditional methods. In addition, reliable and efficient methods for packing microchip columns with the stationary phase, sensitive detectors, and the production of miniaturized high-pressure pumps have spurred the development of microchip chromatography. Furthermore, conventional syringe pumps and ultraviolet (UV) detectors can be used for microchip liquid chromatography, with the advent of effective methods of chip-to-world coupling—the method by which the microchip is attached to instrumentation and devices that are separate from the chip itself. Arguably, the biggest gain in microchip chromatography has been the improved methods for creating stationary phases inside microchannels. Traditional packing materials can be held in place using fritless packing methods, where a wide channel is dramatically narrowed to a width approaching the packing's particle size. When the particles are loaded into the channel, they condense at the narrow region and form a solidly packed bed of stationary-phase material. Alternatively, monolithic columns can be formed from continuous polymer plugs using a variety of polymerization methods. Monolithic columns are attractive because they do not require tedious packing procedures, are formed rapidly, and can be patterned anywhere in the microchannel desired, if the polymerization is initiated by UV radiation.

Microfluidics and flow injection analysis. Although most research in lab-on-a-chip technology and microfluidics to date has focused on separations, there is a growing body of work focused on microfluidics flow injection analysis (FIA). In FIA, a sample is injected into a flowing stream of solution to which various reagents are added, and the reaction product of the sample and the reagents is detected. Microfluidic FIA offers many of the same advantages in chemical analysis that microchip separations offer, including a reduction in sample size and reagent consumption. Furthermore, the use of microfabrication to generate systems allows more function to be built into a device, allowing for both FIA and sample pre- or post-treatment. The area of greatest potential application for microfluidic FIA is in clinical diagnostics, where the measurement of specific markers of disease is essential. The use of microfluidic FIA has the potential to reduce analysis times from minutes to seconds, reduce the required sample from milliliters of blood to a single finger prick, and reduce the cost of analysis. In addition to these benefits, there is potential for developing small, hand-held monitoring devices that will allow doctors to record marker levels in real time at a patient's bedside.

For background information *see* ANALYTICAL CHEMISTRY; CHROMATOGRAPHY; ELECTROCHEMICAL TECHNIQUES; ELECTROPHORESIS; LIQUID CHROMATO-GRAPHY; MASS SPECTROMETRY in the McGraw-Hill Encyclopedia of Science & Technology.

Charles S. Henry

Bibliography. A. J. DeMello, Microfluidics: DNA amplification moves on, *Nature*, 422:28–29, 2003; D. J. Harrison et al., Micromachining a miniaturized capillary electrophoresis-based chemical analysis system on a chip, *Science*, 261:895–897, 1993; S. A. Soper et al., Polymeric microelectromechanical systems, *Anal. Chem.*, 72:642A–651A, 2000; T. Thorsen, S. J. Maerkl, and S. R. Quake, Microfluidic large-scale integration, *Science*, 298:580–584, 2002.

Mobile-phone antennas

Since Heinrich Hertz demonstrated electromagnetic radiation and introduced the concept of an antenna over 100 years ago, antennas have become commonplace, and whip antennas on radios and cars, together with satellite dishes, have become familiar objects. Although the theory and design of antennas is well established, sizable journals continue to be published with numerous antenna innovations. This activity is a result of the demand from users worldwide. Large-scale circuit integration and chip technology continue to shrink the size of electronic packages to such an extent that the antenna is often the most bulky and costly component of a device. The mobile-phone antenna is a significant example of antenna design progress in the past decade. Mobile phones have decreased in volume from 900 cm^3 (55 in.^{3}) to 70 cm^3 (4.3 in.^{3}) in the 16 years from 1987 to 2003 as a result of strong consumer demand, and the associated antennas have followed this downsizing.

Reduction of antenna size and visibility. An obvious user requirement is that the mobile-phone antenna be of small size and preferably not visible at all. It was previously thought that users were very tolerant of whip antennas on radios, but, in fact, most users see the antenna as an encumbrance. Enormous effort and expenditure have been devoted to making the antenna inconspicuous in cellular service applications. An exception is the mobile handset antenna for satellite operation, offering global services. Here the antenna performance is so demanding that the user has been compelled to accept a form of pull-out antenna.

There has always been interest in making antennas of inconspicuous size, but the loss of antenna performance was a deterrent. In general, the antenna size is determined by the operating wavelength because antennas are usually resonant devices, similar in some respects to musical instruments. Some of the known ways of miniaturizing antennas are compressing the structure into a more convenient shape, embedding the antenna in ceramic material, and using superconducting materials to reduce the antenna internal power absorption. Superconductivity is clearly not an option for mobile phones.

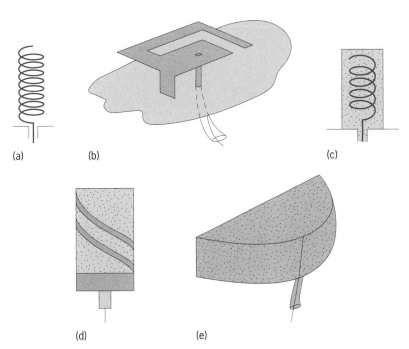

(a)   (b)   (c)

(d)   (e)

Fig. 1. Mobile-phone antennas. (a) Helical coil antenna. (b) Planar inverted F antenna (PIFA). (c) Ceramic chip antenna. (d) Ceramic quadrifilar helix antenna. (e) Dielectric resonator antenna, consisting of a semicircular slab of dielectric material and fed by the inner conductor of a transmission line underneath.

Extensive use is made of the helical antenna (**Fig.** 1*a*), whereby a resonant length of wire is wound into a coil, typically giving a 5-to-1 reduction in height. Some details of the problems facing antenna designers are given in the **table**, comparing a conventional monopole whip antenna and a helical antenna on a ground plane. The antenna height is much reduced but so are the radiation efficiency, bandwidth, and input resistance. Some 20% of the available radiation power is lost internally in the helical antenna structure itself, and this is a direct waste of the battery power, while the smaller input impedance requires additional impedance-matching circuits. When the helical antenna is placed on or within a mobile handset, other design problems arise because the latter represents an electrically small ground plane. Without corrective filters, the handset case itself becomes part of the antenna and cannot be hand-held without absorbing power from the antenna. Manufacturers prefer to use printed antenna technology rather than wound coils, and the planar inverted F antenna (PIFA) [Fig. 1*b*] is another example of a compressed flattened radiating structure, which is installed in many commercial phones.

Both the helical antenna and the planar inverted F antenna are sensitive to the proximity of other components in the handset and the user's hand, while ceramic antennas are less affected. Some examples of ceramic antennas at a research stage are the ceramic chip antenna (Fig. 1*c*), the ceramic quadrifilar helical antenna (Fig. 1*d*), and the dielectric resonator antenna (Fig. 1*e*). Whether or not these ceramic antennas represent the path of future progress for handset antennas will depend on cost factors as well as performance, and printed antennas are very competitive in this respect. Whatever the outcome, the wishes of the user, that the antenna be concealed within the handset case, are likely to remain a high priority.

Reduction of user irradiation. In addition to handling the many design problems, the mobile-phone antenna designer must reduce the amount of electromagnetic radiation absorbed by the user's body and in particular the head. For many years there has been concern about the health effects of electromagnetic radiation from microwave ovens and of high electromagnetic fields near overhead power lines. For mobile-phone users, the situation appears to be more acute because the radiation source is held near the user's brain, with just 1 or 2 centimeters (less than an inch) of space between the head and the antenna. Although evidence of this health risk has yet to be established, mobile-phone manufacturers are giving this issue a high priority.

Clearly a regulatory process is needed that quantifies the extent to which a particular antenna irradiates the user. This irradiation is assessed by the specific absorption rate (SAR), which is defined by Eq. (1).

$$\text{SAR} = \frac{\sigma}{2\rho}|\mathbf{E}|^2 \qquad \text{W/kg} \qquad (1)$$

Here σ is the tissue conductivity, ρ is the tissue density, and $\mathbf{E}$ is the electric field intensity in the tissue. Radio-frequency dosimetry in the human body is very complex due to many factors that affect the absorption rate in tissues, but safety guidelines suggest that the 1-gram-averaged peak SAR should not exceed 1.6 W/kg, while the whole-body-averaged peak SAR should be less than 0.08 W/kg. (The 1-g SAR is a better measure of SAR distribution in the head and represents local variations more accurately.) The frequency of the radiated energy and the duty cycle are among the factors that need to be taken into consideration. Also, it has been established that when a mobile phone is being operated in a confined space

Comparison of the performance of a size-reduced helical antenna with a conventional monopole				
	Resonant frequency, MHz	Bandwidth, MHz	Radiation efficiency, percent	Relative height	Input resistance, Ω
Quarter-wave monopole	150	20	100	1	25
Helical coil	142	1.5	80	1/5	5

such as a car, the user may be exposed to field hot spots that are absent in an open space.

The mobile-phone antenna in itself does not have a SAR rating but must be measured in a natural setting with its handset, together with a model of the human head composed of representative tissue material. Purpose-built rigs are available and assist in standardization, but computer modeling has played a major role in rapidly evaluating different human body and handset scenarios (**Fig. 2**). There is currently much attention on nonthermal tissue damage mechanisms that may take place at very low SAR levels, and for this research the measurement of mobile-phone antenna SAR is especially useful.

Antennas with near-field control. The response of the public to this radiation health concern has been immediate with a demand for hands-free phones and, for some, minimizing of call times. Antenna research has evolved types of antennas that radiate less in the direction of the head, and this development also improves antenna efficiency because any power absorbed in the user's body is wasted from a communication standpoint. Research is ongoing into constructing the mobile handset outer surface from a photonic band-gap material, which can occlude radiation and at least prevent it from being absorbed in the hand. In another design, a metal shield around the antenna base gives some protection to the head, but this solution would appear to have unwanted consequences for antenna operation.

Recently, attention has been focused on creating the definitive low-SAR antenna by a careful understanding of the mechanisms that allow waves to penetrate the human body. Such an antenna would of course have to satisfy all the other handset design requirements, particularly the reduction in size, but, above all, it would establish a design criterion that could increase user confidence. Basic to the discussion of whether such a radiating device is feasible is the definition of the free-space wave impedance, Z_0, given by Eq. (2),

$$Z_0 = \mathbf{E}_T/\mathbf{H}_T = 377 \ \Omega \qquad (2)$$

where $\mathbf{E}_T$ and $\mathbf{H}_T$ are the orthogonal transverse electric and magnetic field intensities, respectively. For the mobile-phone antenna, the potential hazard to the human body lies in the antenna near-field region because the handset is held against the head. In general, antennas are characterized by their electric and magnetic properties, and a whip antenna is predominantly of the electric type whereas a loop antenna is predominantly magnetic. In the near-field region, the wave impedance is no longer 377 Ω: Close-in to an electric antenna, $\mathbf{E}_T/\mathbf{H}_T \gg 377$ Ω, while for a magnetic antenna, $\mathbf{E}_T/\mathbf{H}_T \ll 377$ Ω. It is known that the low-impedance wave of the magnetic antenna will penetrate the human body surface more readily than the high-impedance wave from an electric antenna. In the latter case the wave front is mainly reflected and less body absorption occurs.

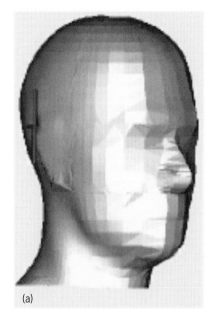

(a)   (b)

Fig. 2. Computer modeling of specific absorption rate (SAR) of a mobile-phone antenna. (*a*) Model of human head with antenna. (*b*) Computer modeling display, showing the regions of highest SAR in the brain.

At present these properties are under investigation together with the measurements on practically realizable electric-type antennas that exhibit low SAR. Ideas on realizability have emerged from recent studies on the use of materials whose relative permability, μ_r, and relative permittivity, ε_r, are both greater than 1. When μ_r and ε_r are approximately equal, the bandwidth and radiation efficiency of the material-embedded antenna are at a maximum but the sought-after electric properties require μ_r to be greater than ε_r. The selection of suitable materials is at present limited for the mobile band of frequencies, but progress in innovative material research, possibly exploiting nanotechnology, suggests that novel materials for the definitive low-SAR mobile-phone antenna will evolve in the near future.

For background information *see* ANTENNA (ELECTROMAGNETISM); ELECTROMAGNETIC RADIATION; ELECTROMAGNETIC WAVE TRANSMISSION; IMPEDANCE MATCHING; MICROWAVE; MOBILE RADIO in the McGraw-Hill Encyclopedia of Science & Technology.

Yiannis C. Vardaxoglou

Bibliography. K. Fujimoto and J. R. James (eds.), *Mobile Antenna Systems Handbook*, 2d ed., Artech House, 2001; J. R. James and J. C. Vardaxoglou, Investigation of properties of electrical-small spherical ceramic antennas, *IEE Electr. Lett.*, 38(20):1160–1162, 2002.

Modular helium reactors

Nuclear reactors are receiving renewed interest in the United States for generation of electric power to help meet future energy requirements. Modular helium reactors (MHRs), a type of high-temperature gas-cooled reactor (HTGR), have high potential for

(a)

(b)   (c)   (d)

Fig. 1. High-temperature gas-cooled reactor (HTGR) fuel. (a) Refractory-coated fuel particle. For HTGRs that use fuel elements in the form of prismatic blocks, (b) coated fuel particles are formed into (c) fuel rods that are inserted into (d) graphite fuel element blocks.

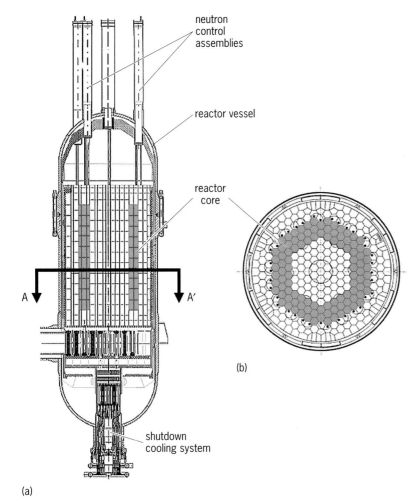

(a)

(b)

Fig. 2. Modular helium reactor design configuration. (a) Vertical section. (b) Horizontal cross section A-A′.

satisfying the need for new energy generation capacity. They can be used to drive Brayton-cycle energy-conversion systems to produce electricity at high efficiency and low cost. Moreover, they can be used for efficient, cost-effective production of hydrogen that would enable further development of a hydrogen energy economy.

HTGR characteristics. The distinguishing characteristics of HTGRs are the use of helium gas coolant, graphite moderator, and refractory-coated particle fuel. The helium coolant is inert and remains single-phase under all conditions; the graphite moderator has high heat capacity, high strength, and stability to high temperatures; the refractory-coated particle fuel retains its integrity at high temperatures. These characteristics combine to provide the capability of HTGRs to have high coolant outlet temperatures, allowing them to be used for high-efficiency electricity generation (higher efficiency than with light-water reactors) or for hydrogen production.

In HTGRs, fission products are retained in the coated fuel particles less than 1 mm (0.04 in.) in diameter (**Fig. 1**). Fissile fuel, in the particle center, is surrounded by a layer of low-density pyrolytic carbon, two layers of high-density pyrolytic carbon, and a layer of silicon carbide sandwiched between the two high-density pyrolytic carbon layers. The fuel particle coatings retain fission products up to temperatures approaching 2000°C (3600°F).

In one type of HTGR, the fuel particles are combined with a graphite matrix and manufactured into fuel compacts that are inserted into fuel channels in hexagonal graphite fuel element blocks (Fig. 1). In an alternative type of HTGR, the particles are combined with a graphite matrix and bonded into tennis-ball-sized spherical "pebble" fuel elements.

HTGR evolution. Seven developmental or demonstration HTGR plants have been built and operated, two in the United States, two in Germany, and one each in the United Kingdom, China, and Japan. Only two are currently operating, the 10-megawatts-thermal High Temperature Reactor (HTR-10) in China and the 30-MWt High Temperature Test Reactor (HTTR) in Japan.

Work on HTGRs in the United States began in the 1950s. The first United States HTGR plant was the Peach Bottom I plant, built in Pennsylvania. It was a small 110-MWt reactor that generated 40 MW of electric power (36% thermal efficiency). The second United States HTGR was the Fort St. Vrain plant in Colorado. Fort St. Vrain had a design power of 330 MWe (842 MWt). Both of these HTGRs had the same basic characteristics. Both used helium as coolant, graphite as moderator, and nuclear fuel in the form of uranium and thorium carbide refractory-coated particles.

In the early 1970s, designs were developed for HTGR plants with significantly larger capacities, 2000 and 3000 MWt. Ten of these large plants were ordered by five different utilities. The oil crisis in the early 1970s and the resulting energy awareness, as well as economic factors, reduced electricity needs to the point that all ten of these HTGR plants, along with many light-water reactor plants, were canceled.

Evolution to the modular helium reactor. In the early 1980s, following the partial-loss-of-core-cooling event at Three Mile Island Unit 2 (in Pennsylvania), HTGR design work focused on designs having enhanced safety characteristics for retention of fission products under all credible conditions. This work resulted in development of the modular high-temperature gas-cooled reactor (MHTGR), now known as the modular helium reactor. In this reactor, the maximum fuel temperatures cannot exceed 1600°C (2900°F) during even worst-case accident events (including complete loss of cooling), using only natural means for rejection of core decay heat (conduction, convection, and radiation). No active systems are required; that is, no working mechanical or electrical equipment is required to ensure that the core does not overheat. Since only passive means are needed to limit the fuel temperature, there is a high assurance that there will be no release of radionuclides beyond normal limits under the worst-case accident scenarios.

The modular helium reactor design incorporates several features to limit fuel maximum temperatures. These design features include a low core power density (about 6.5 W/cm^3), a slender length-to-diameter reactor core aspect ratio, an uninsulated steel reactor pressure vessel for heat rejection from the vessel surface, and an annular active core arrangement utilizing graphite block fuel elements (**Fig. 2**). This design arrangement passively rejects decay heat from the core and maintains maximum fuel temperatures below the 1600°C design limit to ensure that fission product radionuclides are retained within the fuel particles.

High-efficiency electricity generation. The power limit of a modular helium reactor with a core power density of 6.5 W/cm^3 is about 600 MWt based on existing nuclear pressure vessel manufacturing size limitations. Limiting power to 600 MWt adversely impacts economies of scale, but this is partially offset by the passive safety characteristics that reduce the need for complex engineered safety systems. Economic benefits can also be realized by deploying several reactor modules together to form a single

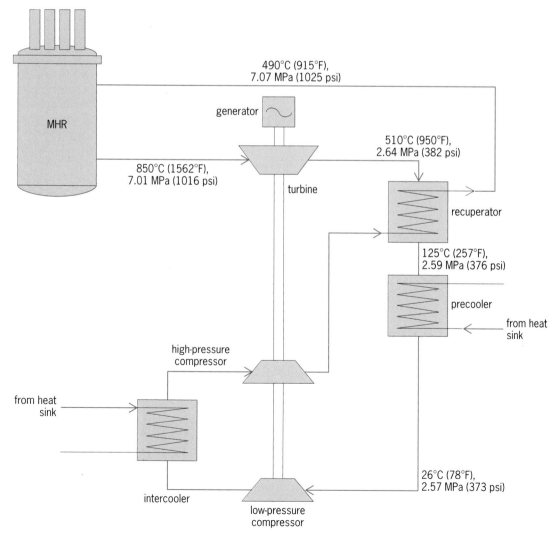

Fig. 3. Direct Brayton cycle for generation of electricity using the high outlet coolant temperature of the modular helium reactor.

plant. But more importantly, substantial economic benefit can be realized, without compromising reactor safety, by using the high-temperature gas coolant to directly drive a gas turbine for the generation of electricity using a Brayton power conversion cycle (**Fig. 3**). The resultant configuration is identified as the gas turbine–modular helium reactor (GT-MHR).

The direct Brayton cycle used by the GT-MHR not only is less complex than the century-old Rankine (steam) cycle used by light-water reactor plants but also has a thermal efficiency of 48%, 50% higher than light-water reactor plants. The substantial improvement in thermal conversion efficiency, coupled with the reduced equipment cost, and reduced costs for operation and maintenance of complex safety and power conversion equipment, results in electricity generation costs for the GT-MHR well below the cost of the most economic generation alternative.

High-efficiency hydrogen production. Concerns about energy security and control of greenhouse gas emissions have prompted interest in production of hydrogen as a primary energy carrier in the United States. Currently, over 11 million tons of hydrogen are produced and consumed yearly in the United States by oil refinery, ammonia production, and chemical process industries. The demand for hydrogen by these users is expected to double by 2010. If hydrogen is used to meet transportation energy needs, the demand for hydrogen would increase severalfold. To meet the growing demand for hydrogen and to provide hydrogen for new, clean-fuel cell vehicles will require substantial increases in hydrogen production.

Virtually all of the hydrogen is currently produced from the reforming of natural gas. The carbon in the natural gas is released as carbon dioxide (CO_2), the primary greenhouse gas, with more than 7 kg of carbon dioxide being released for every kilogram of hydrogen produced. For the wide-scale use of hydrogen as an energy carrier, new sources of hydrogen must be found that do not rely on fossil fuels and that do not pollute the environment. Water splitting, the separation of water into hydrogen and oxygen, can provide a clean source of hydrogen as long as the primary source of input energy into the process is nonfossil, such as solar or nuclear energy. There are two approaches to hydrogen production from water. The first is to use electricity to separate water into hydrogen and oxygen by electrolysis. The second approach is to use heat to drive a thermochemical water-splitting cycle. The use of a thermochemical water-splitting cycle has the potential for greater efficiency than electrolysis owing to the elimination of the intermediate electricity production step.

The most attractive thermochemical water-splitting cycle is the sulfur-iodine cycle (**Fig. 4**). The cycle takes in only water and high-temperature heat and releases hydrogen, oxygen, and low-temperature heat. All reagents are recycled; there are no effluents.

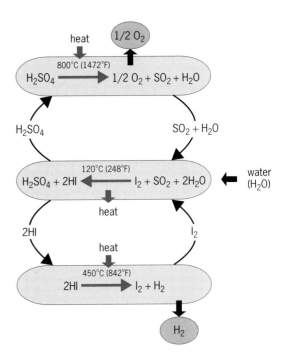

Fig. 4. Sulfur-iodine thermal-chemical water-splitting cycle.

The passively safe modular helium reactor having an outlet coolant temperature capability in the 850–1000°C (1562–1832°F) range is an ideal match for producing the process heat required by the sulfur-iodine cycle. It is estimated that an MHR process heat plant coupled with the sulfur-iodine cycle can achieve 50% hydrogen production efficiency and produce hydrogen cheaply using no fossil fuels and releasing no carbon dioxide or other greenhouse gases. The cost for hydrogen used in fuel-cell-equipped vehicles would be competitive with today's gasoline fuel cost for internal-combustion-engine-equipped vehicles.

For background information *see* BRAYTON CYCLE; FUEL CELL; HYDROGEN; NUCLEAR FUELS; NUCLEAR POWER; NUCLEAR REACTOR in the McGraw-Hill Encyclopedia of Science & Technology.

Walter A. Simon; Malcolm P. LaBar

Bibliography. *Design and Development Status of Small and Medium Reactor Systems 1995*, IAEA-TEDOC-881, International Atomic Energy Agency, Vienna, 1995; C. W. Forsberg and K. L. Peddicord, Hydrogen production as a major nuclear energy application, *Nucl. News*, 44(10):41–45, September 2001; M. T. Simnad (guest ed.), High-temperature helium gas-cooled nuclear reactors: Past experience, current status and future prospects, *Energy Int. J.*, vol. 16, no. 1/2, Pergamon Press, 1991.

Molecular modeling

Chemistry has seen many advances in a relatively short time, driven by scientists who wove seemingly unrelated observations into coherent theorems that explained molecular behavior. Before 1860, the empirical formula for water (H_2O) was thought to be OH, and virtually nothing was known about the structure of organic compounds except that they were composed mostly of carbon and hydrogen. In the intervening years, a great deal of thought was devoted to developing an understanding of molecular behavior, and the results of that work are remarkable. Only a few decades ago, it was common knowledge that no microscope would ever allow human eyes to "see" a molecule. But now we are able to view (and even arrange) individual atoms using the modern techniques of scanning tunneling microscopy (STM) and atomic force microscopy (AFM).

Given this synergistic relationship between increasing scientific insight and new technology, it is no surprise that the fields of computational chemistry and molecular modeling have grown up alongside advances in theoretical chemistry and computer technology. The discoveries and developments in this area have taken place at an accelerated pace in recent years, with molecular modeling becoming an essential part of the molecular design process. In modern academic and industrial laboratories, molecular modeling spans every aspect of chemistry, molecular biology, chemical engineering, and materials science.

Molecular modeling has evolved along with, and adapted to, the changing capabilities of the computational methods and computer systems. For our purposes, molecular modeling methods can be classified into two general areas: correlative structure/property relationship modeling and three-dimensional (3D) structure-based design. Correlative modeling involves the use of machine learning techniques to determine the relative importance of molecular features in order to understand and predict molecular behavior. Structure-based design attempts to calculate the molecular interactions to accomplish a similar goal. Each of these areas has seen advances in the past year, and both subfields rely on having accurate (or at least useful) representations of the molecules under investigation.

Molecular representation. Molecular modelers are almost always presented with difficult problems that involve many variables. For example, why is molecule A better absorbed by the body than molecule B? Or, why does polymer C melt at a lower temperature than polymer D? To answer these kinds of questions, different approaches require alternative kinds of molecular representation.

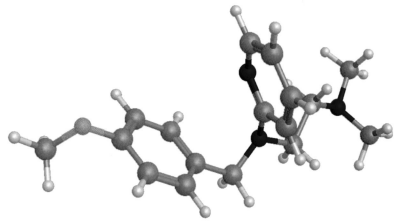

Fig. 2. Pyrilamine 3D structure.

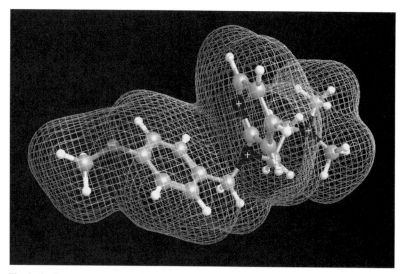

Fig. 3. Pyrilamine van der Waals surface.

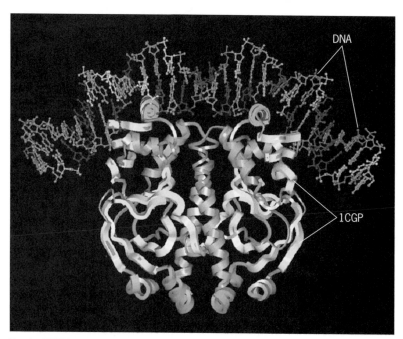

Fig. 4. 1CGP bound to DNA.

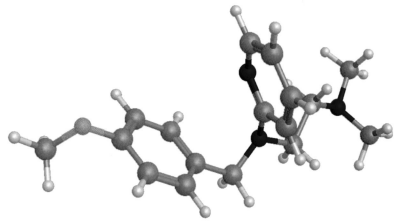

Fig. 1. Pyrilamine 2D structure.

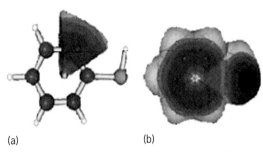

(a)   (b)

Fig. 5. Thiophenol TAE reconstruction. Parts *a* and *b* are described in the text.

Figure 1 shows a two-dimensional (2D) graphical representation of the molecular structure of pyrilamine, a compound with antihistamine properties. This kind of molecular representation is common

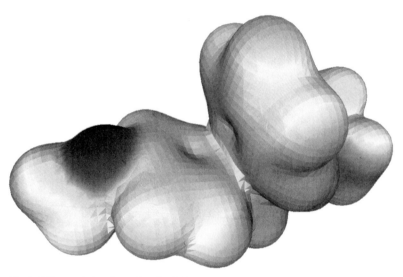

Fig. 6. PIP-encoded surface of pyrilamine.

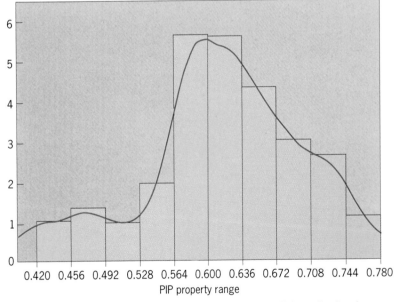

Fig. 7. Histogram representation of the surface distribution of PIP for pyrilamine shown in Fig. 6. Each histogram bin represents the fraction of molecular surface area having the specified range of PIP.

in organic chemistry, and conveys a great deal of information to experienced medicinal chemists. In computational chemistry, this represents a molecular graph—a series of atoms connected by bonds. There is no geometric information here, except what is implied, and a number of the hydrogen atoms have been left off for clarity.

The first step in 3D molecular modeling is to represent molecules in terms of their geometries rather than their 2D structures (**Fig. 2**). In this higher-resolution mode of representation, bond distances, angles, and torsions are explicitly described. At this level of detail, it is possible to capture variations in molecular shapes and conformations. Computational techniques such as molecular mechanics and dynamics can then be used to find the lowest-energy geometries for each molecule and to determine how molecules can interact with each other.

At this point, it is useful to think of molecules as more than a collection of atomic "balls" connected by bond "springs." Realizing that atoms carry electrical charges and are surrounded by clouds of electrons leads to a representation that looks like the van der Waals envelope (**Fig. 3**), where the excess positive charge is shown as light color on the surface mesh, and excess negative charge is shown as dark color. As with the molecular representations shown earlier, the surface in Fig. 3 does not physically exist as shown. It is only a means of understanding how molecules repel or attract each other through electrostatic interactions and appear to other molecules as having finite volumes that cannot overlap. Structure-based calculations are performed using this kind of molecular representation through the use of a force-field calculation that captures these various attractive and repulsive forces. Such explicit force-field calculations are commonly used to explain the behavior of small molecules, and are often used to understand how small, drug-sized molecules might interact with large molecules such as proteins or DNA, or even how large molecules can interact with each other. **Figure 4** shows the interaction of a gene-regulation protein (1CGP) bound to a sequence of DNA that recognizes it specifically.

Transferable atom equivalents. Some of the most exciting recent advances in molecular modeling include the capability of representing molecules in terms of their electron density distributions in a manner fast enough to be a practical tool for molecular design applications. Traditionally, computational approaches for determining molecular properties from their electron densities have been lengthy and expensive undertakings. In recent years, new approximate methods have emerged that allow electron density-derived molecular properties to be obtained quite rapidly.

One such approach involves the use of transferable atom equivalents (TAE). Using the TAE approach, molecular electron densities and their properties are reconstructed from combinations of atomic electron density fragments (transferable atoms). The computer program that does this is called RECON. The piece-wise atomic reconstruction of a molecule using this approach is shown for thiophenol in **Fig. 5**. In Fig. 5*a*, the electron density of a single aromatic carbon atom has been added to the molecular framework. In Fig. 5*b*, the completed electron density reconstruction of thiophenol is shown, including a representation of Politzer's local average ionization potential (PIP), one of the many electron density–derived properties found to be useful for molecular modeling. Introduced by J. Murray and P. Politzer in 1998, PIP is a local quantity defined at every point in the space around a molecule, and is computed by taking a weighted average of the orbital energies. Note that in Fig. 5 the individual atoms have come together to provide a quantitative representation of the electronic properties of the molecule.

Correlative modeling. When the TAE approach is used to reconstruct the electronic properties of the pyrilamine molecule, the PIP-encoded surface shown in **Fig. 6** is obtained. The electron density distributions provided by TAE reconstruction may be used as part of a structure-based molecular modeling approach, or they can be used to generate molecular descriptors, a key component of correlative structure/property modeling. Descriptors are numerical features used to describe molecules in quantitative structure/property relationship studies.

This form of modeling is different from what has been described so far in that molecular structure is only used to produce descriptors, after which the molecules are represented by sets of seemingly abstract numbers derived from their structures and electronic properties. One class of molecular descriptors, shown in **Fig. 7**, can be obtained from the molecular surface property distribution illustrated in Fig. 6. In this case, each histogram "bin" represents the fraction of molecular surface area having a specific range of the PIP property. Since the distribution of this property is intimately related to molecular structure, the bin heights may then be used as molecular descriptors. These descriptors may be rapidly computed (half a million molecules per hour) for small, drug-sized molecules on modest computer systems, making them useful for high-speed property screening applications. This technique has been recently extended to encode both surface property and molecular shape information in the form of 2D shape/property histograms such as the pyrilamine electrostatic potential (EP)-encoded surface shown in **Fig. 8**. The height of each column in Fig. 8 may be used as a shape/property descriptor in correlative structure/property modeling.

In order to use correlative modeling successfully, there must be a set of examples of molecules that elicit various responses from the system being studied. When these responses are provided to the machine learning algorithms along with a set of molecular property descriptors, it is often possible to generate a predictive model for molecular behavior. Significant advances in both descriptor technology and machine learning methods have taken place recently, and the combination of these two new approaches (sparse support vector machine regression and electron density–derived shape/property descriptors) has opened up new levels of virtual screening for use by molecular designers in all areas of chemistry, including drug design, polymer chemistry, protein engineering, and regulation of gene transcription, to name a few.

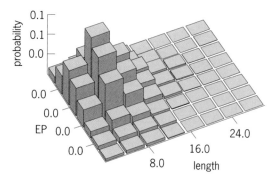

Fig. 8. Pyrilamine electrostatic potential shape/property histogram.

For background information *see* BOND ANGLE AND DISTANCE; CHEMICAL BONDING; COMPUTATIONAL CHEMISTRY; CONFORMATIONAL ANALYSIS; MOLECULAR MECHANICS; MOLECULAR SIMULATION; STEREOCHEMISTRY; VAN DER WAALS EQUATION in the McGraw-Hill Encyclopedia of Science & Technology.

Curt M. Breneman; N. Sukumar

Bibliography. C. M. Breneman et al., Descriptor generation, selection and model building in quantitative structure-property analysis, in J. N. Cawse (ed.), *Experimental Design for Combinatorial and High Throughput Materials Development*, pp. 203–238, Wiley-Interscience, 2002; C. M. Breneman and T. Thompson, Modeling the hydrogen bond with transferable atom equivalents, in D. Smith (ed.), *Modeling the Hydrogen Bond*, pp. 152–174, ACS Symposium Series, Washington, DC, 1993; C. M. Breneman and L. W. Weber, Transferable atom equivalents: Assembling accurate electrostatic potential fields for large molecules from ab-initio and PROAIMS results on model systems, in G. A. Jeffrey and J. F. Piniella (eds.), *The Application of Charge Density Research to Chemistry and Drug Design*, Plenum Press, 1991; C. J. Cramer, *Essentials of Computational Chemistry: Theories and Models*, Wiley, 2002; F. Jensen, *Introduction to Computational Chemistry*, Wiley, 1998; L. B. Kier and L. H. Hall, *Molecular Connectivity in Chemistry and Drug Research*, Academic Press, New York, 1976; L. B. Kier and L. H. Hall, *Molecular Connectivity in Structure-Activity Analysis*, Wiley, 1986; A. Leach, *Molecular Modelling: Principles and Applications*, 2d ed., Prentice Hall, 2001; K. B. Lipkowitz and D. B. Boyd (eds.), *Reviews in Computational Chemistry*, vols. 1–18, Wiley, 1990–2002; J. S. Murray, P. Politzer, and G. R. Famini, Theoretical alternatives to linear solvation energy relationships, *Theochem. J. Mol. Struc.*, 454(2–3):299–306, 1998; R. S. Perlman and K. M. Smith, Novel software tools for chemical diversity, in H. Kubinyi, Y. Martin, and G. Folkers (eds.), *3D QSAR and Drug Design: Recent Advances*, Kluwer Academic, Dordrecht, Netherlands, 1998.

Mycotoxins in food

Mycotoxins are poisonous chemical substances produced by molds during their growth on foods, feeds, and various agricultural commodities such as cereal grains, oilseeds, and nuts. The Food and Agriculture Organization (FAO) estimates that at least 25%, and in some areas as much as 50%, of food crops are affected by mycotoxins, with at least 5–10% of the world food supply lost annually to fungi and mycotoxins. In the United States the estimated costs of fungal and mycotoxin contamination of food and feed are between $418 million and $1.66 billion annually.

The molds that produce mycotoxins are members of the kingdom Fungi and are considered to be microfungi, as opposed to macrofungi such as the toxic mushrooms. Mushroom toxins, or poisons, are not considered mycotoxins because they are an inherent part of the mushroom or fungus; thus the toxic mushrooms are inherently poisonous. Mycotoxins, on the other hand, are excreted into the substrate and are not an inherent part of the mold.

Mycotoxins cause toxicity by acting at the cellular level through diverse mechanisms related to their particular chemical structures. Diseases and disease conditions (states of illness not described as specific diseases) that are caused by mycotoxins are referred to as mycotoxicoses. The effects of mycotoxins on humans and animals are dose-related and include acute and chronic effects and, in the case of some mycotoxins, carcinogenicity and suppression of the immune system. Mycotoxins were first recognized as potential disease agents for humans and

TABLE 1. Mycotoxins of greatest concern and the molds that produce them

Mycotoxins	Molds
Aflatoxins	*Aspergillus flavus, A. parasiticus, A. nomius*
Ochratoxin A	*A. ochraceus, A. carbonarius, Penicillium verrucosum*
Patulin	*Penicillium expansum*
Fumonisins	*Fusarium verticillioides, F. proliferatum, F. subglutinans*
Moniliformin	*Fusarium proliferatum, F. subglutinans*
Deoxynivalenol (Vomitoxin)	*Fusarium graminearum, F. culmorum, F. crookwellense*
Zearalenone	*Fusarium graminearum, F. culmorum, F. crookwellense*
Satratoxins	*Stachybotrys chartarum*

animals in 1960, when more than 100,000 young turkeys (poults) and other young farm animals in England died from poisoning by feed containing contaminated peanut meal imported from Brazil. The disease was first called Turkey X disease because the cause was not readily apparent. Eventually it was shown that the peanut meal was contaminated with a mycotoxin that ultimately became known as aflatoxin, named for the mold that produced the toxin, *Aspergillus flavus*. Ultimately the term "mycotoxin" came to designate a family of compounds.

Although there are numerous molds that may contaminate foods and feeds, the mycotoxins of greatest concern are produced by species found in mainly three genera, *Aspergillus*, *Penicillium*, and *Fusarium* (**Table 1**). The genera *Alternaria* and *Stachybotrys* contain species that produce mycotoxins that are of lesser significance as food contaminants.

Aflatoxins. Aflatoxins consist of a family of approximately 20 related compounds produced by *A. flavus*, *A. parasiticus*, and *A. nomius*; however, only *A. flavus* and *A. parasiticus* are found in food and considered important contaminants. The four main aflatoxins—B_1, B_2, G_1, and G_2—may contaminate foods as a result of mold growth. Two other aflatoxins, M_1 and M_2, are metabolites of B_1 and B_2 that are produced and excreted by animals that have consumed contaminated feed (**Fig. 1**).

Aflatoxins, particularly B_1, are potent liver toxins (hepatotoxins) and carcinogens of the liver (hepatocarcinogens). Aflatoxins can cause acute toxicity, chronic (subacute) toxicity, immunosuppression, and cancers. High doses of aflatoxin cause acute toxicity and often death in both humans and animals. Sublethal doses produce a chronic toxicity that may weaken the immune system and increase susceptibility to bacterial and viral infections. Exposure to very low doses over a period of time produces tumors in test animals and may contribute to liver cancer in humans and animals. The International Agency for Research on Cancer has classified aflatoxin B_1 as a class I human carcinogen. Trout and ducklings are the most susceptible to aflatoxins, with the young of any species more susceptible than older mature individuals of the same species.

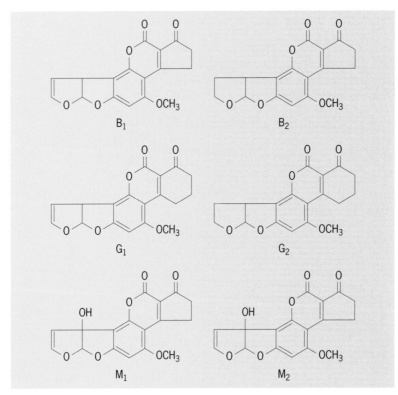

Fig. 1. Chemical structures of aflatoxins.

TABLE 2. U.S. Food and Drug Administration (FDA) and European Union action levels and guidelines*

Mycotoxin	Commodity of food	FDA[‡]	European Union[‡]
Aflatoxin	Human food	20 ppb	2–15 ppb
Aflatoxin M_1	Milk	0.5 ppb	0.05 ppb
Aflatoxin	Dairy feed	20 ppb	
Aflatoxin	Breeding stock feed	100 ppb	
	Mature beef, swine and poultry feed	300 ppb	
Ochratoxin A	Cereal grains[†]		5.0 ppb
	Processed cereal products		3.0 ppb
	Dried fruits, wine		10.0 ppb
Patulin		50 ppb	50 ppb
Deoxynivalenol	Finished wheat products	1.0 ppm	0.5 ppm
	Flour		0.75 ppm
	Raw cereals		0.75 ppm
Fumonisins	Degermed corn meal	2.0 ppm	1.0 ppm
	Whole/partly degermed	4.0 ppm	
	Dry milled corn bran	4.0 ppm	
	Cleaned popcorn	3.0 ppm	
	Cleaned corn for masa	4.0 ppm	

*Canada and other countries have similar guidelines.
[†]Where no levels are given, action levels and guidance levels have not yet been established.
[‡]ppb, parts per billion; ppm, parts per million.

Aflatoxins may be contaminants of nuts (such as peanuts, pecans, Brazil nuts, almonds, and pistachios), corn (maize), cottonseed, figs, and dates. They are very stable compounds and thus can survive most types of food processing, enabling them to contaminate processed and finished food products such as peanut butter, corn meal, and processed foods containing nuts or grains. The food industries in the United States, Canada, and Europe extensively test susceptible foods and commodities for aflatoxin to prevent contaminated products from entering the food supply (**Table 2**).

Ochratoxins. Ochratoxins consist of three metabolites (ochratoxins A, B, and C) produced by *A. ochraceus* and related species, some strains of *A. carbonarius*, *Penicillium verrucosum*, and certain other *Penicillium* species. Ochratoxin A is the most abundant as well as the most toxic of the group (**Fig. 2**). Ochratoxins B and C differ from ochratoxin A in several functional groups, resulting in much lower toxicity and significance. Ochratoxin A is a potent nephrotoxin that causes kidney damage in swine, dogs, and rats. Ochratoxin has been studied as a possible cause of Balkan endemic nephropathy, a disease that occurs in certain families in rural areas of former Yugoslavia, Bulgaria, and Romania, but proof is lacking. Ochratoxin has also been shown to be carcinogenic and may cause kidney tumors in animals. It can be produced in barley by *P. verrucosum* and in green coffee beans, grapes, and raisins by *A. ochraceus* and *A. carbonarius*. Although ochratoxin is destroyed during the roasting of coffee beans, it may persist in raisins, wine, and beer. Ochratoxin A has been found in the blood of Europeans and Canadians, and epidemiological evidence suggests that nearly half of the population of Europe has been exposed.

Patulin. Patulin is a mycotoxin produced by *P. expansum* and other species of *Penicillium*, *Aspergillus*, and *Byssochlamys* (**Fig. 3**). It is a broad-spectrum biocide that is toxic to bacteria, mammalian cell cultures, plants, and animals. However, patulin is rather unstable in most foods (grains, cheese, and cured meats) and has not been proven to cause disease in humans or animals. Its instability stems from its reactions with sulfhydryl groups in sulfur-containing amino acids that render it inactive. However, patulin is of some public health concern because it is a potential carcinogen, as it was shown to cause cancer when injected intradermally in mice (though it did not cause any observed toxicity or carcinogenicity in rats when given orally). In addition, patulin produced by *P. expansum*, the main cause of apple rots, has been found in commercial apple juice and apple sauce, both of which are consumed by infants and young children.

Fumonisins. Fumonisins are common contaminants of corn (maize) that are produced primarily by *Fusarium verticillioides* and *F. proliferatum*, common soil organisms found in all corn-growing

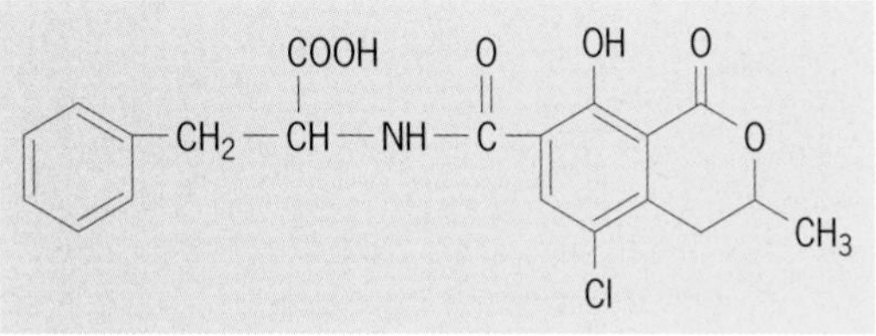

Fig. 2. Chemical structure of ochratoxin A.

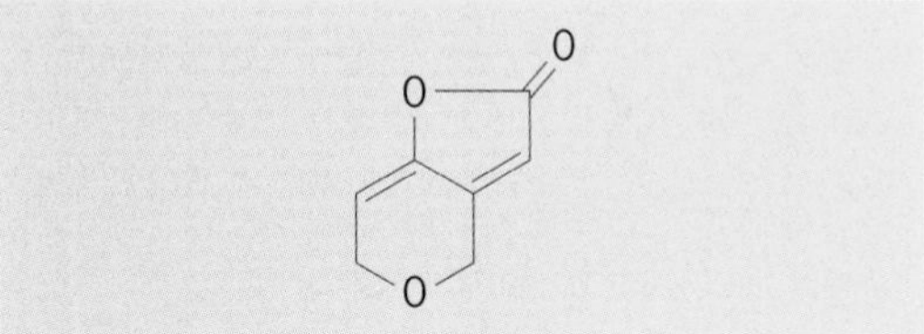

Fig. 3. Chemical structure of patulin.

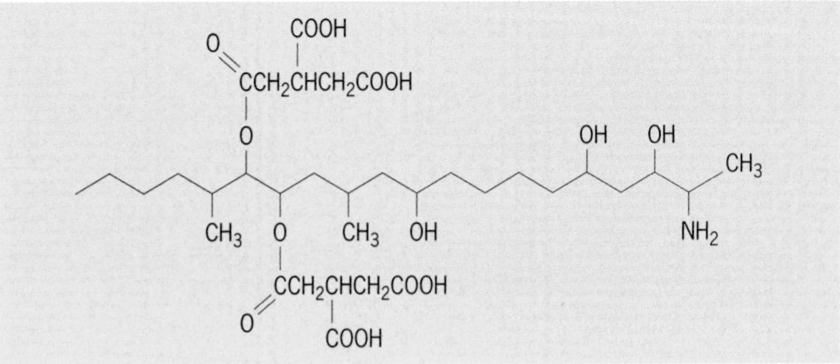

Fig. 4. Chemical structure of fumonisin B₁.

areas (**Fig. 4**). These fungi commonly invade corn plants and contaminate the grain as it develops in the field. Fumonisins have been shown to cause equine leukoencephalomalacia, a degenerative brain disease in horses and other Equidae members (such as donkeys and mules), and an abnormal accumulation of fluid in and around the lungs in swine, resulting in a fatal disease known as porcine pulmonary edema. Experimentally, fumonisins have produced liver cancer in rats and atherosclerosis in monkeys. Fumonisins, specifically the presence of *F. verticillioides* in corn, has been documented and linked to esophageal cancer in humans in the Transkei region of South Africa, northeast Italy, and northern China, where corn is a dietary staple.

Moniliformin. Moniliformin, produced by the corn contaminants *F. proliferatum* and *F. subglutinans*, has been found along with fumonisins in corn and corn-based foods (**Fig. 5**). It is acutely toxic to many

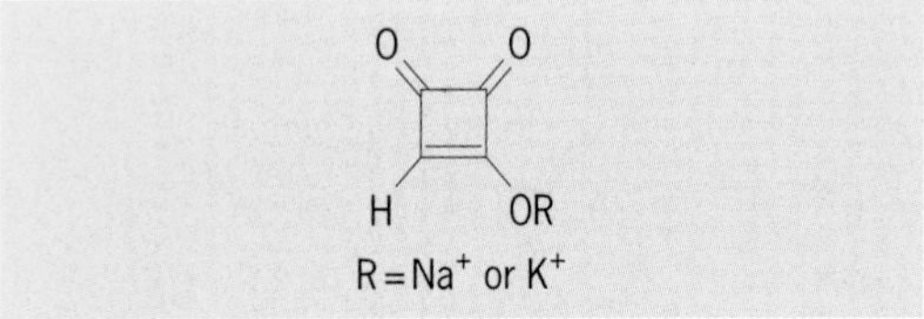

Fig. 5. Chemical structure of moniliformin.

experimental animals, including chickens, ducklings, and rats. Chickens are the most susceptible to moniliformin, which is more toxic to chickens than fumonisins. Moniliformin is a potent cardiotoxic mycotoxin, and subacute toxicosis results in cardiac lesions and enlargement of the heart in rodents and poultry.

Deoxynivalenol. Deoxynivalenol is the most commonly occurring trichothecene (a group of mycotoxins containing an epoxide ring at C-12 and 13 and a double bond at C-9 and 10), and is produced primarily by *F. graminearum* but also by *F. culmorum* and *F. crookwellense* (**Fig. 6**). These organisms infect corn and wheat kernels and produce deoxynivalenol, causing various diseases, particularly Fusarium head blight (scab) in wheat and barley and ear rot in corn.

Deoxynivalenol contamination of grain varies from year to year depending upon the prevalence of Fusarium head blight of wheat and barley and ear rots of corn. The presence of deoxynivalenol in grain also increases its potential to contaminate processed cereal-based foods, and the toxin has been found in breads, pastas, breakfast cereals, infant foods, malt, and beer. Deoxynivalenol causes vomiting in swine, dogs, cats, and humans, and thus is also known as vomitoxin. Swine are the most sensitive to deoxynivalenol, being adversely affected by just 1.0 μg/g (ppm) of feed. Pet foods containing toxic grain have also caused acute toxicities in cats and dogs. Outbreaks of gastroenteritis in humans caused by consumption of moldy wheat contaminated with *F. graminearum* have been documented and reported. Deoxynivalenol is also a proven immunomodulator in animals (and a suspected immunomodulator in humans) and can cause both immunosuppression (increasing susceptibility to infectious diseases) and immunostimulation (which can lead to autoimmune disorders).

T-2 toxin. T-2 toxin, another important trichothecene, is produced by *F. sporotrichioides* and *F. poae* (**Fig. 7**). Although it is more toxic than deoxynivalenol, it occurs less commonly in nature. T-2 toxin inhibits protein synthesis, disrupts deoxyribonucleic acid and ribonucleic acid synthesis, and suppresses the immune systems of animals. It is very toxic to poultry, especially chickens and turkeys, causing oral lesions and necrosis of the mucous membranes of the mouth, tongue, and hard palate. T-2 toxin also causes hemorrhaging in the gastrointestinal tract (resulting in bloody diarrhea), reddening of the skin, and dermal necrosis. In humans, T-2 toxin is believed to be the primary cause of the disease alimentary toxic aleukia, which produces symptoms characteristic of T-2 toxin consumption.

Zearalenone. Zearalenone is a phenolic resorcyclic acid lactone that can be produced by a number of

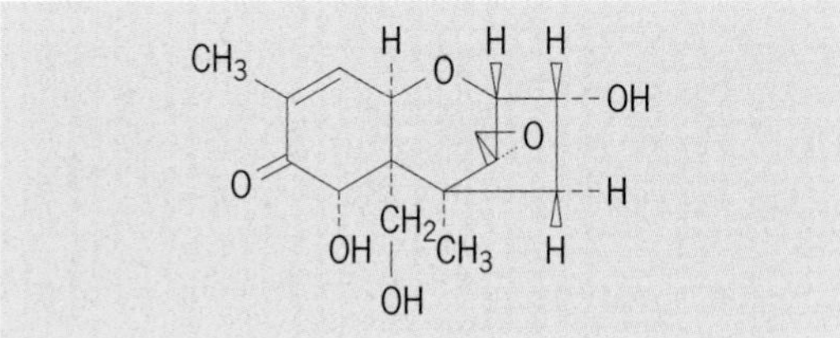

Fig. 6. Chemical structure of deoxynivalenol.

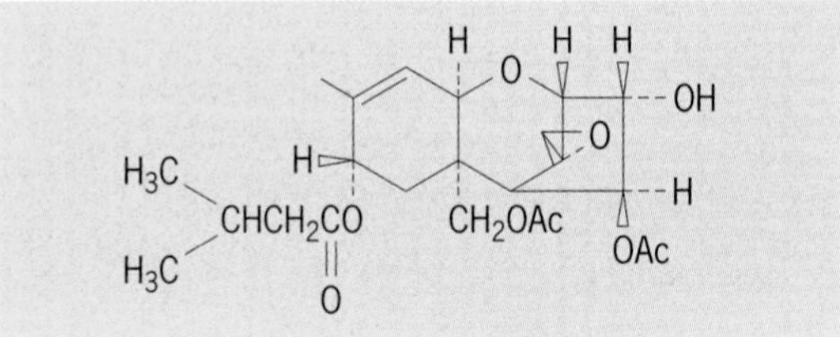

Fig. 7. Chemical structure of T-2 toxin.

Fusarium species (**Fig. 8**). However, it is most commonly produced by *F. graminearum, F. culmorum,* and *F. crookwellense,* the same species that produce deoxynivalenol. Although zearalenone can occur in wheat, barley, rye, and sorghum, it is most frequently associated with high-moisture corn grown in cooler areas of the United States, such as Minnesota, Wisconsin, and Michigan, where early frosts may kill the corn plant before the grain has had a chance to properly dry in the field. Zearalenone affects the reproductive system in animals, especially swine, in which it causes vulvovaginitis (inflammation of the vulva and vagina), infertility, abortion, and feminization of males that results in mammary gland hyperplasia (enlargement) and testicular atrophy. Zearalenone has also been classified as an endocrine disrupter and is considered to be a potential human carcinogen.

Alternaria toxins. *Alternaria* species, especially *A. alternata,* can produce several toxic compounds, including tenuazonic acid, alternariol, alternariol monomethyl ether, altertoxin, altenuene, and AAL toxin. *Alternaria* are common in fresh fruits and vegetables, which provide the high-moisture environment these fungi require for growth. *Alternaria* infections have been observed in apples, oranges, tomatoes, and bell peppers, and the toxins have been detected in commercial apple products, tomatoes, tomato paste, and oranges. *Alternaria alternata* f. sp. (form species) *lycopersici* is a tomato pathogen that produces a host-specific toxin known as AAL toxin, which is structurally similar to fumonisins (**Fig. 9**). When grown on rice and fed to rats, cultures of this strain caused death in 80% of the animals; however, 100 mg of purified AAL toxin did not produce toxic effects in rats, suggesting that additional toxins were present in the rice cultures.

Stachybotrys toxins. *Stachybotrys* toxins are produced by *S. chartarum,* a toxigenic saprophytic, cellulose-decomposing fungus that grows well on high-cellulose materials such as straw, hay, cereal grains, plant debris, and dry wall or wall board in buildings (**Fig. 10**). Although it can be a contaminant of grains, it is considered to be more of a hazard as an indoor airborne fungus. *See* INDOOR FUNGI.

Regulations. Mycotoxins are natural contaminants of foods and various agricultural commodities. The occurrence of mycotoxins is not entirely avoidable, and small amounts of mycotoxins may be expected to occur in human foods and animal feeds. However, mycotoxin contamination can be controlled by good agronomic and good manufacturing practices, which

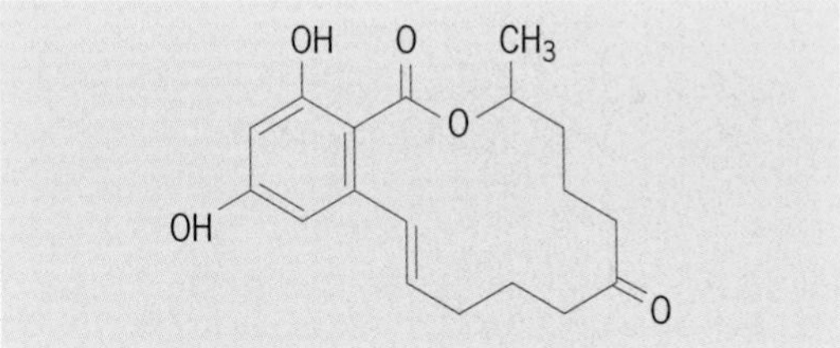

Fig. 8. Chemical structure of zearalenone.

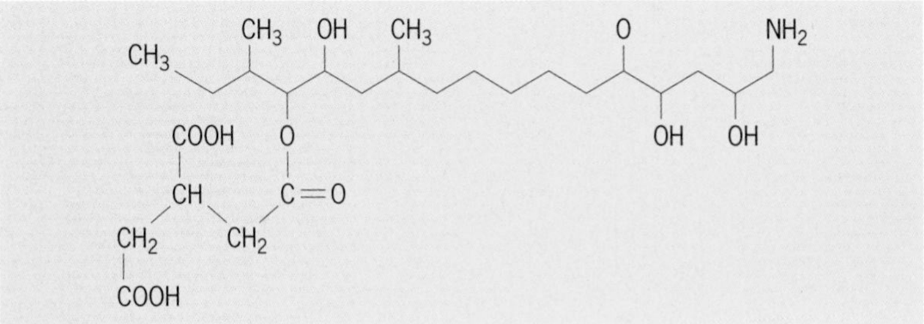

Fig. 9. Chemical structure of *Alternaria alternata* f. sp. *lycoperscici* (AAL) toxin.

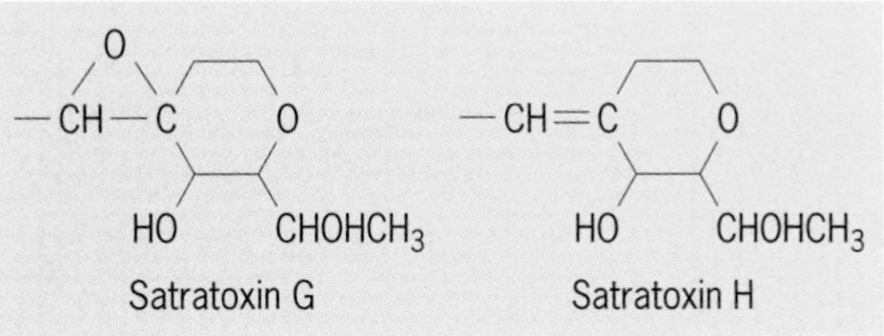

Fig. 10. Chemical structures of major *Stachybotrys* toxins.

consist mainly of preventing mold growth through adequate moisture control in stored products and removal of contaminated products by sorting. The U.S. Food and Drug Administration and other regulatory agencies in the states and other countries try to minimize the amounts of mycotoxins in the food supply by establishing action levels or guidance levels (Table 2) of the amounts of mycotoxins that can be allowed in foods, monitoring the food supply for the presence of mycotoxins, and taking regulatory action (confiscation) against products that exceed the established action levels for a mycotoxin.

For background information *see* AFLATOXIN; FUNGI; MEDICAL MYCOLOGY; MYCOLOGY; MYCOTOXINS in the McGraw-Hill Encyclopedia of Science & Technology.   Lloyd B. Bullerman

Bibliography. L. B. Bullerman, Mycotoxins, *Encyclopedia of Food Sciences and Nutrition,* pp. 4080–4090, Elsevier Science, London, 2003; *Mycotoxins: Risks in Plant, Animal, and Human Systems,* Task Force Rep. no. 139, Council for Agricultural Science and Technology, Ames, Iowa., 2003; J. Richard (ed.), *Mycotoxins—An Overview.* Romer Labs, Inc., Union, MO, 2000; J. E. Smith and M. O. Moss, *Mycotoxins: Formation, Analysis, and Significance,* Wiley, New York, 1985; K. K. Sinha and D. Bhatnagar (eds.), *Mycotoxins in Agriculture and Food Safety,* Marcel Dekker, New York, 1985.

Nanoprint lithography

The ability to reduce the feature dimensions of silicon-based transistors has enabled the semiconductor industry to enhance the performance, increase the speed, and reduce the price of integrated circuits for the past 55 years. In the late 1960s, Gordon Moore predicted that the transistor density for an integrated circuit would double every 18 months (Moore's law). So far, the semiconductor

TABLE 1. Electrically functional inks

Applications (electrical property)	Ink class	Electrically functional inks	Electrical property	Viscosity, pascal second (centipoise)
Conductive layer (conductivity, siemens/cm)	Polymer thick film (PTF) conductor	Silver (Ag), copper (Cu), gold (Au), palladium (Pd), carbon (C), aluminum (Al)	<100 S/cm	>1 Pa · s (>1k cP)
	Conductive polymer	Polyaniline, polyethylenedioxy-thiophene/polystyrenesulfonate	1–100 S/cm	0.01–10 Pa · s (10–10k cP)
	Nanoparticle suspension	Au, Ag, Cu, Pd	1k–10k S/cm	0.001–1 Pa · s (1–1k cP)
Dielectric layer (dielectric constant)	PTF dielectric	Silica (SiO_2), titanium dioxide (TiO_2), alumina (Al_2O_3)	5–50	>1 Pa · s (>1k cP)
	Dielectric polymer	Poly(4-vinylphenol), polyimide, polymethylmethacrylate	2–5	0.01–10 Pa · s (10–10k cP)
	Nanoparticle suspension	SiO_2, TiO_2, Al_2O_3	5–25	0.001–1 Pa · s (1–1k cP)
Resistive layer (sheet resistance, ohms/square)	PTF resistor	C	1–100k Ω/sq	>1 Pa · s (>1k cP)
	Nanoparticle suspension	Carbon nanoscale particles	1–1k Ω/sq	0.001–1 Pa · s (1–1k cP)
Semiconductor layer (mobility, cm^2/ volt second)	Polymer, oligomer	Poly(3-alkylthiophene) [P3αT], poly(dioctylfluorene-bithiophene) [F8T2]	10^{-5} to 10^{-2} cm^2/V · s	<0.02 Pa · s (> 20 cP)
	Nanoscale molecules	Pentacene	10^{-5} to 10^{-2} cm^2/V · s	<0.1 Pa · s (<100 cP)

industry has been able to follow Moore's law by enhancing the resolution attributes of photolithography tools. A photolithography tool is an optical system that projects the image of a mask (representing a layer of an integrated circuit) onto a wafer that has been coated with a light-sensitive material (resist). After exposure, the resist is developed to reveal a pattern which is transferred to the wafer's surface through etching techniques. Semiconductor analysts have theorized that by 2010 photolithography tools with 157-nm light sources will approach their theoretical resolution limit (120 nm).

During the past 10 years, several scientists have reported experimental results that demonstrate the use of graphic-arts printing technologies to create micrometer- and nanometer-sized features. These technologies have created great interest in the semiconductor industry and appear to support Moore's law.

The discovery and the characterization of the fundamental electrical and optical properties of several advanced polymeric materials has led to widespread interest among scientists and engineers to develop printed electronic and optical devices such as

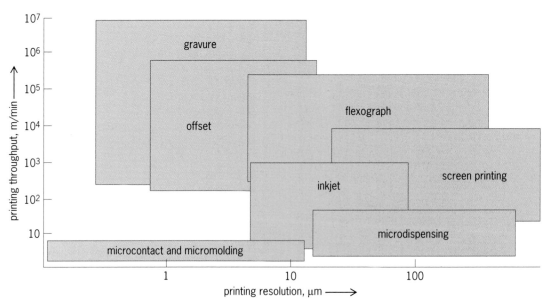

Fig. 1. Printing technologies as functions of printing resolution and throughput.

sensors, organic light-emitting diodes (OLEDs), organic field-effect transistors (OFETs), and micro-optical systems. Many of these polymers can be formulated into inks, enabling the use of traditional and nontraditional low-cost printing technologies. Since product price is largely a function of materials and manufacturing costs, there is great interest in using graphic-arts printing methods and expertise to reduce the cost. By contrast, microelectronics manufacturing technologies such as photolithography, chemical vapor deposition, and electroplating are complex and costly.

Ink technologies. Inks have two fundamental attributes: functionality and printability. A typical graphic-arts printing ink is a multicomponent system that consists of a polymer matrix, filler, binder, solvent, and additives. The polymer matrix or filler provides the functionality (such as color), while the binder, solvent, and additives affect the printability.

As with graphic-arts inks, electrically functional printing inks consist of four components. The functions of the binder, solvent, and additives are similar to those used for graphic-arts inks. Unlike graphic-arts inks, the filler/solute in electrically functional inks are conductive, dielectric, resistive, or semiconductive. Another critical difference is that the electrical purity of the ink must be very high in order to achieve optimal device performance [for example, low ion concentration (less than 10 parts per million)]. This requirement limits the choice of raw materials and requires process steps to purify the inks, thereby increasing the complexity of the ink formulation. **Table 1** lists a few commercially available inks and their reported electrical properties.

Printing technologies. These can be classified as contact and noncontact. **Figure 1** presents the most common printing technologies and their limits.

Noncontact printing. Noncontact printing technologies include inkjet, microdispensing, and laser-based printing (**Fig. 2a**). In noncontact printing, no contact is made between the printing device and the media during printing, and the process is digitally controlled. Since contact is not made, these printing technologies are insensitive to substrate surface roughness, and therefore are chosen when using nonplanar substrates. Also, digital processing enables quick changeover during manufacturing, which is favored for fast prototyping and dynamic information printing.

In early studies, inkjet printing was evaluated for manufacturing passive microelectronic components such as resistors, capacitors, inductors, and antennas. Since then, improvements in controlling droplet formation dynamics and material properties have resulted in the formation of micrometer-sized droplets and consequently the fabrication of smaller structures.

Laser-based printing technologies are currently at different stages of development. The most commonly mentioned systems are (1) laser transfer of materials from a donor film to a substrate, and (2) use of a laser

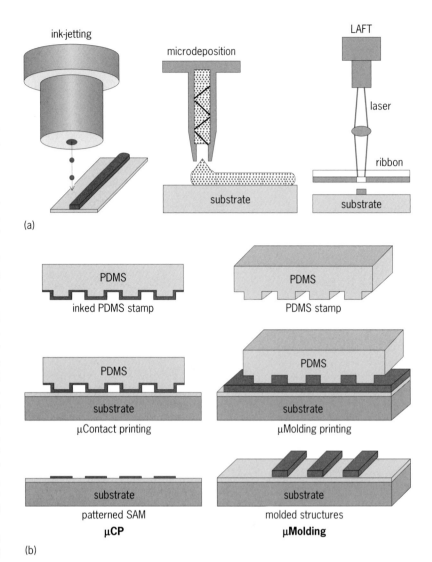

Fig. 2. Printing technologies. (a) Noncontact (LAFT = laser-assisted forward transfer). (b) Soft lithography using polydimethylsiloxane (PDMS) stamps for microcontact-printing (μCP) a self-assembled monolayer (SAM) and micromolding (μMolding).

to melt nanosized materials in flight prior to contact with the substrate. Results have been reported for fabricating a variety of active (field-effect transistors) and passive (capacitors and resistors) devices. In addition, these technologies may be useful for fabricating antennas and electrically conductive traces on, for example, circuit boards.

Contact printing. Contact printing technologies include flexography, lithography, gravure, screen printing, and soft lithography. Among these, screen printing is most commonly used within the electronics industry for printed circuit board fabrication and assembly. In graphic arts printing, contact techniques are capable of printing 40-μm lines/dots; however, finer printing resolutions of 5–10-μm lines/dots are achievable.

A printing technology that has received heightened attention during the past 5 years is soft lithography, which includes microcontact printing (μCP) and micromolding [micromolding in capillaries (MIMIC), microtransfer molding (μTM), soft imprint

TABLE 2. Soft lithography inks for microcontact printing (μCP) and micromolding (μM)

Ink compositions	Substrates	Demonstrated feature dimensions
Alkanethiolates	Gold, silver, copper, palladium, gallium arsenide	>30 nm (μCP)
Alkylsiloxanes	Si/SiO$_2$, Al/Al$_2$O$_3$, glass substrate with $-$OH functional group, plasma-treated polymers	100 nm (μCP)
Alkylphosphonic/carboxylic acids	Aluminum	1000 nm (μCP)
Poly(3,4-ethylenedioxythiophene)/ poly(styrene sulfonate) (PEDOT/PSS)	Gold, indium tin oxide	100/20 μm (μM)

lithography (SIL), and solvent-assisted micromolding (SAMIM)]. Another area of development for these technologies is materials and processes for self-assembled structures. Self-assembly is a spontaneous, noncovalent interaction of the chemical moieties between the printed ink and the receiving media. Most commonly, the self-assembled monolayer will serve as a resist for subsequent chemical patterning processes or as a seed layer (catalyst) for selective material deposition (for example, electroless deposition). To date, the gold/alkanethiolates system appears most often in literature; however, the availability of systems for microcontact printing is limited (**Table 2**). Moreover, subsequent process steps (such as patterning, chemical etching, and electroless deposition) after soft lithography are highly dependent on the quality of the self-assembled monolayer, which directly affects the final product.

Microcontact printing uses a patterned rubber stamp (for example, polydimethylsiloxane) to transfer ink to a substrate (Fig. 2b). In theory, microcontact printing can achieve finer resolution (outside of a cleanroom environment) than traditional wafer printing technologies, with a few researchers reporting approximately 30-nm resolution.

Micromolding has fewer processing-related restrictions than microcontact printing. However, shrinkage during the mold creation must be well characterized to achieve the desired printed dimensions. In the micromolding process, a stamp patterned with micro-sized features is placed in contact with a smooth, planar substrate, creating microchannels between the recessed areas (microfeatures) on the stamp and the substrate's surface. A polymer ink is dispensed along the periphery of the stamp and drawn into the microchannels by capillary forces. After filling, the ink is cured using heat, light, or solvent evaporation.

Recently, Tim Burgin and coworkers studied high-volume manufacturing using soft lithography. This study evaluated the most critical manufacturing issues to achieve high-yield processing of 3-in.-diameter (7.6-cm) wafers with 550-nm features for surface-acoustic-wave devices.

Printed electronics device. An all-printed organic field-effect transistor (OFET) is shown in **Fig. 3**. The OFET structure consists of four printed layers: the gate electrode (polymer thick film), the gate dielectric (polymer thick film), the source and drain electrodes (polymer thick film), and the organic semiconductor (polythiophene). The process begins by contact-printing the gate electrode, which is subsequently covered by the gate dielectric. Next, the source and drain electrodes are contact-printed on top of the gate dielectric. The process is completed by noncontact-printing the semiconductor, which

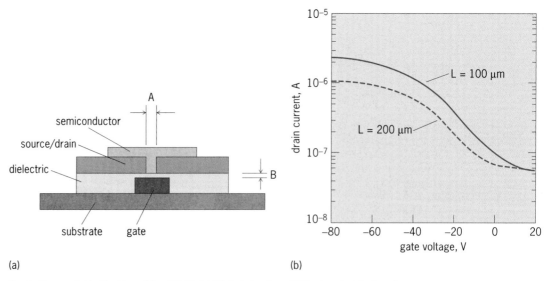

Fig. 3. Organic field-effect transistor (OFET). (*a*) Printed structure. (*b*) Device electrical performance.

must be electrically in contact with, and between, the source and drain electrodes to ensure device operation.

The critical dimensions that affect OFET performance for a given material system are channel length (dimension A in Fig. 3*a*) and dielectric layer thickness (dimension B in Fig. 3*a*). In general, the smaller the channel length, the higher the device's operating frequency and output current; and the thinner the dielectric layer, the lower the operating voltage required. Since the source and drain electrodes are on the same layer (printing plane), the channel length between the source and drain is directly related to the printing resolution. Printing technologies have achieved finer dimensions, thereby enabling the creation of higher-operating-performance OFET devices.

The electrical characteristics of devices with printed OFETs clearly show field-effect transistor behavior (Fig. 3*b*), where channel resistance is a function of the gate bias voltage and increasing the negative gate bias voltage decreases the channel resistance, resulting in an increase of the source/drain current. Figure 3*b* also compares two OFETs with different channel lengths. Their behavior is in accordance with theory such that the smaller the channel length, the higher the output current. Thus, further reducing the channel length will enhance OFET performance, enabling the creation of more advanced printed electronics applications (for example, low-radio-frequency wireless products).

Outlook. Printing technologies have shown promise for manufacturing microelectronic devices. The advantages of using printing technologies for microelectronics will be realized when a stronger understanding of the physical and chemical principles related to printing is established. As scientists continue to investigate printing technologies for microelectronics manufacturing, the focus should be on systems with low cost and a small footprint to reduce factory expenses, high manufacturing yield, high reliability, minimal equipment operator intervention, a modularized platform, and in-line testing capability for quality control.

For background information *see* INK; INTEGRATED CIRCUITS; LIGHT-EMITTING DIODE; MONOMOLECULAR FILM; PRINTING; SEMICONDUCTOR; TRANSISTOR in the McGraw-Hill Encyclopedia of Science & Technology.
Jie Zhang; Daniel Gamota

Bibliography. H. A. Biebuyck et al., Lithography beyond light: Microcontact printing with monolayer resists, *IBM J. Res. Dev.*, 41:159, 1997; T. Burgin et al., Large area submicrometer contact printing using a contact aligner, *Langmuir*, 16:5371, 2000; A. Chen et al., A new method for significantly reducing drop radius without reducing nozzle radius in drop-on-demand drop production, *Phys. Fluids*, 14:1, 2002; T. Granlund et al., Patterning of polymer light-emitting diodes with soft lithography, *Adv. Mater.*, 12:269, 2000; H. Kipphan (ed.), *Handbook of Printing Media*, Springer, 2001; A. Kumar et al., Features of gold having micrometer to centimeter dimensions can be formed through a combination of stamping with an elastomeric stamp and an alkanethiol "ink" followed by chemical etching, *Appl. Phys. Lett.*, 63: 2002, 1993; B. Michel et al., Printing meets lithography: Soft approaches to high resolution patterning, *IBM J. Res. Dev.*, 45:5, 2001; J. Szczech et al., Fine-line conductor manufacturing using advanced drop-on-demand PZT printing technology, *IEEE Trans. Elec. Pack. Manuf.*, 25:26, 2002; J. Tien et al., Microcontact printing of SAMs, *Thin Film*, 24:227, 1998; Y. Xia et al., Soft lithography, *Annu. Rev. Mater. Sci.*, 28:153, 1998; J. Zhang et al., Investigation of using contact and non-contact printing technologies for organic transistor fabrication, *MRS Spring Meeting*, 2002; J. Zhang et al., Material systems used by micro dispensing and ink jetting technologies, *MRS Spring Meeting*, 2000.

Nanosatellites

Nanosatellites, defined as having mass of under 10 kg (22 lb), are being studied to complete numerous missions by the National Aeronautics and Space Administration (NASA), the U.S. Department of Defense, commercial organizations, and other countries. Many of these missions require numerous small spacecraft in a constellation or "swarm." These include orbital communications networks and swarms of small satellites to conduct remote sensing or even spy missions, and to provide unique perspectives on comets or asteroids. For instance, 20 or more small spacecraft could be deployed from a mother ship to observe a comet from many angles. **Figure 1** demonstrates one configuration where a mother ship carries as many as 100 nanosatellites to their final destination in space for deployment.

Fig. 1. Mother ship carrying 100 nanosatellites. (*NASA Goddard Space Flight Center/Peter Rossoni*)

Nanosatellites will see continued and rapid development in all aspects of space exploration. Their small size and rapid build time make them ideal vehicles to respond to rapidly developing challenges and technology developments in a way that is not possible with larger space systems. Their low cost allows many nanosatellites to be deployed and thereby to provide a redundant, reliable space platform for many scientific, military, and commercial missions. Many engineering and science students are training and gaining experience with these systems and their capabilities.

Applications. Several NASA nanospacecraft missions would be made possible by the technology of nanosatellite systems being developed by current research:

1. Nanospacecraft could be made agile by the use of microthrusters and could maneuver close to an asteroid to observe its characteristics. With their small mass, a fleet of them could be sent, with each one carrying a different scientific instrument. There could also be some redundancy so that, in the event of the failure of some of the nanospacecraft, others could take over.

2. Many nanospacecraft could be deployed from a mother craft to view a comet from many different perspectives and map the comet with its dust and plasma tails. Such missions would provide three-dimensional views of these unique objects. A high-performing attitude control and propulsion system could allow these nanospacecraft to change relative perspective and gain valuable information.

3. Nanospacecraft could fly specific biological or materials-science experiments after being deployed by the International Space Station or the space shuttle. The propulsion module could fly the spacecraft away from the crewed vehicle, and rendezvous could take place when the experiment was complete and the samples needed to be returned. Being away from the station, these platforms would provide pristine microgravity environments for research, without the vibrations caused by astronauts and station support equipment.

4. The large number of nanospacecraft capable of being launched on one mission can also allow a swarm of spacecraft to map the Earth's magnetic field or provide remote sensing capabilities.

There is great interest in nanospacecraft in the military and academic communities. Coming improvements in nanosatellite systems would allow more capable spacecraft to accomplish a variety of missions:

1. A nanospacecraft could observe the health and status of a larger military or commercial spacecraft. Microthrusters would allow the spacecraft to fly around and document the condition of the larger spacecraft.

2. Many university groups are building very small nanosatellites of the CubeSat design (discussed below), as well as other nanospacecraft. Micropropulsion modules are currently under development for Cubesat spacecraft. Formation flying experiments would be enabled by this technology.

3. Microthrusters can also be applied to larger spacecraft to provide very fine pointing control for interferometry or laser communication experiments.

System development. Nanosatellites offer extreme challenges in terms of system design. These challenges include the creation of multifunctional structures with high strength but low mass and low volume. Nanosatellites are even more limited than larger ones in terms of mass, power, and volume. For instance, current design practices for spacecraft structures are to provide separate components for thermal management, stiffness, dimensional stability, and strength to withstand launch loads. This practice does not scale well for small spacecraft. Due to their small size and mass, many of the structural components on a small satellite are much thicker or more massive than required. This is due to manufacturing, availability, and tolerance requirements of materials. For instance, the material stiffness needed for a certain minimum deflection and vibration frequency increases in proportion to the structural size raised to the third power, so that the stiffness (and thereby thickness) of nanosatellite structures could be much smaller were it not for the limitations associated with manufacturing capabilities.

A typical nanosatellite has top and bottom decks to which components are mounted, and an octagonal shell sidewall which joins the decks together (**Fig. 2**). Octagonal sidewalls are chosen because they are easier to fabricate than a cylinder and allow more surface area for solar cells. The spacecraft spins about the z axis (the axis perpendicular to the top and bottom decks).

Multifunctionality. To increase the performance of nanosatellite systems, many innovative techniques are being applied to their design. Multifunctionality

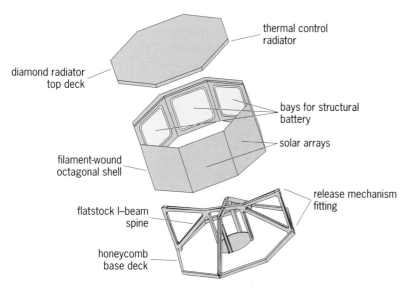

thermal control radiator

diamond radiator top deck

bays for structural battery

solar arrays

filament-wound octagonal shell

release mechanism fitting

flatstock I–beam spine

honeycomb base deck

Fig. 2. Design of a typical nanosatellite. Top deck takes shear and component inertial loads. Octagonal shell takes shear loads and supports solar array and top deck. Flatstock I-beam spine takes compression, lateral, and thruster loads. Base deck takes compression, bending, and thruster loads, and carries ion detector. (*After P. Rossoni and P. V. Panetta, Developments in nano-satellite structural subsystem design at NASA-GSFC, 13th AIAA/USU Conference on Small Satellites, SSC99-V-3, Logan, Utah, August 1999*)

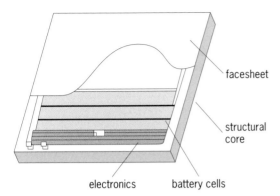

Fig. 3. Example of a multifunctional composite structural member of a nanosatellite. (*After P. Rossoni and P. V. Panetta, Developments in nano-satellite structural subsystem design at NASA-GSFC, 13th AIAA/USU Conference on Small Satellites, SSC99-V-3, Logan, Utah, August 1999*)

is a requirement of future nanosatellite structures. For instance, a battery can be placed inside a composite panel, with the facesheets taking most of the load (**Fig. 3**), and data conduits, propellant lines, and power connectors can be placed inside the core to become a structural member. The facesheet panel can support solar cells, and the core is a flat-cell battery and conditioning electronics. Valve bodies, pressure vessels, and propellant lines could also be integrated into a structure. Filament-wound tanks provide rigidity. Radiation shielding can be accomplished by integrating high-Z (high-atomic-number) materials, such as tungsten, into the structural laminate.

Propulsion. Propulsion systems on board nanosatellites must be small and must consume very little power while still providing velocity boosts similar to those of larger spacecraft. Current research is aimed at decreasing power and mass requirements for high-pressure valves with high response speeds, and fitting propellant tanks into the limited volume of a nanosatellite.

MEMS applications. Micro-electrical-mechanical systems (MEMS) will play a leading role in the development of nanosatellite systems. With MEMS technology a silicon substrate, similar to that used in integrated circuits, is etched into the shape of a rocket engine. In this way, very small rocket engines can be developed for thrusters on nanosatellites. A complete monopropellant thrust chamber has been etched from silicon along with a catalyst bed. The catalytic material coats a series of channels inside the chamber. Challenges include thermal control; fluid flows in such a small nozzle, some of which are only 30 micrometers in diameter; and attaining complete decomposition of the propellant in the catalyst chamber for the monopropellant thruster. *See* MICROFLUIDICS.

MEMS technology is also being used to develop attitude detection systems for nanosatellites. A complete attitude detection system on a chip is being developed. Work is also ongoing to develop rocket engines on a chip, which would allow entire rocket engine modules, perhaps complete with valves, to be fabricated on an integrated circuit chip. This technology offers digital propulsion, in which a propellant is loaded into each individually sealed chamber. When the resistor is energized, the propellant ignites, raising the pressure in the chamber and rupturing the diaphragm. An impulse is imparted as the high-pressure fluid is expelled from the chamber.

University involvement. The small size and typically lower cost of nanosatellites has made them useful vehicles for involving university and other small groups in space exploration. The CubeSat nanosatellite (**Fig. 4**) is a spacecraft in the shape of a 10-cm (4-in.) aluminum cube with a mass of 1–2 kg (2–4 lb). Its basic and simple design allows students to build, test, and fly nanosatellites.

The Arizona State University built and launched *ASUSat-1*, a 5.9-kg (13.0-lb) nanosatellite. Over a 6-year period, over 400 students worked on the project. Its primary mission was Earth imaging, with several secondary missions including orbit determination, amateur-radio communications, passive stabilization techniques, attitude detection, and composite-material research.

The U.S. Naval Academy built and flew the *PCsat* nanosatellite (**Fig. 5**) to begin a series of

Fig. 4. CubeSat nanosatellite. (Stensat Group)

Fig. 5. *PCsat* nanosatellite with antennas, built and flown in space by a team of students at the U.S. Naval Academy. (*From B. Bruninga, B. Smith, and D. Boden, PCsat success! and follow-on payloads, 16th Annual AIAA/USU Conference on Small Satellites, SSC02-I-5, Logan, Utah, August 2002*)

experiments in low-cost spacecraft telemetry and data communications for mobile satellite users. An additional unique feature of *PCsat* was the integration of multiple Internet-linked ground stations, allowing worldwide access to satellite telemetry and communications. This mission demonstrated the capabilities of a constellation of low-Earth-orbiting nanosatellites for worldwide communications on a very limited project budget. Such a constellation has potential to provide Earth point-to-point communications systems for developing nations.

Outside the United States, students and other small groups from the United Kingdom, Canada, Japan, Taiwan, China, and many other nations have flown or are developing nanosatellites.

For background information *see* INTEGRATED CIRCUITS; MICRO-ELECTRO-MECHANICAL SYSTEMS (MEMS); SATELLITE (SPACECRAFT); SCIENTIFIC SATELLITES; SPACE TECHNOLOGY; SPACECRAFT PROPULSION; SPACECRAFT STRUCTURE in the McGraw-Hill Encyclopedia of Science & Technology.   Donald Platt

Bibliography. B. Bruninga, B. Smith, and D. Boden, PCsat success! and follow-on payloads, *16th Annual AIAA/USU Conference on Small Satellites*, SSC02-I-5, Logan, Utah, August 2002; A. Friedman et al., ASUSat1: Low-cost, student-designed nanosatellite, *14th Annual AIAA/USU Conference on Small Satellites*, SSC00-V-2, Logan, Utah, August 2000; H. Heidt et al., CubeSat: A new generation of picosatellite for education and industry low-cost space experimentation, *14th Annual AIAA/USU Conference on Small Satellites*, SSC00-V-5, Logan, Utah, August 2000; S. W. Janson, H. Helvajian, and K. Breuer, *MEMS, Microengineering and Aerospace Systems*, AIAA 99-3802, American Institute of Aeronautics and Astronautics, 1999; D. H. Lewis, Jr., et al., Digital MicroPropulsion, *Sensors and Actuators A: Physical*, 80(2):143–154, Elsevier Science, 2000; M. S. Rhee, C. M. Zakrzwski, and M. A. Thomas, Highlights of nanosatellite propulsion development program at NASA-Goddard Space Flight Center, *14th AIAA/USU Conference on Small Satellites*, SSC00-X-5, Logan, Utah, August 2000; P. Rossoni and P. V. Panetta, Developments in nano-satellite structural subsystem design at NASA-GSFC, *13th AIAA/USU Conference on Small Satellites*, SSC99-V-3, Logan, Utah, August 1999; A. Tsinas and C. Welch, Efficient satellite structural design optimised for volume production, *15th Annual AIAA/USU Conference on Small Satellites*, SSC01-IV-5, Logan, Utah, August 2001.

Networked microcontrollers

Pervasive computing is defined as the seamless integration of computing into everyday life. Some aspects of pervasive computing exist today. For example, some automobiles have 10 to 20 networked microcontrollers coordinating braking and engine control. Automated trains that shuttle passengers at airports also utilize networks of microcontrollers. Modern aircraft could not adequately fly without networked microcontrollers for engine control and navigation. These types of systems have existed for many years, but are only a small fraction of the potential applications of pervasive computing. Recently, the confluence of cheaper hardware and new software has increased the types of pervasive computing systems that will be deployed in the near future. The hardware includes faster microcontrollers, cheaper and smaller static and dynamic random access memory, wireless radio communication devices, and a vast array of sensors and actuators. The software has been packaged in the form of middleware services that support access to and the use of the communication medium. This includes capabilities for message routing, congestion control, group management, clock synchronization, and event services. Together these hardware and software technologies are on the verge of producing a revolution in pervasive computing. *See* ADAPTIVE MIDDLEWARE.

Wireless sensor networks. A key ingredient in this revolution is the next generation of networked microcontrollers. Recent research efforts in networked microcontrollers have focused on wireless sensor networks as a novel underlying structure. The goal is to combine thousands or even tens of thousands of these devices into a coherent sensor network that acts with aggregate behavior. Such networks have a vast array of potential applications. For example, these networks could be deployed from aircraft in hostile terrain. Upon deployment the large collection of devices would self-organize into a cohesive whole and self-locate via the Global Positioning System. They could then monitor the environment for enemy troop movement and inform command-and-control elements of such activity. These networks could be similarly deployed over a city or remote town that suffered an earthquake. The sensor network could quickly assess the situation, identifying gas leaks, fires, and trapped and injured people, and send the information to fire and rescue teams. Wireless sensor networks could also be deployed in a nonrandom and careful manner to help protect airports, make university campuses safer, or improve living conditions in buildings and homes. Wireless networks could provide input to the Internet (for example, by collecting temperature readings and transmitting them for display on a Web site) and to many applications labeled smart, such as smart buildings and smart rooms (to automatically operate such functions as switching lights on and off, and downloading lecture notes when the professor enters the room). Small wireless sensor networks could even be built into clothing. In this case a person's health status could be monitored and reported periodically to primary care providers. With these applications and many other unforeseen possibilities, the future for these systems is tremendously exciting.

At the heart of these systems is the microcontroller. A microcontroller can be thought of as a computer that is simple in comparison to commercial processors such as the Pentium class of machines. It runs most of the computations needed by the device it is supporting. Each device also has memory, sensors, and actuators. The networking among the microcontrollers is provided by a wireless radio device attached to the microcontroller. This frees individual devices from being physically tethered to each other. It is the software that organizes the large collection of tiny devices into a coherent whole. The software research on these systems is producing many novel ideas.

Software capabilities. Current software capabilities for large-scale distributed systems such as the Internet were developed under several key assumptions, including a wired infrastructure, large amounts of available power (just plug the computer into a wall socket), highly reliable devices (such as commercial computers like the Pentium), infrequent network topology changes, non-real-time response, and the importance of each node in the system. The new large-scale wireless sensor networks violate all these assumptions. In particular, they are not wired, power is a precious commodity, each device is not very reliable, topology changes are frequent, real-time response is required, and the aggregate behavior of the network is important, not that of the individual devices. Wireless sensor networks are also very data centric in that they deal with continuous real-world sensed values such as temperature, pressure, or acceleration and are interested in geographic location. For example, a user may query the system to determine the number of people in the front hall of building 5. The user does not care which devices respond as long as they provide the correct answer. These vast differences in assumptions and requirements mean that software research has to develop many new solutions. Some representative examples will be described.

Group management. Networked microcontrollers monitor the environment and upon detection of some event dynamically form a subgroup of devices to assess and act upon that event. For example, suppose an intruder enters an area where a wireless sensor network exists. The devices in the region of the intruder might detect movement, communicate with each other to decide if this is a false alarm, and if not, could cooperate to locate, track, and classify the intruder. The group of devices might also take various actions such as triggering an alarm, turning on floodlights, and informing security. The software to support all these functions includes probabilistic team formation, real-time communication, and sensor-based estimation protocols. Team-formation protocols must support dynamic team membership (meaning that devices may be added to or dropped from the team during an event) and real-time coordination, and must operate in the presence of failures of various types. Teams must follow the objects being tracked or events being addressed. Messages being sent among team members as well as across the sensor network must arrive in time for appropriate action. When the data being transmitted are sensor data, such data must arrive while the data are still fresh. Old data, if used to make (control) decisions,

can cause major failures in the system. Estimation protocols have to perform sensor fusion (combining data from different sensors) and account for noisy and error-prone data. These protocols are different from those found in wired networks, such as a local-area network or the Internet.

Resource management. Resource management issues also take on new dimensions. The software in the wireless sensor network must manage slow processors, small memories, limited communication bandwidth and power, and their trade-offs. For example, these systems could be built to monitor the communication failure rate and, if it is too high, could reduce it by dynamically increasing the radio power transmission levels, or, if a device is mobile, by moving that device closer to the device with which it is communicating. Most wired networks use power from the electric power grid via wall sockets. In wireless sensor networks, devices use power from small batteries or solar cells. They might communicate at short ranges to save power or might even go into various types of sleep mode during periods of inactivity. These approaches can significantly extend the lifetime of these systems.

Spatio-temporal analysis. Wireless sensor networks fundamentally deal with time and space (location). Individual devices must know their location and have synchronized clocks in order to report data to application-level activities (that is, activities for which the system was built), such as the location and time of detected objects or sensed values. The coordinated control of such information from large numbers of devices is very difficult. A number of approaches to this problem are being investigated, including exploring feedback control theory, using learning technology, and employing biological metaphors such as the operation of an ant society. Equally difficult is the analysis of a time-space system, referred to as spatio-temporal analysis, that is, the development of analysis techniques that address both space and time together, rather than just space or just time. Very few analysis results exist so far. Current evaluations of wireless sensor networks are done largely by testing. However, mathematical foundations that extend current analysis are required. Efforts at extending control theory and real-time scheduling theory to support spatio-temporal analysis are underway.

For background information *see* DISTRIBUTED SYSTEMS (COMPUTERS); EMBEDDED SYSTEMS; INTERNET; LOCAL-AREA NETWORKS; MICROPROCESSOR; MICROSENSOR; PROGRAMMABLE CONTROLLERS; SEMICONDUCTOR MEMORIES; TELEMETERING; WIDE-AREA NETWORKS in the McGraw-Hill Encyclopedia of Science & Technology.   John A. Stankovic

Bibliography. M. Addlesee et al., Implementing a sentient computing system, *IEEE Computer*, 34(8): 50–56, 2001; I. F. Akyildiz et al., Wireless sensor networks: A survey, *Computer Networks*, 38:393–422, 2002; M. Weiser, Some computer science issues in ubiquitous computing, *Commun. ACM*, 36(7):75–84, 1993.

Networking chip subsystems

A system on chip (SOC) is a monolithic unit that performs a complex function, such as a personal computer, radar receiver, or cellular phone. Systems on chip are assembled using high-level components, such as processors, controllers, and memory arrays. The communication scheme among these components is key to delivering the desired performance within the prescribed energy consumption budget. For this reason, a large research effort is addressing the design and optimization of on-chip communication. Techniques borrowed from networking technologies are carried over to microelectronic design but have to cope with the advantages and limitations of the manufacturing technology for integrated circuits.

Characteristics of systems on chip. Systems on chip are designed with silicon nanoscale technologies, that is, with transistor gate lengths that are shorter than 100 nm and are decreasing from year to year. Moreover, systems on chip integrate subsystems with a variety of different functionalities (such as digital, analog, and radio-frequency) and a variety of different technologies [such as sensors, optical interfaces, and micro-electro-mechanical systems (MEMs)]. In the coming years, novel technologies may complement or substitute for silicon as a substrate for computational and storage subsystems.

Overall, the device density on the chip is expected to increase, and accommodating the resulting power density and extracting heat rapidly enough to keep the device from overheating will be major design challenges. Energy-efficient design policies will be required. In this direction, voltage levels on the chip will be reduced to the order of a few hundred millivolts. Unfortunately, this voltage reduction will adversely affect signal integrity. Signal delays on the wires will dominate delays in the computational units, and their accurate prediction will be increasingly difficult.

Systems on chip will find application in many embedded systems (such as portable communicators, vehicle control systems, and health monitoring systems) where reliability and robustness are major concerns. Thus, new system-level design methodologies will be driven by the end-application requirements as well as by the physical limitations of the underlying technology.

Two principles will govern systems-on-chip design as a result of its increasing complexity. First, systems on chip will be designed using preexisting components, such as processors, controllers, and memory arrays. Design methodologies will support component reuse in a plug-and-play fashion. Second, reliable operation of the interacting components will be guaranteed by a structured methodology for interconnect design that relies on the application of conventional networking technology to the microelectronic environment.

On-chip networks are also referred to as micronetworks, to distinguish them from local- and wide-area networks. Systems on chip differ from wide-area networks because of the proximity of their components and because they exhibit much less unpredictability. Local, high-performance networks (such as those developed for large-scale multiprocessors) have similar requirements and constraints. A few distinctive characteristics are unique to systems-on-chip networks, namely, energy constraints and design-time specialization.

Whereas computation and storage energy greatly benefit from device miniaturization, the energy for global communication (communication between the chip subsystems) does not scale down with device size. On the contrary, projections based on current delay optimization techniques for global wires show that global communication on the chip will require increasingly higher energy consumption. Hence, communication-energy minimization will be a growing concern in future technologies. Furthermore, network traffic control and monitoring can help in better managing the power consumed by networked computational resources. For instance, the clock speed and voltage of processing elements can be varied according to the available network bandwidth.

Design-time specialization is another facet of the systems-on-chip network design. Whereas macroscopic networks emphasize general-purpose communication and modularity, in systems-on-chip networks these constraints are less restrictive. The communication network fabric is designed on silicon from scratch. Standardization is needed only for specifying an abstract network interface for the processing and storage elements, but the network architecture itself can be tailored to the application, or class of applications, targeted by the systems-on-chip design.

Network architectures and protocols. Network design entails the specification of network architectures and control protocols. The architecture specifies the topology and physical organization of the interconnection network, while the protocols specify how to use network resources during system operation.

Architectures. The current dominant on-chip communication paradigm is shared-medium, in which the same transmission medium is shared by all the communications, as exemplified by bus-based architectures. Several bus standards (such as AMBA) are currently used, but their effectiveness is likely to fade as more components are interconnected, making bus standards slow and energy-inefficient communication means.

The direct or point-to-point network is a network architecture that overcomes the scalability problems of shared-medium networks. In this architecture, each node is directly connected with a subset of other nodes in the network, called neighboring nodes. Nodes are on-chip computational units, but they contain a network interface block, often called a router, which handles communication-related tasks. Each router is directly connected with the routers of the neighboring nodes. In contrast to shared-medium

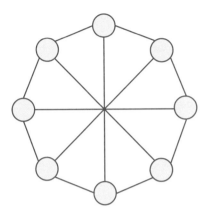

Fig. 1. Topology of the octagon network.

architectures, as the number of nodes in the direct network increases, the total available communication bandwidth also increases. Direct interconnect networks are therefore very popular for building large-scale systems.

An octagon network is an example of a direct network on chip. It was designed for network processors. In an octagon network, eight processors are connected by an octagonal ring and four diameters. Messages between any two processors require at most two hops (**Fig. 1**). If one node processor is used as the bridge node, more octagons can be tiled together.

Indirect or switch-based networks are an alternative to direct networks for scalable interconnection design. In these networks, a connection between nodes has to go through a set of switches. The network adapter associated with each node connects to a port of a switch. Switches do not perform information processing. Their only purpose is to provide a programmable connection between their ports, or, in other words, to set up a communication path that can be changed over time. As an example (**Fig. 2**), SPIN is an indirect network on chip with a fat tree topology. Messages reach the processing elements by traveling up and down the routing tree.

Protocols. On-chip global wires are the physical support for communication and embody the physical network architecture. Global wires can be seen as noisy channels. In micronetworks, noise is the abstraction of the signal disturbances, such as timing errors, cross talk, and electromagnetic interference.

Network protocols are designed in layers. The data-link layer abstracts the physical layer as an un-

reliable digital link. The main purpose of data-link protocols is to increase the reliability of the link up to a minimum required level, and to regulate the access to the network, where contention for a communication channel is possible.

Error detecting and correcting codes are used in different ways to provide for signal transmission reliability. When only error detection is used, error recovery involves the retransmission of the faulty bit or word. When using error correction, some (or all) errors can be corrected at the receiving end. Error detection or correction requires an encoder-decoder pair at the channel's end, whose complexity depends on the encoding being used. Obviously, error detection is less hardware intensive than error detection and correction. In both cases, a small delay has to be accounted for in the encoder and decoder. Data retransmission has a price in terms of latency. Moreover, both error detection and correction require additional (redundant) signal lines.

An effective way to deal with errors in communication is to packetize data, because error containment and recovery is then easier. Indeed, the effect of errors is contained by packet boundaries and error recovery can be carried out on a packet-by-packet basis. In this case, the redundant data lines can be avoided by adding the redundant information at the tail of the packet, thus trading off space for delay.

As an example, the SPIN micronetwork defines packets as sequences of 36-bit words. The packet header fits in the first word. A byte (8 bits) in the header identifies the destination (hence, the network can be scaled up to $256 = 2^8$ terminal nodes), and other bits are used for packet tagging and routing information. The packet payload can be of variable size. Every packet is terminated by a trailer, which does not contain data but a checksum for error detection (**Fig. 3**).

The protocol network layer implements the end-to-end delivery control in advanced network architectures with many communication channels. Key tasks are switching (for example, circuit, packet, or cut-through switching) and routing (for example, deterministic or adaptive routing). The protocol transport layer decomposes messages into packets at the source. It also resequences and reassembles them at the destination. Packet size is a critical design decision, because the behavior of most network control algorithms is very sensitive to this parameter. While in most macroscopic networks packets are standardized to facilitate internetworking, in micronetworks

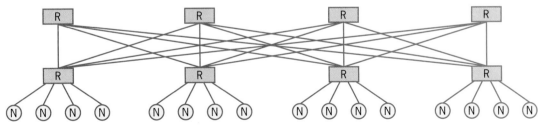

Fig. 2. Topology of the SPIN network. R = router; N = node.

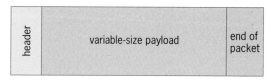

Fig. 3. Packet in the SPIN micronetwork.

the packet size can be tuned to optimize performance or energy consumption.

Micronetwork protocols have to interface to the systems-on-chip middleware, that is, to the layer of system software which is cognizant of the underlying network architecture and which supports the execution of application programs. The design of effective middleware for networked systems on chips can make use of the experience in distributed operating systems, but must satisfy specific requirements of systems on chips such as low communication latency. *See* ADAPTIVE MIDDLEWARE.

In summary, the micronetwork protocols are designed in layers and abstract the network and its controls to the software application programs. Some applications require guaranteed services by the underlying layers, while best-effort services suffice in some other cases. Thus, the overall strategy in choosing network architectures and protocols depends on the specific systems-on-chip function.

For background information *see* DATA COMMUNICATIONS; INFORMATION THEORY; INTEGRATED CIRCUITS; LOCAL-AREA NETWORKS; MICROPROCESSOR; PACKET SWITCHING; WIDE-AREA NETWORKS in the McGraw-Hill Encyclopedia of Science & Technology.

Giovanni De Micheli

Bibliography. L. Benini and G. De Micheli, Networks on chips: A new SoC paradigm, *IEEE Computers*, pp. 70–78, January 2002; D. Bertozzi, L. Benini, and G. De Micheli, Low-power error-resilient encoding for on-chip data busses, *DATE, International Conference on Design and Test Europe*, Paris, pp. 102–109, 2000; P. Guerrier and A. Grenier, A generic architecture for on-chip packet-switched interconnections, *Design Automation and Test in Europe Conference*, pp. 250–256, 2000; F. Karim, A. Nguyen, and S. Dey, On-chip communication architecture for OC-768 network processors, *38th Design Automation Conference Proceedings*, 2001; T. Theis, The future of interconnection technology, *IBM J. Res. Dev.* 44(3):379–390, May 2000; J. Walrand and P. Varaiya, *High-Performance Communication Networks*, Morgan Kaufman, 2000.

Nobel prizes

The Nobel prizes for 2002 included the following awards for scientific disciplines.

Chemistry. The chemistry prize was awarded for the development of methods for analyzing and identifying large biological molecules, specifically proteins. John B. Fenn of Virginia Commonwealth University and Koichi Tanaka of Shimadzu Corp. (Japan)

shared half the prize for developing ionization methods for mass-spectrometric protein analysis. Kurt Wüthrich of the Swiss Federal Institute of Technology and The Scripps Research Institute (La Jolla, California) received the other half for developing nuclear magnetic resonance (NMR) spectroscopy for determining the three-dimensional structure of proteins. In recent years, these techniques have led to a greater understanding of how proteins function in cells.

Proteins are biological macromolecules made up of various α-amino acids that are joined by peptide bonds. A peptide bond is an amide bond formed by the reaction of an α-amino group (—NH$_2$) of one amino acid with the carboxyl group (—COOH) of another.

In biological systems, proteins provide structural support (for example, muscles and tendons), act as catalysts (for example, enzymes), regulate physiological processes (for example, hormones), and transport other substances (for example, hemoglobin and lipoproteins). In addition, chromosomes and viruses consist largely of nucleoproteins, which are proteins combined with nucleic acids.

Mass spectrometers are used to identify molecules, often in very small quantity, according to their masses. A mass spectrometer has an ion source, a mass analyzer that separates ions according to their mass/charge ratios, and a detector. For many years, researchers were unable to produce protein ions because conventional techniques either could not ionize molecules with high molecular masses or were too energetic and would destroy the proteins. Fenn and Tanaka overcame these problems, around the same time, by very different methods.

In 1988, Fenn reported the identification of intact protein molecules with molecular masses of 40,000 daltons (Da) using a technique called electrospray ionization. Electrospray ionization produces charged droplets of protein molecules in water (or a water/solvent mixture) which, due to repulsive electrostatic forces as the solvent evaporates, break up and release gas-phase protein ions with varying charges. These charged molecules are detected as a set of mass/charge peaks from which the molecular mass is calculated with very high accuracy, using a signal-averaging method.

During 1987–1988, Tanaka showed that a low-energy (330-nanometer) nitrogen laser could be used to vaporize and ionize proteins, such as carboxypeptidase A (34,472 Da) and cytochrome *c* (12,384 Da), embedded in an inert matrix material without breaking them apart. In this technique, called soft laser desorption, the matrix material absorbs energy from a laser pulse and transfers it to the protein molecules. Protein ions ejected from the matrix then are accelerated by an electric field into a mass analyzer to obtain their molecular mass.

Today, electrospray ionization and soft laser desorption are standard techniques used in proteomics, the study of the complete set of proteins present in the various cells of an organism.

To understand how a protein functions, researchers need an accurate picture of its structure. For nearly 50 years, x-ray crystallography has been used to visualize the three-dimensional molecular structure of proteins. In 1985, Wüthrich determined the first complete three-dimensional protein structure, in solution, by NMR spectroscopy. Solution-NMR more closely approximates the conditions of living cells, and has become an important tool for studying protein folding and the dynamic interactions of proteins with other biomolecules. For interpreting the NMR spectrum, Wüthrich developed a method, called sequential assignment, to assign resonances in an NMR spectrum to the correct nuclei in the protein. In addition, he developed the computational techniques for translating the NMR data into a three-dimensional structure.

For background information *see* AMINO ACIDS; MASS SPECTROMETRY; NUCLEAR MAGNETIC RESONANCE (NMR); PEPTIDE; PROTEIN; X-RAY CRYSTALLOGRAPHY in the McGraw-Hill Encyclopedia of Science & Technology.

Physiology or medicine. Sydney Brenner of The Molecular Sciences Institute, Berkeley, California; H. Robert Horvitz of the Massachusetts Institute of Technology, Cambridge; and John E. Sulston of The Wellcome Trust Sanger Institute, Cambridge, England, were jointly awarded the Nobel Prize for Physiology or Medicine for their discoveries of key genes involved in the regulation of organ development and programmed cell death.

The life cycle of multicellular organisms begins with a unicellular stage, the fertilized egg, from which all other cells in the body descend. Through cell division and cell differentiation, the mechanism by which cells in a multicellular organism become specialized to perform specific functions in a variety of tissues and organs, the organism grows and becomes more complex and takes on its characteristic form. However, normal life also requires programmed cell death. A fine balance between the processes of cell division and cell death is essential for maintenance of the appropriate number of cells in tissues and organs. The discoveries of these three prize winners concerning the genetic regulation of organ development and programmed cell death have made significant contributions to our understanding of these processes as well as the pathogenesis of many diseases, including cancer.

In the early 1960s, Brenner realized that the short generation time and transparency of the 1-mm-long nematode *Caenorhabditis elegans* made it an ideal model organism for the genetic and microscopic study of cell differentiation and organ development in multicellular organisms. In 1974, he used the chemical compound ethyl methane to induce specific mutations in the genome of *C. elegans* and showed that the different mutations could be linked to specific genes and to specific effects on organ development.

Building on Brenner's work, in 1976 Sulston showed that every nematode undergoes exactly the same program of cell division and differentiation; that is, the cell lineage is invariant. These findings led to his major discovery that specific cells in the cell lineage always die through programmed cell death. He described the visible steps in the cellular death process and demonstrated the first mutations of genes participating in programmed cell death, including the *nuc-1* gene, which encodes a protein required for degradation of the deoxyribonucleic acid of the dead cell.

Continuing the work of Brenner and Sulston, Horvitz conducted a series of experiments beginning in the 1970s in which he used *C. elegans* to investigate the genetic control of cell death. In 1986, he identified the first two "death genes," *ced-3* and *ced-4*, which are necessary for the execution of cell death. Later he showed that another gene, *ced-9*, interacts with *ced-4* and *ced-3* to protect against cell death. Horvitz went on to show that the human genome contains a gene similar to *ced-3*. Indeed, today we know that most of the genes that are involved in regulating cell death in *C. elegans* have human counterparts.

The contributions made by these Nobel laureates—the establishment of *C. elegans* as an experimental model system, the mapping of cell lineage, and the identification of genes controlling cell death—have contributed to many disciplines, ranging from developmental biology to medicine. We now know that both increased and decreased cell death underlies many diseases. For example, acquired immune deficiency syndrome, neurodegenerative disorders, stroke, and myocardial infarction are due to excessive cell death. However, autoimmune disorders and many types of cancer are characterized by a reduction in cell death, leading to the survival of cells normally destined to die. Research on programmed cell death in cancer is very active. In fact, treatments directed at increasing programmed cell death are already in use. For example, both tumor irradiation and cancer chemotherapy lead to programmed death of cancer cells, resulting in slower tumor growth. It is conceivable that in the near future researchers will have the knowledge necessary to cure some types of cancer and to develop better treatments for other diseases in which programmed cell death plays a role.

For background information *see* CANCER (MEDICINE); CELL DIFFERENTIATION; CELL DIVISION; CELL LINEAGE; CELL SENESCENCE AND DEATH; DEVELOPMENTAL GENETICS in the McGraw-Hill Encyclopedia of Science & Technology.

Physics. The physics prize was awarded for advances in detecting cosmic particles and radiation. Raymond Davis, Jr., of the University of Pennsylvania and Masatoshi Koshiba of the International Center for Elementary Particle Physics at the University of Tokyo shared one-half of the prize for the detection of cosmic neutrinos. The other half was awarded to Riccardo Giacconi of Associated Universities Inc. in Washington, DC, for contributions which have led to the discovery of cosmic x-ray sources.

According to the presently accepted theory, the Sun's energy source is a series of nuclear reactions that transform hydrogen into helium. These reactions should release huge numbers of neutrinos, but detecting them is very difficult because they react very weakly with matter. To detect solar neutrinos, Davis placed a tank containing 615 metric tons (678 tons) of the cleaning fluid tetrachloroethylene (C_2Cl_4) deep in a mine (to reduce interference from cosmic rays). A solar neutrino was detected through its reaction with a chlorine-37 nucleus to form argon-37, a radioactive isotope with a half-life of about 50 days. Approximately every 2 months, helium was bubbled through the tank to extract an average of 17 argon-37 atoms (from among 2×10^{30} chlorine atoms in the tank), whose decays were detected by a proportional counter. Davis's experiment, which began in 1967 and ran almost continually from 1970 to 1994, was the first to detect solar neutrinos, but at only about one-third the predicted rate; the deficit became known as the solar neutrino problem.

Koshiba's experiment was conducted in the Kamioka mine in Japan and used a large tank of water. The Kamiokande experiment was originally designed to detect proton decay by measuring the Cerenkov light emitted by the decay particles, using photomultipliers surrounding the tank. In 1986, an upgraded experiment, Kamiokande II, began detecting solar neutrinos through their elastic scattering of electrons. This experiment indicated the arrival times and directions of the incoming neutrinos, enabling Koshiba's group to prove that the neutrinos came from the Sun. (About half the predicted number of neutrinos was observed.) In February 1987, 12 neutrinos were observed from Supernova 1987A, 170,000 light-years away (out of 10^{16} neutrinos reaching the detector), confirming the theory that enormous numbers of neutrinos are emitted in supernova explosions. (Such neutrinos were also observed by the Irvine-Michigan-Brookhaven experiment in the United States.)

A much larger detector built by Koshiba's group, Super Kamiokande, began operating in 1996. In 1998, it observed neutrinos emitted in the atmosphere and found evidence for neutrino oscillations, in which one type of neutrino transforms into another. Neutrino oscillations are probably the solution to the solar neutrino problem, but they imply that neutrinos have mass, requiring modification of the standard model of elementary particles. The work of Davis and Koshiba is the basis of a new discipline, neutrino astronomy.

Since cosmic x-rays are completely absorbed by the Earth's atmosphere, they can be studied only from balloons, rockets, and satellites. Moreover, imaging x-rays requires special optics, involving total reflection at grazing angles of incidence. Starting in 1959, Giacconi and the late Bruno Rossi worked out principles for the construction of an x-ray telescope. In 1962, Giacconi's group launched a rocket with nonimaging x-ray detectors that discovered the first x-ray source outside the solar system, as well as a uniform x-ray background. These unexpected discoveries stimulated interest in x-ray astronomy. Giacconi remained at the forefront of the field until 1981, leading the development of the first orbiting x-ray observatory, *UHURU*, launched in 1970, and the first orbiting x-ray telescope, the *Einstein Observatory*, launched in 1978. Both observatories extended knowledge of the x-ray sky far beyond previous observations. Beginning in 1976, Giacconi led efforts to develop an *Advanced X-ray Astrophysics Facility (AXAF)*, which was finally launched in 1999, renamed *Chandra*, and has provided resolution comparable to the Hubble Space Telescope. X-ray astronomy, pioneered by Giacconi, has greatly expanded our view of the universe, revealing a variety of processes in which great amounts of energy are released on short time scales in connection with incredibly compact objects such as neutron stars and black holes.

For background information *see* ASTROPHYSICS, HIGH-ENERGY; NEUTRINO; NEUTRINO ASTRONOMY; SOLAR NEUTRINOS; SUPERNOVA; X-RAY ASTRONOMY; X-RAY TELESCOPE in the McGraw-Hill Encyclopedia of Science & Technology.

Noble gas MRI

Magnetic resonance imaging (MRI) has been a key tool in medicine and biomedical research for more than two decades. In conventional MRI, two-dimensional images of thin slices of tissue are created by measuring the nuclear magnetization of protons (the nuclei of hydrogen atoms) in tissue in the presence of static and oscillating magnetic fields. Generally, a strong static field, typically 1.5 tesla, produces the nuclear magnetization, which is due to a small excess of protons with spin-up (parallel to the static magnetic field) compared to spin-down (opposite the static magnetic field). This excess, called the nuclear polarization, is only about 10 parts per million but, combined with the high concentration of protons in tissue, produces the remarkable images used by physicians and scientists. Nuclear magnetization, the product of nuclear polarization and concentration, sufficient for MRI with much lower concentrations characteristic of gases can also be produced with lasers. This laser technique, called optical pumping, has led to new methods of imaging called laser-polarized noble gas MRI or hyperpolarized gas MRI.

Laser-polarized gases for MRI. Two gases are predominant in laser-polarized gas MRI: helium-3 (^{3}He) and xenon-129 (^{129}Xe). The nucleus of the isotope helium-3 has two protons and one neutron and is very rare compared to the dominant isotope, helium-4 (^{4}He). Helium is not found in quantity in the Earth's atmosphere because the average thermal velocity of helium atoms exceeds the escape velocity from the Earth's gravitational field. The isotope helium-4 is abundant in the Earth's crust because it is produced by radioactive alpha-decay of heavy nuclei. Natural gas and oil wells are sources of helium. The isotope helium-3 is produced by decay of the radioactive hydrogen isotope tritium (^{3}H), which is produced for thermonuclear weapons and for use as a radioactive tracer. The world's total supply of helium-3 is limited to a few hundred kilograms.

Xenon gas is found in the Earth's atmosphere, and the isotope xenon-129 has a natural abundance of about 27%; that is, in a sample of atmospheric xenon, 27% is the isotope ^{129}Xe. Xenon can be easily condensed from the atmosphere, though enrichment of specific isotopes significantly increases the cost. Xenon-129 is also found in meteorites because the iodine-129 component of the rocks decays to xenon-129. The abundance, cost, and availability of the two noble gas MRI isotopes are important considerations for future applications to medicine and biomedical research.

Preparation of laser-polarized gases. Laser-polarized noble gas MRI brings together two distinct technologies: laser optical pumping and magnetic resonance. Research into optical pumping techniques and noble gas polarization has been motivated most strongly by high-energy and nuclear physics experiments that use helium-3 to measure the short-range properties of the neutron. Special lasers developed for that work have made it routine to produce the large quantities of laser-polarized noble gases used for imaging.

Optical pumping makes use of light of a specific wavelength, corresponding to transitions in atoms of helium or rubidium. For helium, the wavelength of the light is 1083 nm, and for rubidium the wavelength is 795 nm. The light is circularly polarized by passing it through special optical elements so that the photons have a specific projection of angular momentum parallel or opposite to their direction of motion and the static magnetic field. When the atoms absorb the light, the atomic electrons pick up some of the angular momentum, producing electron polarization.

Two separate approaches are used: For helium, metastable triplet states populated in a discharge maintained in helium-3 gas are optically pumped by the 1083-nm light, and the helium-3 nuclei of metastable atoms are polarized by the hyperfine interaction. In a subsequent collision, a metastable atom with a polarized nucleus can transfer its metastability but not its nuclear polarization to a ground-state atom. The result is a ground-state atom with nuclear polarization and a metastable atom with no nuclear polarization. The newly metastable atom can then be optically pumped and its nucleus polarized. As these processes continue, the polarization is spread among ground-state atoms.

Optical pumping of rubidium vapors with 795-nm light produces very high spin polarization of the unpaired atomic electrons. If the rubidium vapors have been mixed with xenon-129 or helium-3 gas, the electron spin of the rubidium atoms is transferred to the nuclei of the noble gas atoms during the brief duration of atomic collisions. The electron-nuclear spin exchange is also mediated by the hyperfine interaction, with a rate constant that is proportional to the duration of the overlap of the quantum-wave functions of the rubidium electron and the noble gas

nucleus. For helium-3 this duration is on the order of picoseconds, but for xenon-129 a weakly bound van der Waals molecule can form if a third body carries away momentum and energy. Since the lifetime of the van der Waals molecule can be much longer than the duration of a collision, the overlap of the rubidium electron with the noble gas nucleus is greatly enhanced for xenon-129. The time scales necessary to produce high nuclear polarization of noble gas atoms are much different: many hours for helium-3 and only a few minutes for xenon-129.

For xenon-129, the time constant for decay of nuclear polarization can be increased from minutes to about 1 hour if the xenon is frozen at liquid nitrogen temperatures in a magnetic field. At lower temperatures, the time constant increases significantly. Freezing of xenon is used to collect sufficient quantities of

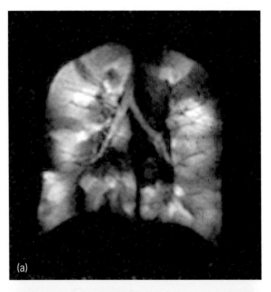

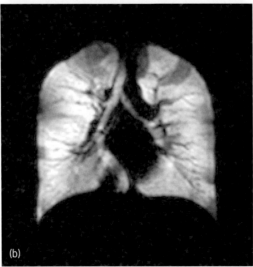

Fig. 1. Magnetic resonance imaging, using laser-polarized ^{3}He gas, of the lungs of a patient with severe asthma symptoms, (*a*) before and (*b*) after bronchodilator use. The images indicate the effects of bronchodilator use in clearing ventilation defects, and show the quality and resolution possible with laser-polarized noble-gas air-space imaging. (*Courtesy of Dr. Talissa Altes, University of Virginia*)

laser-polarized xenon-129. About a liter at standard temperature and pressure is useful.

Administration of laser-polarized gases. For magnetic resonance imaging, the laser-polarized gas must be administered to a patient or subject. For human subjects, the gas is transported to the vicinity of the MRI scanner and blown into a plastic or rubber bag before the subject inhales one or more breaths. Gas can also be dissolved in liquids such as normal saline solution (0.9% saline) for injection, and recently the gas has been encapsulated in small bubbles that can be safely injected into a patient for imaging blood vessels. Laser polarized gas images show where the gas is carried by breathing or by the blood, and it thus serves as a magnetic tracer. For animal studies, the gas can be introduced by injection or by ventilation of an anesthetized subject.

Magnetic resonance imaging. Magnetic resonance imaging makes use of data from the signals induced by the nuclear magnetization as it is manipulated and evolves in applied magnetic fields. Sequences of pulses of oscillating fields at the nuclear magnetic resonance frequencies and time-dependent gradients of the applied field are used to disperse the signals before and during data acquisition. The data for each step in time and each step of the applied gradients are processed using Fourier transform techniques to produce signal data as a function of frequency and phase. The applied gradients are then used to convert frequency and phase to two orthogonal spatial dimensions. Contrast in proton MRI is provided by the tissue dependence of the magnetic properties of proton nuclear spin time constants. Contrast may be enhanced by introducing materials that have an effect on these time constants, for example chelated gadolinium, which is used in tumor detection.

Studies. A wide range of unique studies is possible with noble gas MRI.

Lung air-space imaging. The most basic studies enabled with these new techniques are static and dynamic images of the lung air space. Proton MRI is not used for lung air-space imaging due to the low concentration of protons in the air space and due to magnetic effects at the gas-tissue interface that confound the imaging. Very detailed images are routine with both helium-3 (**Fig. 1**) and xenon-129 gas. Ventilation images show where the inhaled air travels in the lungs and how effectively it flows or diffuses through airways from the trachea to the alveoli. Any parts of the lung not ventilated in normal breathing are readily apparent. Ventilation images are not unique to MRI. Other modalities, specifically imaging of gamma-ray photons from radioactive xenon-133 (^{133}Xe), have been used to indicate large ventilation defects; however, the advantages of noble gas MRI include the absence of any radioactive dose to the patient and its much higher spatial resolution. The higher spatial resolution has been useful in studies that show small ventilation defects in the lungs of asthma patients and the effects of a bronchilator in clearing them (Fig. 1). A variant of ventilation imaging called restricted diffusion imaging can be used to study the size and distribution of the alveoli that terminate the bronchial passageways. Study of

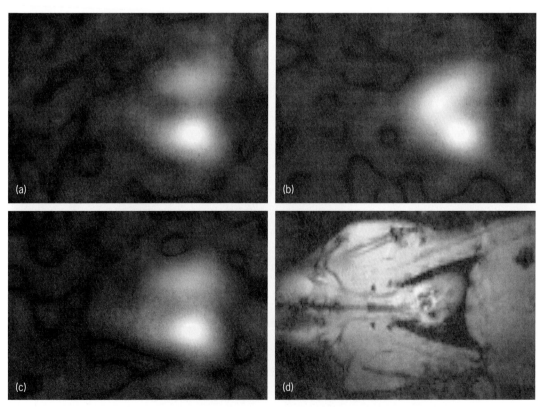

Fig. 2. Coronal magnetic resonance images of the thorax of a rat. (*a*) Laser-polarized xenon-129 in the gas phase, confined to the lung air space. (*b*) Xenon-129 dissolved in blood. (*c*) Xenon-129 dissolved in tissue. (*d*) Conventional proton image. (*University of Michigan*)

 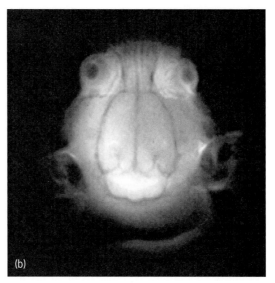

Fig. 3. Magnetic resonance imaging of the head of a rat. (*a*) Laser-polarized xenon-129 image. (*b*) Conventional proton image. The two images are coregistered so that the xenon image can be superimposed onto the proton image to show where xenon has been carried by blood flow. (*University of Michigan*)

alveolar size and distribution in patients with chronic obstructive pulmonary disease shows that alveolar size varies significantly more than in a healthy subject.

Imaging of xenon dissolved in blood and tissues. Imaging of noble gases dissolved in tissues is possible only with xenon-129, which is highly soluble in the blood and tissues. The solubility of xenon in circulating human blood is 17%, while the solubility of helium-3 is less than 1%. The solubility of xenon in tissues is in many cases even higher. The isotope xenon-129 has been used widely in nuclear magnetic resonance (NMR) because of the large range of sensitivity of the NMR frequency to the chemical environment. These chemical shifts allow spectroscopic separation of the portions of xenon-129 dissolved in various tissues from one another and from the xenon-129 in the gas phase. Chemical shift imaging allows images of distinct phases of xenon-129 so that an entirely new dimension can be added to MRI tracer techniques. The most extensive work has been done with rats, because it is currently difficult to produce enough laser-polarized xenon-129 for effective human imaging.

Figure 2 shows coronal images of xenon-129 in the thorax of a rat at three different chemical shifts, as well as a conventional proton image. The chemical shifts correspond to xenon-129 in the gas phase (confined to the lung air space), and to xenon-129 dissolved in blood and in tissue. The images can be overlaid to correlate anatomical features in the proton image with the functional information revealed by the xenon images. These images indicate that the xenon breathed by the subject crosses the blood-gas barrier through the parenchyma of the lungs, dissolves in the blood, and is carried to the heart where it dissolves in the cardiac tissue.

Brain imaging. Xenon-129 has also been imaged in the kidney and in the brain of the rat. **Figure 3** compares a laser-polarized xenon-129 image and a conventional proton image of the head of a rat. The proton image shows features of the rat's anatomy, including significant musculature in the jaws and the olfactory lobe at the front of the brain. The xenon image indicates significant xenon dissolved in the tissue of the cerebrum but little in the cerebellum. This is consistent with significant cerebral blood flow but much less flow to the cerebellum.

NMR spectroscopy study of the brain has revealed that there are at least four distinct chemical shifts, presumably corresponding to different tissues, as well as the identified blood dissolved phase. Research that assigns these chemical shifts to specific tissues, possibly layers of the cerebral cortex, will open a new dimension of study of the function of the brain using magnetic tracers to map blood flow. Magnetic tracer techniques are being developed that use the dependence of MRI images on time and NMR parameters.

For background information *see* DYNAMIC NUCLEAR POLARIZATION; HELIUM; INERT GASES; LASER; MAGNETIC RESONANCE; MEDICAL IMAGING; NUCLEAR MAGNETIC RESONANCE (NMR); OPTICAL PUMPING; XENON in the McGraw-Hill Encyclopedia of Science & Technology.   Timothy E. Chupp; Scott D. Swanson

Bibliography. T. A. Altes et al., Hyperpolarized ³He MR lung ventilation imaging in asthmatics: Preliminary findings, *J. Magnet. Reson. Imaging*, 13:378–384, 2001; V. Callot et al., Perfusion imaging using encapsulated polarized ³He, *Magnet. Reson. Med.*, 46:535–540, 2001; T. E. Chupp and S. D. Swanson, Medical imaging with laser polarized noble gases, *Advances Atom. Mol. Opt. Phys.*, 45:41–97, 2001; H.-U. Kauczor and K. F. Kreitner, Contrast-enhanced MRI of the lung, *Eur. J. Radiol.*, 34:196–207, 2000; S. D. Swanson et al., Distribution and dynamics of laser polarized xenon magnetization in vivo, *Magnet. Reson. Med.*, 42:1137–1145, 1999; T. Walker and W. Happer, Spin-exchange optical pumping of noble-gas nuclei, *Rev. Mod. Phys.*, 69:629–642, 1997.

Nuclear magnetic resonance logging (petroleum engineering)

In order to manage and optimize production, petroleum engineers are interested in reservoir fluids (water or hydrocarbon) as well as the properties of the reservoir rock, such as the void space that can contain fluids (porosity) and the capacity to transmit fluids (permeability). Logging techniques are used to measure a rock formation's fluid, porosity, and permeability properties with depth, either during well drilling or after drilling, before the well is lined (cased) with a steel pipe. Nuclear magnetic resonance (NMR) logging, also known as magnetic resonance (MR) logging, is a formation evaluation method that exploits the properties of the hydrogen nucleus—a proton. Magnetic resonance is one of the most complicated, yet powerful formation evaluation methods available. The MR measurements are similar to those used in a wide variety of applications, including medical imaging and spectrographic analysis in chemistry. Borehole MR is a small subset of these applications. MR logging is limited to measuring fluids, and the only commonly sensed atom is hydrogen.

Theory. A proton possesses electric charge and spin and, as a result, has both a magnetic field and angular momentum due to its spinning mass. One can think of the proton as a tiny bar magnet spinning about its axis. A proton will wobble in a cone (precess) about an external magnetic field like a disturbed mechanical top spinning in the Earth's gravitational field. The frequency of proton precession, called the Larmor frequency, is proportional to the strength of the external magnetic field and is on the order of 2 MHz.

Magnetic resonance is a cyclical measurement, alternating between the polarization of protons with a permanent magnet and the destruction of this polarization using radio-frequency electromagnetic waves. A MR measurement begins by aligning or polarizing the protons with a powerful permanent magnet, just as a compass needle aligns with a magnetic field. The increase of polarization with time is described by a time constant of an exponential function, the longitudinal relaxation time T_1. The polarization rate depends on the fluid type and whether the fluid is in contact with the rock matrix grains. For fluids in rocks, the polarization rate cannot be described by a single value for the time constant T_1. A distribution of time constant values, or a T_1 distribution, is required.

After polarization, the protons are manipulated using radio-frequency pulses from an antenna. The protons induce a very weak signal (echo) in the antenna, which also serves as the receiver. Measurement continues by alternating transmitting pulses and receiving echoes until the echo signal has decayed. Tool output at each depth measured is a series of echoes whose amplitude decreases with time. Through a series of computations, this echo decay train is transformed into a T_2 (transverse relaxation) time constant distribution.

Interpretation. The shape of the polarization buildup curve and the echo decay train, described by the T_1 and T_2 time-constant distributions, provides information on the physical properties of the rock (petrophysical information). The area under the T_2 distribution corresponds to the total porosity of the formation. As the measurement progresses, echo amplitude decay occurs due to three relaxation mechanisms: surface, bulk, and diffusion relaxation.

Surface relaxation is due to interactions at the interface of the fluid and the pore walls. Small pores have fast decay times (short T_2 times), whereas large pores decay slowly with long T_2 times. The pore size distribution provides permeability and bound-fluid information. Small pores contain large amounts of capillary- and clay-bound fluids that are not producible (do not contribute to permeability). Large pores contain free fluids, that is, producible fluids (do contribute to permeability). Many reservoirs have high water saturation but, on production, do not yield undesirable water because the water is bound in small pores.

Bulk relaxation occurs when fluids are not in contact with the pore walls, and is due to interactions of protons within the bulk fluid. For example, bulk relaxation occurs in oil drops surrounded by water in a rock, where water wets the rock surfaces. The frequency of these interactions is related to the fluid's viscosity and therefore provides oil viscosity information.

Diffusion relaxation arises because the permanent magnet varies in strength (gradient) in the region of investigation, and the fluid molecules move with random motions characterized by a diffusion coefficient. By generating sets of MR data with various time increments between the pulsing of the tool, the diffusion coefficients can be measured and saturation and fluid viscosity information can be derived (**Fig. 1**).

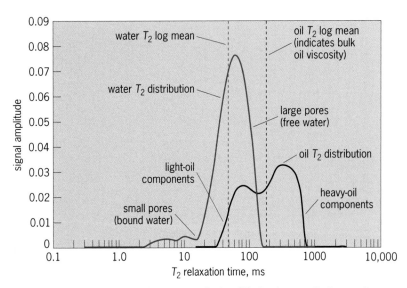

Fig. 1. Oil and water T_2 distributions extracted using diffusion-in-a-gradient magnetic-field resonance principles. The shape of the water distribution provides information on bound- and free-water components in the formation. The shape of the oil distribution provides oil viscosity information. The area under each curve corresponds to the oil-water saturation, and the total porosity is indicated by the sum of the two areas. (*Courtesy of Schlumberger*)

Well logging. Magnetic resonance logging measurements can be recorded at the well site on a continuous basis as the tool assembly ascends or descends in the borehole. A more detailed laboratory MR analysis can be done on reservoir rock core samples taken during drilling operations. These measurements can be used for calibration of the borehole MR response or as a basis for drilling future wells in the field. The disadvantages of laboratory MR measurements are that the rock is no longer in its natural condition, and measurements are not available in time to make well completion (production preparation) decisions.

The value of MR data is greatest when recorded in real time [logging-while-drilling (LWD) mode]. The latest MR and communication technologies enable key formation data to be transmitted as the well is drilled from downhole LWD tools to the surface and then to oil companies' offices around the world for immediate analysis and instant decision-making. This allows for borehole trajectory optimization based on petrophysical parameters. For example, the most permeable layers of a formation can be tracked or zones with high water production potential can be avoided.

Magnetic resonance data have many uses. Completion engineers use permeability information from MR measurements to design hydraulic fracture treatments for reservoir stimulation. Reservoir engineers assess rock quality to locate thin, high-porosity "pay zones" that might otherwise be bypassed or to find vertical permeability barriers, enabling better production management. Geologists and petrophysicists use MR logs to enhance their understanding of pore geometry for depositional (facies) analysis. In addition, interpretation of MR logs in combination with other logging measurements enables more accurate characterization of reservoir hydrocarbons, improving assessment of well producibility.

Tools. Magnetic resonance was discovered in the 1940s, and oil field applications and patents soon followed. Early tools, however, suffered from poor reliability and a low signal-to-noise ratio.

The NUMAR Corporation developed the first commercially successful MR logging tool, the MRIL®, which is run centralized in the borehole. The latest version, the MRIL®-Prime, incorporates improvements to increase logging speed and efficiency. This is accomplished through prepolarizing magnets that prepare the proton alignment in advance of the antenna, and through measurement capability that allows for increased data density.

The Combinable Magnetic Resonance (CMR, mark of Schlumberger) tool shown in **Fig. 2**, introduced in 1995, is run pressed against the borehole using a bowspring. A short directional antenna placed between a pair of compound magnets focuses the CMR measurement on a 6-in. (15-cm) vertical zone located 1.1 in. (2.8 cm) inside the formation and separated from the tool by a 0.5-in. (1.27-cm) blind zone. This design allows for a lightweight and compact tool capable of operating in large holes and salty mud

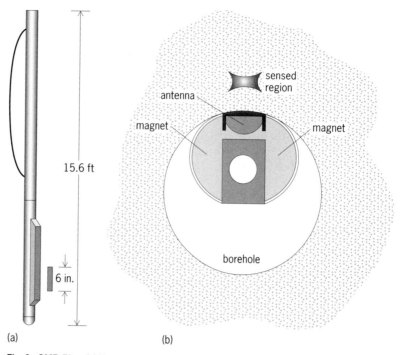

Fig. 2. CMR-Plus. (*a*) Tool with a bowspring for centering, and the 6-in. (15-cm) measurement aperture indicated. (*b*) Cross section. The tool's permanent magnets and radio-frequency antenna produce a concentrated sensed region where the magnetic resonance measurement is made. (*Courtesy of Schlumberger*)

environments, while delivering precision measurements with high vertical resolution.

In the mid-1990s, the introduction of higher rate and more sophisticated pulsing techniques extended the capabilities of these tools for characterizing fluid mobility and measuring the smallest pore sizes. More recently, substantial improvements in magnet design and data acquisition capabilities have led to significant increases in logging speeds.

A new tool, CMR-Plus (mark of Schlumberger), enables faster data acquisition through a new magnet design with a longer prepolarizing field that increases logging speeds to 3600 ft/h (1097 m/h) in fast-relaxation environments. Higher logging speeds enable economic acquisition of data over longer intervals, including potentially productive zones that were initially of little interest. The tool can log in smaller holes ranging down to a 5.9-in. (15-cm) diameter. A new pulse-acquisition sequence and an upgraded electronics package increase the signal-to-noise ratio and improve measurement precision.

Outlook. Presently, MR logging technology provides many key answers, including total porosity independent of mineralogy, the spectrum of pore sizes present, bound-fluid volume, permeability, fluid type, quantitative oil-water-gas saturation, and oil viscosity. The future for NMR logging is likely to reflect two trends: continued improvements in NMR technology and the method's wider acceptance by the oil and gas industry. Increasingly, this innovative technology is seen not simply as a high-tech niche application but as a tool providing answers not available from standard logging methods.

For background information *see* BIOT-SAVART LAW; MAGNETIC RESONANCE; NUCLEAR MAGNETIC RESONANCE (NMR); OIL AND GAS WELL COMPLETION; OIL AND GAS WELL DRILLING; PETROLEUM ENGINEERING; PETROLEUM GEOLOGY; WELL LOGGING in the McGraw-Hill Encyclopedia of Science & Technology.

James Kovats

Bibliography. D. Allen et al., How to use borehole nuclear magnetic resonance, *Oilfield Rev.*, 9(2):34–57, 1997; D. Allen et al., Trends in MR logging, *Oilfield Rev.*, 12(3):2–19, 2000; G. Coates, L. Xiao, and M. Prammer, *NMR Logging Principles and Applications*, Gulf Publishing, 2001; T. Farrar (preface) and E. Becker (preface), *Pulse and Fourier Transform NMR: Introduction to Theory and Methods*, Academic Press, 1978; R. Freedman et al., A new NMR method of fluid characterization in reservoir rocks: Experimental confirmation and simulation results, Society of Petroleum Engineers, Pap. SPE 75325, *SPE J.*, pp. 452–464, December 2001; B. Kenyon et al., Nuclear magnetic resonance imaging: Technology for the 21st century, *Oilfield Rev.*, 7(3):19–33, 1995; H. Smith and F. Ranallo, *A Non-Mathematical Approach to Basic MRI*, Medical Physics Publishing, 1989.

Organic light-emitting devices

Organic light-emitting devices (OLEDs) are an attractive technology for use in a new generation of flat-panel displays because of their wide viewing angle, ease of fabrication, and compatibility with a wide variety of substrates, including plastics, textiles, and paper.

Although light emission from organic materials upon electrical excitation has been demonstrated in the past, it was not until the pioneering work at Kodak (small-molecule OLEDs) and the University of Cambridge (polymer OLEDs) that the commercial prospects of such technology were envisioned.

Materials and device structure. The organic materials used in OLEDs are either small molecules or polymers (**Fig. 1**). A typical polymer OLED consists of a single polymer layer sandwiched between two electrodes (**Fig. 2a**). The polymer layer both transports the charges and emits light. OLEDs based on small molecules have more than one organic layer placed between the two electrodes. For example, the bilayer OLED in Fig. 2b has a hole transport layer (HTL) and an electron transport layer (ETL). One of these layers must emit light. Otherwise, a light-emitting layer must be placed between the hole transport layer and the electron transport layer, increasing the complexity of the device. In all cases, the thickness of each organic layer (small molecules or polymer) is several tens of nanometers.

Fabrication. One attractive feature of OLEDs is their ease of fabrication. Small-molecule OLEDs can be fabricated using high-vacuum thermal deposition techniques. In this case, the organic materials and the electrodes are evaporated (or sublimed) in vacuum (about 10^{-5} to 10^{-7} torr or 10^{-3} to 10^{-5} pascal). In some cases, the electrodes may be sputter-coated [a process whereby energetic ions eject materials from a target (source) which are then deposited as a thin layer on a substrate]. However, care must be taken not to damage the organic layers while sputtering the electrode materials on top of them.

Due to their relatively high molecular weight, polymers cannot be vacuum-deposited. They have to be dissolved and applied as a solution. Polymer solutions lend themselves to low-cost coating (such as spin coating) and printing techniques (such as inkjet, gravure, and screen printing). Spin coating is the most mature process in polymer OLEDs fabrication. The lifetime of devices having a spin-coated layer is better than those in which the polymer is printed. However, recent progress and understanding of the polymer printing process is closing this performance gap.

Operational mechanism. Although the exact operating mechanisms of OLEDs are more complex, a general description of how OLEDs function is as follows. Upon application of a direct-current (dc) voltage in a molecular OLED, holes are injected from the anode into the highest occupied molecular orbital (HOMO) of the hole transport layer, and electrons are ejected from the cathode into the lowest unoccupied molecular orbital (LUMO) of the electron transport layer (**Fig. 3a**). In polymer OLEDs, the polymer layer transports the charges (Fig. 3b).

The charges drift under the influence of the electric field and recombine on a molecular site in the

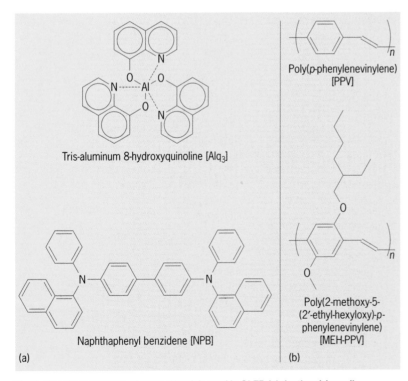

Poly(p-phenylenevinylene) [PPV]

Tris-aluminum 8-hydroxyquinoline [Alq₃]

Naphthaphenyl benzidene [NPB]

Poly(2-methoxy-5-(2′-ethyl-hexyloxy)-p-phenylenevinylene) [MEH-PPV]

(a) (b)

Fig. 1. Chemical structure of some materials used in OLED fabrication: (a) small molecules and (b) polymeric materials. Alq₃ is normally used as an electron transport layer; it can also emit green light around 520 nm. NPB is used as a hole transport layer. The polymers shown transport charge and emit light.

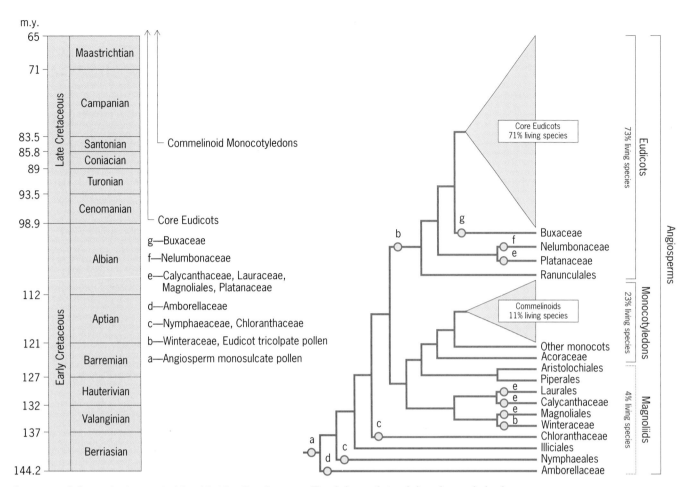

Age versus phylogenetic placement of the oldest fossil angiosperms. The phylogenetic tree is based on analysis of morphological and molecular data. A general congruence between the age of fossils and their phylogenetic placement is observed during the early diversification of angiosperms. The oldest reliably identified angiosperm fossils belong to early-differentiated branches in the angiosperm phylogenetic tree; the first fossil representatives of derived (more advanced) groups such as the core eudicots (a large group including spinach, mistletoe, eukalypt, legumes, and astors among many others) and commelinoid monocots (including palms, gingers, bananas, sedges, and grasses) occur in younger sediments. *(After J. A. Doyle and P. K. Endress, Morphological phylogenetic analysis of basal angiosperms: Comparison and combination with molecular data, Int. J. Plant Sci., 161(suppl.):S121–S153, 2000)*

feature associated with eudicot angiosperms, a group that includes the majority of living angiosperm species (see illus.). Pollen in distinctive tetrads (groups of four grains resulting from sexual division) belonging to Winteraceae (winter's bark), one of the early differentiated angiosperm lineages, occurs in Barremian-Aptian palynofloras (121 m.y.). Mesofloras of Barremian–Aptian age contain the oldest angiosperm flower remains. These include a small bisexual flower of Nymphaeaceae (water lily), belonging to the second-differentiated branch in the angiosperm phylogenetic tree, and small female flowers, fruits, and other dispersed reproductive remains belonging to Chloranthaceae, another early differentiated group sometimes resolved as the fourth branch in the angiosperm phylogenetic tree. Illiciales (star anise), belonging to the third branch within angiosperms, and monocotyledons are possibly represented in Barremian-Aptian mesofloras. Macrofloras of Aptian age (121–112 m.y.) contain leaf remains with an epidermal pattern similar to

that unique to Amborellaceae, the first-differentiated branch within angiosperms.

Mesofloras of early Albian age (112–108 m.y.) contain bisexual flowers unequivocally belonging to Calycanthaceae (*Chimonanthus*) and to Lauraceae (cinnamon), two closely related, early-differentiated angiosperm groups, and to Magnoliales, an early-differentiated lineage that includes Magnoliaceae (magnolia), Myristicaceae (nutmeg), and Annonaceae (sweetsop), among others (see illus.). Early Albian mesofloras also contain leaves, male and female inflorescences, flowers, and other dispersed flower parts of Platanaceae (London plane), one of the earliest-differentiated lineages within the eudicots. Other early-differentiated eudicot branches are represented in slightly younger paleofloras, including leaves and fruits of Nelumbonaceae (sacred lotus) from a middle Albian (108–103 m.y.) macroflora, and male and female inflorescences and flowers of Buxaceae (boxwood) from a late Albian (103–99 m.y.) mesoflora.

Late Cretaceous. Late Cretaceous paleofloras (99–65 m.y.) contain a significantly greater diversity of angiosperms, including core eudicots and monocotyledons, which together constitute the vast majority of living angiosperm species. Most major angiosperm groups had differentiated by the end of the Cretaceous, but many diversified during the Tertiary (65–1.8 m.y.). Some large core eudicot groups, such as Gentianales (gentian), Solanales (tomato), Lamiales (mint), and Asterales (sunflower), differentiated much later, in the Eocene (55–37 m.y.) or maybe as late as the Oligocene (37–26 m.y.). Several present-day angiosperm-dominated ecosystems, such as grasslands and lowland tropical rain forests, originated during the Tertiary.

Fossil flower and pollen features. The wealth of unequivocal data provided by mesofossils results in a revised and enriched concept of ancestral flowers. Flowers preserved in ancient mesofloras are very small. While their minute size may partially result from reduction during fossilization, it is very unlikely that they were large during life. Most are few-parted, and both unisexual and bisexual flowers are present, although unisexual forms prevail. Floral organs are usually arranged in whorls (concentric circles). Epigynous flowers (floral organs inserted above the base of the ovary) co-occur with hypogynous forms (floral organs inserted below the base of the ovary). Many flowers lack a perianth (sterile floral envelope), but whether they were truly naked or lost their perianth during preservation cannot be reliably determined. When a perianth is present, it is simple and consists of a small number of tepals (undifferentiated perianth organs) that are free from each other. The number of stamens (organs where pollen is formed) per flower is usually small, and they are arranged in a single whorl, although multiwhorled and apparently spirally set stamens also occur. Ancient stamens exhibit a weak differentiation into filament (basal sterile part) and anther (distal fertile part). Anthers have four pollen sacs arranged in two thecae (pollen sac pairs), an organization preserved in the vast majority of living angiosperms. Each theca opens either by a longitudinal slit or by laterally hinged valves.

An extensive diversity of pollen types occurs in association with angiosperm reproductive organs in the oldest mesofloras. Most correspond to monosulcate or monosulcate-derived types, characteristic of magnoliid and monocotyledonous angiosperms (see illus.). Flowers usually have one or few carpels (organs where seeds are formed and become a fruit or part of a fruit at maturity) most frequently free from each other, and arranged in a whorl or sometimes spirally set on an elongated floral receptacle. Stigmas (areas of pollen reception and germination) are nonspecialized and sessile. Carpels usually contain a single seed or very few seeds. Most seeds are anatropous (the micropyle faces the site of seed attachment), and few are orthotropous (the micropyle is opposite the site of seed attachment), and apparently derived from ovules with two integuments (ex-

ternal layers). A wide variety of unspecialized fleshy and dry fruit types are present.

Comparison to early-differentiated living angiosperms. The image of ancestral flowers provided by recent fossil discoveries is consistent with that derived from phylogenetic analysis of molecular and morphological data from living and extinct plants. Features of living representatives of early divergent angiosperms complete the picture, indicating that ancestral carpels were most probably of the ascidiate (sac-like) type and that carpel sealing was not achieved by fusion of morphological surfaces, as found in most living angiosperms, but rather by the presence of a dense liquid inside the cavity of the carpal. The picture emerging from integrating characters of ancient fossil flowers with phylogenetic analysis is a consistent and significantly improved concept of the attributes of ancestral flowers. Character distribution in younger fossil flowers and among later-differentiated angiosperm lineages suggests that traits that appeared subsequently in flower evolution include a bipartite perianth (with two whorls of morphologically distinct organs), fusion and functional interaction among organs in one or several whorls, and carpels with long styles (carpel extension that usually bears the stigma).

Overview. The diversity and abundance of angiosperm remains in paleofloras of different types and ages document the affinity (relationship to living species) and ecological traits of predominant angiosperm groups during the early phases of their radiation. Wind-pollinated plants, which produce vast amounts of pollen, are much better represented in palynofloras than insect-pollinated plants, which produce smaller amounts of pollen. Deciduous trees, which abundantly produce and seasonally shed their leaves, are much better represented in macrofloras than evergreen or herbaceous plants, which produce fewer leaves. The physical and chemical processes leading to formation of mesofossils allow preservation of delicate organs such as flowers and flower parts, including those of insect-pollinated and herbaceous plants. These vegetation types are rarely represented in palynofloras or in macrofloras. Mesofloras of Early Cretaceous age contain a variety of flower types and pollen associated with reproductive organs that greatly exceeds the abundance and diversity of angiosperms found in other types of paleofloras of equivalent age. The observed differences suggest that herbaceous, insect-pollinated forms dominated early phases of angiosperm diversification. Ancient mesofloras contain an extremely high diversity of floral types belonging to extinct groups of probable general magnoliid affinity. (Present-day magnoliids include magnolias, avocado, and cinnamon.) This unexpectedly high diversity reveals that the earliest phases of angiosperm diversification proceeded very rapidly. The taxonomic composition of these earliest floras, dominated by magnoliid types, was very different from that of living floras, in which eudicot angiosperms are usually predominant. Floral specialization displayed by magnoliid lineages in ancient paleofloras is equivalent to that of living representatives, suggesting that magnoliid pollination mechanisms became established very early during their diversification. On the other hand, fossil magnoliid fruit types are simple and nonspecialized, suggesting that seed dispersal strategies displayed by living forms evolved subsequently, probably in association with animal groups that diversified during the Cretaceous and Tertiary.

For background information *see* PALEOBOTANY; PALYNOLOGY; PLANT KINGDOM; PLANT PHYLOGENY in the McGraw-Hill Encyclopedia of Science & Technology.   Susana Magallón

Bibliography. P. R. Crane et al., The origin and early diversification of angiosperms, *Nature*, 374:27–33, 1995; P. R. Crane and P. S. Herendeen, Cretaceous floras containing angiosperm flowers and fruits from eastern North America, *Rev. Palaeobot. Palynol.*, 90:319–337, 1996; E. M. Friis et al., Reproductive structure and organization of basal angiosperms from the early Cretaceous (Barremian or Aptian) of western Portugal, *Int. J. Plant Sci.*, 161(suppl.):S169–S182, 2000; E. M. Friis et al., Fossil history of magnoliid angiosperms, in K. Iwatsuki and P. H. Raven (eds.), *Evolution and Diversification of Land Plants*, pp. 121–156, Springer, Tokyo, 1997.

Paper manufacturing (thermodynamics)

The annual production of paper and paperboard is approximately 300 million metric tons (330 million tons) worldwide. Paper products include newspapers, magazines, books, soft tissues, office paper, and boxboard. Paper is made from natural fibers (mostly from wood) which are processed with water, chemicals, minerals, and heat. The principles of papermaking have been known for 2000 years, and the process today is highly technical. A pulp and paper mill is a huge complex, with many different stages for grinding raw wood into small particles, separating fibers from raw materials (mechanical or chemical pulping), bleaching, washing, evaporating water, drying, and processing into paper. The typical production rate of dry cellulose pulp in a paper mill is around 400,000–600,000 metric tons (440,000–660,000 tons) per year. The amount of raw wood needed is several times more—the practical equivalent of 300–500 truckloads arriving daily at a mill. The best available technology is needed to produce an environmentally acceptable, energy-efficient process, which requires an ever-smaller input of chemicals and wood, and uses water with an increased amount of recirculated matter.

Thermodynamic applications. In the pulp and paper industry, the application of thermodynamics includes classical heat and energy-efficiency studies, surface phenomena, colloidal science, solubility, reactions, and chemical affinity. Chemical thermodynamics focuses on energy changes in chemical reactions and provides a fundamental and widely applicable tool for studying multiphase solutions in the papermaking processes. Chemical equilibrium shows the natural boundaries of the physical and chemical interactions in different chemical environments, temperatures, and pressures. Thermodynamics answers the question by how much and in what direction the system responds to chemical and physical changes. Modern processes recycle the water, thus reducing the amount of fresh water needed in paper manufacturing. As the aqueous process solutions become more concentrated with dissolved organic compounds and ions, the solids precipitation has to be controlled in the process. Thermodynamics provides a practical tool for estimating the chemical states of the pulp and paper solutions. Such a fundamental approach is related to the chemical energy, chemical reactions, solubility of gases and salts, and pH, an important process parameter. The thermodynamic data, combined with the results of the separately measured fiber and solution properties, is needed for the development, optimization, and control of the pulp and papermaking processes.

Bleaching. Lignin is the noncarbohydrate portion of the cell wall of the plant material, while cellulose (a high-molecular-weight, chemically linear polysaccharide) is the main solid constituent of woody plants. Bleaching is the process of removing the lignin, a dark material, from wood by means of chemicals and heat, leaving cellulose, which is white. As the lignin is removed, the brightness of the pulp increases. The most common bleaching agents are chlorine (Cl_2), chlorine dioxide (ClO_2), hypochlorite (ClO_3^-), hydrogen peroxide (H_2O_2), oxygen (O_2), and ozone (O_3). The choice of bleaching agent depends on the wood material, chemical and energy costs, environmental issues, and the desired quality of the bleached fibers.

For environmental reasons, peroxides have been used as a replacement for chlorine gas. Hydrogen peroxide, together with pressurized oxygen gas, has been studied for totally chlorine-free (TFC) bleaching. Oxygen-pressurized peroxide bleaching is done at moderate oxygen pressures of 0.1–2 MPa (1–20 bar). Pressurizing allows reaction to take place near and above 100°C (212°F). An alkaline (NaOH) bleaching solution dissociates hydrogen peroxide into proton (H^+) and the reactive perhydroxide ion (OOH^-). The dissociation reaction in alkaline conditions can be expressed by reaction (1) or reaction (2), where aq is the aqueous phase. The

$$H_2O_2(aq) \longleftrightarrow H^+ + OOH^- \qquad (1)$$

$$OH^- + H_2O_2(aq) \longleftrightarrow OOH^- + H_2O \qquad (2)$$

change in Gibbs free energy, $\Delta_r G$, of dissociation reaction (1) can be expressed as a sum of standard change of Gibbs free energy, $\Delta_r G°$, and a logarithmic term including the activity ratio Q [Eq. (3)], where

$$\Delta_r G = \Delta_r G° + RT \ln Q \qquad (3)$$

R is the gas constant and T is the absolute temperature. The equilibrium constant K is expressed by means of the activities a_i in equilibrium composition of the system, which are written with composition terms of molalities b multiplied by activity coefficients γ_i, giving the activity, Eq. (4), with

$$a_i = \frac{\gamma_i b_i}{b^\circ} \qquad (4)$$

$b^\circ = 1$ mol/kg for each individual species i in the solution. The equilibrium constant is Eq. (5). At equi-

$$K = \frac{a_{H^+}\, a_{HO_2^-}}{a_{H_2O_2}} \qquad (5)$$

librium, where the Gibbs free energy is at its global minimum, $\Delta_r G$ is zero. Then one obtains the important relationship in chemical thermodynamics, Eq. (6).

$$\Delta_r G^\circ = -RT \ln K \qquad (6)$$

The numerical value of $\Delta_r G^\circ$ for any reaction can be obtained from thermochemical tables by means of enthalpy H, entropy S, and heat capacity C_P. In calculations for aqueous solutions, a temperature-dependent standard data and activity coefficient model are needed to describe the total Gibbs energy for the solution. Process solutions involve complex reactions and electrochemical equilibrium between the fiber phase and bulk ionic solution, precipitates, dissolved gases, and organic compounds. Because of the complex surface reactions on the fibers, alkaline and peroxide are consumed, organic acids are formed, and a pH change is observed. The thermodynamic model gives useful information on the initial pH of the reactive bleaching solution at different alkaline and peroxide concentrations. The pH of the bleaching solution in the process is important because it influences the degree of dissociation of hydrogen peroxide and the final pulp quality. For optimal results, the pH should be around 10.3–11.3. Both chemical changes and physical properties are observed by sampling during the fiber bleaching reaction. The fibers' physical properties, such as brightness, kappa number, and fiber strength, are measured by standardized methods. The kappa number measures the degree of delignification. A decrease in the kappa number corresponds to greater bleaching and an increase in pulp brightness. For the pulp samples in **Fig. 1**, the kappa numbers of the samples from left to right are 7, 4, and 10.

Hydrogen peroxide is a highly reactive chemical, and in the presence of metal it will decompose into water and oxygen, reaction (7), where g is the gas

$$H_2O_2(aq) \longleftrightarrow H_2O + 1/2\ O_2(g) \qquad (7)$$

phase. In order to avoid the decomposition reaction of peroxide in the bleaching solution, the content of naturally present metal ions Mn^{2+}, Cu^{2+}, Fe^{3+}, Ca^{2+}, and Mg^{2+} is lowered by chelation, a chemical reaction or process for removing undesired metal ions. Ethylenediaminetetraacetic acid (EDTA) and diethylenediaminepentaacetic acid (DTPA) are used to

Fig. 1. Laboratory-bleached pulp samples with kappa numbers 7, 4, and 10.

chelate transition metals before starting the peroxide bleaching process.

Multiphase calcite chemistry. Acid-base chemistry plays an important role in the papermaking process because controlling the process pH at a constant level has a major effect on the paper production rate. Calcium carbonate ($CaCO_3$), also known as calcite, is used as filler in high-grade paper. The dissolution and precipitation processes of aqueous $CaCO_3$ solutions are controlled by the acid-base chemistry. Thermodynamics gives quantitative information on what the state of the multiphase $CaCO_3$ solution will be if the temperature, pressure, or chemical composition is altered. Sulfuric acid (H_2SO_4) has commonly been used for pH control in $CaCO_3$-buffered solutions. It dissociates in water, forming proton (H^+), bisulfate (HSO_4^-), and sulfate (SO_4^{2-}) ions. As the ions build up in the water (which is recirculated to reduce the amount of wastewater), precipitation of unwanted salts can occur. Carbon dioxide (CO_2) gas can be used as an alternative, acidifying reagent in calcite-buffered solutions.

By altering the carbon dioxide pressure, one can alter the pH and control the calcium carbonate precipitation and dissolution (**Fig. 2**). Natural sources of CO_2 from biological activity (the irreversible decomposition process of organic matter) may also

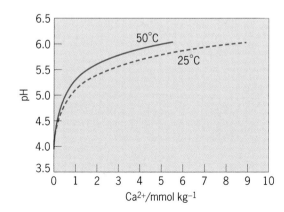

Fig. 2. Change in pH as calcite is added in distilled water that is in equilibrium with 1 atm (1.013 bar or 0.1013 MPa) carbon dioxide partial pressure. The end points of the line represent the saturation point of calcite. Note that the solubility of calcite is less at 50°C (122°F) than at 25°C (77°F).

interface quantum states into the conduction band of the host material. The electrons are accelerated to high energy and eventually collide with a host or dopant atom. This can result in either excitation of another electron into the conduction band (charge multiplication) or impact excitation of a LC atom. It is the impact excitation of LCs that leads to light output by radiative relaxation. The functional difference between the powder and thin-film phosphors is the source of the electrons. In powder phosphors electrons come from the surfaces of small grains of a separate phase, typically copper sulfide, whereas in thin-film devices electrons come from the interfaces of the insulating layers. Unlike the dc phosphors, high-field ac phosphors rarely generate light by electron-hole recombination, since the high fields rapidly separate any electron-hole pairs before they can recombine.

Design of phosphors. There are several requirements of a phosphor host material. The first is that the host must not absorb the light being emitted. This means that the bandgap of the host must be larger that the energy of the light that is to be generated. The reason can be seen in Fig. 1. If the energy levels of the LC were farther apart than the bandgap, the electrons would occupy the conduction band rather than the localized quantum state on the LC atom. In addition, the LC atom must fit in the host's crystalline lattice. Atoms that are too large often will not fit into the host crystal and will separate into other nonluminescent phases, decreasing the efficiency of the phosphor.

There are several other requirements of the host material that can vary according to the application. For photoluminescent phosphors, it is important that the host material efficiently absorbs the UV light or the energy will be wasted. It is also important that the host and LC have some overlap of energy levels so that the excitation can be efficiently transferred to the LC. Cathodoluminescent phosphor hosts must be stable under electron bombardment. Electroluminescent phosphors must have the ability to efficiently transport high-energy electrons.

Luminescent centers can be either line (single-wavelength) emitters or broadband emitters. Line emitters have very well defined, sharp energy levels involved in the radiative relaxation. Ions such as the rare-earth elements (lanthanum, europium, praseodymium, neodymium, and so on) are generally line emitters because their electrons involved in the light emission are shielded from the electrons of the atoms surrounding them. Broadband emitters (such as Mn^{3+}) have radiative transitions involving the same electrons that interact with the surrounding atoms. Since there can be variation in the nature of the surroundings, the energy levels occupy a broad range rather than narrow energy levels. Line emitters give off light whose color is relatively independent of the host. In contrast, the color of light from broadband emitters can vary widely, depending on the host material. For example, manganese gives yellow light in a zinc sulfide host, but it gives green light in a zinc germanate host.

In addition to the LC, other atoms (sensitizers) may be added to increase light emission. These sensitizers may increase the efficiency of the energy transfer to the LC. One example is the addition of cerium to zinc sulfide phosphors doped with terbium. Cerium has an excited level near that of terbium so it can transfer energy to terbium, which then gives off more green light.

Another type of modifier affects the electrical properties of electroluminescent host materials. One example is the addition of potassium chloride to sputter-deposited thin films of zinc sulfide doped with manganese. This enhances the hole (positive charge left behind when the electron is absent) mobility and thus reduces charging in the thin-film phosphor. This ultimately enhances the efficiency of the phosphor.

Degradation of phosphors. A big technical challenge in the development of phosphors for advanced displays is minimizing the effects of degradation of the phosphor with time. The eye can perceive a 5% change in brightness of adjacent areas. Display phosphors cannot have a brightness change of more than 5% over the typical 20,000-hour design lifetime for a commercial display. The brightness can change during operation as a result of numerous effects, including heating, migration of dopants or defects, and chemical reaction with the surrounding ambient. Some of these effects are exacerbated by the method in which the phosphor is used. For example, chemical effects can be worse for cathodoluminescent phosphors where the electron beam creates highly reactive species from the background gases, a process known as electron-stimulated surface chemical reaction. Design of the phosphor to minimize aging effects can include phosphor particle coating and careful engineering of the defects to prevent migration.

For background information *see* BAND THEORY OF SOLIDS; CATHODOLUMINESCENCE; ELECTROLUMINESCENCE; ELECTRON-HOLE RECOMBINATION; FIELD EMISSION; HOLE STATE IN SOLIDS; LIGHT-EMITTING DIODE; LUMINESCENCE; RARE-EARTH ELEMENTS; SEMICONDUCTOR in the McGraw-Hill Encyclopedia of Science & Technology.   Mark R. Davidson; Paul H. Holloway

Bibliography. G. Basse and B. C. Grabmaier, *Luminescent Materials*, Springer, New York, 1994; Y. A. Ono, *Electroluminescent Displays*, World Scientific, Singapore, 1995.

Photodetachment microscopy

Since lasers became widely tunable in the 1970s, they have become a popular tool for the excitation of atoms, molecules, and their ions. Photons emitted by a laser are essentially monoenergetic, since the laser wavelength can be set very precisely. This makes it possible to selectively excite the electrons bound in atoms and molecules to higher energy levels. If the photon energy is sufficiently high, it can eject an electron from an atom or molecule to form a positive ion. This process is called photoionization. Negative

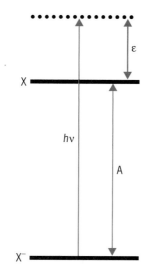

Fig. 1. Energetics of photodetachment. Absorption of a photon of energy $h\nu$ by a negative ion X⁻ leads to ejection of an electron of energy ε. The detachment energy of the negative ion A, which is the electron affinity of the neutral atom X, can be measured as the difference $A = h\nu - \varepsilon$.

ions are systems that are weakly bound due to the fact that the outermost electron moves in the short-range field of a neutral atom or molecule. The binding energies of these electrons in negative ions are significantly smaller than those in atoms or molecules. Photodetachment is a process analogous to photoionization that ejects an electron from a negative ion. The energetics of photodetachment is shown in **Fig. 1**. It can be seen that for monoenergetic photons the detached electron has a well-defined energy.

Motion of the detached electron. In a typical photodetachment experiment, the total detachment probability is measured as a function of the laser wavelength. Important information can also be determined by measuring the kinetic energies and angular distributions of the detached electrons. In general, the electrons are emitted over a range of angles. In the presence of an electric field, however, the electrons can be guided into a detector. The

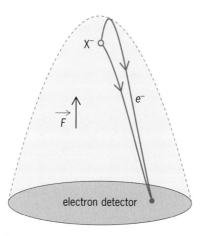

Fig. 2. Trajectories of an electron detached from a negative ion X⁻ in the presence of an electric field F. The detection probability is distributed within a circle on the electron detector. Since every point within this circle can be reached by a pair of trajectories, interference between the trajectories is expected to occur.

effect of an electric field is to cause pairs of trajectories to rejoin at the detector (**Fig. 2**). In 1981 the theorists Yu. N. Demkov, V. D. Kondratovich, and V. N. Ostrovskii pointed out that electron interference fringes, or more precisely rings, should be observable provided a detector with high spatial resolution was used. The authors coined the name "photoionization microscope" for such an apparatus.

Properties. It would appear that making a microscope based on the photodetachment of electrons from negative ions would be easier than making one based on the photoionization of electrons from neutral atoms or molecules. This is because the outermost electron in a negative ion is less tightly bound than in an atom or molecule and can be detached by relatively low-energy photons. Once free, the motion of the detached electron is relatively unperturbed, since it moves in the field of a neutral particle. As a result, the trajectories of the detached electron in the presence of an electric field are the simple parabolas shown in Fig. 2. In contrast, the motion of an ionized electron is more complicated due to the presence of a long-range Coulomb field, and the trajectories of such an electron are no longer parabolas.

There are some practical conditions that must be met in order to observe interference rings in a photodetachment microscope. For example, the interval between adjacent rings must be larger than the resolution limit of the detector. The maximum phase difference must also be large enough to make the number of rings greater than unity. The spectral width of the excitation source must also be sufficiently narrow so as not to wash out the ring interference pattern. The combination of these conditions requires that the detached electron energy be of the order of 100 microelectronvolts and the electric field strength be a few hundred volts per meter.

Figure 3 shows an example of an interference pattern arising from the photodetachment of a negative oxygen ion (O⁻) taken under these conditions. The intense outer ring is due to the accumulation of classical trajectories at the maximum radius that the electron can reach on the detector. The inner dark rings and the central dark spot are the interference maxima that are associated with the quantum nature of the electron motion. The deviation from a perfect circular shape is due to inhomogeneities in the electric field. These long-range inhomogeneities do not, however, influence the number of rings in the interferogram.

Qualitatively, the interference patterns show that the coherence of the detached electron wave can survive propagation over macroscopic distances (about 0.5 m or 1.5 ft). The fringes are also among the largest ever obtained using matter waves, with a period up to 0.3 mm (0.012 in.). The measured interferograms are exactly what would be expected in the case of the propagation of an electron wave from a pointlike source in the presence of an electric field. The agreement between the measured data and a theoretical propagator formula gives confidence that the interference patterns can be used for quantitative measurements.

Photodetachment spectrometry. The most important parameter that comes out of the fitting of the experimental data with a propagator formula is the exact energy ε at which the electron is ejected. This energy determines both the size of the central spot (proportional to $\varepsilon^{1/2}$) and the number of interference fringes (proportional to $\varepsilon^{3/2}$). Because of the latter dependence, ε is essentially constrained by the number of rings observed in the interferogram. This is fortunate since the number of rings can be measured more accurately than the absolute radius of the central spot. The measurement of a radius entails determining the pixel size in a separate measurement. The energy of the detached electron, ε, can thus be determined with an interferometric accuracy to less than 1 μeV. This makes the photodetachment microscope the most accurate electron spectrometer ever built, as far as the absolute uncertainty is concerned. This technique has been applied to measurements of the electron affinities of silicon, oxygen, and fluorine, with an accuracy of a few microelectronvolts. Part of the uncertainty arises in determining the laser wavelength. Thus, photodetachment microscopy is able to accurately measure electron affinities without scanning the wavelength of the laser, which is necessary in the competing technique of laser photodetachment threshold spectrometry.

Photodetachment microscopy also has the advantage of being a remarkably sensitive technique. Electrons produced at energies higher than those of the slow electron detached from the ground-state negative ion are effectively discriminated against. Such electrons could arise from parasitic sources or from internally excited ions. The slow electrons of interest concentrate in an area proportional to ε, which more than counteracts the tendency of the detachment cross section to decrease as $\varepsilon^{1/2}$ in the vicinity of an s-wave detachment threshold. Faster electrons get spread over a larger surface. The signal-to-noise ratio is sufficiently good that it is possible to detect the electrons detached from minor $^{17}O^-$ and $^{18}O^-$ isotopes and to measure the electron affinities of all their parent atoms, even though the $^{17}O^-$ beam current is not detectable, due to masking by a large unwanted $^{16}OH^-$ ion current.

Molecular photodetachment microscopy. The fact that all the photodetachment microscopy images for atomic negative ions or anions were similar encouraged researchers to extend the same imaging technique to a molecular anion, OH^-. A typical result is shown in **Fig. 4.** With molecular anions there is an additional level of complexity. They can rotate and store internal energy in integral multiples of a quantum of rotation, which is inversely proportional to the moment of inertia of the molecular system. The series of large rings in Fig. 4 corresponds to increases in the rotational quantum number associated with both the initial OH⁻ and the final OH molecule. Since the moments of inertia of these two systems are almost the same, the detachment threshold is nearly independent of the rotational state, and the rings, which are classical features, can merge together in a band head.

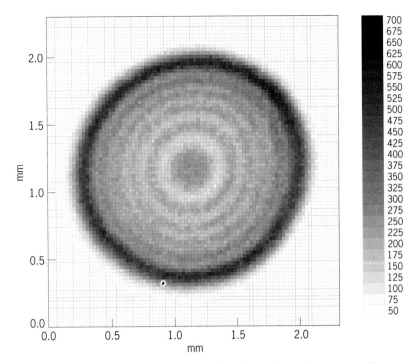

Fig. 3. Interference pattern obtained by photodetaching O⁻ ions in the presence of a 423-V/m electric field. By counting the observed interference rings, one can determine that the electron exited from the negative ion with a kinetic energy of 162 μeV. The intensity scale on the right indicates the number of electrons received per pixel (after 15 minutes of accumulation).

The small spot in the middle of Fig. 4 is analogous to the spot in Fig. 3. The two rings contained in the spot are quantum interference rings. Both are produced by electrons of the same detachment energy ε. The detachment energy is significantly smaller than

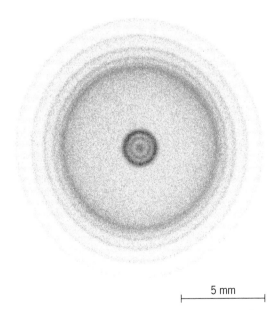

5 mm

Fig. 4. Photodetachment of OH⁻ ions in the presence of a 396-V/m electric field. The interferogram of interest is the central spot, which corresponds to the true detachment threshold of OH⁻ (transition from the ground state of OH⁻ to that of OH). The interference pattern appears quite similar to the atomic one. The external rings are not interference fringes, but classical rings corresponding to different energies associated with the rotational excitation of the molecule.

that corresponding to the outer rings due to the fact that the rotational quantum number increases by one unit during the detachment process. Remarkably, despite the rotational structure of the molecule and its polar character (it is not known whether this polar nature influences the motion of the outgoing electron), the interference pattern for molecular anions remains similar to that obtained using atomic anions. This similarity makes it possible to measure the electron affinity of OH.

Prospects. Photodetachment microscopy can be applied to many atomic anions whose detachment energies are either unknown or known with accuracies no better than a few millielectronvolts. It is expected that negative ions will continue to serve as unique atomic systems for testing short-range effects, either in collisions or in detachment processes, free of the overwhelming long-range influence of an ionic core. There are plans to investigate the influence of a magnetic field on photodetachment microscopy in the near future. Experiments involving crossed electric and magnetic fields can serve as tests of special relativity.

For background information *see* ELECTRON AFFINITY; MOLECULAR STRUCTURE AND SPECTRA; NEGATIVE ION; PHOTOIONIZATION; QUANTUM MECHANICS; SUPERPOSITION PRINCIPLE in the McGraw-Hill Encyclopedia of Science & Technology.   Christophe Blondel

Bibliography. T. Andersen, H. K. Haugen, and H. Hotop, Binding energies in atomic negative ions: III, *J. Phys. Chem. Ref. Data*, 28:1511–1533, 1999; C. Blondel, C. Delsart, and F. Dulieu, The photodetachment microscope, *Phys. Rev. Lett.*, 77:3755–3758, 1996; C. Bracher et al., Three-dimensional tunneling in quantum ballistic motion, *Amer. J. Phys.*, 66:38–48, 1998; S. J. Buckman and C. W. Clark, Atomic negative-ion resonances, *Rev. Mod. Phys.*, 66:539–655, 1994; Yu. N. Demkov, V. D. Kondratovich, and V. N. Ostrovskii, Interference of electrons resulting from the photoionization of an atom in an electric field, *JETP Lett.*, 34:403–405, 1981; H. Massey, *Negative Ions*, 3d ed., Cambridge University Press, 1976.

Plant pathogen genomics

Plants face enemies that cost the world's agricultural industry billions of dollars each year and threaten the secure food supply essential for social and economic progress. These enemies can include insects, animals, and other plants, but are usually microbial pathogens such as bacteria, fungi, oomycetes (funguslike organisms), and viruses. The danger that agricultural terrorists might seek to spread virulent plant pathogens lends an additional urgency to research on plant diseases. Over the last 20 years, scientists have increasingly turned to molecular biology to understand the complex biological mechanisms involved in plant pathogenesis and to develop novel strategies to help plants fight disease. Standard molecular approaches traditionally have focused on one or several of the genes involved in these mechanisms. The recent addition of genomics (a scientific method that deals with nearly complete sets of genes derived from the entire sequence of an organism's genetic material) to the toolkit of molecular biology has dramatically increased our understanding of how plants and pathogens interact.

Genomic analysis of plant pathogens. Genomic information about plant pathogens helps scientists understand the biology of the organisms, making it possible to develop novel biological, genetic, or chemical control methods. This information will also facilitate rapid and precise diagnosis of plant diseases through tests that can identify unique deoxyribonucleic acid (DNA) sequences from a particular pathogen. These tests can alert farmers to the presence of the pathogen early, before disease is apparent, so that measures can be taken to control the disease and prevent its spread to other fields.

Viruses. Hundreds of viruses are known to infect plants. Viral pathogens are unique because they consist essentially of only nucleic acids [DNA or ribonucleic acid (RNA)] and protein. They require a host cell for replication and cause disease because their rampant replication disrupts normal host developmental and metabolic functions. The DNA sequences of numerous plant-virus genomes have been determined since the 1980s. In comparison with living organisms, viral genomes are extremely small and simple, and may possess only a few genes (see **table**). These small genomes typically encode proteins for replicating the viral genome and for moving between adjacent plant cells. Cell-to-cell movement is critical for systemic infection by the viruses; hence, plant defenses focus on blocking this movement.

Bacteria. Bacteria are single-celled microorganisms that cause a variety of diseases in plants. Plant pathogenic bacteria manifest many types and growth characteristics, but genomic analyses have focused on those of greatest economic importance. The sequencing of the genomes of at least 16 diverse bacterial plant pathogens is underway; although only a fraction of the size of plant genomes, each bacterial genome likely encodes several thousand proteins (see table). With these sequence data and the development of large-scale approaches to discovering the function of genes (functional genomics), novel strategies have been devised to identify genes relevant to pathogenicity. For example, we know that many bacterial pathogens secrete their toxic proteins (known as effector or, historically, avirulence proteins) directly into the cells of the host plant. Recent studies of bacterial genomes suggest that plant pathogenic bacterial strains may secrete more than 30 different proteins into plant cells. In addition, these pathogenic bacteria possess more than 20 proteins that function either as toxins or as degraders of plant cell walls, and they possess up to 93 proteins that may serve to attach the bacteria to plant surfaces prior to infection. Identifying the bacterial genes encoding the secretory systems or the specific proteins targeted to the plant will improve our understanding of the mechanism of bacterial infection, and could lead to improved methods for controlling these pathogens.

Genome sizes of several plants and plant pathogens

Organism	Approximate average size*	Example	Example genome size in DNA bases	Number of genes
Virus	10,000	Tobacco mosaic virus	6,400 (made of RNA)	4
Bacteria	3,500,000	*Xanthomonas* (black rot)	5,100,000	2,700
Fungi	100,000,000	*Magnaporthe* (rice blast)	40,000,000	12,000
Plants	10,000,000,000	*Arabidopsis*	125,000,000	28,000
		Rice	450,000,000	50,000 (?)
		Corn	2,500,000,000	50,000 (?)

*Average size for each type of organism, indicated in base pairs of DNA, does not accurately indicate the wide variability that exists across diverse species; the genome sizes may vary from the average by up to a factor of 100.

An essential component of genomics is bioinformatics, which involves the use of computer programs to analyze an organism's DNA sequence and make predictions about its biology. Bioinformatic analyses can reveal information about evolution, gene function, and gene regulation. One use of bioinformatics is in comparative analyses of bacterial genomes; the comparisons reveal the sets of proteins that are shared across diverse bacterial strains and may be important for pathogenicity. Proteins encoded uniquely in one genome may be adapted for the specific host or mode of infection in that strain of bacteria.

Eukaryotic pathogens. Genomic studies of eukaryotic plant pathogens, which include fungi, the funguslike oomycetes, nematodes, insects, and even pathogenic plants like mistletoe, have progressed more slowly than those of bacteria and viruses, primarily because these are more complex organisms. The complexity of these organisms may correlate with genome size—the genomes and the number of genes in them may be 10 to 100 times larger than those in bacteria, and are often on the scale of those in the host (see table). One of the first eukaryotic plant-pathogen genomes to be extensively analyzed will be that of the rice blast fungal pathogen, *Magnaporthe grisea* (**Fig. 1**). Sequencing of this 40 million base-pair genome is underway, and preliminary analyses suggest that it contains more than 12,000 genes, many of which are likely to be necessary for pathogenesis. Among plant pathogens, the sequences for many actively used genes (expressed genes) are available for at least seven fungi or oomycetes and for at least seven species of nematodes. Whole-genome sequencing projects are currently being initiated for several of these organisms, including the oomycete *Phytophthora infestans*, the cause of Ireland's 1845–1846 late-blight potato epidemic.

While the general themes involved in eukaryotic versus bacterial infection of plants are likely to be similar (for example, recognition and attachment to a host cell, degradation of the cell wall, secretion of toxins or infection-related effector or avirulence proteins, and subsequent colonization of the host tissue), the specific mechanisms and proteins are likely to be substantially different. Because little is known so far of the molecular details of eukaryotic pathogenesis in plants, further study should yield exciting biological discoveries.

Genomic analysis of plant defense and immune systems. Because plants are sessile organisms (fixed in place), they cannot evade pathogens. And unlike animals, plants have no circulatory system, so defense molecules produced in one part of the plant cannot be easily transported to another part to fight an infection. Therefore, plants must possess a wide range of mechanisms and responses to identify and repel potential parasites, and each plant cell must be individually protected. The defense may be prepared in advance; for instance, the plant may produce antimicrobial compounds and a waxy cuticle to set up a passive barrier to infection. Or the defense may be reactive; that is, the plant may produce detoxifying enzymes after a pathogen attack or block the spread of the pathogen by introducing barriers of dead or reinforced cellular material.

Most studies indicate that the defense mechanisms of diverse plant species are similar. Research has therefore focused on the model plant *Arabidopsis* (**Fig. 2**), which is easily studied in the laboratory. The recent availability of whole-genome sequences for *Arabidopsis* and for rice has led to a better

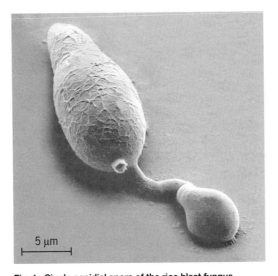

5 µm

Fig. 1. Single conidial spore of the rice blast fungus *Magnaporthe grisea* forming an appressorium (foreground) at the site of contact with the surface. In this image, the surface is the synthetic material Mylar. The appressorium is used to penetrate the plant host tissue. *Magnaporthe grisea* is the first eukaryotic plant pathogen for which the genome has been completely sequenced. (*Courtesy of Richard J. Howard*)

Fig. 2. The genome of *Arabidopsis thaliana* was completely sequenced in 2000. The compact genome, relative to many crop plant genomes, and the small size of the plant make it an ideal laboratory model system for plant molecular biology research.

understanding of the nature and complexity of plant defense systems. The availability of the genome sequences for *Arabidopsis* and rice has also accelerated one of the most time-consuming aspects of genetic research: the molecular identification of the genes that cause a specific characteristic (such as resistance to a pathogen).

R-proteins. Molecular genetic experiments over the last 10 years have demonstrated that plants employ several overlapping sets of defense systems. Research has focused on genes used for the last century by traditional plant breeders to confer resistance; these produce resistance proteins (R-proteins, encoded by R-genes) which are usually part of a switch that turns on a complicated defense system involving hundreds of proteins, small molecules, and cellular and physiological changes. The R-proteins recognize specific pathogens and trigger plant responses and are now known to be receptorlike proteins with similarities to animal and other plant proteins. The function of the related animal proteins is to bind specific molecules and then commence signaling. The plant R-proteins may bind secreted pathogen proteins to signal pathogen presence and activate plant defenses. The result of the recognition is localized cell death in the plant tissue, observed as a spot of dead tissue, called the hypersensitive response (HR). The HR surrounds and isolates pathogens with dead tissue, effectively blocking systemic infections.

NBS-LRR proteins. Genomic studies using the available plant whole-genome sequences have described the total number of plant genes that may constitute the plant immune system. Analyses of the nearly complete *Arabidopsis* and rice sequences have created new opportunities for experimental

work on these genes and the proteins that they encode. Most plant R-proteins have been found to be predicted nucleotide binding site–leucine rich repeat proteins (NBS-LRR proteins, named after the subunits of the protein), which function to detect a wide range of pathogens, including viral, bacterial, fungal, oomycete, and animal pathogens, using widely divergent methods of infection. More than 150 genes encoding NBS-LRR proteins have been found in the *Arabidopsis* genome (out of a total of some 28,000 genes) and may play direct roles in pathogen recognition, with a large number of additional proteins playing secondary roles. In contrast, the rice genome may encode more than 500 NBS-LRR proteins potentially involved in pathogen recognition (out of some 50,000 genes). While other types of proteins can function as R-proteins, they are not expected to be as prevalent as NBS-LRRs. Other plant genomes likely contain a similar proportion of R-proteins to rice and *Arabidopsis*. Genomic information, however, is more than just a catalog of proteins; it also allows scientists to investigate changes in the activity and regulation of plant genes when the plant is attacked. Such studies of transcriptional changes as well as of signaling dynamics and pathways are currently underway.

Conclusion. Genomic approaches to plant-pathogen interactions hold great promise for understanding host-pathogen interactions and the molecular basis of recognition, infection, and defense. This knowledge is likely to lead to novel tactics for control of pathogens, including the enhancement of genetically based plant defenses and the development of more effective, yet environmentally benign chemical treatments. The recent availability of bacterial and plant genomic sequences has advanced the field of plant pathology by revealing the full complement of proteins involved on both sides of the interaction and by providing clues about the functions of these proteins. Genomics is only one of many experimental tools, but the use of genomics data in genetic and biochemical experiments designed to test specific scientific hypotheses will help protect plant agricultural products from natural enemies.

For background information *see* AGRICULTURAL SCIENCE (PLANT); BACTERIAL GENETICS; FUNGAL GENETICS; GENETIC MAPPING; PATHOTOXIN; PLANT PATHOLOGY; PLANT VIRUSES AND VIROIDS in the McGraw-Hill Encyclopedia of Science & Technology.

Blake C. Meyers

Bibliography. J. L. Dangl and J. D. Jones, Plant pathogens and integrated defence responses to infection, *Nature*, 411(6839):826–833, 2001; J. Glazebrook, E. E. Rogers, and F. M. Ausubel, Use of Arabidopsis for genetic dissection of plant defense responses, *Annu. Rev. Genet.*, 31:547–569, 1997; R. W. Michelmore and B. C. Meyers, Clusters of resistance genes in plants evolve by divergent selection and a birth-and-death process, *Genome Res.*, 8(11):1113–1130, 1998; M. A. Van Sluys et al., Comparative genomic analysis of plant-associated bacteria, *Annu. Rev. Phytopathol.*, 40:169–189, 2002.

Plant vulnerability to climate change

The distribution and physiology of plants are constantly changing over different time scales as the plants are affected by changes and fluctuations in the environment. Recently, climate change has received much attention, and there is increasing evidence that it is already having an effect on some species, not only on their distribution but also on the timing of key events, such as leafing and flowering. With global temperatures predicted to increase by 1.4–5.8°C (2.5–10.4°F) over the period 1990–2100, and global precipitation predicted to increase (but with regional variation), it is important to assess the implications of such changes and to identify which ecosystems and plants might be particularly vulnerable.

Vulnerability to climate change is the degree to which a system is susceptible to, or unable to cope with, such adverse effects. It is a function of the character, magnitude, and rate of climate change to which the system is exposed, the system's sensitivity to that change, and its ability to adapt to the changes (see **table**). This vulnerability will be influenced by both the physiology of the species and the geography of the ecosystem and species.

Direct impacts of climate change. The vulnerability of species will depend on their sensitivity to the changes in temperature and precipitation, which affects their metabolism and functioning. If the changed conditions are well within the tolerance range of the species, the species should not be directly vulnerable to climate change. But if conditions shift the species toward its limit for one of these variables, it may start to experience indirect effects, such as becoming less productive or competitive. This could lead to changes in the abundance of the species within the community, which could affect the community's structure and functioning. In the western United states shortgrass steppe (prairie), for example, blue grama (*Bouteloua gracilis*) could decrease in productivity as spring temperatures warm, allowing exotic and native forbs (broadleaved, non-woody plants, not grasses) to increase and making the ecosystem less tolerant of drought and grazing.

If, however, the new conditions take a species outside its tolerance range, individuals will start to die. This is already seen in coral reefs during El Niño years, when the symbiotic algae die if there is a greater than 1°C (1.8°F) increase in sea surface temperatures over the seasonal maximum. This mortality response, like the temperature increase, is usually temporary, but with prolonged or higher periods of raised sea surface temperatures the entire coral ecosystem will be threatened.

Populations in the warmest part of a species' range (for example, southern populations in the Northern Hemisphere) are likely to be the most vulnerable to climate change, as they are nearest to their upper thermal limit, and local extinctions are likely. This is occurring in the Edith's checkerspot butterfly (*Euphydryas editha*) in western North America. At the same time, northern populations of species may start to expand and the distribution of a species will move poleward. Vulnerability also will occur where there is little or no overlap between present and potential future distributions; that is, little or none of its current distribution will be suitable under future climate change and thus dispersal will be important for the species' survival. The New Zealand kauri tree (*Agathis australis*), for example, could lack such an overlap and the ability to reach its new areas of suitable climate space.

Latitude and altitude. Geographically, ecosystems and species at high latitudes and altitudes will be particularly at risk as they run out of space for poleward or altitudinal movement. In Arctic regions, for example, the increased temperatures mean that many species, particularly mammals such as polar bears

Vulnerable ecosystems	
Ecosystem	Reason for climate change vulnerability[*]
Terrestrial	
Polar (for example, tundra)	Great changes in temperature and precipitation (C), lack of opportunity for poleward migration (A)
High mountain	Lack of opportunity for altitudinal migration in response to temperature increases (A)
Islands	Sea-level rise [on small islands] (C), barriers to dispersal (A)
Wetlands	Increased drought (C)
Karoo–S. Africa	Increased frequency of drought (C), changing fire regimes (C), loss of specialist pollinators (A)
Cape Floral Kingdom (Fynbos)–S. Africa	Increased frequency of drought (C), changing fire regimes (C), lack of space for altitudinal or latitudinal migration (A)
Coastal	
Mangroves	Sea-level rise (C), changing sediment flux (C), lack of opportunity for inland/poleward migration (A)
Sea grass beds	Sea-level rise (C), changing sediment flux (C)
Coral reefs	Temperature increases (C), CO_2 (C)

[*] (C) climate change–related factor, (A) adaptation–related factor.

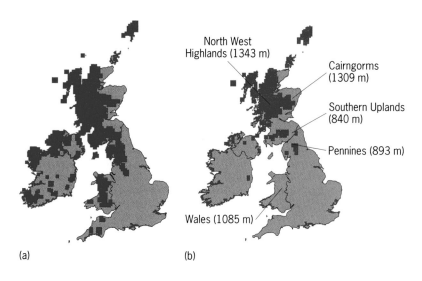

(a) (b)

Modeled distribution of suitable climate space for Dwarf willow (*Salix herbacea*), an Arctic-alpine plant, in Britain and Ireland (*a*) under current climate (1961–1990) and (*b*) under the UKCIP98 2050s High scenario, with maximum altitude in the upland areas in dicated. The Dwarf willow could become vulnerable in Wales and in the southern part of its range, through loss of climate space, and have to move to higher altitudes elsewhere. (*Adapted from P. A. Harrison, P. M. Berry, and T. P. Dawson, eds., Climate Change and Nature Conservation in Britain and Ireland: Modelling Natural Resource Responses to Climate Change (The MONARCH project), UKCIP Tech. Rep. Oxford*)

and walrus, will be negatively affected as the tundra and ice cover contract. For plants, the impact will depend on their physiological response to the warming, including the associated increased nutrient supply. The latter, for example, could adversely affect mosses, lichens, and evergreen shrubs, while warming in the winter and spring may encourage premature growth, so subsequent frost could lead to damage. In upland Britain, as in mountain ecosystems elsewhere, the lack of suitable climate space at higher elevations will make Arctic-alpine species vulnerable. The **illustration** shows the potential retreat of such a species poleward and to higher altitudes.

Barriers. Species that need to move poleward to adjust to climate change could be vulnerable if they encounter barriers to their movement, such as mountains or oceans. One concern is the Cape Floral Kingdom, South Africa, which is an area of high species richness and where there is little opportunity for such movement, except into intensively managed farmland on the tip of South Africa. Island ecosystems could be similarly threatened.

Dispersal ability. In order for ecosystems or species to adjust to changes in suitable climate space and fulfil their new potential range, dispersal will be required. This may be difficult if the species have restricted distributions, have small populations, are slow-growing, or are of limited dispersal capability. Many species restricted to a specified region or locality (endemic species) meet the first two constraints. In Costa Rica, for example, which has a high level of biodiversity, the montane (mountainous) forests are very sensitive to changes in precipitation and under some climate change scenarios certain zones may disappear. These forests are often associated with high levels of

endemism, and thus their associated species would be very vulnerable. Similarly, the Lake Baikal ecosystem, which contains many endemic species, is sensitive to changing patterns of ice formation and, given its isolated location, is highly vulnerable.

Forest ecosystems will have a slow response time to climate change because of the longevity of trees, their dominant species. This could mean that fast-growing, early-successional species will replace late-successional species in old-growth forest, if the latter start to die because their climate tolerance is exceeded. Species with limited dispersal capacity, especially those relying on clonal (asexual) reproduction, such as mountain avens (*Dryas octapetala*) and the sedge *Carex bigelowii*, could also be vulnerable. In the Arctic, although many species currently reproduce clonally because of the environmental limitations, with increased temperatures some may be able to produce flowers and set seed more frequently in the future.

Timing and interactions. The vulnerability of plants and ecosystems could also be affected by the disruption of the synchrony in the timing of life-cycle events (phenology) or of species' interactions. For example, it is thought that warmer temperatures and dry summers could lead to an increased density of eastern spruce budworm moth in the northern spruce and fir forests of the eastern United States and Canada. In certain areas, these climatic changes may mean that the moth is no longer affected by some of its parasitoid and bird predators, such as wood warblers, which will become overwhelmed if moth numbers become too great. If this were the case, the forests would become vulnerable to attack. This example indicates some of the biological complexities of identifying vulnerable species and ecosystems. Generally, it is thought that the stresses produced by climate change may make species more susceptible to pathogens. Similarly, many species have specialist relationships with other organisms, and as these relationships are inherently vulnerable it is possible that they could be disrupted by climate change. If a fig tree (*Ficus* spp.) which had a mutually dependent relationship with a specific fig wasp were to respond differently to climate change than its associated wasp, it could lose its pollinator or the wasp its fruit for egg laying.

Indirect impacts of climate change. Vulnerability can also occur because of the direct effects of rising carbon dioxide (CO_2) levels, which are the main cause of climate change, or indirect effects of climate change such as rising sea level. The latter could particularly affect coastal wetlands, such as salt marshes and mangroves, especially where the presence of sea defenses or human habitation make migration inland unfeasible. The Sundarbans of Bangladesh and India and the Port Royal mangrove wetlands in Jamaica are thought to be particularly vulnerable.

Other impacts on vulnerability. Identifying the possible vulnerability of ecosystems and plants is difficult as there are various physical drivers of environmental change and biological interactions, as well as other factors, such as human land use/land cover change and habitat destruction which will impact upon vulnerability. It is thought that in the short term (1–10 years) these factors will have a more significant impact on the distribution and survival of species than climate change. Biodiversity hot spots, for example, are areas that have high species numbers, including many endemic species, but many also are experiencing large habitat losses. The Mediterranean and savanna hot spots are thought to be particularly vulnerable to climate change, increasing CO_2, and habitat losses. *See* BIODIVERSITY; HOT SPOTS.

Ecosystems and plants are already vulnerable to decreases in their abundance and range, which could lead to extinction, because of human activities. In the short term, these other pressures may have a greater local impact on vulnerability, but climate change will contribute to longer-term stresses on plants and ecosystems and lead to changes in their distribution and abundance. Climate change is not an isolated factor, and an integrated approach is needed in order to understand ecosystem and plant vulnerability.

For background information *see* BIOGEOGRAPHY; CLIMATE MODIFICATION; ECOSYSTEM; EL NIÑO; FOREST ECOSYSTEM; GLOBAL CLIMATE CHANGE; GRASSLAND ECOSYSTEM; PLANT GEOGRAPHY in the McGraw-Hill Encyclopedia of Science & Technology.

Pam Berry

Bibliography. F. S. Chapin III, O. E. Sala, and E. Huber-Sannwald (eds.), *Global Biodiversity in a Changing Environment*, Springer, New York, 2001; M. R. Malcolm et al., Habitats at risk: Global warming and species loss in globally significant terrestrial ecosystems, *IPCC Climate Change and Biodiversity*, IPCC Tech. Pap. ed. by V. Gitay et al., World Wildlife Fund for Nature, Gland, Switzerland, 2002; *IPCC Climate Change 2001: Impacts, Adaptation and Vulnerability—Summary for Policymakers*, Intergovernmental Panel on Climate Change, Cambridge University Press, Cambridge, 2001.

Pluto

Pluto has been an intriguing object ever since its discovery in 1930, lying in relative darkness nearly 31 astronomical units (at present) from the Sun. It is difficult to study because of its great distance from the Earth and the Sun; however, observational efforts since the 1970s have revealed it to be an interesting place. In 1978 Pluto was discovered to have a satellite, Charon, about half its diameter, which orbits Pluto with a period of 6.37 days—the same as Pluto's rotation period. Spectral measurements reveal Pluto to have surface ices of methane (CH_4), carbon monoxide (CO), and nitrogen (N_2), and the gaseous phase of these molecules forms a tenuous atmosphere that stretches at least 100 km (60 mi) above the planet's surface.

Pluto resides in a special mean-motion resonance with Neptune that precludes the collision of the two planets, despite the fact that it comes closer to the

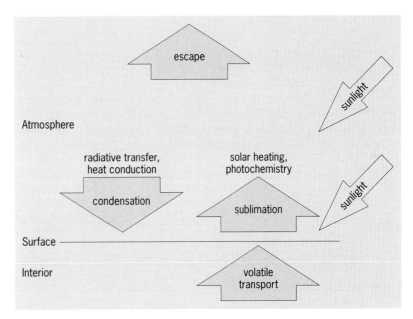

Fig. 1. Processes in Pluto's atmosphere. The horizontal line represents the body's surface. The arrows indicate the source of the effect and the region where the process is at work, in the interior of Pluto or within its atmosphere. (*After J. L. Elliot and S. D. Kern, Pluto: Pluto' atmosphere and a target occultation search for other bound KBO atmospheres, Earth, Moon, and Planets, 2003*)

Sun than Neptune for a brief time during its 248-year orbit. Pluto's average distance from the Sun is 39.4 astromical units, where it is in the midst of a large population of small, icy bodies discovered in 1992 and termed Kuiper Belt objects. Pluto is perhaps the "king" of this region, having enough gravity to pull itself into a spherical shape and to hold an atmosphere. Charon is not known to have an atmosphere at this time, but the option has not been ruled out. Spectral observations of Charon reveal a clear signature of water ice (H_2O) but show no evidence for CH_4. Measurements continue to be made of both Pluto and Charon to better understand what happens as they move away from the Sun.

Atmospheric processes. Pluto is a cold, icy body having a density near 2.0 g/cm^3 with an interior of rock and ice. Its atmosphere is derived from surface ices, and a number of processes are at work there (**Fig. 1**). Within the atmosphere, sunlight is absorbed by CH_4 molecules, and their resulting heat is conducted to the surface. Absorbed sunlight also drives certain photochemical reactions that create a wide variety of molecular products. Ices on the surface sublimate into the atmosphere when hit by sunlight. Areas of ice that are not in sunlight are cooler and provide locations where frost can condense onto the surface. Condensation releases heat that warms the ice in shaded regions. Thus, the processes of sublimation and condensation act as a thermostat to keep the surface-ice temperature nearly the same all around the body. These processes also transport N_2 frost (via the atmosphere) from the equatorial to the polar regions. At the top of the atmosphere, some molecules have enough energy to overcome Pluto's gravity and escape into space. These molecules are replenished by a net sublimation of ice. Over Pluto's

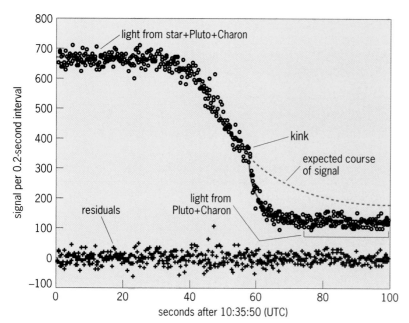

Fig. 2. Light signature from a star occulted by Pluto observed from the Kuiper Airborne Observatory (KAO) in 1988. The signal gradually fades as the star is occulted by Pluto's atmosphere; the reappearance of the star is not shown. The expected signal, based on a clear atmosphere of constant temperature with no haze layer, is also plotted. The residuals are the differences between the data and the signal that would be expected from a clear atmosphere of constant temperature above a haze layer, which may be responsible for the kink in the light signature. The fluctuations in the light signature and the residuals are due to the noise in the data. (*After J. L. Elliot et al., Pluto's atmosphere, Icarus, 77:148–170, 1989*)

lifetime it has been estimated that about 3 km (2 mi) of surface ice has been lost in this manner, but the lost ice is replenished by the transport of ices from the subsurface.

Occultations. Pluto and other solar system bodies are observed with several techniques, including imaging, spectroscopy, and occultations. An occultation occurs when a planet's path passes between the observer and a star, blocking the starlight from view. The light from the star diminishes, and the signature of this decrease reveals properties of the occulting body's atmosphere (if it has one). To observe an occultation, it is necessary to be in the correct location on Earth when the body's shadow, cast in starlight, passes over. This is similar to the circumstances surrounding a solar eclipse, in which the shadow of the Moon, cast in sunlight, passes over a limited portion of the Earth.

Recent occultations. Because Pluto's shadow covers only about one-fifth of the Earth's diameter, coverage with fixed telescopes is not always possible. Furthermore, the telescope must be in night when the event occurs, with the Moon not too bright or close to Pluto. Weather is also a critical factor—clouds can ruin the best opportunities.

A Pluto occultation was observed in 1988. The light failed to drop abruptly, instead it faded gradually (**Fig. 2**)—the telltale sign of an atmosphere. The slow drop was due to the refraction of starlight in Pluto's atmosphere as Pluto "moved" in front of the star. However, instead of continuing its expected course, the signal dropped unexpectedly to zero starlight, leaving only the light from Pluto and Charon. This abrupt change in the slope of the light curve has been called the "kink," and indicated that there was something interesting going on in this region. Unfortunately, no other occultations were observed for over 14 years.

Two prospects finally presented themselves in July and August 2002. **Figure 3** shows the occultation paths for the stars P126A and P131.1. Since what can be learned from an occultation depends on the

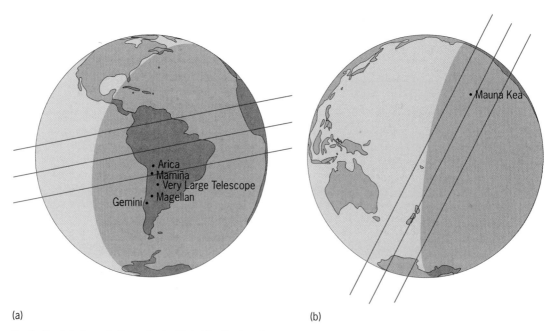

(a)   (b)

Fig. 3. Predicted occultation paths for Pluto. The Earth is shown as viewed from the direction of Pluto, and the shaded areas are the parts of the Earth that are in night. The straight lines are the boundaries and centerlines of the occultation paths. Locations of telescopes are also shown. (*a*) Occultation of the star P126A on July 20, 2002 (*after* http://occult.mit.edu/research/occultations/Candidates/Predictions/P126.html). (*b*) Occultation of the star P131.1 on August 21, 2002 (*after* http://occult.mit.edu/research/occultations/Candidates/Predictions/P131.1.html).

amount of light that can be collected, it was desired that the largest telescopes in the shadow path be used. This would give the best opportunity to answer questions posed from the results of the 1988 occultation.

The star in July 2002 was bright ($R = 12.3$, where R refers to the magnitude of a celestial object at red wavelengths on a logarithmic scale), but the Moon was also nearly full and close to Pluto in the sky, which made the observations difficult. In addition, the path of the occultation proved to be slightly north of the large telescopes in Chile (Gemini, Magellan, and the Very Large Telescope) that were hoping to observe the event (but that were clouded out in any case). Measurements were carried out with small portable telescopes near the towns of Mamiña and Arica in northern Chile. These measurements indicated that the kink in the original light curve (Fig. 2) had changed or gone away, but additional data would be required for a definitive result.

The star P131.1, occulted by Pluto in August 2002, was about 25 times fainter ($R = 15.7$) than the one occulted in July, but the occultation path was perfectly placed for measurements by many large telescopes on Mauna Kea in Hawaii. The event was also visible from telescopes in the extreme western United States. Ten individual data sets were obtained for analysis. **Figure 4** images Pluto and Charon as they passed in front of this star.

Two alternative scenarios for Pluto's atmosphere had earlier been proposed to explain the kink. One suggested that the kink was caused by an atmospheric haze, and the other suggested that it was caused by a sharp temperature drop in the atmosphere. The points across the bottom of Fig. 2, labeled residuals, are the differences between the data and the signal that would be expected from a clear atmosphere of constant temperature above a haze layer, the first of these scenarios. The clustering of the residuals around the value 0 demonstrated that the data followed this model fairly closely except for the effect of noise, but although this agreement was consistent with the first scenario it did not exclude the second. In order to determine which interpretation is correct, observations of an occultation with measurements at two different wavelengths of light (visible and infrared) could provide insight. The observations made in summer 2002 were targeted to test the models proposed to explain the 1988 atmospheric measurement by taking observations with multiple telescopes using instruments in both the visible and infrared.

Results of 2002 occultations. In addition to being more extensive, the August observations proved to be of much higher quality than those secured in July. The new data confirmed that the kink seen in the 1988 measurement has greatly diminished or gone away completely. They also showed that Pluto's atmosphere has expanded, with the pressure increasing by a factor of 2 since 1988. This result is perhaps opposite what many expected. Since Pluto has been moving away from the Sun since 1989, causing the incident solar illumination to decrease, many antic-

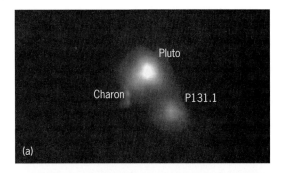

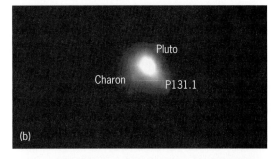

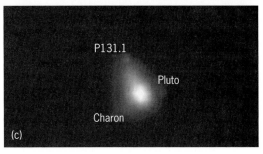

Fig. 4. Images of Pluto-Charon and the star P131.1.
(a) 32 min before, (b) 24 min before, and (c) 41 min after the occultation on August 21, 2002. (*Williams College Occultation Expedition*)

ipated that Pluto's atmosphere would collapse onto its surface. However, distance is not the only factor that determines how much sunlight is absorbed by a planet's surface. A dark surface absorbs more sunlight and heats up more than a brighter surface. Hence, an alternative explanation for Pluto's warming is that its surface has become darker in recent years. This view is consistent with brightness measurements over the past decade, which show this to be the case, although other effects may also be contributing to the surface warming.

Preliminary analysis of the multiple-wavelength observations indicates that the haze model is more likely than the sharp thermal-gradient interpretation to be the correct explanation of the 1988 measurements. However, some ambiguities remain, and additional measurements are needed to properly interpret this result.

Future observations. The next occultation opportunity for Pluto has not been specifically identified, but several should occur within the next few years, since Pluto is now moving so that its position is projected against the star-rich central bulge of the Milky Way Galaxy. The new Stratospheric Observatory for Infrared Astronomy (SOFIA), a 747-SP aircraft containing a 2.5-m (98-in.) telescope which is expected to go into service in 2005, can be used to observe these occultations, even if they occur over water and/or remote locations on Earth. Observers can fly directly to the location of interest and get above whatever weather might cause problems for telescopes on the surface. SOFIA is being fitted with two instruments, HIPO and FLITECAM, that have high time resolution and allow for simultaneous measurements in the visible and infrared.

In addition, a NASA New Horizons mission will fly by the Pluto-Charon system in 2015 (if it is launched in 2006 as presently scheduled). The primary mission goals include mapping the surface appearance of these two bodies, studying the surface composition with spectra in the near-infrared, and probing the atmosphere with ultraviolet spectrometers. An occultation of the spacecraft's radio signal will probe the atmospheric structure, and the mission seeks to measure the interaction of Pluto's atmosphere with the solar wind. From these data a more comprehensive picture will be acquired for the Pluto-Charon system.

For background information *see* ASTRONOMICAL OBSERVATORY; ECLIPSE; KUIPER BELT; OCCULTATION; PLUTO; TELESPCOPE in the McGraw-Hill Encyclopedia of Science & Technology.   James Elliot; Susan Kern

Bibliography. J. L. Elliot and C. B. Olkin, Probing planetary atmospheres with stellar occultations, in G. W. Wetherill et al. (eds.), *Annu. Rev. Earth Planetary Sci.*, 24:89–123, 1996; J. L. Elliot, M. J. Person, and S. Qu, Analysis of stellar occultation data, II. Inversion, with application to Pluto and Triton, *Astron. J.*, vol. 126, 2003; S. A. Stern and J. Mitton, *Pluto and Charon: Ice Worlds on the Ragged Edge of the Solar System*, Wiley, 1997; S. A. Stern and D. J. Tholen (eds.), *Pluto and Charon*, University of Arizona Press, Tucson, 1997.

Polyelectrolyte multilayers

Polyelectrolytes are charged, usually water-soluble polymers which may be handled and processed at ambient conditions using materials of very low toxicity. It has been known for many years that solutions of oppositely charged polymers form complexes through multiple electrostatic interactions. Although many applications have been proposed for these complexes, few have been realized since the complexes are intractable once formed. A technique for producing polyelectrolyte complexes as uniform ultrathin films, some with internal structuring, was described in the early 1990s. Since then, numerous morphologies, compositions, and applications derived from complexed multicomposite (many different materials) films have been explored.

Layer-by-layer assembly. Thin films of charged organic materials are conveniently prepared on a variety of substrates by exposing the surfaces in an alternating sequence to oppositely charged materials (**Fig. 1**). The film is built in a layer-by-layer

fashion, with nanometer thickness control (**Fig. 2**). On each adsorption step, material of the opposite charge interacts with the surface, reversing the surface charge and priming it for the next step. The multilayering of polyelectrolytes in this way represents a general approach to film formation, where any charged, well-dispersed species (organic or inorganic) may be used. Thus, molecular and nanoparticulate materials may be combined to yield rugged, yet versatile thin films. Although electrostatic forces are the interaction of choice when building multilayers, it is also possible to rely on weaker forces, such as hydrogen bonding, charge transfer, and hydrophobic effects—multiple interaction points that work cooperatively to enhance the overall binding energy. The use of aqueous solutions (synthetic or natural polyelectrolytes) or suspensions (colloidal charged particles) is a very attractive feature of this technology, making it amenable to biological materials and applications.

Multilayer components may be applied from dilute solutions by alternate immersion, spraying, or flushing (for internal coating of tubes or channels). A solvent rinse (usually water) removes the excess material. If adsorption is performed with this rinse, the amount of material adhering to a surface is self-limiting. Variables in the multilayer process include polymer or particle composition, solvent, exposure time, concentration, temperature, pH (for weak acids/bases), and molecular weight. Most multilayers are made at ambient conditions with aqueous solutions of high-molecular-weight polymers. The addition of a low-molecular-weight salt (in most cases NaCl) controls the layer thickness for all-polymer multilayers. Salt swells the multilayer, allowing more polymer to penetrate. If particles are used, parameters such as pH, salt concentration, and particle concentration must be more specifically refined to promote particle adhesion. Particle deposition tolerates a narrower range of pH and salt concentration compared with polymer multilayering because particle surface charge tends to be of the weak acid type and the mixing is less intimate.

Multilayers may be grown on virtually any surface that is relatively clean, and since the amount of material adsorbed is self-limiting, there are no restrictions on the size or shape of article to be coated—every surface contacted by the coating solution grows a uniform film. Particles are almost as straightforward to coat as are planar surfaces. In many cases, the particle on which the multilayer is deposited may be dissolved or degraded, leaving a continuous hollow shell or capsule (template) suitable for transporting and releasing various chemical species. This templating approach has been used for biological particles, such as blood cells. An enormous palette of materials suitable for multilayering leads to a notion of synthesis on the supramolecular scale that presents broad opportunities for creativity on the molecular and nanocomposite level. For polyelectrolytes, extensive interaction between components allows for mixing or blending on a molecular level. Molecular mixing gives materials that are uniform in composition at the nanometer level.

Properties. The binding of polymer segments is cooperative, up to a point, so the overall interaction energy between molecules is strong and complexes are essentially irreversible. Multilayers are nonequilibrium structures, with reversible rearrangements permitted at the very local level (as in swelling). Irreversibility is an essential prerequisite to formation, since dispersed quasisoluble particles of the complex are thermodynamically more favored. Because of the "hit and stick" propensities for polyelectrolyte adsorption, they are easy to form but still remain somewhat permeable. In addition, molecular conformations frozen by nonequilibrium conditions make it possible to obtain layered materials.

Some properties of the multilayers reflect those of the starting materials, and others are unique to the complexes formed. For example, individual polyelectrolytes are water-soluble and contract on addition of salt, but are insoluble and swell in salt when complexed. In all-polymer polyelectrolyte multilayers (PEMUs), neutron or x-ray analysis methods show the films to be locally amorphous due to intimate molecular mixing or interpenetration of polyelectrolytes from adjacent layers, but structure is observed if the alternating materials are separated by several layers of a different material. Additional structural hierarchy is introduced if rigid, inorganic particles participate in the multilayers. Clay minerals, for example, serve as effective barriers for isolating individual layers. And in the final film, incorporated nanoparticles will confer the particular property (structural, magnetic, or optical) for which they have been designed.

Bulk properties should be differentiated from surface or interfacial properties. Reversible swelling of multilayers is influenced by salt concentration or pH. All multilayers contain water under ambient conditions, favoring those applications exploiting their soft and compliant nature but confounding some electronics applications. Certain properties emphasize the nonequilibrium conditions under which multilayers have been formed, such as the preservation of bulk noncentrosymmetric orientation of non-linearly optically active polymers.

Surface properties of PEMUs are dictated by the charge and hydrophobicity of the final layer. For example, a multilayer terminated by a positive polyelectrolyte will impart a net positive charge to the surface, encouraging the adsorption of negative species (small and macromolecular) to the surface. A multilayer terminated with a hydrophilic polymer will be wetted more efficiently by water. Various related hydrodynamic phenomena, such as electroosmotic flow, electrophoretic mobility, and streaming potential, depend on the polarity and magnitude of the surface charge. The broad range of polarity available from the constituent polyelectrolytes facilitates the design of compatible interfaces between dissimilar materials, such as polymer and glass, or polymer and metal.

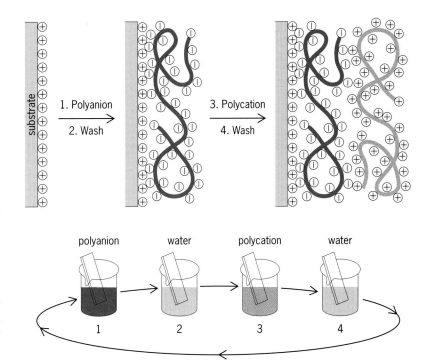

Fig. 1. Layer-by-layer deposition process. (*Courtesy of Gero Decher*)

Applications. Because the composition and properties of multilayered composites may be tuned to such a high degree, the suite of potential applications is very diverse. Some early PEMUs showed potential as antistatic coatings when one of the polyelectrolytes used was an electrically conducting conjugated (alternating carbon-carbon double and single bonds) polymer. Undoped conjugated polymers were also built into multilayers for organic light-emitting diodes. Such electronics applications are affected by the multilayers' water and salt content, which tend to degrade electrical contacts under applied electric fields.

More recent and promising applications include PEMUs for interfacing "hard" systems to aqueous environments. For example, chemical responsiveness may be built into thin films for biological applications. Extensively swollen by water, multilayers are gentle and effective hosts for incorporating enzymatically active molecules. In bioreactor applications,

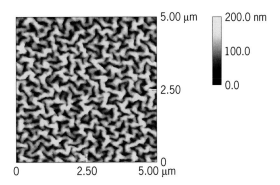

Fig. 2. Image taken with an atomic force microscope, revealing the surface morphology of a multilayer approximately 200 nm thick.

active enzymes incorporated in multilayer films have an additional benefit: they are protected from proteases and other molecules that degrade them.

Because they are so thin, multilayers exhibit very high permeation rates when used for membrane filtration. Membrane or analytical separation of charged species, and even optically active isomers, has been demonstrated. Programming of multilayer permeability under the influence of pH changes has proven useful in controlled drug release, especially with templated capsules. Separation of gaseous mixtures by pervaporation (the species being separated is in the vapor phase on at least one side of the membrane) through PEMUs has yielded separation factors on the order of 10^4. Permeability to ions facilitates the color-changing dynamics of multilayers used as electrochromic displays and in sensors.

Chemically induced instability (for example, by changing pH) of PEMUs has been used in soft lithography to pattern surfaces or to separate multiple membranes. In some cases, an internal change in the charge density causes phase separation into microporous material of controlled (reduced) refractive index, which is useful for antireflection coatings.

PEMUs are well suited to controlling interfacial energies and interactions, where acceptable performance is observed even from the thinnest of films. PEMUs offer much promise in the biomedical field for controlling protein adsorption and cell adhesion. For example, they have been used to advantage in coating articles to be implanted in vivo, and to direct cell growth modes and directions on patterned surfaces. Wetting and lubrication of PEMU surfaces depends strongly on the nature of the "top" layer. A hydrophilic polyelectrolyte promotes efficient wetting. Contact lenses coated with PEMUs were the first commercial product employing multilayers to be manufactured on a large scale. It is envisioned that future applications of PEMUs will emphasize bioengineering applications, as well as polymer/inorganic composites, which are currently limited by the lack of large-scale material sources and coating facilities.

For background information *see* ADSORPTION; COLLOID; ELECTROSTATICS; INTERMOLECULAR FORCES; MONOMOLECULAR FILM; SUPRAMOLECULAR CHEMISTRY in the McGraw-Hill Encyclopedia of Science & Technology.   Joseph B. Schlenoff

Bibliography. P. Bertrand et al., Ultrathin polymer coatings by complexation of polyelectrolytes at interfaces: Suitable materials, structure and properties, *Macromol. Rapid Commun.*, 21:319–348; G. Decher, Fuzzy nanoassemblies: Toward layered polymeric multicomposites, *Science*, 277:1232–1237, 1997; G. Decher and J. B. Schlenoff (eds.), *Multilayer Thin Films: Sequential Assembly of Nanocomposite Materials*, Wiley-VCH, Weinheim, 2003; T. R. Farhat and J. B. Schlenoff, Doping controlled ion diffusion in polyelectrolyte multilayers: Mass transport in reluctant exchangers, *J. Amer. Chem. Soc.*, 125:4627–4636, 2003; R. K. Iler, Multilayers of colloidal particles, *J. Colloid Interface Sci.*, 21:569–594, 1966.

Population growth and sustainable development

The global human population is approaching 6.4 billion, with 80 million people added each year. This staggering growth reflects an average of 4.1 births and 1.8 deaths each second. Fortunately, such rapid growth is not expected to continue forever due to declining family sizes in most nations. Depending on the level of optimism assumed, the United Nations predicts that the population will level out at 7.7 to 11.2 billion people between 2030 and 2100 (**Fig. 1**).

Humans have an enormous impact on the Earth and its resources. It was recently estimated that we have altered 39–50% of the Earth's surface, with 16–23% of the world's land area converted for agriculture. Although it is difficult to measure land conversion precisely, it is clear humans have altered a large portion of the planet's land surface. Furthermore, we divert 70% of the readily available fresh water for agricultural and other human uses. The exploitation of land and fresh water by the human population means that there are fewer resources and less habitat available to other species. Humans also deplete species directly through harvest. For example, roughly two-thirds of fish species are being harvested at unsustainable rates. Humans exert additional pressures on the environment in the form of pollution, introduction of species to areas outside their natural range (upsetting natural population dynamics), and deforestation and desertification (causing global food shortages and climate change). Fossil fuel combustion, soil cultivation, biomass burning, and tropical deforestation have increased atmospheric carbon dioxide levels, causing more heat to be trapped in the atmosphere and resulting in a 0.6°C increase in mean global temperature over the last century.

Impact of future growth. The human population is not evenly distributed across the globe; 80% of the world's population is found in less developed countries (Fig. 1). Projected future population growth is also expected to be concentrated in those nations. In particular, the United Nations Population Reference Bureau calculates that the population of less developed countries will increase by 57% before 2050,

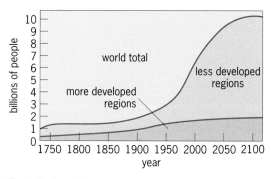

Fig. 1. Projected human population growth. (*Reprinted with permission from W. P. Cunningham and M. Cunningham, Principles of Environmental Science: Inquiry and Applications, McGraw-Hill, New York, 2002*)

whereas more developed countries are expected to increase by only 3%.

Population density and biodiversity. The areas with the most rapid human population growth and highest population density generally harbor the highest diversity of species. For example, the areas of highest species diversity (biodiversity hotspots, such as the Mediterranean Basin, the Tropical Andes, and the Caribbean) have a human population growth rate 1.8 times higher than the world average, and a population density 71% greater than the world average. The fact that population density and future growth are highest in the most biodiverse regions will likely accelerate the ongoing extinction crisis. *See* BIODIVERSITY HOTSPOTS.

Resource consumption. Consumption of resources is also not distributed evenly throughout the world. However, the global distribution of consumption is virtually opposite that of population growth, with the vast majority of resource use occurring in more developed nations. For example, roughly 16% of the world's population is responsible for using 80% of the available resources, and an American citizen uses 30 times more resources on average than a citizen of India. As poorer countries continue to develop, their consumption is projected to increase dramatically.

Given that human populations will continue to grow and that individuals in poorer nations will likely increase their per capita consumption of resources over time, it is critical to assess potential ways to minimize the impact that current and future human populations will exert on the environment. A variety of strategies are currently being used to balance population growth, economic development, and environmental health. These strategies for sustainable development include such diverse goals as changing farming practices, providing new kinds of tourist experiences, and designing more environmentally friendly cities.

Sustainable development. There are currently 3 billion people living on the equivalent of less than $2 US per day. Poverty often promotes environmental destruction, as poorer people rely heavily upon extraction of natural resources. At the same time, environmental degradation worsens poverty, resulting in a vicious cycle of environmental degradation and human suffering. Combating poverty is clearly an important step in the struggle to preserve the environment and ecosystems.

In 1992, participants at the United Nations Conference on Environment and Development drafted the Rio Declaration, which established that although individual countries are entitled to use their natural resources for economic development, each nation should ensure that those resources and a healthy environment will be available for both current and future generations. Below are examples of how the concept of sustainable development has been used to help reduce the impacts of human societies and promote conservation.

Sustainable agriculture. Each year, 500 million tons of pesticides are utilized worldwide. At the same time, nearly 800 million people suffer from a lack of nourishment. Sustainable agriculture seeks to bolster food production while reducing the environmental degradation associated with conventional farming methods. Sustainable farming emphasizes the prevention of erosion, conservation of water resources, and retention of soil nutrients. It also advocates decreased dependence on synthetic fertilizers and pesticides and fossil fuels to reduce both environmental toxicity and farmer expense.

To achieve these goals, some farmers have adopted intercropping, which involves growing multiple species of crop or noncrop plants intermixed in a single field. The presence of additional plant species can reduce water loss and erosion. Another strategy involves planting cover crops during nongrowing seasons. Cover crops can prevent erosion, improve soil quality and nutrient levels, and improve crop performance during the growing season. For the control of insect pests and plant diseases, sustainable agriculture emphasizes the use of natural pesticides and the attraction or release of predatory insects.

As human populations grow, environmental problems associated with agriculture also increase. For example, the amount of land dedicated to food production has increased 400% since 1700. Shifting to more sustainable growing methods can help us to balance environmental health with the nutritional demands of future generations.

Sustainable energy. As poorer nations are beginning to industrialize and acquire consumptive habits similar to those of developed nations, there is an increasing need to shift toward sustainable sources of energy. Currently, national economies are largely dependent on nonrenewable sources of energy, especially fossil fuels (oil, coal, and natural gas), which make up 86% of world consumption (**Fig. 2**). An important goal of sustainable development is to help societies shift to renewable sources of energy, particularly those that minimize pollution. The United

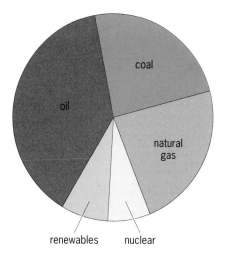

Fig. 2. World energy consumption by source. (*Data from Energy Information Administration's International Energy Annual 2001***,** http://www.eia.doe.gov/emeu/iea/contents. html**)**

Nations Development Program provides financial assistance to catalyze sustainable energy technology and assist developing nations with implementing sustainable energy programs. Although many nations are designing programs to establish solar and wind energy use, much work remains to be done. As of 2000, only 7% of global energy use was derived from sustainable sources (Fig. 2).

Ecotourism. Widespread implementation of sustainable practices is often inhibited by a lack of funding. One solution is ecotourism, which allows tourists the opportunity to explore relatively pristine natural areas without damaging the environment. The profits generated from ecotourism benefit the local culture and economy, thereby providing an incentive for the local people to preserve their surrounding ecosystem. There are ecotourism programs available in many locations, including Australia, Greenland, Costa Rica, and Nepal. Currently, ecotourism represents 7% of the $425 billion spent yearly by international travelers, and future trends look promising: whereas overall tourism revenues are increasing by 4% each year, ecotourism is expected to increase by 10–30%.

Planning environmentally friendly cities. The impact of humans on the environment is related to the spatial distribution of the population. For example, a river or stream system becomes seriously degraded when more than 10% of the surrounding land has been converted for human use. Thus, a more concentrated population would generally be much less harmful to a watershed than urban sprawl. High-density living has many environmental benefits, including a decrease in air and water pollution as people commute shorter distances and use more public transportation. Thus, it is important to design new urban centers in ways that will minimize impacts on the surrounding environment.

Environmentally friendly planning has played a major role in the development of many large cities, including Portland, Oregon. In 1973, Portland developed an urban growth boundary to limit the city's expansion. The urban growth boundary is surrounded by a green belt, an area of rural land where little or no urban growth is allowed. Green belts can help prevent urban sprawl and force inhabited areas to become more densely populated. However, despite the many benefits of high-density living, the proportion of people living in dense urban areas has decreased through recent years. Many people associate dense living with crime, lack of privacy, and traffic congestion. Fortunately, some densely populated cites are attractive and can serve as models for future urban development.

Future outlook. The human population is large and growing rapidly. Increased awareness of the overpopulation problem and implementation of family planning measures are essential in the effort to control human population growth. However, regardless of how many people inhabit the planet, the impacts of the human population on the environment and on biodiversity could be either large or small. Impacts depend not only on population size but also on how much each individual consumes and pollutes. Hope for the future may lie in the development of technologies that are less harmful to the environment, and in reduced per capita consumption.

For background information *see* ADAPTIVE MANAGEMENT; CONSERVATION OF RESOURCES; ECOLOGY, APPLIED; ENVIRONMENTAL MANAGEMENT; HUMAN ECOLOGY; INDUSTRIAL ECOLOGY; POPULATION ECOLOGY in the McGraw-Hill Encyclopedia of Science & Technology.   Sabrina West

Bibliography. R. P. Cincotta et al., Human population in the biodiversity hotspots, *Nature*, 404:990–992, 2002; R. Livernash and E. Rodenburg, Population change, resources and the environment, *Pop. Bull.*, no. 53, 1998; Population Reference Bureau, *2002 World Population Data Sheet*, 2002; United Nations Environment Programme, *Rio Declaration on Environment and Development*, 1992; P. M. Vitousek, Human domination of earth's ecosystems, *Science*, 277:494–499, 1997.

Precise earthquake location

Earthquakes occur when the friction that prevents a fault from slipping is overcome by the gradual accumulation of stress acting across the fault and starts the material on the two sides of the fault slipping past each other. As slip occurs, frictional resistance drops, and the resulting rapid and unstable slip generates seismic waves that radiate outward in all directions from the initiation point, or hypocenter. These seismic waves are recorded by seismographs deployed on the Earth's surface, and their arrival times are the information that is used to determine the hypocenter. The distribution of earthquake hypocenters is one of the principal tools that seismologists use to understand earthquake behavior. Thus, it is important to minimize the uncertainties in determining earthquake hypocenters. Because networks of seismographs are set up to locate and catalog earthquakes automatically, refinement of earthquake locations is often referred to as relocation.

The interior of the Earth has a complex and varied geology, particularly in the Earth's crust where most earthquakes occur. Variations in the geology lead to variations in the velocity of seismic-wave propagation—the speed at which waves propagate through the Earth's interior. The Earth's subsurface velocity structure varies strongly with position and is only incompletely known. Because the arrival time of seismic waves depends on this velocity structure, discrepancies between the assumed velocity structure and the true velocity structure translate into errors in earthquake locations. This is the primary source of location errors for earthquakes that occur within an established seismic network. Another important cause of earthquake mislocation is error in the measurement of seismic-wave arrival times. Recent progress in earthquake relocation has focused on strategies that reduce both of these error sources.

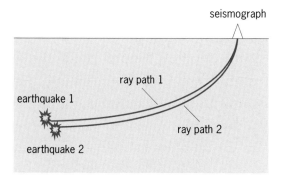

Fig. 1. Earthquakes 1 and 2 are located close to one another. The arrival times of seismic waves at a seismograph on the Earth's surface will be influenced by the structure that the rays (connecting the earthquakes and the seismograph) encounter as they propagate through the Earth's subsurface. Because the two rays follow nearly the same path, they will be influenced by nearly the same velocity structure. Thus, measuring the difference in arrival times of the seismic waves will largely eliminate the effect of this part of the velocity structure from the problem. The arrival time difference will be primarily sensitive to the spatial separation of the two earthquakes.

Relocating of groups of earthquakes. Relocating groups of earthquakes simultaneously can help to eliminate much of the error that arises from incomplete knowledge of the Earth's subsurface velocity structure. In many cases, precision (relative earthquake locations) is of more interest than accuracy (absolute earthquake locations). For example, earthquakes can be located using arrival time differences, rather than the absolute arrival times, of seismic waves. Measuring differences in arrival times from nearby earthquakes (**Fig. 1**) has the virtue of removing the contribution of unmodeled velocity structure that is shared by the two adjacent ray paths. The result is a measurement that is primarily related to the relative location of the earthquake hypocenters.

Measuring arrival times using waveform cross-correlation. The other principal source of errors in earthquake locations are measurement errors in the arrival times of seismic waves. Measurement errors can be reduced by taking advantage of the similarity of seismic waves from earthquakes that occur near one another (**Fig. 2**) using a technique known as waveform cross-correlation, in which very small differences in arrival times are translated into differences in location and, as a result, precise relocation. Under these conditions waveform cross-correlation easily yields subsample precision in relative arrival time measurements for similar waveforms. For typical seismographic networks such as those deployed across California to detect and locate earthquakes, the uncertainty in relative arrival time measurements for similar earthquakes can be reduced from several tenths of a second to several thousandths of a second.

Combining cross-correlation–based differential arrival times with large-scale simultaneous relocation reduces errors in relative earthquake locations by up to 2 orders of magnitude. This approach to relocating earthquakes has blossomed in recent years due to the widespread availability of digital data from earthquake data repositories and the computational ability to solve large systems of equations simultaneously. Initially these techniques were applied to modest numbers of earthquakes from relatively small source volumes, but more recently they have been applied to tens of thousands of earthquakes over distances approaching 100 km (62 mi).

High-resolution locations. Improved earthquake locations lead to the ability to resolve the fine details of active faults at depth as they are expressed by microearthquake activity. The images of seismicity resulting from high-precision relocation have led to new discoveries and insight into the fundamentals of the earthquake process. For example, precise relocations allowed J.-L. Got and coworkers to demonstrate that much of the seismicity near Kilauea volcano, Hawaii, could be attributed to slip on a nearly horizontal surface deep under the volcano that is thought to separate the erupted edifice

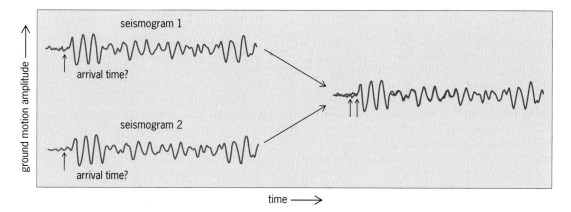

Fig. 2. Seismograms from earthquakes 1 and 2 are very similar to one another due to their proximity. Seismograms at left show possible picks of first arrival time of seismic waves (arrow). Because these signals emerge from background seismic noise, some error enters into this measurement. By overlaying the waveforms as on the right, it is possible to align the two seismograms, and to measure the relative arrival times very precisely, often to within a few milliseconds. This alignment is done mathematically using cross-correlation. Misalignment of the two arrows shows that the measurement error when seismograms are considered individually is substantial, approximately 0.1 second in this case. This measurement error would be mapped into errors in earthquake location when hypocenters are estimated from such measurements.

of Hawaii from the much older sea floor beneath. Observations of repeating microearthquakes have led to the conclusion that friction on creeping faults at depths where earthquakes are generated follow friction laws derived from laboratory observations, and that the state dependence also found in laboratory friction laws applies on real faults as well. Recurrence intervals of repeating earthquakes have also been used to argue for scale-dependent stress drops, although there is a strong counterargument that is itself based on the distance dependence of precisely located earthquake pairs. High-precision earthquake locations have resolved streaks of seismicity that appear to follow the direction of the local slip vector. These streaks remain enigmatic. In some cases they appear to be isolated stuck areas on otherwise continuously slipping faults, whereas in others they may delineate the boundary between slipping and locked areas.

Joshua Tree aftershock sequence. On April 24, 1992, a magnitude-6.1 earthquake occurred near the San Andreas Fault in the Mojave Desert of Southern California. Known as the Joshua Tree earthquake, it was a right-lateral strike-slip earthquake (fracture where the far side is moving right and horizontally relative to the near side) on a nearly vertical fault that was preceded by a foreshock sequence and ruptured primarily to the north. Like many earthquakes, the Joshua Tree earthquake was followed by many smaller earthquakes in the source region. A notable aspect of the Joshua Tree aftershock sequence was how prolific it was. The Joshua Tree earthquake was followed by more than 7000 aftershocks within 2 months of the mainshock. Many of the aftershocks in the Joshua Tree sequence occurred off the mainshock fault plane, which is somewhat uncommon. Despite these interesting aspects, the Joshua Tree earthquake has received relatively

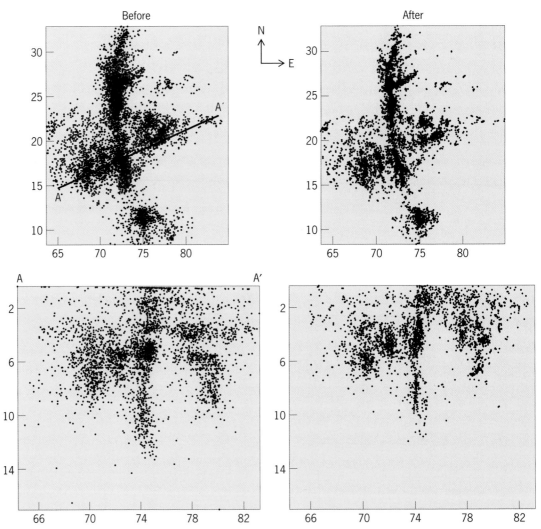

Fig. 3. Catalog locations (left) compared with relocations (right) for the 1992 Joshua Tree earthquake aftershock sequence. Each dot represents one earthquake. Upper panels show a map view with a distance scale in kilometers. Lower panels show the cross section A-A′ from the upper left panel. The horizontal axis shows distance in the A-A′ direction, and the vertical coordinate indicates depth (km). The mainshock rupture plane is marked by concentrated seismicity and trends generally northward, with some curvature and segmentation. In cross section, the mainshock plane is the vertical lineation of seismicity at 0–10 km depth at 74 km in the horizontal. Note the emergence of detailed small-scale structure in the relocations that are obscured by location errors in the original locations.

little attention because it was followed just miles away by the much larger M-7.2 June 28, 1992, Landers earthquake.

Shown in **Fig. 3** are 7439 earthquakes of the Joshua Tree aftershock sequence that were recorded by at least four seismographs. The panels on the left show the original (catalog) locations, which were obtained by measuring the arrival times one seismogram at a time and locating each earthquake individually. The locations in the panels on the right were obtained by using cross-correlation to derive precise relative arrival times and then relocating all earthquakes simultaneously using the double-difference relocation technique. The double-difference relocation technique determines the relative position of a group of earthquakes by using the measured difference in the arrival times from two earthquakes (that is, by subtracting the arrival time of a wave from one earthquake from the arrival time of the same wave from another earthquake) to solve for the difference in position of the two earthquakes. The data used to solve for these locations consist of a total of 3.8 million observations or earthquake arrival times, which were used to solve for nearly 30,000 unknown hypocentral parameters.

The errors in location for the standard earthquake catalog shown on the left side of Fig. 3 are about 1 km (0.6 mi) in the horizontal position and 2 km (1.2 mi) in the vertical. After relocation, errors are estimated to be approximately 50 m (164 ft) in the horizontal and 100 m (328 ft) in the vertical. The latter set of uncertainties reflects the relative, rather than the absolute, error. However, there is no doubt that the images of relocated seismicity on the right side of Fig. 3 are much clearer than those on the left and better meet our expectation that earthquakes occur due to slip on planar faults.

For background information *see* EARTHQUAKE; FAULT AND FAULT STRUCTURES; SEISMOGRAPHIC INSTRUMENTATION; SEISMOLOGY in the McGraw-Hill Encyclopedia of Science & Technology.

Gregory C. Beroza; E. E. Zanzerkia

Bibliography. D. A. Dodge, G. C. Beroza, and W. L. Ellsworth, Detailed observations of California foreshock sequences: Implications for the earthquake initiation process, *J. Geophys. Res.*, 101:22,371–22,392, 1996; A. Douglas, Joint epicentre determination, *Nature*, 215:47–48, 1967; J.-L. Got, J. Fréchet, and F. W. Klein, Deep fault plane geometry inferred from mulitplet relative location beneath the south flank of Kilauea, *J. Geophys. Res.*, 99:15,375–15,386, 1994; G. Poupinet, W. L. Ellsworth, and J. Fréchet, Monitoring velocity variations in the crust using earthquake doublets: An application to the Calaveras fault, California, *J. Geophys. Res.*, 89:5719–5731, 1984; A. M. Rubin, D. Gillard, and J.-L. Got, Streaks of microearthquakes along creeping faults, *Nature*, 400:635–641, 1999; C. G. Sammis, R. M. Nadeau, and L. R. Johnson, How strong is an asperity, *J. Geophys. Res.*, 104:10,609–10,619, 1999; D. P. Schaff et al., High resolution image of Calaveras fault seismicity, *J. Geophys. Res.*, 107:633, 2002; J. E. Vidale et al., Variations in rupture process with recurrence interval in a repeated small earthquake, *Nature*, 368:624–626, 1994; F. Waldhauser and W. L. Ellsworth, A double-difference earthquake location algorithm: Method and application to the northern Hayward fault, *Bull. Seismol. Soc. Amer.*, 90:1353–1368, 2000; F. Waldhauser, W. L. Ellsworth, and A. Cole, Slip-parallel seismic lineations along the northern Hayward fault, California, *Geophys. Res. Lett.*, 26:3525–3528, 1999.

Predictive genetics of cancer

The completion of the Human Genome Project in April 2003 marked the beginning of a revolution in predictive medicine. In the most important development since the landmark discovery of the deoxyribonucleic acid (DNA) double helix 50 years prior, the complete sequence of the 3 billion bases that make up the human genome is now known. With this knowledge, some of the most complex biological problems will be unraveled and understood at the most fundamental level, the gene. The genetic contributions to common but complex diseases (those due to variation in several, perhaps many, genes) such as asthma, diabetes, heart disease, mental illness, and cancer will be discovered. New technologies to sequence the entire genome of one person in one day will transform the traditional practice of medicine. Genetic testing, the detection of disease-causing mutations in the sequence of an individual's genome, will become routine. Clinicians will use genetic tests to determine a person's risk of developing diseases for preventive medicine. The hope is that predictive genetic testing will ultimately reduce the incidence and severity of common diseases.

Cancer and genetics. All cancer is genetic, a result of mutation or changes in the DNA sequence of genes. DNA mutations are inherited or occur spontaneously when mistakes are made as cells replicate their DNA in preparation for cell division. DNA mutations may also be induced as a result of exposure to environmental agents such as radiation or harmful chemicals.

Cancer develops slowly. For most solid tumors, there is a 20-year interval from the time of the first mutation to clinical detection of a tumor. Initially, one gene in one cell undergoes a mutation that gives it a growth advantage over other cells. This cancer-prone cell passes this mutation to all cells that it produces. Over time, these cells acquire additional mutations in other genes that allow them to divide, invade, and metastasize (invade and destroy normal tissue). It is thought that a cell must undergo at least five to seven successive mutations for it to become a cancer cell.

Classification of cancer genes. Genes that are mutated in cancer cells are usually genes that control the orderly replication of cells. When damaged, these genes allow the cell to reproduce without restraint. Cancer genes are classified according to their biological function. Proto-oncogenes normally function to encourage cell growth and proliferation at the appropriate time. When mutated, they function as oncogenes and continually signal cellular growth and division. Conversely, tumor suppressor genes normally function to halt cellular growth. Mutated tumor suppressor genes allow cells to proliferate uncontrollably. An important subset of tumor suppressor genes are those that repair damaged DNA. Mutation of DNA repair genes hinders the cell's ability to detect and/or repair mutations, thus accelerating damage to other genes.

Inherited cancer syndromes. Although all cancer is the result of mutation, a small fraction, less than 10%, is inherited. In inherited forms of cancer, the mutation is passed down from parent to child in the germline. In families with inherited cancer, disease is the result of a single mutation in a single gene. Cancer genes in these families are usually dominant and highly penetrant (a mutation that is 100% penetrant would give rise to cancer in all people who inherit the gene). An individual who inherits one mutated copy of a highly penetrant gene is "predisposed" to cancer; that is, the person is more likely to develop cancer than the average person. These individuals often develop cancer at an earlier age than the population average for a particular cancer type.

Cancer susceptibility genes. Tumor suppressor genes that are inherited include genes that predispose to breast, ovarian, and colon cancer. Germline mutations in other tumor suppressor genes cause rare inherited cancers such as retinoblastoma (childhood eye cancer), Wilms' tumor (childhood kidney cancer), Li-Fraumeni syndrome (variety of cancers), neurofibromatosis (connective tissue and brain tumors), renal cancer, skin cancer (basal cell carcinoma and melanoma), gastric cancer, chondrosarcoma (cartilage cancer), and endocrine tumors (see **table**). Inherited mutations in oncogenes that cause medullary thyroid cancer and renal cancer have also been identified.

Tests for mutations in over 25 known cancer genes are available in many research laboratories, and genetic testing for the more common hereditary breast and colon cancer genes is available commercially. Clinicians can identify disease-predisposing mutations that may cause breast and colon cancer, long before any symptoms appear.

Breast cancer. Mutations in the genes *BRCA1* (breast cancer 1) and *BRCA2* (breast cancer 2) predispose individuals to breast and ovarian cancer. In families affected by hereditary breast and ovarian cancer, the genetic test is complicated. *BRCA1* and *BRCA2* are extremely large genes, and hundreds of different mutations have been detected in different families. Normal, innocuous DNA sequence variations also occur in the *BRCA1* and *BRCA2* genes. Thus, discerning which variants cause cancer is not always possible. However, once a cancer-causing mutation in an affected individual is identified, other members of the family can be tested for the specific "family" mutation. Together, mutations in *BRCA1* and *BRCA2* ac-

count for 20–25% of breast cancer that clusters in families. Mutation carriers have a lifetime breast cancer risk approaching 85% and a lifetime ovarian cancer risk of 20–40%. Over 750 different mutations in *BRCA1* and *BRCA2* have been identified.

Colon cancer. Hereditary non-polyposis colon cancer (HNPCC) is the most common form of hereditary colon cancer, accounting for 5–8% of all colon cancers. Inherited mutations in at least five different DNA repair genes predispose individuals to colon cancer in high-risk families. The lifetime risk of colon cancer is 80% in mutation carriers.

Predictive genetic testing in cancer. The expectation that the completion of the human genome will generate a list of cancer genes has generated great enthusiasm for predictive genetic testing, especially in families with multiple affected individuals. For most *BRCA1*, *BRCA2*, and HNPCC mutation carriers, the risk of developing breast or colon cancer is high, making predictive genetic testing appropriate for blood relatives of known mutation carriers. However, the vast majority of cancer is not due to a single mutation in a specific gene. This is true even for cancer patients that have several relatives with the same cancer type. For example, breast cancer is so common that some families have more than one affected member purely by chance. Nonetheless, an individual with a first-degree relative with cancer will have twice the risk of developing cancer compared with the general population, suggesting that common cancer susceptibility genes do exist. These inherited alterations are likely subtle sequence variations in low-penetrance (not all gene carriers manifest disease) genes rather than inactivating mutations in highly penetrant genes.

Individuals that carry predisposing mutations in relatively common, low-penetrance cancer genes will have a much lower risk of developing cancer. Their cancer risk will increase only in the presence of other genetic and environmental factors. Unaffected members of these families may benefit from predictive genetic screening once the risk of developing disease for each factor is known.

Benefits and risks of genetic testing. The clinical benefits of knowing that one is at high risk of developing cancer remain uncertain and ultimately depend on the effectiveness of modifying behavior and preventive treatments. If no successful behavior modifications or preventive treatments exist, predictive genetic testing for cancer will be of little value. Most people with a mutation in an HNPCC gene will develop colon cancer unless they increase cancer surveillance through annual colonoscopy and remove precancerous lesions. A woman who inherits a mutation in *BRCA1* may reduce her risk of developing breast cancer by a combination of more frequent screening, chemoprevention, prophylactic surgery, and lifestyle modifications. Whether these approaches actually reduce cancer risk is controversial and will be resolved only by studying many high-risk individuals who choose various combinations of preventive measures.

Inherited cancer syndromes

Syndrome	Gene	Tumors
Familial breast and ovarian cancer	BRCA1	Breast cancer Ovarian cancer
	BRCA2	Breast cancer Ovarian cancer Male breast cancer
Hereditary non-polyposis colorectal cancer	MLH1 MSH2 MSH6 PMS1 PMS2	Colorectal cancer Endometrial cancer
Retinoblastoma	RBI	Retinoblastoma Osteosarcoma
Li–Fraumeni syndrome	TP53 CHK2	Soft tissue sarcoma Breast cancer Brain tumors Leukemia
Familial polyposis	APC	Colorectal cancer
Wilm's tumor	WT1	Kidney cancer
Von Hippel–Lindau syndrome	VHL	Renal cell cancer Hemangioblastoma Retinal angioma Pheochromocytoma
Basal cell nevus syndrome	PTC	Basal cell cancer
Cowden syndrome	PTEN	Breast cancer Thyroid cancer Endometrial cancer
Tuberous sclerosis	TSC1 TSC2	Renal carcinoma Hamartomas Rhabdomyoma
Neurofibromatosis 1	NF1	Neurofibroma
Neurofibromatosis 2	NF2	Acoustic neuroma Meningioma Schwannoma
Juvenile polyposis	SMAD4 BMPR1A	Hamartomatus polyps Colorectal cancer
Familial gastric cancer	CDH1	Gastric cancer Lobular breast cancer
Familial melanoma	CDKN2A CDK4	Melanoma Pancreatic cancer
Multiple endocrine neoplasia type 1	MEN1	Parathyroid hyperplasia Pancreatic tumors Pituitary tumors
Multiple endocrine neoplasia type 2	RET	Medullary thyroid cancer Pheokromocytoma
Familial renal cancer	MET	Papillary renal cancer
Peutz–Jegher's syndrome	LKB1	Gastrointestinal tract cancer Breast cancer Testicular cancer
Carney complex	PRKAR1A	Pituitary cancer Testicular cancer Thyroid cancer
Hereditary paraganglioma and phaeochromocytoma	SDHD SDHC SDHB	Paraganglioma Phaeochromocytoma

The physical risks of predictive genetic testing are associated with the risk of drawing a blood sample. More significantly, there are potential risks involved in the way the results of either a positive or negative test impact a person's life. Knowing that one is genetically predisposed to cancer can have severe psychological consequences. In addition, ensuring genetic confidentiality is a major concern. These issues should be discussed with a genetic counselor before an individual consents to predictive genetic testing.

Genetic testing and public health. It is estimated that 1 in 300 women carry germline mutations in breast cancer susceptibility genes, the same proportion of

Americans that inherit mutations in colon cancer susceptibility genes. Because breast and colon cancer are so common in the general population, even a small fraction of the total is a relatively large number of predisposed individuals and therefore a public health concern. However, there is little clinical value in screening large populations because costs (using current sequencing technology) would be enormous for the relatively few mutation carriers identified in this way. Furthermore, for the majority of familial cancer cases, the common cancer susceptibility genes are not known. In the future, large-scale population-based predictive cancer screening will benefit public health when economical mutation detection strategies are developed and the risk associated with each cancer gene is understood. Only then will predictive cancer testing and preventive medicine reduce cancer morbidity and mortality in the general population.

Conclusion. The science of predictive cancer genetics is rapidly evolving. With the completion of the sequencing of the human genome, new cancer-predisposing genes will be discovered, offering important opportunities for diagnosis and risk assessment. Genetic tests that determine risk of common cancers will be developed and will be used in routine clinical practice. The development of specific strategies to reduce cancer risk in genetically susceptible individuals will be a major focus of medicine. Ideally, predictive genetic testing for cancer will allow the patient and clinician to establish an effective strategy for prevention.

For background information *see* BREAST DISORDERS; CANCER (MEDICINE); DEOXYRIBONUCLEIC ACID (DNA); GENE ACTION; HUMAN GENETICS; HUMAN GENOME PROJECT; MUTATION; ONCOGENES; ONCOLOGY; TUMOR in the McGraw-Hill Encyclopedia of Science & Technology.   Piri L. Welcsh

Bibliography. L. B. Andrews, *Future Perfect: Confronting Decisions about Genetics*, Columbia University Press, New York, 2001; W. B. Colman and G. J. Tsongalis (eds.), *The Molecular Basis of Human Cancer*, Humana Press, 2001; D. H. Hamer and P. Copland, *Living with Our Genes: Why They Matter More than You Think*, Anchor Books, 1999; K. Offit, *Clinical Cancer Genetics: Risk Counseling and Management*, Wiley-Liss, New York, 1998; B. Volgelstein and K. W. Kinzler (eds.), *The Genetic Basis of Human Cancer*, McGraw-Hill, 1997; J. D. Watson and A. Berry (contributor), *DNA: The Secret of Life*, Knopf, 2003.

Primate conservation

Across the globe, primate species in natural environments are under increasing pressures from a growing human population. According to the International Union for the Conservation of Nature (IUCN), 35% of living primates (62 of 177 species) are currently considered either "critically endangered" (defined as "facing an extremely high risk of extinction in the wild in the immediate future") or "endangered" (defined as "facing a very high risk of extinction in the wild in the near future"), while an additional 51 species are considered "vulnerable" to extinction in the medium-term future (see **table**). Most living primates are restricted to tropical and subtropical forests, and the vast majority of critically endangered and endangered primates are either found in or are endemic to 11 of the 25 biodiversity "hotspots" recently identified by Conservation International as priority areas for global conservation. In fact, all critically endangered primate species, and over half of the endangered species, can be found in just six hotspots that together comprise just over 0.5% of the Earth's surface, suggesting very clear priority habitats where immediate conservation efforts should be focused (**Fig. 1**). *See* BIODIVERSITY HOTSPOTS.

Threats to primates. The most serious anthropogenic threats facing primates come from habitat loss due to deforestation and conversion of forests to agricultural land, habitat fragmentation, habitat degradation and modification associated with logging and other forms of natural resource extraction, and subsistence and commercial hunting. All of these processes are ultimately tied to human population growth, which tends to be greatest in developing countries where most of the world's primates are found. *See* POPULATION GROWTH AND SUSTAINABLE DEVELOPMENT.

Deforestation and habitat conversion. Over the last 8000–10,000 years, human activity has reduced the moist tropical forests of Africa, Southeast Asia, and South America that are home to most primates to less than one-half of their former extent. Deforestation and the conversion of forested lands to agricultural use have been particularly dramatic during the past several decades: tropical forests were deforested at an average rate of nearly 0.9% per year globally during the 1980s and at a rate of 0.52% per year between 1990 and 1997. Depending on the precise data used, it has been estimated that up to 12.3 million hectares of tropical forest were lost annually to deforestation throughout the 1990s, a figure only slightly lower than in preceding decades. The rate of tropical deforestation and conversion varies with geographic region and is highest in Southeast Asia, where human population density is greatest. Some of the highest rates of deforestation are seen in precisely those areas of the tropics identified as biodiversity hotspots and containing large numbers of threatened primates (Fig. 1). Indeed, habitat loss is often considered to be the primary risk factor for future primate extinctions.

Habitat fragmentation and modification. In a number of regions of the world, primates are threatened not just by habitat loss but also by fragmentation and modification of their natural environments. Such changes may coincide with large-scale deforestation but may also arise as incidental effects of less drastic habitat disturbances. For example, "selective" logging and other putatively sustainable extractive forestry practices typically require the building of a network of roads, which can fragment existing forest patches or

Number of primate species in various categories of extinction risk in each major geographic region where living primates are found				
Geographic region	Critically endangered	Endangered	Vulnerable	Lower risk or data-deficient	Total
Africa	2	12	7	23	44
Asia	6	16	15	30	67
Americas	9	9	21	4	43
Madagascar	2	6	8	7	23
Total	19	43	51	64	177

make previously unexposed areas of forest accessible to colonization by humans thus facilitating subsequent deforestation and hunting.

Although the effects of fragmentation and modification per se on the risk of extinction of local primate populations are poorly understood, based on ecological theory and limited empirical studies it is clear that a number of factors influence whether a particular primate species can persist in such a habitat. In general, primate populations are more likely to persist in larger fragments and in fragments that are closer to potential source populations from which recolonization is possible. Additionally, primates with less specialized dietary and other ecological needs, those with smaller home range requirements, and those that are predominantly folivorous (leaf-eating) rather than frugivorous (fruit-eating) appear more likely to survive in fragmented or selectively modified habitats.

Another general consequence of fragmentation is the more rapid loss of genetic diversity due to inbreeding and stochastic (random) genetic drift. This loss of diversity may compromise a population's ability to respond adaptively to changing environmental conditions and thus increase its risk of extinction. Today some of the most highly endangered primates, such as the muriquis and lion tamarins endemic to Brazil's Atlantic Forest, exist only in fragmented and modified habitat landscapes (**Fig. 2**).

Hunting. Hunting is also a major threat to primate populations. Humans have been important predators of primates for tens of thousands and perhaps hundreds of thousands of years, and human hunting has been suggested as a direct causal factor in the extinctions of a number of primate taxa (for example, multiple species of lemurs on Madagascar following human colonization of the island 1500 to 1000 years ago; and Miss Waldron's red colobus, a large red-and-black monkey, from western Africa during the twentieth century). Recent studies have demonstrated the dramatic effect that even small-scale subsistence hunting can have on primate populations. For example, surveys of vertebrate biomass in 25 primary rainforest sites across western Amazonia have revealed that the density of large-bodied ateline primates—the prehensile-tailed howler, spider, and woolly monkeys—is as much as 10 times lower in areas subject to intense subsistence hunting than

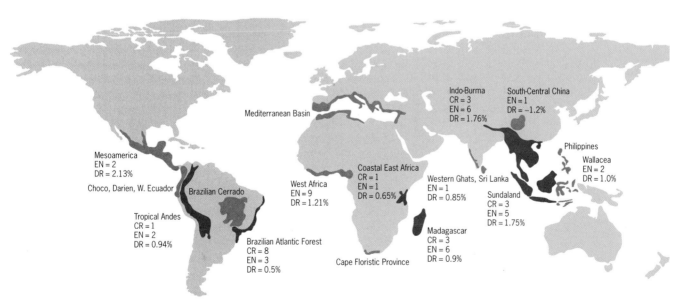

Fig. 1. Sixteen global biodiversity hotspots in which nonhuman primates are found. Eleven of these hotspots harbor critically endangered and/or endangered primates, and six (darkly covered) contain all 19 critically endangered species and over 50% of endangered species. CR = number of critically endangered primate species found in the hotspot, EN = number of endangered species, and DR = estimated rate of deforestation during the 1990s. (*Adapted from N. Myers et al., Biodiversity hotspots for conservation priorities, Nature, 403:853–858, 2000*)

Fig. 2. The critically endangered Northern muriqui, *Brachyteles arachnoides*, one of the largest living South American primates. It is estimated that about 500 mature Northern muriquis remain in the wild, divided among several remnant and regenerating forest patches in a highly fragmented landscape. Although muriquis still face multiple risks from humans, international conservation efforts have been effective in improving their chances for survival. (*Photo by Luiz G. Dias, courtesy of K. B. Strier*)

in sites facing little or no hunting pressure. Throughout Amazonia, these three primate species are often cited as the preferred prey of local subsistence hunters, and it has been estimated that between 1 and 2.6 million individual atelines (yielding between 6800 and 16,700 tons of meat) are harvested annually by hunters in the Brazilian Amazon alone.

Today commercial hunting is thought to pose one of the most significant threats to local primate populations, particularly in parts of West Africa where bushmeat is a highly desirable food resource and where many species of primates are being hunted for market at rates that significantly exceed estimates of maximum sustainable yield. Beyond being hunted for food, primates are commonly killed as agricultural pests in some parts of the world, a trend that can only be expected to increase as human population density increases. Finally, many primates are hunted as part of the trade in exotic pets and in animal parts used as souvenirs or in traditional medicines, as well as in the endeavor to acquire subjects for biomedical research, although international restrictions on the trade in endangered species (for example, legislation of the Convention on International Trade in Endangered Species) have effectively lessened some of these pressures over the past few decades.

Importance. There are a number of reasons why primates are critical focal taxa for wildlife conservation efforts. First, it is likely that primates play crucial ecological roles in some of the ecosystems in which they are found, and the loss of primates from these ecosystems may have dramatic and far-reaching effects. For example, in many tropical forests primates are important dispersers of seeds for numerous species of plants. In the neotropics (southern Mexico, Central and South America), up to 90% of woody plant species depend on frugivores for dispersal of their seeds; primates constitute a major portion of the total biomass of neotropical frugivores and are known to swallow and disperse in their feces enormous numbers of viable seeds from hundreds of different plants. Moreover, the germination success of seeds is often significantly improved following passage through the digestive tracts of primates. Consequently, frugivorous primates have the potential to significantly impact the recruitment patterns (number and likelihood of dispersed seeds surviving to adulthood) of many plant species, including those of economic value to local human populations, and can thus influence the long-term health and persistence of forested ecosystems. Recent studies have found that seedling recruitment is markedly lower in forest fragments lacking primate seed dispersers.

Second, given their long lifespans and slow reproductive rates, primates may be some of the most sensitive indicators of anthropogenic stress on an area; successful efforts to conserve primates are thus likely to also serve the broader goal of preserving many other elements of the same ecosystems. This may be an especially compelling reason to focus primate conservation efforts on those tropical forest hotspots that harbor considerable plant and animal biodiversity beyond just primates. Moreover, primates may be particularly charismatic and effective "flagship species" around which to rally local and international conservation interest.

Finally, primates tend to be better studied than many other vertebrates in the tropical ecosystems in which they are found; as a result, their ecological requirements may be well understood, thus making management plans designed to meet these ecological requirements more likely to be effective.

Approaches. Central to most conservation efforts is the prioritization of species or habitat areas for conservation attention. Once priority habitats or species have been identified, there are many tactics that may be employed to meet conservation goals. The most common of these involve setting aside specific areas as protected reserves and trying to promote the sustainable use of primates and their forest habitats among the local people.

Prioritization of habitats or species. The biodiversity hotspot concept is an example of a habitat prioritization approach, identifying critical areas for conservation action based on the enormous diversity in plant and animal life that these various areas support. Particular primate species are often prioritized for conservation based on an evaluation of their risk of extinction, and one of the workhorse techniques for such risk assessment is a set of procedures known as population and habitat viability analysis (PHVA). PHVA typically uses computer simulation to evaluate the likelihood that a particular population will persist in an area for a specified period of time given a set of demographic parameters defining the population (for example, fertility and mortality schedules, and the age and sex composition) and the specific extrinsic forces operating on both the population and local people (for example, stochastic fluctuations in environmental conditions, rates of habitat loss and fragmentation, and the availability of alternative economic opportunities).

Protected reserves. In the last two decades, much has been written about general principles of reserve design, but it is important to note that the efficacy of reserves is often limited by problems of enforcement; in many developing countries, there is little funding to support enforcement and, as a result, many reserves are protected in name only. This has led some conservationists to suggest that reserves should be designed with the additional explicit criterion of minimizing their accessibility via roads or river courses.

Promoting sustainable use. In the last 10 years, many conservationists have advocated the use of integrated conservation and development projects (ICDPs) as a strategy for promoting the sustainable use of primate populations and their habitats. Such projects aim to provide an economic incentive to local human populations for participating in conservation efforts, and are born of the realization that local people are unlikely to embrace conservation actions imposed by governments or external agencies without some clear benefit to themselves. While theoretically compelling, in practice ICDPs have seldom lived up to their promise, in part because they often fail to take into account the fact that successful projects may provide an economic incentive for increased human migration into an area, thus further taxing the natural resource base they aim to protect.

Other approaches. A number of other tactics have been used in recent primate conservation efforts with varying degrees of success, including breeding animals in captivity with the implicit goal of restocking wild populations, translocating animals between different natural areas to manage genetic diversity in situ, establishing stricter legal obstacles to trade in animal parts, and promoting ecotourism as an alternative source of revenue for local human populations. What appears clear is that no single conservation approach can be applied across the board; instead, effective conservation strategies vary depending on the demographic characteristics and ecological requirements of the species targeted for conservation and on the economic values that humans place on these species.

For background information *see* BIODIVERSITY HOTSPOTS; ECOLOGICAL MODELING; ECOSYSTEM; ENDANGERED SPECIES; POPULATION VIABILITY; PRIMATES in the McGraw-Hill Encyclopedia of Science & Technology.   Anthony Di Fiore

Bibliography. F. Achard et al., Determination of deforestation rates of the world's humid tropical forests, *Science*, 297:999–1002, 2002; J. Baillie and B. Groombridge (eds.), *1996 IUCN Red List of Threatened Animals*, IUCN, Gland, Switzerland, 1996; T. M. Brooks et al., Habitat loss and extinction in the hotspots of biodiversity, *Conserv. Biol.*, 16:909–923, 2002; G. Cowlishaw and R. Dunbar, *Primate Conservation Biology*, University of Chicago Press, 2000; Food and Agriculture Organization of the United Nations, *State of the World's Forests 2001*, FAO, Rome, 2001; C. Hilton-Taylor (ed.), *2002 IUCN Red List of Threatened Animals*, IUCN, Gland, Switzerland, 2002; W. R. Konstant et al., The world's 25 most endangered primates, *Neotropical Primates*, 10:128–131, 2002; N. Myers et al., Biodiversity hotspots for conservation priorities, *Nature*, 403:853–858, 2000; C. A. Peres, Effects of subsistence hunting on vertebrate community structure in Amazonian forests, *Conserv. Biol.*, 14:240–253, 2000.

Punctuated equilibria (evolutionary theory)

The term "punctuated equilibria" refers to Stephen Jay Gould and Niles Eldredge's 1972 proposal regarding the nature of biological data as preserved in the fossil record and the implications that those data have for evolutionary theory. Eldredge and Gould proposed that the history of life was not characterized by morphologically "connecting together all the extinct and existing forms of life by the finest of graduated steps," as hypothesized by Darwin in 1859 and subsequently portrayed by mid-twentieth-century textbooks (**Fig. 1**). Rather, Gould and Eldredge envisioned evolutionary history as a network characterized by long periods of morphologic stability (stasis mode) punctuated here and there by rapid events of speciation (punctuation mode) [**Fig. 2**]. When first proposed, the concept of punctuated equilibria caused considerable controversy because aspects of the punctuated-mode pattern were promoted as falsifying some long-held beliefs about Neo-Darwinian evolutionary theory—termed "phyletic gradualism" by Gould and Eldredge. Later, punctuated equilibria also came to be used by creationists (incorrectly) to throw doubt on the entire theory of evolution. Today, the punctuated equilibria model is regarded as a valid characterization of a number of fossil lineages, but is no longer regarded as challenging contemporary evolutionary theory.

Punctuation mode: allopatric speciation. Drawing on the work of scientists from the mid-1800s to the mid-1900s, most evolutionary biologists and paleontologists had come to regard the speciation process as reflecting the accumulation of a number of small, progressive changes in organismal populations. The progressive nature of this transformation was thought to allow these populations to acquire

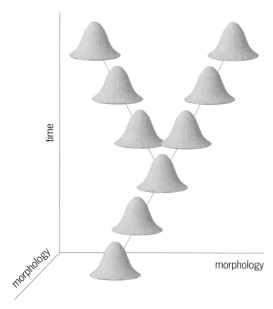

Fig. 1. Gradualist model of morphological change through time. In this conceptualization a series of fossil populations are represented as morphological frequency distributions moving to the right across a morphological change field (bottom of diagram) at a uniform rate. The speciation event occurs in the middle of the diagram when a branch of the original population splits off from the main lineage and begins moving off on a new trajectory (to the left), once again at a uniform rate. Note that consistent trends in morphological change are present within species as well as between species under the gradualist model. This figure, which is similar to the illustrations produced in paleontological textbooks of the 1950s and 1960s, was criticized by N. Eldredge and S. J. Gould because it implies substantial levels of morphological change within species (between speciation events). Defenders of the gradualist model, however, argue that such illustrations are grossly oversimplified caricatures of Neo-Darwinian evolutionary theory.

novel attributes while retaining their overall fitness. This model, in turn, led to the concept of large-scale evolutionary change arising as a direct extrapolation of these small-scale modifications.

In conjunction with this work, it was realized that small populations characterized by limited interbreeding with other populations and located on the periphery of a species' geographical range (where they might be subjected to unusual environmental conditions) would be favorable candidates within which morphological variations might spread and become established, producing new species. Such populations are termed allopatric. Many evolutionary theorists, including Eldredge and Gould, believe allopatric speciation is a dominant mode of species production.

Although Eldredge and Gould accepted the Darwinian mode of species transition by finely graduated steps within allopatric populations, they were the first to document examples of the punctuated speciation pattern—drawn from their own paleontological research (on snails and trilobites, respectively)—and argue that such a pattern was the expected signature of allopatric speciation in the fossil record.

Two factors are thought to be responsible for this. First, the small size and isolated nature of the allopatric population reduces the probability that it will be preserved in the fossil record. This means that for many (if not most) species direct evidence of the gradual morphological transition is simply not there to be found. New species seem to appear suddenly in the fossil record because these appearances record the migration of fully formed species into local areas, not evolutionary transformation in situ. Second, the short duration of such speciation events (approximately 5000–10,000 years) relative to the long intervals over which species exist (approximately 5 million to 10 million years) means that, even if a paleontologist was fortunate enough to happen upon the fossil record of an isolated population in which a new species was formed, the speciation event would be compressed into a very small physical interval, perhaps as small as a single depositional layer or bed. Given the realities of sampling the fossil record, this entire transition could easily be subsumed within a single rock sample and appear as a single point on a graph or table. Thus, the punctuation-mode aspect of punctuated equilibria theory does not conflict with the speciation mechanism of standard Neo-Darwinian evolutionary theory. Instead, it adjusts scientists' expectations of what an allopatric speciation event would look like in the fossil record.

Equilibrium mode. Theoretical justification for Gould and Eldredge's second pattern—the stasis or equilibrium mode of morphological change—has proven more controversial. From the beginning, both authors were struck by the manner in which

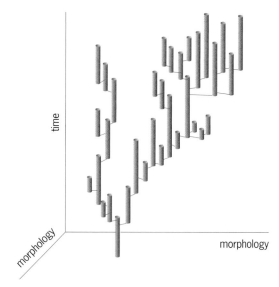

Fig. 2. Punctuated equilibria model of morphological change through time (*after S. J. Gould and N. Eldredge, 1972*). In this conceptualization of evolutionary change, species are represented as straight rods (symbolizing that morphological change within species is limited to random variation about a static average) and their pattern of ancestor-descendent relations is represented by short, horizontal ties (symbolizing rapid morphological changes occurring in small, isolated populations).

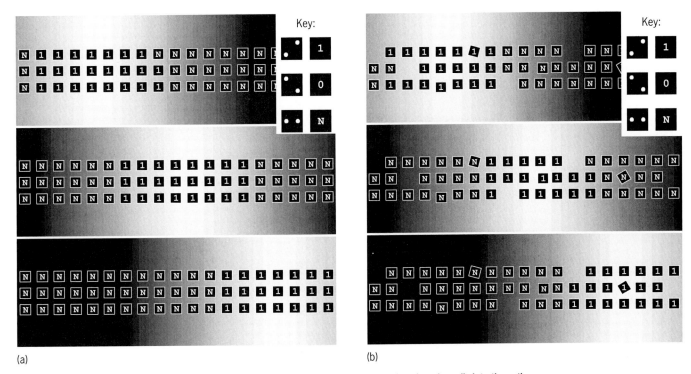

(a)   (b)

Fig. 3. Clocked molecular QCA. Lighter shaded regions are those where the electric field pushes the cells into the active state. Snapshots at different times show the information packet moving to the right. (*a*) Perfectly ordered triple-wide wire. (*b*) Disordered wire. The triple redundancy provides immunity to many kinds of disorder.

have been proposed which could be clocked using locally produced fields from clocking wires buried in the substrate on which the molecules are attached. This would not entail making individual contact to each molecule.

Much work remains to be able to reliably position QCA molecules on a surface and construct circuits from them. The cell-cell interaction in QCA provides significant resilience against defects in positioning molecules. In fact, using three-wide QCA wires (**Fig. 3**) could provide built-in error recovery. The potential for achieving truly molecular functional densities, up to 10^{14} devices per square centimeter, makes this a very attractive avenue for further research.

Power gain and dissipation. One crucial element for any robust digital technology is power gain. As information is moved from stage to stage, some of the energy in the signal path is inevitably lost to the environment. Devices with power gain allow the restoration of this energy so that signal integrity is maintained as the information is moved and processed. With conventional devices, this is accomplished by current from the power supply. With QCA, the restoring energy comes from the clock. Power gain in QCA has been examined theoretically and demonstrated experimentally in metal-dot circuits. It is a crucial feature that distinguishes QCA from many other nanodevice proposals.

Power dissipation will likely be the chief limiter of high-performance nanoscale devices. QCA circuits are clocked adiabatically; that is, the potentials are changed gradually enough that each cell is always

very close to its instantaneous ground state. Because the cells are never excited energetically much above their ground states, they do not have much energy to dissipate to their environment as heat. QCA therefore generates extremely low levels of power dissipation, approaching the theoretical limits imposed by fundamental considerations of logical and physical reversibility.

Extensions. QCA has been implemented in magnetic systems, and spin-based proposals exist at the theoretical level. QCA is not quantum computation in the sense that it uses only classical degrees of freedom (charge configuration) to encode information. However, if one makes the generous assumption (as is usually done) that switching can be accomplished without breaking quantum-mechanical phase coherence across the entire cellular array, then one can show that quantum computation could be accomplished with a QCA circuit.

Architecture. To exploit QCA cells as computational devices requires rethinking computer architecture from the ground up on a basis different than transistors. This process has begun, and circuits as complex as a simple microprocessor have been designed and simulated. The connection between layout and timing is the most profound new element in QCA architecture. Much work remains to explore the implications.

For background information *see* AUTOMATA THEORY; INTEGRATED CIRCUITS; IRON; LIGAND; MEDICAL IMAGING; MESOSCOPIC PHYSICS; METALLOCENES; NUCLEAR MAGNETIC RESONANCE (NMR); OPTICAL PUMPING; QUANTIZED ELECTRONIC STRUCTURE

(QUEST); TRANSISTOR; TUNNELING IN SOLIDS in the McGraw-Hill Encyclopedia of Science & Technology.

Craig S. Lent

Bibliography. I. Amlani et al., Digital logic gate using quantum-dot cellular automata, *Science*, 284(5412): 289-291, 1999; C. S. Lent, Bypassing the transistor paradigm, *Science*, 288:1597-1598, 2000; C. S. Lent et al., Quantum cellular automata, *Nanotechnology*, 4:49-57, 1993; C. S. Lent, B. Isaksen, and M. Lieberman, Molecular quantum-dot cellular automata, *J. Amer. Chem. Soc.*, 125:1056-1063, 2003; A. O. Orlov et al., Realization of a functional cell for quantum-dot cellular automata, *Science*, 277(5328): 928-930, 1997; J. Timler and C. S. Lent, Power gain and dissipation in quantum-dot cellular automata, *J. Appl. Phys.*, 91(2):823-831, 2002.

Renewable energy sources

With the exception of hydropower, renewable energy sources such as wind, biomass, geothermal, and solar have so far played only a minor role in worldwide electricity production. The primary reason for this delay in market penetration of these energy sources is their high initial cost. However, increasing awareness about the considerable external costs of producing electricity from fossil fuels can help justify the high capital costs for renewable energy projects (see **table**). Another drawback for some of the renewable energy sources, namely solar and wind, is the intermittent nature of the resource. Due to the increasing availability of storage technologies, however, these barriers are slowly being surmounted.

Electricity derived from nonrenewable energy sources (fossil fuels) produces emissions such as nitrogen, sulfur, and carbon dioxides. Not only do these emissions add to climate change via the greenhouse effect, but they also contribute to acid rain. In comparison, renewable "green energy" sources such as hydropower, wind, geothermal, and solar energy create little to no pollution and are relatively noninvasive. One further advantage of these renewable energies is that they have an almost limitless supply of power. In view of all these advantages, commercial-scale electricity production from renewable energy

sources is increasing at a much faster rate than that from conventional nonrenewable sources.

Hydropower. Hydropower is energy that is generated by the flow of water. The electrical generators in hydroelectric plants have the same operating principles as their counterparts in fossil fuel–fired power plants, except that the turbine is water-driven rather than steam-driven. The turbine is attached to a shaft which connects to a generator through a series of devices called speed increasers. Hydroelectric installations can be placed in two categories: those that have reservoirs upstream from the generation station, and those that are run-of-river (relying only upon the flow of the river to keep the plant running). In projects with reservoirs, the plant operator can use the dam to store water when there is a lot of river flow and can release the water when there is little river flow, to stabilize the water power captured by the turbines. In many cases the reservoir is used for flood control purposes as well.

At the end of 2000, approximately 740 gigawatts of hydropower contributed about 21% of electricity generation capacity worldwide. Opportunities for new and large hydropower projects are very limited in most developed countries, primarily because most economically exploitable sites have already been developed. There are many potential hydropower sites in the developing world, but they would require flooding large tracts of land (in order to store water behind dams), and concern about the damage to wildlife habitats and the risk of uprooting indigenous peoples places severe restrictions on many large hydropower development sites.

Wind energy. Wind power uses kinetic energy that is extracted from the wind and converted into electric energy using turbines and generators. In horizontal-axis wind turbines, a tower is used to hold the blades and nacelle (an enclosure containing the electric generating equipment) at a certain height to catch the wind. The nacelle houses the gearbox which connects to the blades, electrical generator, and the control mechanism to convert the kinetic energy of the wind to electrical energy. The nacelle is free to rotate around on a vertical axis to always face the incoming wind as closely as possible. The pitch angle of the rotor blades can be controlled to optimize the energy capture and to protect the turbine against damage during high winds (see **illus.**).

Even though the concept of mechanical power from wind energy dates back many centuries, commercial-scale wind electricity has been produced only for the last two decades and has become commercially competitive only in the last few years. At the end of 2002 the worldwide wind energy capacity was 32,037 megawatts, of which 23,832 MW was in Europe; the United States capacity was 4674 MW; and the capacities in India and China were 1702 MW and 473 MW respectively.

Modern wind turbines are for the most part capable of generating electricity at all times, when adequate wind is available, with typical availabilities exceeding 98%. On the average, they produce

Relative cost of electricity generation		
Resource	Generation cost, US cents/kWh	External cost of generation,* US cents/kWh
Coal	3.11–3.41	1.94–14.6
Gas turbine	2.53–3.41	0.97–3.89
Nuclear	3.31–5.74	0.19–0.58
Good wind site	5.84	0.05–0.24
Optimal wind site	3.89	0.05–0.24

*The estimated costs to society and the environment due to their operation, not including nuclear waste and decommissioning costs.
SOURCE: International Atomic Energy Agency, ExternE, and *Wind Power Monthly* (*Wall Street Journal*, August 27, 2002).

electricity at the rate of 35–40% of full capacity in good wind resource areas. In comparison, modern-day coal-fired power plants have a capacity factor of about 70%. The audible noise from wind turbine generators is of concern to communities living nearby. As a result, more and more such generators are being placed away from habitable areas and off-shore. Today, large new wind farms at excellent wind sites generate electricity at a cost of 4–6¢/kWh. That places the cost of power from the United States' most efficient wind farms in a range that is competitive with that of electricity from new coal-fired power plants, but without any harmful greenhouse gas emissions.

Electricity from biomass. Biomass comes from organic materials produced by plants, such as leaves, roots, and stalks. Common sources of biomass include agricultural wastes, wood materials, municipal waste, and energy crops (grown specifically to produce energy). Biomass power is one of the largest sources of renewable electricity, with a worldwide capacity of 24.7 GW at the end of 2000. The United States was the largest biomass electricity generator in the world, producing over 11 GW during that year. Biomass can be converted into electricity by one of several processes. The majority of biomass electricity is generated using a steam cycle, where biomass material is burned to generate steam in a boiler. The resulting steam is then used to turn a turbine which is connected to a generator. Biomass can also be used with coal in a furnace to run a boiler. This co-firing process helps to reduce harmful air emissions (such as sulfur dioxide) from coal-fired power plants. Another way to generate electricity is by using chemical processes to convert solid biomass into a fuel gas such as methane. The fuel gas can then be used in a piston-driven engine, a high-efficiency gas turbine generator, or a fuel cell. Fuel gas can also be integrated into industrial manufacturing plants for combined heat and power (CHP). Although energy derived from biomass could be thought of as green energy in that it is renewable, it does contribute to increased carbon dioxide levels, exacerbating the greenhouse effect.

The cost to generate electricity from biomass depends on the cost of the biomass fuel supply, the type of technology used, and the size of the power plant, which can range from a few kilowatts up to 80 MW. In direct-fired biomass power plants in the United States, generation costs are about 9¢/kWh. It is expected that advanced technologies such as gasification-based systems could bring this cost down to as little as 5¢/kWh.

Geothermal power. Geothermal power comes from the energy contained in the Earth's hot subsurface rock layers. Water is used to absorb heat from the hot rock and transport it to the Earth's surface, where it is converted to electrical energy through turbine-generators. In a flash steam power plant, water from high-temperature (>240°C, or 464°F) reservoirs is partially flashed to steam (steam is produced by reducing pressure on the water). The heat is converted

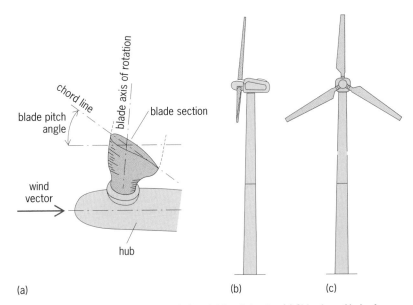

Horizontal-axis wind turbine with upwind, variable-pitch rotor. (*a*) Side view of hub of rotor, showing base of one blade. (*b*) Side view. (*c*) Front view.

to mechanical energy by passing the steam through low-pressure steam turbines, which in turn drive a generator to produce electricity. A small fraction of geothermal energy generation worldwide is generated using a heat exchanger and secondary working fluid to drive the turbine. Exploitable geothermal reservoirs exist in high-temperature, highly permeable, fluid-filled rock within the Earth's upper crust, typically in areas associated with young volcanic rocks. Driven by heat loss from underlying magma, hot fluids rise along preexisting zones of high permeability.

As of 1999, the electricity generating capacity from geothermal sources worldwide was 8 GW. During that year geothermal power plants generated 49 billion kWh of electricity. These plants are highly capital intensive because enough steam-supply wells have to be drilled up front to provide the full plant capacity at startup. Nonetheless, after they are put into use, these wells continue to supply steam for 20 years, resulting in low operating costs.

Solar energy. Energy transmitted from the Sun in the form of electromagnetic radiation can be captured and stored to produce electricity. There are two systems—solar thermal and solar photovoltaic—that are used to generate commercial-scale electricity worldwide. In the solar thermal system, solar heat is generally converted to electricity by concentrating the incoming sunlight to raise the temperature of a working fluid enough to produce steam and run a turbine. In the United States, commercial power plants generated 410 MW of solar thermal electricity at the end of 2000. In the photovoltaic process, a region of the light spectrum imparts enough energy to create electron-hole pairs in a semiconductor material to directly convert light into electricity. The fundamental unit of a photovoltaic panel is the cell. The main material of the cell is a semiconductor material such as gallium arsenide, crystalline silicon, or amorphous silicon. Electricity is collected and transported by metallic contacts placed on both surfaces of the cell. The world's largest photovoltaic power plant—a 2-MW system—is operated by the Sacramento (California) Municipal Utilities District. Electricity generated from this system during the day is used in homes or businesses. As with other privately owned renewable sources, any unused electricity goes into the utility grid for net metering (turning the customer's meter backward). During the night, power is drawn from the grid, similar to a large battery backup system. At present, commercially available photovoltaic modules convert sunlight into electricity with efficiencies ranging 6–20%. Average photovoltaic modules cost under $5 per peak watt, and produce electricity from 25–40¢/kWh depending on the technology (and sunshine conditions). At the end of 2000, worldwide solar photovoltaic capacity stood at over 1000 MW.

For background information *see* BIOMASS; CONSERVATION OF RESOURCES; DAM; ELECTRIC POWER GENERATION; ENERGY SOURCES; GEOTHERMAL POWER; PHOTOVOLTAIC CELL; RENEWABLE RESOURCES; SOLAR ENERGY; WIND POWER in the McGraw-Hill Encyclopedia of Science & Technology.   Saifur Rahman

Bibliography. R. Dipippo, *Geothermal Energy as a Source of Electricity: A Worldwide Survey of the Design and Operation of Geothermal Power Plants*, 2000; W. Harris, *Water Makes Hydropower (From Resource to Energy Source)*, 2003; D. L. Klass, *Biomass Renewable Energy, Fuels, and Chemicals*, 1998; T. Markvat (ed.), *Solar Electricity*, 2d ed., 2000; S. Rahman, Green Power—What is it and where can we find it, *IEEE Power Energy Mag.*, vol. 1, no. 1, 2003.

Retinoids and CNS development

The developing central nervous system (CNS) crucially depends on the maintenance within embryonic cells of precise levels of retinoic acid (RA), a low-molecular-weight, lipophilic compound derived from vitamin A. The level of RA must not rise or decrease excessively during pregnancy, or CNS damage will result. Although the broad effects of excess or insufficient levels of RA on the nervous system have been known for some time, it has been only within the last 10 years, through the use of anatomically specific molecular markers, that we have begun to understand the specific role of retinoic acid in body patterning and CNS development.

Retinoic acid. Retinoic acid acts at the level of the cell nucleus to establish or maintain patterns of gene activity, including gene activity in the developing CNS. Vitamin A, or retinol, is obtained from the diet in the form of carotenoids (from plant sources) or retinyl esters (from animal sources), and the family of molecules derived from retinol is known as the retinoids. Cells that require RA metabolize it from retinol, which travels in the blood linked to retinol-binding protein after its release from liver stores. In the cytoplasm, retinol binds to cellular retinol-binding protein (CRBP), which facilities its enzymatic conversion to RA. There are two types of enzymes that perform this conversion. The first type, the retinol dehydrogenases or alcohol dehydrogenases, convert retinol into retinaldehyde; the second type, the retinaldehyde dehydrogenases, convert retinaldehyde into RA (**Fig. 1**). RA is then bound in the cytoplasm by cellular retinoic acid–binding protein (CRABP).

There are several different forms of active RA: all-*trans*-RA and 9-*cis*-RA, which are the ligands for the nuclear retinoic acid receptors and didehydro-RA, which is found predominantly in bird embryos. Subsequently, these active retinoic acids are catabolized to compounds such as 4-oxo-RA, 4-OH-RA, and 18-OH-RA by a group of cytochrome P450 enzymes called CYP26 (Fig. 1).

After synthesis in the cell cytoplasm, all-*trans*-RA and 9-*cis*-RA enter the nuclei and bind to ligand-activated (or ligand-dependent) transcription factors, the retinoic acid receptors (RARs) and the retinoid X receptors (RXRs), which form part of the gene superfamily that includes the steroid hormone receptors (**Fig. 2**). There are three of each receptor type: RARα, RARβ, RARγ and RXRα, RXRβ, RXRγ. Each of these six receptors is encoded by a specific gene from which several isoforms can be generated by transcriptional modifications such as the use of multiple promoters (DNA sequences that indicate where transcription should begin) and posttranscriptional changes such as differential splicing (the use of alternative exons, or coding regions, in mRNA). The ligand for the RXRs is 9-*cis*-RA, whereas the RARs bind both 9-*cis*-RA and all-*trans*-RA. These receptors act as ligand-dependent transcription factors by recognizing consensus sequences known as retinoic acid response elements (RAREs), which are present in the enhancer sequences (DNA sequences that increase transcription) of RA-responsive genes. The RARs and RXRs do not act alone, but as heterodimers; for example, the proteins RARα and RXRβ are paired to form a functional protein complex.

Developing embryo. The importance of maintaining the correct level of RA in the embryo has been established from many studies stretching over the twentieth century, which examined the effects of excess or insufficient RA on the development of animal embryos. Excess RA, for example, causes severe defects of the nervous system, including hydrocephalus (abnormal accumulation of cerebral spinal fluid within the skull), anencephaly (malformation of the skull with little or no brain present), exencephaly (formation of brain outside the skull), anophthalmia (absence of eyes), microphthalmia (smallness of one or both eyes), defects of the retina, and abnormal neural crest migration (which results in defective cranial sensory nerve formation). Several of these abnormalities, such as cerebellar and cranial nerve defects, have been seen in infants born to mothers who have taken Accutane® (13-*cis*-retinoic acid) during pregnancy.

Experimental results also strongly suggest a role for RA in neuronal differentiation. For example, when embryos were treated with RA at tail bud stages, ectopic neural tubes were produced within the tail bud, showing that RA can induce the formation of neural tissue from mesoderm (middle germ layer). RA has also repeatedly been shown to induce differentiation of neurons and glia in tissue cultures of embryonal carcinoma cells. When dissociated or explanted neuronal cells are used, RA induces more and/or longer neurites (axons or dendrites) than would be normally observed. Hundreds of genes have now been identified which are regulated by RA during this process of neuronal differentiation.

The opposite type of study, in which RA is made unavailable to the embryo by dietary deprivation, has also highlighted the CNS. For example, an early study found that RA deprivation induced anophthalmia in pig embryos. Subsequent experiments revealed that excess and deficient amounts of RA produced a remarkably similar range of embryonic defects. In the CNS, these defects included hydrocephalus, spina bifida, anophthalmia, and microphthalmia.

These early observations, in which rather gross defects to the nervous system were reported, did not really give precise information about the role of RA in CNS development. However, in the last 10 years, the study of the ability of RA to induce embryonic defects has been revisited for two reasons. First, there are now anatomically specific molecular markers, such as the *Hox* genes (which regulate body patterning in developing embryos), that can be used to determine the precise positional changes of body segments in embryos that either are retinoid-deficient or have been treated with excess RA. Second, a great deal has been learned about the molecular biology of RA action via the RARs, and the defects caused by retinoid deprivation have now been recapitulated in mice genetically altered to have mutations in RAR genes.

Excess retinoic acid. Administration of excess RA has two observable effects on the normal development of the CNS (**Fig. 3***a*). The first effect involves the anterior CNS regions, the forebrain, and the second is centerd on the hindbrain.

Forebrain effects. When embryos of experimental organisms (ranging from fish to mammals) are treated with RA at early stages of development, such as mid/late primitive streak stages in mice or rats, the forebrain and eyes are lost (Fig. 3*b*). This involves the downregulation (decreased expression) of genes such as *Otx2*, *XCG-1*, *Emx1*, and *Dlx1*. As a result of this loss of anterior tissue, the remaining hindbrain and spinal cord seem to expand to compensate, and the expression of posterior genes such as *Krox20*, *Pax2*, and various *Hox* genes are upregulated (increased).

Hindbrain effects. Effects centered on the hindbrain are seen when excess RA is administered at slightly later stages of development or at lower doses. The

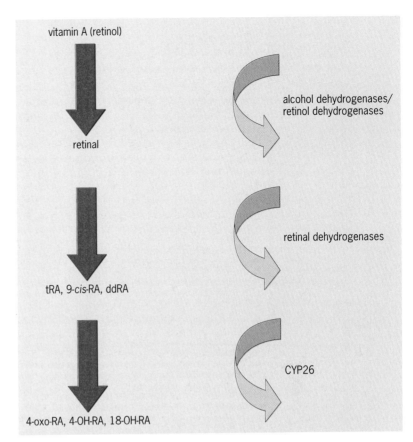

Fig. 1. Metabolic pathway from vitamin A (retinol) to retinoic acid and its subsequent catabolism. Retinol is first oxidized to retinal via the action of the medium-chain alcohol dehydrogenases (there are at least five of these enzymes) and the short-chain retinol dehydrogenases (there are at least eight). Retinal is then metabolized to various forms of retinoic acid (all-*trans*-retinoic acid, 9-*cis*-retinoic acid, and didehydroretinoic acid) by the retinal dehydrogenases (there are at least four). These retinoic acids bind to the retinoic acid receptors (see Fig. 2) and are subsequently catabolized to inactive metabolites by a class of P450 enzymes called CYP26 (there are at least three of these enzymes).

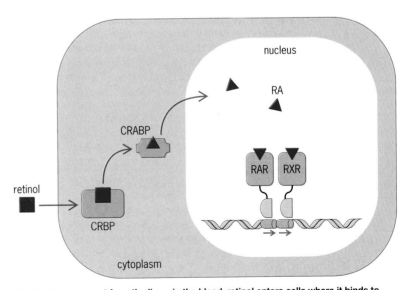

Fig. 2. After transport from the liver via the blood, retinol enters cells where it binds to cellular retinol-binding protein (CRBP) and is metabolized to retinoic acid (RA) under the action of the enzymes described in Fig. 1. RA is then bound in the cytoplasm by cellular retinoic acid-binding protein (CRABP). The RA generated then enters the nucleus, where it binds to two classes of retinoid receptors, the retinoic acid receptors (RARs) and the retinoid X receptors (RXRs). These two receptors heterodimerize and activate gene transcription. All-*trans*-retinoic acid can bind only to the RARs, whereas 9-*cis*-retinoic acid can bind to both the RARs and RXRs.

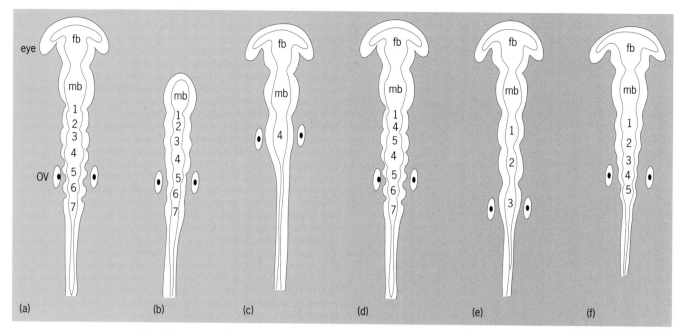

Fig. 3. Examples of the abnormalities of brain structure caused by an excess or a deficiency of RA. (*a*) Representation of a normal developing brain structure of a chick embryo. At the top is the forebrain (fb) with a bulge either side from which the eyes will develop. Next is the midbrain (mb) and then the hindbrain, which is composed of seven bulges or rhombomeres. Below the hindbrain is the spinal cord. Beside rhombomere 5 are two ovals which represent the otic vesicles (OV) from which the ears will develop. (*b*) Effect of a high level of excess RA at an early stage of development. The forebrain and eyes are missing. (*c*) Effect of a lower dose of RA at a slightly later stage than *b*. The anterior hindbrain has become one large rhombomere 4, and the remaining tissue has lost its segmental structure. The otic vesicles have moved forward along with the loss of anterior hindbrain tissue. (*d*) An alternative effect to *c* is shown, whereby rhombomeres 2 and 3 are transformed into a 4 and 5. (*e*) Effect of removing all the RA from the embryo or inhibiting all the RA signaling. In this case there are only three rhombomeres instead of the normal seven, and they enlarge to fill the available hindbrain space. (*f*) Effect of decreasing RA signaling (not completely inhibiting it). Here rhombomeres 6 and 7 have been lost. As inhibition of signaling increases, more rhombomeres are lost.

normal developing hindbrain consists of a series of repeating bulges known as rhombomeres, each expressing a distinct set of marker gene expressions (Fig. 3*a*). Each region of the hindbrain gives rise to discrete motor neuronal populations and cranial sensory ganglia. After RA treatment, the hindbrain *Hox* genes are induced and ectopically expressed in more anterior domains. Then they regress, leaving behind altered expression patterns that result in altered anatomies. Some embryos lose the anterior rhombomeres (numbers 1, 2, and 3), which become fused into a single large rhombomere, number 4, according to gene expression data. The remaining rhombomeres (numbers 5, 6, and 7) lose their boundaries (Fig. 3*c*). In other embryos, the gene expression patterns that are left behind suggest a transformation of rhombomere phenotype rather than a loss of rhombomeres. In these cases, the normal rhombomeric sequence of 1, 2, 3, 4, 5, 6, 7, is transformed into 1, 4, 5, 4, 5, 6, 7 (Fig. 3*d*). This means that instead of the trigeminal nerve arising from rhombomere 2 and the facial nerve arising from rhombomere 4, the trigeminal nerve is gone and is replaced by a duplicated facial nerve.

Deficiency of RA or decreasing RA signaling. Detailed studies on the effects of dietary deficiency of RA have been performed on the embryos of quails and rats. Various other methods have been utilized to create a deficiency of RA or decrease RA signaling. For exam-

ple, the RARs and RXRs have been mutated singly, doubly, or triply to prevent RA signaling through the receptors. The enzyme that plays the most significant role in synthesizing RA in the embryo, RALDH2, has also been mutated. Another method involves culturing chick embryos in synthetic retinoid-like compounds, which irreversibly bind to the RARs to prevent signaling.

These various experimental manipulations reveal that the absence of RA or RA signaling is associated with four major defects in the CNS:

Myelencephalon fails to develop. First, the posterior part of the hindbrain, known as the myelencephalon, completely fails to develop in the absence of RA. Instead of the normal seven rhombomeres, numbers 4, 5, 6, and 7 are missing, and the remaining three rhombomeres expand in size to compensate (Fig. 3*e*). Thus, the anterior hindbrain becomes attached to the spinal cord. When RA signaling is gradually reduced in a stepwise fashion rather than removed altogether, the posterior hindbrain rhombomeres are gradually lost one by one as the reduction occurs, demonstrating that the posterior hindbrain is gradually built up by increasing levels of RA (Fig. 3*f*).

Neurites fail to extend. The second defect seen in the complete absence of RA in quail embryos is the failure of the majority of neurites to extend from the developing neural tube into the periphery. As a result, motor neurons fail to form, and within the neural tube there are some neurons that manage to extend neurites, but their trajectories are completely chaotic and abnormal. These results clearly complement the findings of studies of embryonal carcinoma cells in which RA induced the differentiation of neurons. This phenomenon is similar to the effect of RA on primary neurons in amphibian embryos. These neurons, which are the first to differentiate and are unique to fish and amphibians, are responsible for getting the embryo moving after hatching and, thus, are important for survival of the young fish or amphibian embryo. The addition of RA to these embryos induces excess primary neurons to differentiate; conversely, decreasing the signaling of RA or the levels of RA reduces their number. These effects on neuronal differentiation are mediated by the control that RA exerts over prepattern genes such as *X-ngnr-1* and *X-MyT1* and neurogenic genes such as *X-delta-1.*

Apoptosis of neural crest cells. The third defect seen in RA-deficient embryos of quails and rats is the death of neural crest cells by apoptosis, which inhibits development of the peripheral sensory nervous system and associated ganglia as well as the sympathetic nervous system and the enteric nervous system (a network of nerves located in the gastrointestinal wall), which are also derived from the neural crest. This defect is lethal for the embryo and complements data in vitro which showed that cultured neural crest cells require RA for their survival.

Abnormal shape and growth of spinal cord. Fourth, the shape in transverse section of the developing spinal cord and its subsequent growth is abnormal, suggesting a role for RA in the proliferative expansion of the CNS. In addition, several of the neuronal subtypes present in the dorsoventral axis of the spinal cord are abnormal. In the normal spinal cord, subsets of sensory neurons are present in the dorsal part, subsets of interneurons are present in the central part, and subsets of motor neurons are present in the ventral part. This precise patterning is set up by two concentration gradients: a concentration gradient of a molecule called sonic hedgehog originating at the ventral part of the cord and decreasing dorsally, and a concentration gradient of bone morphogenetic proteins originating in the dorsal part of the cord and decreasing ventrally. RA is somehow involved in the establishment of these two gradients because in its absence the ventral gradient is expanded, the dorsal gradient is decreased, and some of the classes of interneurons that form in between these two gradients are missing. Later in development, RA signaling is involved in the differentiation of one of the subsets of motor neurons, the lateral motor column (LMC) neurons. The enzyme RALDH2 is expressed in these motor neurons, and when it is ectopically expressed, ectopic motor neurons are generated.

Clinical implications. Since studies have shown that excessive levels of RA are so deleterious to the developing CNS, this must be avoided at all costs during pregnancy. As mentioned above, the oral drug Accutane® (13-*cis*-retinoic acid), a treatment for severe cystic acne, when inadvertently ingested during pregnancy has resulted in several of these CNS abnormalities appearing in newborn infants. Examples include cerebellar defects, a decrease in size of the forebrain, motor and sensory developmental delays, and severe mental retardation. Children who have been exposed but do not display these major abnormalities often show cognitive impairment. Conversely, a maternal diet deficient in RA could lead to CNS abnormalities in infants, such as cognitive impairment. Although maternal RA deficiency is perhaps unlikely in the developed world, in the developing world childhood blindness caused by vitamin A–deficient maternal diets is a serious problem.

For background information *see* ANIMAL MORPHOGENESIS; CENTRAL NERVOUS SYSTEM; DEVELOPMENTAL BIOLOGY; DEVELOPMENTAL GENETICS; EMBRYONIC DIFFERENTIATION; EMBRYONIC INDUCTION; NERVOUS SYSTEM (VERTEBRATE); NEUROBIOLOGY; VITAMIN A in the McGraw-Hill Encyclopedia of Science & Technology.   Malcolm Maden

Bibliography. V. Dupe and A. Lumsden, Hindbrain patterning involves graded responses to retinoic acid signalling, *Development*, 128:2199–2208, 2001; H. Kalter and J. Warkany, Experimental production of congenital malformations in mammals by metabolic procedure, *Physiol. Rev.*, 39:69–115, 1959; E. J. Lammer et al., Retinoic acid embryopathy, *New Eng. J. Med.*, 313:837–841, 1985; J. Langman and G. W. Welch, Effect of vitamin A on development of the central nervous system, *J. Comp. Neurol.*, 128:1–16, 1967; M. Maden, Retinoid signalling in the development of the central nervous system, *Nature Rev. Neurosci.*, 3:843–853, 2002; R. E. Shenefelt, Morphogenesis of malformations in hamsters caused by retinoic acid in relation to dose and stage at treatment, *Teratology*, 5:103–118, 1972; O. Wendling et al., Roles of retinoic acid receptors in early embryonic morphogenesis and hindbrain patterning, *Devleopment*, 128:2031–2038, 2001.

Ring-opening polymerization

In ring-opening polymerization, macromolecules are formed from cyclic monomers such as cyclic hydrocarbons, ethers, esters, amides, siloxanes, and sulfur (eight-membered ring). Thus, ring-opening polymerization is of particular interest, since macromolecules of almost any chemical structure can be prepared.

Polymerization is initiated with the breaking of a single (sigma) bond in the cyclic monomer. The driving force of ring-opening polymerization comes from the strain of the rings. The major source of the ring strain is the angular strain, with the release of the strain being energetically favorable.

As the ring size increases, ring strain decreases, although this change is not absolute. Some cyclic monomers, such as six-membered cyclic ethers, are virtually strain-free, and their polymerization to high-molecular-weight polymer is not possible. By contrast, six-membered cyclic esters and siloxanes (for example, hexamethylcyclotrisiloxane) polymerize easily, due to their particular conformation. Conformational strain (related to the opposition of hydrogen atoms across the rings) is responsible for the strain for some of the larger cyclic monomers.

The majority of heterocyclic monomers polymerize by ionic mechanisms. That is, either anions or cations located at the end of a growing macromolecule attack the monomer molecule, breaking a bond between the heteroatom (for example, nitrogen or oxygen) and the adjacent atom so that both bonding electrons remain with one of the atoms (heterolytic cleavage).

Anionic polymerization. In the anionic polymerization of cyclic ethers or esters, mostly alkoxide anions are involved. Examples of monomers and propagating species are shown in reactions (1)–(3), where Mt^+ denotes metal cations, mostly Na^+ or K^+, and less often divalent cations.

tions in biomedicine, such as drug delivery systems and gels (highly swelling in water) used for healing wounds.

Polysiloxanes are very versatile polymers used in engineering applications due to their outstanding heat resistance, as well as in biomedicine because of their biocompatibility. Silicone rubber (sealant) is another popular product. It can be made as a liquid which solidifies on contact with air.

Another technically important family of polymers made by anionic ring-opening polymerization is based on ε-caprolactam. Polyamides derived from this monomer are used as fibers for textiles known as nylon (for example, Nylon 6).

Coordinated polymerization. The most recently developed living polymerizations of cyclic ethers and cyclic esters are based on multicenter concerted mechanisms of propagation. This mechanism of coordinated polymerization was elaborated in Lodz, Poland, and is described in reaction (4) for the

Cyclic ethers

$$\dots - (CH_2CH_2O)_2CH_2CH_2O^\ominus Mt^\oplus \quad (1)$$

Cyclic esters
(lactones)
[except four-membered rings]

$$\dots - C(CH_2)_xCH_2OC(CH_2)_xCH_2O^\ominus Mt^\oplus \quad (2)$$

Cyclic siloxanes

$$\dots - [Si(CH_3)_2O]_3 - [Si(CH_3)_2O]_2Si(CH_3)_2O^\ominus Mt^\oplus \quad (3)$$

(Ground state of the deaggregated,   (ε-Caprolactone)
actually growing species)

(Transition state)

$$(4)$$

(Ground state)

The polymerization of some of these monomers (for example, ethylene oxide or some lactones) are called "living polymerizations" because there is no termination to stop chain growth.

Polymers of ethylene oxide find numerous applications. Very high molecular weight polymer, up to several million, is used to increase the fluidity of water. Water containing a fraction of a percent of poly(ethylene oxide) can be pumped (for example, by firefighters) with much less energy and therefore for much longer (higher) distances. Lower-molecular-weight products are used as water-soluble surface-active agents. In addition, poly(ethylene oxide) is biocompatible, and it is finding applica-

polymerization of ε-caprolactone initiated with dialkylalkoxyaluminum (R and $R' = C_2H_5$), where in a concerted manner two bonds (1 and 1') are broken and simultaneously two new bonds (2 and 2') are formed. The coordinated propagating species are less reactive than ions and therefore are much more selective. Alkoxides of iron (Fe), tin (Sn), titanium (Ti), and several others polymerize cyclic esters in the same way. Only with these initiators is it possible to avoid formation of the unwanted cyclic oligomers, which notoriously contaminate the desired linear macromolecules.

The most comprehensively studied cyclic ester is L,L-dilactide, giving the novel industrial polymer poly(L-lactide) [reaction (5)]. It meets several criteria

L-Lactic acid → L,L-Dilactide →

Poly(L-lactide) (5)

of future polymers; namely, it is based on renewable starting materials, and its biodegradation products are water and carbon dioxide. Municipal waste is used as the starting material and converted in a high yield into L,L-dilactide. L,L-dilactide is a six-membered ring with two chiral centers in the molecule, and is prepared from L-lactic acid, a product of bacterial fermentation of mono- or polysaccharides.

The most common initiator is tin octoate [Sn(Oct)$_2$ or Sn(OC(O)CH(CH$_2$CH$_3$)(CH$_2$)$_3$CH$_3$]. It has recently been shown that Sn(Oct)$_2$ is not an initiator by itself, and like other carboxylates reacts with impurities present in monomer (for example, with lactic acid or water) and gives tin alkoxide, which initiates and then propagates in the same way, as described above for ROAlR$_2$ (that is, on the —SnOR bond).

Although L,L-dilactide is a six-membered cyclic compound that has relatively low ring strain, its thermodynamics of polymerization are favorable enough to ensure high monomer conversion at the polymer-monomer equilibrium. The polymerization proceeds with almost complete retention of configuration (at the stereocenter), in contrast to the anionic polymerization, in which substantial loss of optical activity (racemization) occurs.

The mechanical properties of poly(L-lactide) [containing approximately 5% of D,L-dilactide units] resemble those of polystyrene. Poly(L-lactide) is colorless and transparent, and is mostly used as a commodity plastic for films and fibers. When prepared with special precautions and at the high purity conditions, poly(L-lactide) can be used as a biocompatible material (for example, surgical sutures) whenever bioresorption is required.

In the recent years, the number of published works on, and related to, poly(L-lactide) has been fast increasing, reaching 500 papers in the year 2002.

Cationic ring-opening polymerization. A majority of the monomers that polymerize by anionic mechanisms can also be polymerized cationically. However, several monomers that undergo cationic ring-opening polymerization do not polymerize anionically, such as cyclic acetals, some cyclic ethers (oxetanes, tetrahydrofurans and larger rings), cyclic

1,3-oxaza monomers (for example, oxazolines), and cyclic imines (for example, aziridines). Cationic ring-opening polymerization can be initiated by protonic acids, carbocations, onium ions, or Friedel-Crafts catalysts, most often initiating in the form of complex acids with water. Polymerization proceeds with onium ions as active species, shown in reactions (6) and (7). Polyoxazolines upon hydrolysis can be converted into the linear polyamines.

Polytetrahydrofuranium ion — from — Tetrahydrofuran (6)

Polyoxazolinium ion — from — Oxazoline (7)

Cationic ring-opening polymerization is used in industry mostly for polymerization of 1,3,5-trioxane and its copolymerization with ethylene oxide or 1,3-dioxolane. This is the largest single polymeric material of technical importance made by cationic ring-opening polymerization. It has a number of applications as a high-melting polymer with excellent mechanical properties and high dimensional stability, such as electronic components and automotive parts.

The poly(1,3,5-trioxane) [that is, polyformaldehyde] hemiacetal end groups resulting directly from polymerization are unstable, and on heating produce formaldehyde by an unzipping reaction (8).

$$\ldots -CH_2OCH_2OCH_2OCH_2OH \longrightarrow \ldots -CH_2OCH_2OCH_2OH + CH_2O \quad (8)$$

The homopolymer is stabilized by blocking the end groups, thus increasing the energy of activation for the first step of depolymerization. Another way to increase the polyacetal stability is to copolymerize 1,3,5-trioxane with ethylene oxide (or 1,3-dioxolane). At the end groups or at their immediate vicinity are units, $\ldots CH_2CH_2O-$, that are stable toward depropagation, since removal of the monomer molecule from the chain's end would require much more energy—in this instance a highly strained three-membered ring would have to be formed.

Another industrially important product is polytetrahydrofuran (polyTHF), mostly used as an oligomer with a molecular weight in the range of a few thousand. The α,ω-dihydroxypolyTHF are used as soft, elastic blocks in elastoplastic multiblock copolymers, namely with aromatic polyesters or polyamide rigid blocks, as well as in elastic fibers. Elastic fibers (such as spandex), which are additives to many modern textiles, are polyurethanes which contain soft and elastic poly(THF) blocks.

Activated monomer mechanism. Cationic ring-opening polymerization of heterocyclic compounds is plagued by inevitable side reactions. Recently a novel mechanism of cationic ring-opening polymerization was elaborated. It was based on the observation that formation of cyclic side products by backbiting is suppressed in the presence of alcohols or water. Backbiting is the only way (except end-to-end biting) of macrocyclics formation when onium ions' active centers are located at the chain ends [reaction (9)]. In the traditional process, a macromolecule

$$\text{Growing chain with an active center*} \quad \longrightarrow \quad \left[\text{~~~~*} \right] \xrightarrow[\text{cyclic oligomer}]{\text{elimination of the}} \text{~~~~*} + \text{~~~~} \quad (9)$$

having active cationic species at its end ($\sim$*) attacks the nucleophilic unit in the chain (for example, an ether or an ester bond) and expels an oligocyclic compound in the next step.

The observed absence of macrocyclics when polymerization is conducted in the presence of alcohols led to the formulation of a mechanism in which cationic species are no longer at the end of the macromolecules, but are located on the monomer molecules. This is called the activated monomer mechanism (AMM). The simplest system, namely AMM in the cationic ring-opening polymerization of ethylene oxide, is as follows.

1. Protonation of the monomer [reaction (10)].

$$\text{"H}^{\oplus}\text{"} + \underset{\text{(Ethylene oxide)}}{CH_2-CH_2} \longrightarrow CH_2-CH_2 \quad (10)$$

("H$^{\oplus}$" is a proton, linked ionically to a counterion; counterions are omitted, however.)

2. Initiation: addition (S_N2 substitution) of the initiator (for example, an alcohol molecule ROH) to the activated (that is, protonated) monomer molecule [reaction (11)].

$$ROH + \underset{O^{\oplus} H}{CH_2-CH_2} \longrightarrow R-O-CH_2-CH_2-O^{\oplus} \quad (11)$$

3. Transfer of a proton from the protonated end group to the next monomer molecule, followed by addition of thus activated monomer molecule to the . . . —OH end group in the growing macromolecule.

Propagation proceeds by repeating this sequence. The tertiary oxonium ions at the end of macromolecules are not formed, eliminating formation of macrocyclics.

For background information *see* MANUFACTURED FIBER; POLYACETAL; POLYAMIDE RESINS; POLYETHER RESINS; POLYMER; POLYMERIZATION in the McGraw-Hill Encyclopedia of Science & Technology.

S. Penczek

Bibliography. G. Allen and J. C. Bevington (eds.), *Comprehensive Polymer Science*, vol. III, Chap. 31–37 (Anionic ROP) and 46–52 (Cationic ROP), Pergamon Press, Oxford, 1984; A. Duda and S. Penczek, Mechanisms of aliphatic polyester formation, Chap. 12 in Biopolymers, vol. 3b: *Polyesters II—Properties and Chemical Synthesis*, edited by A. Steinbüchel and Y. Doi, Wiley-VCH, Weinheim, 2002; K. J. Ivin and T. Saegusa (eds.), *Ring-Opening Polymerization*, vols. I–III, Elsevier Applied Science, London, 1984; S. Penczek, P. Kubisa, and K. Matyjaszewski, *Cationic Ring-Opening Polymerization*, Springer, Berlin, vol. I, 1980, vol. II, 1985; M. Szwarc and M. van Beylen, *Ionic Polymerization and Living Polymers*, Chapman & Hall, New York, 1993.

RNA interference

In 1990, the first indication for the existence of nucleic acid sequence guided gene silencing came from scientists who were introducing additional copies of a gene responsible for the darkening of flower color into the *Petunia* genome. In addition to creating darker flowers, the insertion of multiple copies of the gene created white flowers or flowers with patches of white mixed with patches of color (variegated). The white and variegated plants had recognized the newly introduced transgenes as foreign and marked them as well as the endogenous homologous gene for silencing—a process that became known as cosuppression. Subsequent experiments showed that ribonucleic acid (RNA) transcribed from the transgenes was the silencing trigger. Andrew Fire and colleagues discovered in 1998 that injection of double-stranded RNA (dsRNA) into the nematode worm *Caenorhabditis elegans* caused sequence-specific degradation of cytoplasmic messenger RNAs (mRNAs) containing the same sequence as the dsRNA trigger. This phenomenon was termed RNA interference (RNAi), and was soon related to the cosuppression events described earlier in plants.

Biochemical data from plants and the fruit fly *Drosophila melanogaster* revealed that the true mediators of RNAi were short interfering RNAs (siRNAs) of distinct length and structure—one of two types of small RNA produced from cleavage of long dsRNA. RNAi was rapidly developed as a tool to study gene function and was found to naturally occur in protozoa and all higher eukaryotes tested. However, it was not until the discovery of siRNAs that RNAi could be readily applied to mammals. Genetic and biochemical investigations of the mechanisms guiding RNAi in many different organisms revealed conservation of a cellular machinery that cleaves long dsRNAs into small dsRNAs that guide mRNA degradation, translational, inhibition, and chemical modifications of deoxyribonucleic acid (DNA) and its physically associated proteins (chromatin).

General mechanism. Most of the work investigating the mechanism of RNAi has utilized *Arabidopsis thaliana* (a small weed), *C. elegans*, *D. melanogaster*, and humans. Although some species-specific characteristics of RNAi have evolved, the basic steps are conserved. The core pathway can be divided into initiation and effector steps (see **illus.**). Initiation is mediated by the Dicer enzyme that cleaves long dsRNA molecules into 20- to 30-nucleotide (nt) RNA duplexes (siRNAs) that contain 2-nt 3′ hydroxyl overhangs and 5′ phosphates. In the effector step, these siRNAs guide multiple rounds of sequence-specific cleavage of mRNA by the RNA-induced silencing complex (RISC). (Synthetic double-stranded siRNAs introduced into cells bypass Dicer processing and directly enter the effector step.) Dicer pro-

cessing of long dsRNA occurs asymmetrically from both ends and determines which strand of a siRNA will guide the RISC effector complex to the complementary target mRNA. Active RISC contains only one strand of the siRNA, allowing the single-stranded siRNA to basepair with its complementary sequence in the target mRNA. RISC cleaves a target RNA only once, 10 nucleotides from the 5′ phosphate of the antisense (noncoding) guide RNA. It is not well understood how siRNAs are incorporated into RISC or what enzyme or enzymes catalyze removal of one strand of the siRNA, although Dicer has been implicated in both processes.

Components of RISC. The only components of RISC identified thus far in several different species by both genetic and biochemical data are members of the

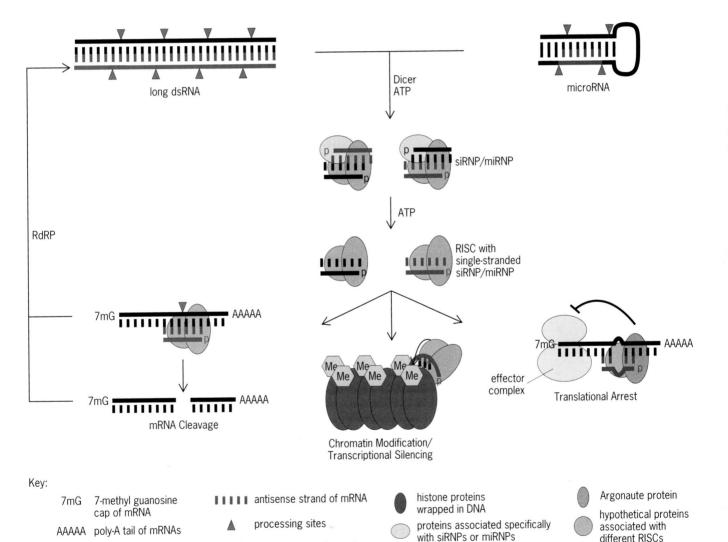

Key:

7mG	7-methyl guanosine cap of mRNA	
AAAAA	poly-A tail of mRNAs	
∥∥∥∥	sense strand of mRNA	

antisense strand of mRNA

▲ processing sites

Me methyl groups directed to DNA and histone proteins

histone proteins wrapped in DNA

proteins associated specifically with siRNPs or miRNPs

Argonaute protein

hypothetical proteins associated with different RISCs

Three mechanisms by which dsRNA leads to sequence-specific gene silencing. dsRNA is cleaved by Dicer enzyme into 20- to 30-nt small interfering RNA duplexes (siRNAs) or microRNAs (miRNAs) which are incorporated into protein complexes, forming ribonucleoproteins termed siRNPs or miRNPs, respectively. RNA-induced silencing complexes (RISCs) containing a single strand of siRNA direct mRNA degradation, translational inhibition, or chromatin modification. In some organisms, RNA-directed RNA polymerase (RdRP) resynthesizes long antisense RNA strands from cleaved mRNA, to amplify siRNA production.

Argonaute family of proteins. *Drosophila melanogaster* Argonaute1 and Argonaute2 and their human homologs, EIF2C1 and EIF2C2, copurify with the RNA cleavage activity in biochemical experiments. Argonaute proteins share conserved protein tht sequence motifs, termed PIWI and PAZ domains, but their functions are unknown. The PIWI domain is present in all members of the Argonaute family, while the PAZ domain is found in both the Argonaute and Dicer families. Two other proteins that copurify with human RISC under certain conditions are Gemin3 (a putative RNA helicase) and Gemin4 (implicated in RNA processing). Two other *D. melanogaster* proteins implicated as components of RISC are dFXR (*Drosophila* fragile X-related protein) and VIG (Vasa intronic gene). VIG contains a motif known to bind RNA, and the dFXR human homolog regulates the translation of several mRNAs via its association with the noncoding RNA named BC1. It is apparent that a family of related RISC complexes must exist that mediate sequence-specific targeting via a noncoding RNA.

Amplification and systemic spread of RNAi. Plants and *C. elegans* have evolved additional complexities in their RNAi processes. Both can amplify the amount of long dsRNA in a cell to increase production of siRNAs and degradation of mRNAs. This mechanism utilizes an RNA-directed RNA polymerase (RdRP), and an mRNA template to synthesize long antisense RNA strands (see illus.). These organisms have also evolved mechanisms to spread the silencing systemically throughout the body and to their unborn progeny. It is generally hypothesized that these spreading mechanisms evolved as an antiviral response that allows a single cell infected with a dsRNA virus to protect the rest of the organism. Indeed, several plant viruses and at least one animal virus have evolved mechanisms to evade such a defense mechanism by actively repressing different steps of RNAi, including the systemic silencing. Oddly, there are no known viruses that infect *C. elegans*, although the *C. elegans* genome does contain many classes of retroviral-like transposons (mobile DNA elements which have sequences resembling the genomes of retroviruses).

MicroRNAs (miRNAs). The RNAi machinery has functions beyond protection against invaders. Plants and animals have genes that produce noncoding RNAs that are processed by components of the RNAi pathway to regulate expression of endogenous genes. The primary products of these noncoding RNA genes are ~70 nt in length and form hairpin structures (a single strand loops back, aligning two complementary sequences) that are substrates for Dicer (see illus.). The Dicer products are ~22-nt microRNAs, the second type of small RNA, derived from one arm of the hairpin. To date, more than 200 miRNAs have been identified in vertebrates. Many miRNAs are conserved across species, suggesting that they mediate essential functions. In many cases, several different miRNA genes are transcribed in one long primary transcript molecule that is processed in the nucleus to release multiple hairpin precursors, which are subsequently processed by Dicer in the cytoplasm.

Most miRNAs are partially complementary to sequences in the 3′ untranslated regions of their target mRNAs and inhibit translation of the target mRNAs by unknown mechanisms. It is postulated that these miRNAs guide effector complexes that are closely related to or identical to RISC. Consistent with this idea, miRNAs that are fully complementary to their targets cause cleavage and degradation of the target mRNA. While many plant miRNAs display nearly perfect complementarity to target mRNAs and cause RNA degradation, all animal miRNAs identified thus far are not fully complementary to any known mRNAs, suggesting that the function of most animal miRNAs is to directly inhibit protein synthesis and not to cause mRNA degradation. Interestingly, the mouse homolog of the RISC-associated dFXR protein binds the noncoding RNA named BC1 to direct translational repression.

Transcriptional gene silencing/chromatin modification. Genetic screens to identify mutant organisms that are defective in RNAi or cosuppression (silencing of transgene and homologous endogenous gene) were conducted in *A. thaliana*, *Neurospora crassa* (a fungus), *C. elegans*, and *D. melanogaster*. These screens revealed links between RNAi and transcriptional silencing. Several mutations that disrupt RNAi in *C. elegans* activate expression of previously silent transposable elements (mobile segments of DNA that can move from one chromosomal site to another). Dicer cleavage products containing transposable element sequences have been cloned from *A. thaliana*, *C. elegans*, *D. melanogaster*, and *Trypanosoma brucei* (flagellated protozoan). In *A. thaliana*, long (24 to 26 nt) and short (21 to 23 nt) classes of siRNAs are produced from dsRNA. The long class is dispensable for mRNA degradation but is needed for systemic spread of silencing and for transcriptional silencing associated with changes in chromatin structure. These chromatin changes include DNA methylation and posttranslational modifications of histone proteins (see illus.). Small RNAs corresponding to transposable elements from *A. thaliana*, *D. melanogaster*, and *T. brucei* are all in the long class, suggesting that 24- to 26-nt RNAs guide silencing of transposable elements in these organisms.

H3-K9 methylation. Mutations disrupting some members of the Argonaute family affect transcriptional silencing. For example, *A. thaliana* Argonaute4 mutants relieve transcriptional repression of certain genes and reduce methylation of both DNA and histone H3 at the ninth lysine residue (H3-K9). These mutants also fail to accumulate 24- to 26-nt RNAs homologous to the AtSN1 retrotransposon (transposon that moves throughout the chrosome via an RNA intermediate). Similarly, *D. melanogaster* mutants that have mutations in the PIWI Argonaute gene fail to accumulate small RNAs homologous to silenced transgenes, and are defective for transcriptional silencing of certain transgenes.

A relationship between RNAi and methylation of histone H3 is also found in the unique DNA-elimination mechanism in *Tetrahymena thermophila* (ciliated protozoan). In this case, H3-K9 methylation of genomic loci to be eliminated correlates with accumulation of small RNAs and depends on the TWI Argonaute protein. Further supporting a role for RNAi-related mechanisms in this process is the observation that dsRNA introduced into *T. thermophila* at certain developmental stages induces the elimination of DNA homologous to the introduced dsRNA.

Heterochromatin formation. The role of RNAi mechanisms in modulating chromatin structure appears to be phylogenetically widespread. Although the fission yeast *Schizosaccharomyces pombe* does not have a complete RNAi pathway, it contains homologs of Argonaute, Dicer, and RdRP. Genetic evidence indicates that these proteins help form heterochromatin-like structures (condensed chromatin regions) at *S. pombe* chromosome centromeres. The model proposes that siRNAs produced from degradation of long centromeric dsRNA transcripts target an unknown effector complex (presumably containing Argonaute) to specific genomic loci, leading to an increase in the local methylation of histone H3 (see illus.). Many aspects of *S. pombe* heterochromatin structure are similar in plants and animals, but it is yet to be determined if the role of RNAi-related mechanisms in centromeric heterochromatin formation is also conserved.

Applications of RNAi. Classical genetic approaches identify gene mutations that disrupt the function or pathway being studied. Recovery and mapping of mutations affecting phenotypes is time-consuming and not always easily applicable to mammalian systems. Reverse genetic approaches involve disruption of a gene with unknown or suspected function to see the effect on a function or pathway. In many cases this is also expensive and time-consuming. Now that many genomes of key model organisms have been largely sequenced, RNAi provides a new method to investigate gene function that in many cases is simpler than the classic approaches. Indeed, several near-genome-wide "RNAi screens" have already been used to study processes in *C. elegans*, *D. melanogaster*, and human cells.

Application of RNAi to mammalian cells was facilitated by investigation of the RNAi mechanism. Introduction of large dsRNA kills many mammalian cells through activation of a potent antiviral interferon response. Elucidation of siRNA structure, however, led to the discovery that the siRNAs can effectively reduce specific gene expression in mammalian cells without activating the host antiviral response that shuts down all host protein synthesis. The siRNA approach holds great therapeutic promise—siRNAs are naturally used by cells to regulate gene expression and thus are nontoxic and highly effective. Moreover, several studies demonstrate that siRNAs can act in mice to fight viral infections and induce tumor regression. Currently the main obstacle facing development of siRNA-mediated therapies is the scarcity of efficient methods for introducing siRNAs into cells of different tissues. This, however, is a problem for all gene therapy approaches and is the focus of intensive study.

For background information *see* DEOXYRIBONU-CLEIC ACID (DNA); GENE; GENE ACTION; MOLECULAR BIOLOGY; NUCLEIC ACID; NUCLEOPROTEIN; RIBONU-CLEIC ACID (RNA) in the McGraw-Hill Encyclopedia of Science & Technology.   Y. Dorsett; T. Tuschl

Bibliography. E. Bernstein et al., Role for a bidentate ribonuclease in the initiation step of RNA interference, *Nature*, 409:363–366, 2001; S. M. Elbashir et al., Duplexes of 21-nucleotide RNAs mediate RNA interference in mammalian cell culture, *Nature*, 411: 494–498, 2001; A. Fire et al., Potent and specific genetic interference by double-stranded RNA in *Caenorhabditis elegans, Nature*, 391:806–811, 1998; S. M. Hammond et al., Argonaute2, a link between genetic and biochemical analyses of RNAi, *Science*, 293:1146–1150, 2001; M. Lagos-Quintana et al., Identification of novel genes coding for small expressed RNAs, *Science*, 294:853–858, 2001; T. A. Volpe et al., Regulation of heterochromatic silencing and histone H3 lysine-9 methylation by RNAi, *Science*, 297:1833–1837, 2002.

Robotic controlled drilling

Robotic controlled drilling significantly reduces oil and natural gas well construction costs. It increases production by directing the wellbore and its exposure to the hydrocarbon reservoir (pay zone), where oil or gas migrates, or is forced, into the wellbore for transport to the surface. The greater the wellbore exposure to the pay zone, the greater the production from each well.

Early oil wells were drilled vertically to reach reservoirs just below the surface. As easily exploitable reservoirs became more difficult to find, drilling moved offshore. Offshore drilling platforms are expensive, so as many wells as possible are drilled from each. By the 1980s, as many as 60 wells were drilled radially from a single platform to cover as much of the reservoir as possible. About 20 years ago, horizontal drilling became the norm, whereby the well was drilled to a target entry point in or near the pay zone and then drilled along the length of the reservoir. Today horizontal wells are routinely drilled with 3–6 km (1.9–3.7 mi) of reservoir exposure, and it has been estimated that a single horizontal well serves the purpose of five to ten conventional wells at less than twice the cost of a conventional well.

Because oil and gas migrate upward, greater recovery results when the horizontal wellbore is placed close to the top of the reservoir. However, hydrocarbon reservoirs often are not uniform structures. They may be broken into separate pockets by geological faults or exist on several different levels. As a result, well sections may be orientated in nonradial directions to maximize recovery. Over the last 20 years, well drilling has become a process of analyzing (for example, seismic) data to determine the "lie" of the reservoir where the greatest hydrocarbon reserves are, and analyzing how best to maximize their recovery.

Reservoir feedback methods. Thirty years ago, there was little or no feedback from the reservoir to tell drillers whether or not they were drilling in the correct direction or depth until the well was prepared for production (completed). Instead, the well was drilled to a geometrical target based on seismic data, along with geological depth-related information obtained from exploration wells.

Measurement while drilling. Since the late 1970s, tools developed for measurement while drilling (MWD) have been used to transmit geometrical information (the inclination and direction at the bottom of the well) to the surface. Since the distance along the wellbore from the start point is known, the position (Cartesian coordinates) of the bottom of the wellbore can be calculated. If the well drilling drifts off course, it can be detected and corrected sooner.

Formation evaluation. Each rock type in a reservoir has characteristic properties such as resistivity (for example, if the rock is wet with hydrocarbons its electrical resistance will be low) or radioactivity. This allows geologists to determine what rock is currently being drilled, its characteristics, and usually at what depth it is. This in turn allows the geologist to ascertain some clearly identifiable layer of rock (geological marker) that will allow further refinement of a seismic map.

Measurement while drilling and formation evaluation while drilling have made it possible to determine quickly a reservoir's location and alter the drilling path as needed. These processes approach real-time speeds, allowing decisions to be made in hours instead of days or weeks. They were an essential precursor to robotic drilling.

Rotary steerable tool. The development that preceded and allowed for robotic drilling was the rotary steerable tool. Initiated and controlled from the surface, it could change direction with a minimum delay. Several methods of steering wells in the desired direction already existed, but these processes were comparatively slow. Since the middle of the 1980s, the mainstay of directional drilling was the steerable downhole motor. However, it required that the drill pipe stop rotating while the bit was redirected and allowed to drill for a short distance in order to change the well's direction. The process could not be automated. Another problem was that as wells became longer and more complex, forward progress could be made only if the whole drill string (all the elements that connect the drilling bit at the bottom to the power source on the surface) were continuously rotating. Otherwise, the drill string would not slide down the hole. Rotary steerable systems overcame these and many other obstacles.

Robotic drilling systems. The robotic drilling systems available today are capable of drilling a predetermined well path without human intervention. There are circumstances, of course, where human judgment is required. In manual mode, inclination and direction (azimuth) measurements are transmitted to the surface and monitored by the driller who maintains or changes the tool's course. In automated mode, the tool monitors its own direction and inclination and makes automatic corrections to maintain the desired well path, which is generally straight and flat to maximize reservoir production. It is relatively easy to program such a tool to drill from one fixed point in space to another along a predetermined path without human intervention. However, the economic advantages of this are limited.

For the most advanced systems, directional control is achieved by deflecting the drive shaft, which runs inside a stiff housing, to point the bit in the desired direction and to the desired degree (**Fig. 1**). Thus, both tight- and wide-radius curves can be generated, each with a constant curvature. The response speed and the severity of response can be preprogrammed or reprogrammed while drilling. Less sophisticated systems generate a single degree of curvature, and the resultant path is a series of tight turns and straight lines of various lengths to give the desired average curvature.

The aim generally is to reach the reservoir with as smooth a wellbore path as possible, and to align the reservoir section of the wellbore with the lie of the reservoir, which may not be exactly horizontal. It has proved possible to drill a series of branching wellbores (as though the wellbores were tributaries of a river system draining the hydrocarbons and feeding them to the surface) such that all the wellbore is confined within a 1-m-thick (3-ft) vertical slice close to the reservoir's top (**Fig. 2**).

Other robotics are used within the drilling system to monitor and compensate for wear and to

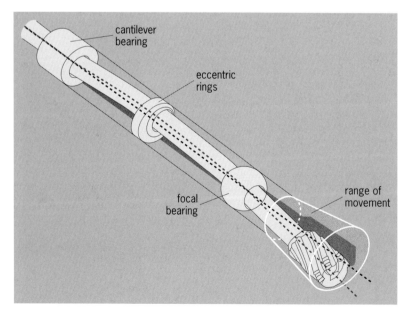

Fig. 1. Point-the-bit rotary steerable tool showing the principle of operation.

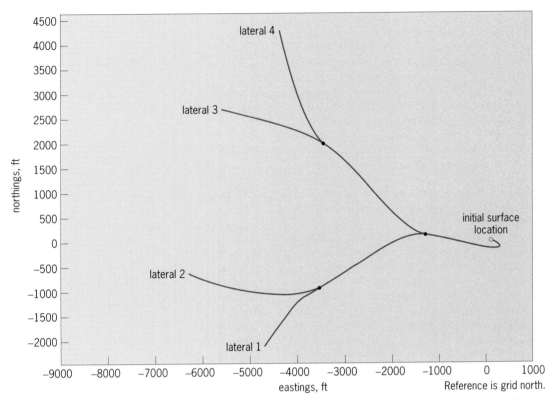

Fig. 2. Flat drainage system more than 1.5 km (1 mi) below the sea floor, showing a four-branch well as viewed from above.

recognize the onset of adverse conditions leading to loss of directional control.

While these technologies and methodologies for well drilling and placement have revolutionized reservoir exploitation, they still rely almost entirely on steering the system via geometrical measurements, while success is measured by formation evaluation methods. What is important is how much oil or gas such a well produces, and for how long. In formation evaluation while drilling, the maximum production from a wellbore is related to particular values, such as the porosity, resistivity, and density of the rock being drilled. These are normally measured, and the raw data processed, by computers downhole. The values are coded (binary is the simplest option) and transmitted via pressure pulses through the drilling fluid to the surface. There they are decoded, and the values derived are compared with known or expected values. If necessary, drilling is redirected to where the optimum production zone is found.

Outlook. The ultimate robotic drilling system will seek and maintain a path that maximizes reservoir production without human intervention. This is technically achievable today, and a commercial product is expected in one or two years. Such a system is well suited for remote operation from onshore drilling data centers where all available data (for example, from seismic surveys, and modeling based on exploration wells) will be used to construct a three-dimensional (3D) Earth model which is updated continuously by real-time formation evaluation while drilling. Drillers, geologists, and geophysicists will be able watch a 3D visual representa-

tion of the part of the Earth's crust they are drilling and, if necessary, redefine the wellpath as it is being drilled. Although this is not yet a routine operation, an advanced point-the-bit robotic drilling system has been steered remotely with complete success from an onshore data collection center while the robotic drilling system itself was located 4 km (2.5 mi) below the ocean surface.

For background information *see* OIL AND GAS, OFF-SHORE; OIL AND GAS FIELD EXPLOITATION; OIL AND GAS WELL COMPLETION; OIL AND GAS WELL DRILLING; SEISMIC EXPLORATION FOR OIL AND GAS; WELL LOGGING in the McGraw-Hill Encyclopedia of Science & Technology.   Thomas Macaulay Gaynor

Bibliography. A. Yaliz et al., *Case Study of a Quad-Lateral Horizontal Well in the Lennox Field: Triassic Oil-Rim Reservoir*, Society of Petroleum Engineers, SPE 75249, 2002; T. Yonezawa et al., *Robotic Controlled Drilling: A New Rotary Steerable Drilling System for the Oil and Gas Industry*, Society of Petroleum Engineers, SPE 74458, 2002.

Salmonellosis in livestock

Salmonella is a genus of gram-negative bacteria of the family Enterobacteriaceae, which also includes *Escherichia coli*. *Salmonella* bacteria usually reside in the gastrointestinal tract and feces of reptiles, amphibians, birds, humans, and livestock (such as cattle, swine, chickens, and turkeys). *Salmonella enterica* species is a bacterial pathogen to humans and animals that can cause severe illness and death.

Environments associated with infected animals are also usually contaminated with *Salmonella*. However, it can persist with no adverse effects in a host. A subclinically infected host can be especially contagious. For example, apparently healthy animals infected with *S. enterica* can infect others' farms or enter the food chain and contaminate food packing and processing areas, kitchens, and final foodstuffs. *Salmonella* is one of the leading causes of food-borne illness, costing the United States an estimated $2.3 billion annually, due to illness and lost productivity.

Serovars. Nearly 2500 serovars (strains or subspecies) of *S. enterica* are recognized, exhibiting subtle to extreme differences in host affinity and pathogenicity. Serovar typing is based on identifying recognizable antigenic properties of the bacterium's surface. It can be useful for identifying the source of infection or contamination. Some serovars have unique host affinities, such as Enteritidis in poultry, Choleraesuis in swine, and Dublin in cattle. However, these and many other serovars can also cause food-borne illness in humans. Typhimurium, Enteritidis, and Newport are the three most commonly reported serovars in human cases of salmonellosis. Typhimurium is common in all livestock species. Enteritidis is most often found in poultry and can be deposited inside a normal-looking table egg. Newport, often associated with dairy cattle, is increasingly important due to the current strains' resistance to multiple antibiotics.

Pathogenesis in livestock. *Salmonella* in livestock can cause significant morbidity and mortality, particularly in young or immune-suppressed animals. Chronically and latently infected animals harbor the organism in the deep crypts of the intestine, in abdominal lymph nodes, or in parts of the liver. *Salmonella* can cause septicemia (blood poisoning) or debilitating enteritis (inflammation of the small bowel) within 24–48 hours of exposure. Chronic infection may result in chronic enteritis with periodic bouts of diarrhea. However, chronic and latent infections usually cause no illness (mostly because they involve small numbers of bacteria in isolated locations within the body). Most latent infections appear to be short-lived (1–2 weeks); however, a few may last months to years.

Infected animals intermittently shed the organism. A number of factors may cause the latently infected animal to begin shedding *Salmonella*, increasing exposure to other livestock and the environment. These factors include parturition (birth), changes in feed, disruptions in feeding, antibiotic therapy, transportation, and animal mixing. The increased shedding may result in outbreaks of clinical disease or increased prevalence of the bacteria in animals. Increased prevalence at the time of slaughter increases the pathogen load on the processing plant, and the risk of food-borne infection.

Epidemiology in livestock. The epidemiology of this pathogen in livestock is extremely complex due to its diversity, latency capabilities, wide host distribution, and environmental persistence. Various *Salmonella* serovars are distributed throughout the livestock population. For intensively raised animals such as poultry, swine, and feedlot cattle, 50% or more of United States flocks and herds are positive for *Salmonella* at some time during the production cycle. However, within an infected farm, 5–15% of animals might be shedding at any one time. Pasture-raised animals, such as beef cows and sheep, have less *Salmonella*; the bacteria are found in only 1% of cows and 10% of cow herds.

Animal-to-animal transmission is usually through the fecal-oral route, although respiratory infection is possible. Thus prevalence estimates are usually based on detection of the organism in feces. However, most methods do not detect low numbers of *Salmonella*. Additionally, latently infected animals do not shed bacteria in feces. Therefore, nondetected animals are likely present on tested farms, making the true prevalence rate higher than current published estimates.

Environment. The fact that the environment is a significant source of *Salmonella* infection has been underappreciated until recently. *Salmonella* is as much an environmental organism as it is an animal pathogen. It can survive for long periods (weeks to months) in any moist or dry, hot or cold location, or in material that is not exposed to direct sunlight. It can be transmitted by farm buildings that previously held infected animals, vehicles traveling from infected premises, dust blown within or between buildings, water running or sprayed between pens, manure or human sewage residue spread on pastures, water tanks or bowls, and feed.

Feed. In many cases, feed is the source of new infections on a farm. Studies have traced the appearance of new serovars (for example, Agona) to feed ingredients such as fish meal. Animal feed necessarily contains animal and vegetable by-products as protein or energy sources. A U.S. Food and Drug Administration survey of animal and plant protein processors demonstrated that 56.4% of the animal protein and 36% of the vegetable protein products taken from 124 processors were positive for *Salmonella*.

Animal reservoirs, vectors, and fomites. Another important source of original farm contamination and transmission are reservoirs such as rodents, cats, and birds; vectors such as insects; and fomites (inanimate objects), such as tools, boots, and equipment. For example, researchers have shown that workers' boots are the most likely place from which to recover *Salmonella* on swine farms. *Salmonella* has been recovered from feed trucks, livestock trucks, and farm vehicles. Although the bacteria also have been recovered from many birds trapped around livestock premises, the direct transmission of *Salmonella* between birds and animals, measured by recovering the same serovar from both sources, is not always strong. However, rodents and cats present a significant risk, as both freely deposit *Salmonella*-contaminated feces into the feed. Cats are not a healthy form of rodent control, particularly on swine farms, where they

of Earth: The Future of American Space Power, University of Kentucky Press, Lexington, 2001; J. Oberg, *Star-Crossed Orbits: Inside the U.S.-Russian Space Alliance*, McGraw-Hill, New York, 2002; R. Saxer et al., Evolved expendable launch vehicle system: The next step in affordable space transportation, *Program Manager* (Defense Systems Management College), 31(2):2–15, March-April 2002.

Satellite observation of clouds

Clouds and precipitation are two of the most important atmospheric topics for understanding and quantifying correctly the issues of global climate change. Since most of the current political and societal debate is centered on global warming and greenhouse gases, there is a general tendency to forget that clouds play a relevant role in modulating the Earth's climate.

There is growing scientific awareness of the crucial role played by clouds on climate. Recent findings suggest that injection of large quantities of aerosols (suspensions of solid or liquid particles in air) into the Earth's atmosphere produces brighter clouds, which are less efficient at releasing precipitation. Aerosol sources such as forest fires, industrial pollution, and desertification are major drivers of climate.

Small- and large-scale cloud systems originate because of favorable meteorological conditions, including atmospheric dynamics, availability of water vapor, aerosol content, solar radiation, and electric fields. A cloud system is a very complex portion of the atmosphere where condensation, collision coalescence, drop breakup, icing, electrification, and other microphysical processes take place and often compete with or counteract each other.

The satellite meteorology community is working to make sure that the cloud physics processes (which were studied in the laboratory during the last

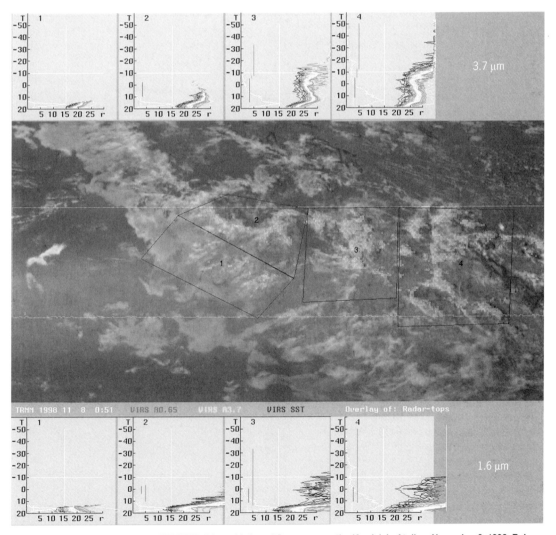

Fig. 1. Multispectral imagery from TRMM Visible and Infrared Scanner over the Kwajalein Atoll on November 8, 1998. Rain areas as detected by the Precipitation Radar on board the satellite are shown as stippled boxes. The graphs at the top and the bottom of the image report cloud top temperature *T* (°C) versus the retrieved effective radius *r* (μm). The 3.7-μm product (top) correctly identifies the cloud top structure and delimits the various microphysical zones. On the contrary, the 1.6-μm channel response (bottom) suffers from contamination from the lower levels in the cloud, and the graphs are much noisier and the effective radii too high. Numbers on the graphs refer to the numbered boxes in the image. (*Courtesy of D. Rosenfeld, Hebrew University, Jerusalem*)

Fig. 2. Channel 2 image (0.725–1.00 μm) of the National Oceanic and Atmospheric Administration's *NOAA16* satellite Advanced Very High Resolution Radiometer (AVHRR) of a multicell storm over northern Texas on June 1, 2001. The relatively high resolution (1.1 × 1.1 km²) allows the identification of all relevant storm features. At the center and in the southeastern side of the cloud system, the growing towers of the cumulonimbus cloud updrafts appear as round structures. Wave trains depart from the updrafts, indicating a pulsating vertical motion of the storm during its growth stage. East and outside of the storm main body, an arched thin cloud band and several other low-level cloud lines delimit the outflow boundary of the storm, that is, the front of cooler air associated with the storm downdraft. (*Courtesy of M. Setvák, Czech Hydrometeorological Institute*)

30 years) are correctly identified as a basis for improving rainfall measurements on a global scale and quantifying the cloud mechanisms that influence global climate changes. Moreover, the World Meteorological Organization (WMO) has recognized that water resources are one of the world's most crucial problems. Local and global changes in the water cycle will affect entire populations and the quality of life on the planet.

Satellite sensors. From the first meteorological satellites in the 1960s up to the 1990s, cloud observations relied upon broadband channels (wavelength bands) in the visible and infrared portion of the electromagnetic spectrum. Observation of the cloud motion and atmospheric dynamics was privileged, with only little capability for cloud physics and structure studies, including precipitation. The use of near-infrared channels, such as the one at 3.7 micrometers, however, opened the door to the use of multispectral information at cloud top.

From 1999 to 2002, several satellites were launched with a new generation of multispectral radiometers (electromagnetic radiation sensors) conceived for cloud, aerosol, and land applications, including the Moderate Resolution Imaging Spectroradiometer (MODIS) on NASA's *Earth Observing System* (*EOS*) satellite, among the high-resolution [250–1000-m (820–3300-ft) footprint] sensors, and the Spinning Enhanced Visible and Infrared Imager (SEVIRI) on board Europe's *Meteosat Second Gen-*

eration (*MSG*) satellite, among the geosynchronous instruments.

Satellite imagery. It is very challenging to make quantitative cloud observations from altitudes of 800–36,000 km (500–22,000 mi), depending on the orbit. The radiance fields measured by a sensor depend on a number of factors such as illumination conditions, satellite-sun angles, homogeneity of the cloud top, thermodynamic phase (ice/water) of the hydrometeors, and the composition of the atmospheric column between the sensor and the cloud top. The physical interpretation of the sensor radiances is therefore a complicated process that involves a number of assumptions and a radiative transfer model that simulates the atmosphere-cloud-radiation interactions.

Measurements include the water droplet/crystal size (a mean quantity that characterizes the size of particles that are present in a certain size distribution at the cloud top), the cloud optical depth, and the thermodynamic phase (ice/water content). D. Rosenfeld and I. M. Lensky proposed an effective method that links the imagery of measured quantities to color tables, where the colors indicate cloud types or characteristics, such as thick/thin, maritime/continental, and convective/stratiform, and allow the monitoring of their evolution. **Figure 1** shows an example observation of oceanic rain clouds using visible and infrared channels.

The most important contributions of satellite cloud observations are the identification of the raining potential and quantification of the impact on the global climate. Precipitation and climate are strictly interconnected by positive/negative feedback mechanisms affecting the delicate equilibrium of the water cycle. The global monitoring of cloud types is starting to nail down numbers for global climate models and for assimilating data into numerical weather prediction (NWP) models. The International Satellite Cloud Climatology Project (ISSCP) is the most important of such efforts together with the Global Precipitation Climatology Project (GPCP).

Apart from a better understanding of cloud formation/dissipation mechanisms and the support given to rainfall measurements, there are other fundamental problems which satellites can be used to solve. For example, a major unknown is the influence of high-altitude cirrus clouds on global warming, including the contribution from aircraft contrails. It is currently believed that a 4% increase in the total amount of these semitransparent, thin clouds could lead to a surface temperature rise of about 2°C (3.6°F). In reality, while it is known that cirrus clouds allow a large part of the radiation to come through to Earth and act as shields for the outgoing thermal infrared radiation, the total amount of these clouds and their ice content are unknown. These unknown parameters are essential for predicting the overall influence of cirrus clouds on climate change.

Another problem concerns the formation mechanisms of deep convective clouds formation (such as those of tropical and midlatitude storm systems), which act as giant engines for the vertical transport of humidity into the upper troposphere and the lower stratosphere. An example of such events is given in **Fig. 2**.

Cloud physics. The retrieval of cloud top microphysical data is one practical example of cloud observation from satellite sensors. Cloud structure and severe weather events are of interest for their impact on weather predictions. Numerical weather prediction models are striving to include cloud microphysics in order to describe the actual observations for data assimilation for creating initial conditions and to incorporate the physical mechanisms of clouds and precipitation formation.

Figure 3*a* shows a multicell, deep convective storm over southeastern Spain. The image of the cloud top reflectivity at 3.7 μm depicts an area of high reflectivity in the form of a fan in the northern portion of the storm. These structures have been observed many times in Europe and the United States and are associated with deep convection such as for cumulonimbus clouds that occupy the entire tropospheric column and reach nearly to the tropopause or even overshoot it. The satellite sensors were the first to document the occurrence of high reflectivity patterns at the cloud top. Observations and radiation modeling over the years have shown that these reflectivity patterns are plumes of very small ice crystals that are injected above the storm top, as is the case in Fig. 3*a*.

The proof of the existence of such small crystals is given in Fig. 3*b*, where the curves of simulated (modeled) reflectance at 3.7 μm are plotted versus the brightness temperature (K) at 10.8 μm for the plume in Fig. 3*a*. The calculations are made as if the slab of particles consisted of bullet rosette crystals with four branches of varying radii. The simulations perfectly reproduce the measurements using these two sensor channels. These observations show how the satellite helps in understanding cloud structure in a very active meteorological environment, opening the way for identifying potential indicators of severe weather.

Outlook. Single-sensor, single-mission strategies for observing clouds and precipitation are a closed chapter, and truly international missions are planned that concentrate on cloud physics and meet the requirements of nowcasting, weather forecasting, and climate and hazard management. Over the next 5 years, new experiments promise advances in the understanding of the clouds surrounding the Earth.

In 2004, NASA will launch the *CloudSat* spacecraft as part of a five-satellite constellation called the A-train. *CloudSat* will fly the first space-borne millimeter-wavelength radar, which will observe, along with the other satellites, cloud condensate and precipitation and provide profiles of these properties with a vertical resolution of 500 m (1640 ft). The driving idea behind the project is to go beyond the simple cloud top observations and measure vertical profiles of cloud properties with the highest possible

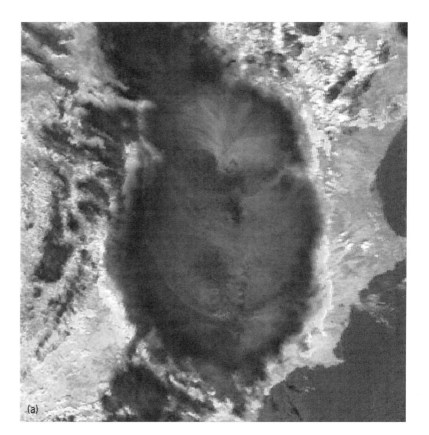

(a)

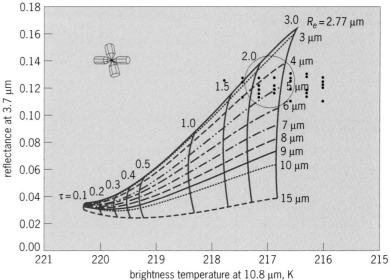

(b)

Fig. 3. Multicell storm over southeastern Spain as observed by the *NOAA 11* AVHRR instrument on August 28, 1991. (*a*) The area is about 440 × 440 km² (270 × 270 mi²). In the northernmost part of the storm, a fanlike feature spreads over the storm top, indicating the presence of small ice crystals injected from the storm in the atmosphere above. (*b*) Simulated reflectance versus brightness temperature curves with superimposed measurements from the AVHRR sensor (filled circles), indicating that rosette ice crystals are strong candidates for the composition of ice crystal plumes above the storm. Horizontal curves refer to various cloud top effective radii (R_e), and vertical curves to varying cloud optical thicknesses (τ).

resolution. The other satellites in the A-train include the *EOS Aqua* and *Aura*; a second NASA satellite, *CALIPSO*, that flies an aerosol lidar (light detection and ranging); and the French Space Agency (CNES) satellite, *PARASOL*, carrying the POLDER (polarization and directionality of the Earth's reflectances) polarimeter.

The Tropical Rainfall Measuring Mission (TRMM) has contributed to quantifying how much precipitation falls in the tropical atmosphere. However, at present we cannot estimate (within a factor of 2) the mass of water and ice in these clouds and how much water and ice is converted to precipitation, or with any certainty what fraction of global cloudiness produces precipitation that falls to the ground. The Global Precipitation Measurement (GPM) mission scheduled for 2008 will consist of a mother ship with an advanced double-frequency precipitation radar and a passive microwave radiometer that will fly in formation with seven or eight drones, each hosting at least a radiometer. The drones will calibrate their rainfall estimates with the precise measurements of the mother ship, thus widening the single-satellite swath on the Earth's surface which will ensure a global coverage of precipitation every 3 hours.

For background information *see* AEROSOL; CLOUD; CLOUD PHYSICS; INFRARED RADIATION; METEOROLOGICAL SATELLITES; MICROWAVE; PRECIPITATION (METEOROLOGY); RADIOMETRY; REMOTE SENSING; SATELLITE METEOROLOGY; TERRESTRIAL RADIATION; TROPICAL METEOROLOGY; WEATHER; WEATHER FORECASTING AND PREDICTION in the McGraw-Hill Encyclopedia of Science & Technology.   Vincenzo Levizzani

Bibliography. S. Q. Kidder and T. H. Vonder Haar, *Satellite Meteorology: An Introduction*, Academic Press, San Diego, 1995; V. Levizzani and M. Setvák, Multispectral, high-resolution observations of plumes on top of convective storms, *J. Atm. Sci.*, 53:361–369, 1996; V. Ramanathan et al., Aerosol, climate, and the hydrological cycle, *Science*, 294:2119–2124, 2001; D. Rosenfeld and I. M. Lensky, Spaceborne sensed insights into precipitation formation processes in continental and maritime clouds, *Bull. Amer. Meteorol. Soc.*, 79:2457–2476, 1998; W. B. Rossow and R. A. Schiffer, ISCCP cloud data products, *Bull. Amer. Meteorol. Soc.*, 72:2–20, 1991; G. L. Stephens et al. and the CLOUDSAT science team, The Cloudsat mission and the A-Train, *Bull. Amer. Meteorol. Soc.*, 83:1771–1790, 2002.

Scramjet technology

Traveling from New York to Sydney, Australia, in about 2 hours or spending a weekend at the Space Station or the Moon may sound like science fiction, but supersonic combustion ramjet (scramjet) technology in development today is expected to make global rapid travel and affordable orbit access a reality within the next 25 years.

The scramjet engine that could make such expedient travel possible is designed to be air-breathing and has recently attained some milestones in development. A breakthrough was realized in the spring of 2003 when a flight-weight scramjet engine was run

at 4.5 and 6.5 times the speed of sound (Mach 4.5 and 6.5) using conventional hydrocarbon fuel. More significantly, this represents the first flight-weight, hydrocarbon-fueled scramjet engine. This test was conducted by the Pratt & Whitney Space Propulsion Hydrocarbon Scramjet Engine Technology (HySET) team, in partnership with and sponsored by the U.S. Air Force Research Laboratories (AFRL) Propulsion Directorate.

Scramjet operation. A ramjet is a jet engine that relies on the compressing or ramming effect on the air taken into the inlet while the aircraft is in motion. The geometry of the engine compresses the air rather than a piston, as in an internal combustion engine, or a compressor, as in a gas turbine. In a ramjet, the compressed air is slowed to subsonic speeds (less than Mach 1) as it is routed into the combustion chamber, where it is mixed with fuel, burned, and expanded through a nozzle to generate thrust.

A scramjet normally operates at speeds above Mach 4. To get a missile or aircraft to Mach 4, a rocket or jet engine is typically used for the initial acceleration. As the vehicle is traveling through the atmosphere on its way to Mach 4, the ignition condition for the scramjet, air is being compressed at hypersonic (greater than Mach 5) and supersonic (Mach 1–5) speeds on the lower surface of the aircraft at high pressure (**Fig. 1**). From the scramjet inlet, the air flows at supersonic speed into the combustor, where fuel is mixed with the air, burned, and expanded to make thrust. The mixing, burning, and expansion process takes place in less than 0.001 second. Engineers who developed this technology have referred to scramjet combustion as attempting to "light a match in a hurricane."

Advantages. In their operating speed regime, scramjets offer several advantages over conventional jet and rocket engines. Although not as efficient as a gas turbine engine, a scramjet has the ability to operate at much faster speeds. While burning hydrocarbon fuel, a scramjet engine can operate at speeds up to Mach 8. Using hydrogen as the fuel, a scramjet engine can operate at speeds in excess of Mach 12. Because of its air-breathing nature, the performance (specific impulse) of a scramjet is significantly higher than that of the best-performing rocket engines. One big advantage of a scramjet engine is the minimal number of moving parts required for operation. Since it relies on the shape of its flow path to compress the air by dynamic means, a scramjet engine does not require the use of high-pressure pumps, compressors, or turbines as do gas turbine or rocket engines. While reducing moving parts increases engine reliability and durability, benefits such as high speed, relative fuel efficiency, and range are the reasons that scramjets will revolutionize the aerospace industry.

The development of scramjet propulsion technology will enable affordable and reusable space transportation systems with operation much like a commercial airliner. Scramjet-powered access-to-space

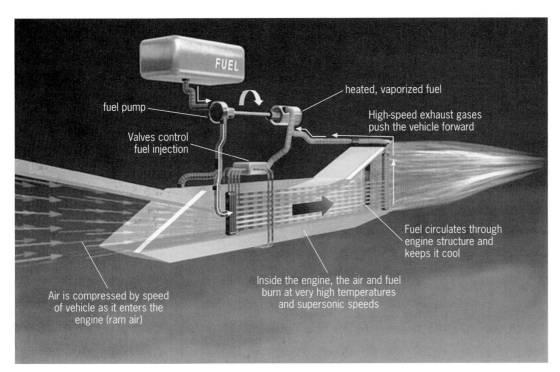

Fig. 1. Operation of a scramjet.

systems that will take off and land horizontally like commercial airplanes are being conceptualized today. These systems will be designed to achieve orbit while employing relatively low acceleration profiles, potentially opening access to orbital flight to a much larger section of the population.

Scramjets will be used in conjunction with either gas turbine engines or rockets in combined cycle engines (CCEs). In a turbine-based combined cycle (TBCC) engine, a turbine engine will accelerate the aircraft from takeoff to a scramjet takeover speed (Mach 3–4), at which point the scramjet will be ignited and the turbine will be shut down. In a rocket-based combined cycle (RBCC), small rockets embedded in the flow path of the scramjet will first be used as ejectors to accelerate the vehicle to ramjet and then to scramjet operation. In both cases, the CCEs will be used to accelerate the space vehicle to escape velocity while flying through the atmosphere using the oxygen in the air. As the vehicle gains altitude and escapes the Earth's atmosphere, propulsion will transition to rockets since air will not be available to the scramjets.

Future air-breathing access-to-space systems could increase the payload fraction by two-to-five times that of conventional rocket-powered systems. Since the scramjet engine takes its oxidizer from the air, a scramjet-powered vehicle does not require an oxidizer tank or a supply control system. This reduction in oxidizer volume and weight can be translated into smaller vehicles and reduced launch costs. Current goals are to be able to place 1 lb (~0.5 kg) of payload into low Earth orbit for $100 by 2025.

Initial technology developments for reusable applications will concentrate on developing variable geometry to optimize scramjet performance and extend the Mach number envelope. Since a hypersonic vehicle will spend prolonged periods traveling at high speed in the atmosphere, its surface will require significantly more thermal protection than the expendable launch systems in use today. Structural and thermal durability will be increased to allow multimission capability. Additionally, scientists and engineers today are defining scalability factors that must be taken into consideration when designing the larger engines required for space applications.

Development. The U.S. Air Force established the Hypersonic Technology (HyTech) Program in 1995 after the National Aero-Space Plane's development was terminated. In 1996, Pratt & Whitney won a $48 million contract for HySET to demonstrate the performance and structural durability of a hydrocarbon-fueled scramjet engine. As a result of the HySET Program's success the X43-C, a joint USAF/NASA program, has emerged. The goal of this program is to flight-test a three-module derivative of the HySET engine simulating elements of a space mission in 2007. In parallel and as risk reduction for the X43-C Program, the Air Force is also considering flying in 2006 a propulsion flight demonstration program: the Endothermically Fueled Scramjet Engine Flight Demonstrator (EFSEFD) Program. This flight demonstration will use a single HySET engine derivative module and will focus on the military application of the scramjet.

Building block approach. Once the hypersonic vehicle mission requirements were translated into scramjet engine component requirements, the HySET team generated the computational fluid dynamic codes. These codes define combustion products and

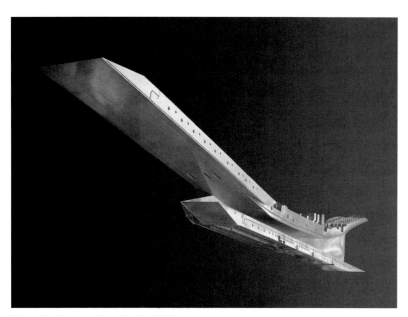

Fig. 2. Performance Test Engine (PTE).

optimize the fueling locations and concentrations required for top performance. Their results were used to design the hardware, allowing optimization of the engine design prior to test.

From subscale testing, the team updated codes and defined the Performance Test Engine (PTE). This full-scale test of a heat sink engine validated the computer codes and demonstrated engine performance and operability. The PTE is a heavy-weight, heat-sink engine fabricated in solid copper (**Fig. 2**). The use of conventional jet fuel provided a major challenge to the team. Prior to this program, most developmental scramjets used hydrogen, which is much easier to mix and burn than jet fuel. In addition, the team developed a special fuel-conditioning system to simulate the fuel temperatures that would be generated by a fuel-cooled engine in flight.

The PTE was operated at Mach 4.5 and 6.5 consistent with scramjet takeover and cruise conditions for a missile application. The team tested and examined 95 test points, demonstrating excellent operability. The engine achieved 110% of its performance test objectives at Mach 4.5 conditions and 100% of its test objectives at Mach 6.5. Based on the performance and operability of the engine during test, the team went on to design and fabricate a more sophisticated full-scale, fuel-cooled engine.

As part of the building block approach, the team built on lessons learned, and used test results from the PTE to calibrate and improve analytical tools leading to the evolution of the Ground Demonstration Engine (GDE). With every test result, the analytical tools used for the design were calibrated, resulting not only in the ability to understand the behavior of the hardware at test but also in the definition of a design set of tools to be employed in subsequent iterations. Along with the development of the scramjet's operability and performance, the HySET

team used the building block approach for developing the thermal and structural technologies. Multiple thermo-structural components with increasingly larger sizes were designed, built, and tested in scramjet operating conditions. In 2001, this culminated in the creation of the first hydrocarbon-fueled scramjet engine, GDE-1, weighing less than 150 lb (59 kg) and using readily available nickel-based alloys for its fabrication.

Operation. In the GDE-1 engine, the fuel must enter the combustor as a vapor in order to facilitate mixing and burning with the airflow. This is achieved through regenerative cooling of the hardware using a revolutionary endothermic cooling approach. The liquid fuel enters the front of the engine and is routed through the structure to the rear of the engine. The fuel cools the metallic structure as it passes through very small longitudinal passages. The passages in the structure are coated with a catalyst. While the fuel is absorbing the heat from the structure, the fuel temperature and pressure rise, causing the catalyst to "crack" the fuel in an endothermic reaction; that is, the fuel breaks down into gaseous lighter hydrocarbon components such as hydrogen, ethylene, and methane. This process gasifies the fuel for easier burning while allowing the fuel to absorb a significantly larger amount of heat from the scramjet structure than otherwise would be possible. The process is akin to a flying fuel refinery. At the rear of the engine structure, the cracked fuel is collected and metered into the scramjet combustor and burned. The combustion process expands through a nozzle, where it produces positive net thrust.

Testing. In the spring of 2003, the GDE-1 was tested initially at low speed (Mach 4.5) and then at Mach 6.5 (**Fig. 3**). As in the prior component tests, the results were used to calibrate and validate the design tools. By July 2003, GDE-1 had completed about 60 tests, successfully validating the thermo-structural, performance, and operability predictions. The results of

Fig. 3. GDE-1 test at Mach 4.5.

GDE-1 testing have proven that with today's technology scramjets fueled by hydrocarbon fuels can generate the thrust required to support hypersonic flight.

The HySET team is already working on the next scramjet engine, GDE-2. It will also be fuel-cooled and flight-weight as was GDE-1. However, this will be a full system test and will introduce control hardware and software so that the engine runs as a complete closed-loop system. The HySET team plans to test the GDE-2 in the second half of 2004.

Outlook. With the HySET program's current success, the next step for hypersonic propulsion is flight. Building on the lessons learned during PTE, GDE-1, and the future GDE-2 ground testing, the HySET team's goal is flight demonstration of this technology. The X43-C and EFSEFD programs will provide the first opportunities to demonstrate the operation of a hydrocarbon-fueled scramjet engine in flight. The lessons learned during these flight experiments will pave the way for hypersonic flight.

For background information *see* ESCAPE VELOCITY; HYPERSONIC FLIGHT; MACH NUMBER; RAMJET; SPACE SHUTTLE; SUPERSONIC FLIGHT in the McGraw-Hill Encyclopedia of Science & Technology.   Joaquin Castro

Bibliography. E. T. Curran and S. N. B. Murthy (eds.), *Developments in High-speed Vehicle Propulsion Systems*, vol. 165, AIAA, 1996; E. T. Curran and S. N. B. Murthy (eds.), *High-speed Flight Propulsion Systems*, vol. 137, AIAA, 2000; E. T. Curran and S. N. B. Murthy (eds.), *Scramjet Propulsion*, vol. 189, AIAA, 2000.

Seed morphological research

Seeds, the defining organs of the seed plants, or spermatophytes, protect and disperse the next generation of plants, represented by the young sporophytes (embryos). Being the result of millions of years of evolution and representing the most complicated organs a seed plant will ever produce in its life, seeds can tell us a lot about how primitive or advanced a given species is in evolutionary terms, about its natural relationships, and its dispersal mode and germination behavior.

Seed structure. Generally, a seed consists of three genetically different regions: (1) the seed coat, representing diploid maternal tissue; (2) the haploid (in gymnosperms, for example, cycads and conifers) or triploid (in angiosperms, that is, flowering plants) nutritive tissue called endosperm; and (3) the diploid embryo, as the product of the fusion of the egg cell with a spermatic nucleus from the pollen (**Fig. 1**).

Despite this simple-sounding general composition of a seed, the anatomy of the seed coat as well as the internal morphology of the seed shows a tremendous diversity and can reach high levels of sophistication.

Ovule types. Seeds develop from fertilized ovules and at the time of fertilization consist of the nucellus, which contains the embryo sac, and one or two enveloping layers called integuments which develop

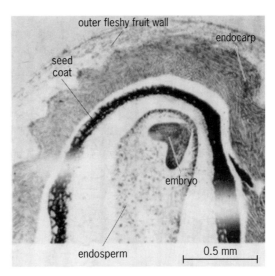

Fig. 1. Section through portion of young raspberry drupelet (one fruit segment). (*From R. M. Reeve et al., Seed, McGraw-Hill Encyclopedia of Science & Technology, 9th ed., 2002*)

into the seed coat. At the apex of the ovules, the integuments have an opening called the micropyle, which allows the pollen tube access to the egg cell. Water and nutrients are supplied to the ovule via the funicle, a kind of umbilical cord connecting the ovule to the placenta (**Fig. 2**). The simplest type of ovule is called orthotropous. It has the funicle entering from its base (chalaza), with the micropyle lying in a straight line opposite. Such ovules are characteristic of gymnosperms, but are found in only a

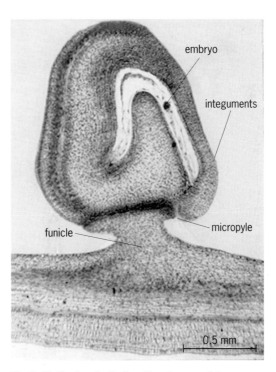

Fig. 2. Median longitudinal section of a campylotropous ovule of a pea shortly after fertilization, showing attachment to pod tissues. (*From R. M. Reeve et al., Seed, McGraw-Hill Encyclopedia of Science & Technology, 9th ed., 2002*)

few angiosperms. The second type of ovule is called anatropous and is basically an orthotropous ovule in which the nucellus is rotated 180°. The funicle "fuses" along the side with the ovule to form the raphe so that the point where the ripe seed later detaches from the funicle comes to lie next to the micropyle. Chalaza and micropyle still lie in a straight line opposite each other. This type of ovule is found in most angiosperms. The third ovule type is called campylotropous, and refers to all ovules with bent longitudinal axes, that is, curved embryo sacs (Fig. 2). They represent a derived state with the clear advantage that the embryo is folded and thus can reach twice the length of the seed, giving rise to a larger seedling.

Embryo types. The internal morphology of seeds varies tremendously with respect to the relative size of the endosperm and the relative size, shape, and location of the embryo. Based on these criteria, in 1946 A. C. Martin distinguished 12 types of embryos: broad, capitate, lateral, peripheral, linear, dwarf, micro, spatulate, bent, folded, investing, and rudimentary. Martin further classified these 12 embryo types into two major divisons (except for the rudimentary embryo): peripheral (embryo is small and within the lower half of the seed or elongate and contiguous to the seed coat) and axial (embryo is central, following the longitudinal axis of the nucellus), the axial type having three subdivisions (**Fig. 3**). The phylogenetic significance of the internal morphology of seed is demonstrated by the fact that, for example, the lateral type is found only in the Poaceae (grasses), and the peripheral type is found in almost the entire order Caryophyllales (carnation family and relatives).

Seed coat characters. The most distinctive character of the seed coat lies in the position and structure of the main mechanical layer, which is composed of thick-walled cells. This mechanical layer can be one or more cells thick and consists of radially elongate palisades, horizontally elongate fibers, or cuboid cells, with either evenly or unevenly thickened walls. Unevenly thickened walls of the cells of the outermost layer of the seed coat are the main reason for an intricate and highly characteristic sculpturation of the seed surface, as in the Cactaceae and Caryophyllaceae.

The anatomical structure of the seed coat and the micromorphology of the seed surface can be excellent indicators of natural relationships. Seed characters provide phylogenetically significant information for the natural (monophyletic) circumscription of families and even genera. For example, less derived (more primitive) angiosperms generally have medium-to-large seeds with complicated multilayered seed coats, whereas the most derived flowering plants, such as the Orchidaceae in the monocotyledons (having one seed leaf) and the Asteridae in the dicotyledons (having two seed leaves), have small, simplified seeds with few-layered or even just single-layered seed coats.

Since the majority of seeds do not show any strong anatomical or morphological adaptations to the environment (because ovules and seeds are enclosed within the ovary and not immediately exposed), the crucial characters of their seed coats can be considered as having evolved in a strictly phylogenetic way. This is the reason why a certain type of seed coat anatomy is often characteristic for whole families, entirely independent from the different environments inhabited by the plants.

One such example of an extremely widespread and ecologically diversified group is the Fabaceae which, despite inhabiting almost all climates and habitats, have a principal seed coat structure that shows hardly any obvious environmental adaptations. The members of this family are almost always characterized by an exotesta (seed coat outer layer) of Malpighian cells (macroscleids) showing a light line (linea lucida) and an outer hypodermis of hourglass cells, or osteoscleids.

The phylogenetic significance of seed characters is also demonstrated by the recent dismemberment of the long-established family Euphorbiaceae. Based on the differences he found in the anatomy of the seed coats, E. J. H. Corner suggested in 1976 that the family is probably polyphyletic and that the three uniovulate subfamilies obviously represent a separate entity from the two biovulate subfamilies (**Fig. 4**). This fact has recently been supported by the phylogenetic analysis of deoxyribonucleic acid (DNA) sequences.

The strongest influence on the formation of the seed coat is exerted by a hard indehiscent pericarp or endocarp (a fruit wall that remains attached to the seed when mature) as it is often developed by uniovulate ovaries, where there is obviously no need for dehiscence (bursting open to release seeds). In such seeds the coat is reduced, since it no longer serves as the mechanical protection of the seed. It

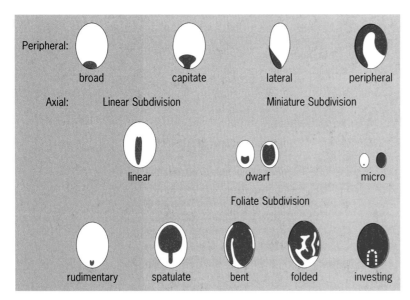

Fig. 3. Twelve seed types defined by Martin based on shape, size, and position of the embryo. (*Reprinted with permission from A. C. Martin, The comparative internal morphology of seeds, Amer. Midland Naturalist, 36:520, 1946*)

then usually fails to develop the anatomical details that normally provide the phylogenetic clues.

Adaptations. Obvious adaptations sometimes refer to seed germination (such as opercula, pluglike structures in the micropylar area which become disjoined during germination by circular dehiscence, in palms, begonias, and gingers). More often the adaptations refer to the biotic or abiotic dispersal of the seeds, such as a brightly colored, fleshy covering (sarcotesta) to attract potential dispersers such as birds, a mucilaginous surface facilitating the passage through dispersing animals' intestines, hairs or wings to assist wind dispersal, or a variety of special appendages.

Dust and balloon seeds. The syndrome of minute wind-dispersed seeds (dust seeds and balloon seeds, generally smaller than 0.3 mm) has arisen independently in monocotyledons and dicotyledon families with remarkable convergences. Without any obvious adaptive advantage, it is striking that such seeds are found in families in nutrient-poor environments or otherwise nutritionally specialized plants (such as parasitic and carnivorous plants).

Appendages. Other adaptations, such as the formation of wings, hairs, and arils, are clearly linked to the abiotic or biotic dispersal of the seeds. Wings and hairs are mostly formations of the seed coat and usually indicate wind dispersal and are found in a wide range of unrelated taxa. Arils, seed appendages which assist dispersal by animals, are generally fleshy and often brightly colored. These appendages, sometimes called elaiosomes ("oil bodies" because they are often rich in oil), are formed by the funicle (funicular arils), by various regions of the seed coat (localized arils), or derived from a combination of several such locations (complex arils).

Funicular arils are frequently encountered in the Fabaceae, Passifloraceae, and some Caryophyllaceae, in which they attract ants (in smaller seeds) or birds and even mammals (in larger seeds) for dispersal. One example of an edible aril is the durian, a fruit of the family Bombacaceae which is very popular in Malaysia and other parts of Southeast Asia. Designed to attract mammals for their dispersal, the spiny fruits emit a foul smell with a strong fecal and onion tang and open to expose seeds, which are entirely wrapped in a pleasant-tasting aril derived from the funicle.

A localized aril developing from an area of the seed itself is the exostome aril (or caruncle), a small swelling of the testa (seed coat) around the micropyle. These caruncles are usually rich in oil and meant to attract ants, which carry away the seeds to later feed on the aril. Similar elaiosomes can also be the result of a proliferation of the raphe or the chalaza with the same function of attracting potential animals for dispersal (most often ants). The extraordinary aril of the bird-dispersed seeds of *Strelitzia reginae* (bird-of-paradise flower) consists of a tuft of thick, deep-orange-colored hairs produced by both the funiculus and the testa in the micropylar region.

High-resolution x-ray computed tomography. Recently, high-resolution x-ray computed tomography

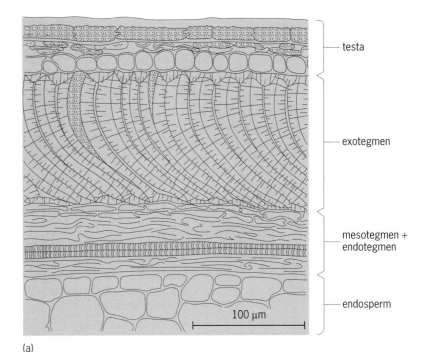

(a)

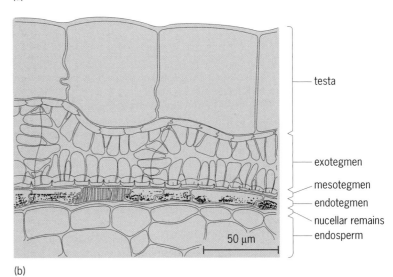

(b)

Fig. 4. Longitudinal section of seed coat. (*a*) *Beyeria viscosa*, Euphorbiaceae, showing the different layers of the seed coat, including the characteristic mechanical layer (the exotegmen) of the seed coat composed of bent palisade-like sclereids, and the endosperm. (*b*) *Phyllanthus muellerianus*, Phyllanthaceae, showing the exotegmen composed of longitudinally elongated, sclerenchymatic cells. Until recently, both *Beyeria* and *Phyllanthus* were considered members of the same family (Euphorbiaceae). The assumption, based on significant differences in their seed coat anatomy, that the two genera actually belong to two different families has recently been confirmed by molecular phylogenetic studies. *Phyllanthus* and related genera are now classified in their own family, Phyllanthaceae. (*Drawings by W. Stuppy*)

(HRCT) has been used to study plant organs and tissues. HRCT produces two-dimensional images, or "slices," that display differences in x-ray absorption arising mainly from differences in density within an object. From these slices, which have a certain finite thickness, a continuous three-dimensional map of the density variations in an object is created by digitally restacking slices separated by the proper distance. The examination of a fruit of *Syagrus flexuosa*, a Brazilian palm species, for example,

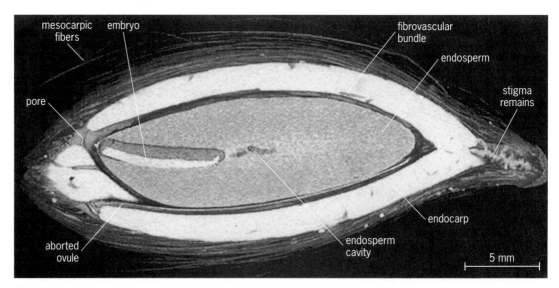

Fig. 5. Three-dimensional cutaway reconstruction of the fruit of the South American palm *Syagrus flexuosa*, obtained from high-resolution x-ray computed tomography (HRCT). The object displays sufficient differences in x-ray attenuation to show all relevant morphological details: peripheral mesocarpic (from the middle layer of the fruit wall) fibers, endocarp (inner layer of fruit wall) with fibrovascular bundles, pores at the base of the fruit which represent openings in the hard endocarp to allow the germinating embryo to escape from the hard endocarp, one of the two aborted ovules, and the endosperm including the central cavity, and embryo (the outer peel of the fruit, the epicarp, was removed prior to the investigation). (*Reprinted with permission from W. H. Stuppy et al., Three-dimensional analysis of plant structure using high-resolution X-ray computed tomography, Trends Plant Sci., 8(3):2–6, 2003*)

has shown that there can be sufficient differences in x-ray attenuation between embryo, endosperm, and covering structures (seed coat or fruit wall) to allow a detailed noninvasive study of the internal morphology of fruits, seeds, and other plant parts (**Fig. 5**).

The great advantage over traditional serial sectioning is that it requires no fixing, sectioning, or staining and it is completely noninvasive. The three-dimensional reconstructions provided by HRCT can be manipulated digitally to perform a large array of accurate and reproducible measurements and visualization tasks, including not only longitudinal and cross sections of the very same specimen but also views of arbitrarily oriented sections and selective extraction of features of interest.

Conclusion. Comparative seed morphology and anatomy is still a rather unexplored discipline in the field of plant science, offering many exciting research opportunities. The seed structure of many species is yet unknown and thus represents a vast untapped source of ecological and phylogenetic information that can contribute toward a better understanding of the evolution of seed plants on our planet.

For background information *see* DORMANCY; MAGNOLIOPHYTA; PLANT EVOLUTION; PLANT PHYLOGENY; PLANT TAXONOMY; SEED in the McGraw-Hill Encyclopedia of Science & Technology.

Wolfgang Stuppy; John Dickie

Bibliography. F. D. Boesewinkel and F. Bouman, The seed: Structure, in B. M. Johri (ed.), *Embryology of Angiosperms*, pp. 567–610, Springer, Heidelberg, 1984; F. D. Boesewinkel and F. Bouman, The seed: Structure and function, in J. Kigel and G. Galili (eds.), *Seed Development and Germination*, pp. 1–24, Marcel Dekker, New York, 1995; J. B. Dickie and W. Stuppy, *Seed and Fruit Structure: Significance in Seed Conservation Operations*, 2003; E. Werker, *Seed Anatomy*, Borntraeger, Berlin, 1997.

Self-healing polymers

Materials that have the ability to heal automatically have been a dream of engineers and scientists for a long time. After long periods of use cracks will form and propagate in materials, eventually leading to mechanical failures, some of which (such as airplane crashes) could be catastrophic. In addition, microcracks are always hidden deep within materials. For decades, scientists worldwide have been looking for materials that can heal themselves after crack formation, especially polymers.

When separate molecules or polymer chains are connected by covalent bonds, they are called cross-linked. Highly cross-linked polymers have been widely studied and used as matrices for composites, foamed structures, structural adhesives, insulators for electronic packaging, and so on. Their densely cross-linked structure provides superior mechanical properties such as high modulus, high fracture strength, and solvent resistance. However, highly cross-linked polymers are susceptible to irreversible damage by high stresses due to the formation and propagation of cracks, which may lead to a dangerous loss in their load-carrying capacity.

There has been an intense search to find methods to heal cracks of linear polymers, also known as thermoplastics. Hotplate welding and crack healing of thermoplastics is well established, where intermolecular noncovalent interactions (chain entanglements) at the crack interface are responsible for mending. Small-molecule-induced crack healing has been studied for thermoplastics; in this mechanism low-molecular-weight solvents (such as ethanol) are supplied to the interfaces, followed by heating to heal the cracks.

Research for re-mending and self-healing of cross-linked polymeric materials has been increasingly exciting in recent years. However, because it is impossible to cause chain entanglement in highly cross-linked polymers, scientists did not obtain significant results until two new design concepts for polymeric materials recently were revealed. In 2001, a method was introduced for making polymer composites embedded with encapsulated "healing" chemical reagents. And in 2002, a concept was introduced for a re-mendable polymeric matrix that can be healed multiple times by simple thermal treatment.

Self-healing polymer composites. With some specific additional ingredients, self-healing polymer composites can be prepared. To the best of our knowledge, the concept of self-repair was first introduced to heal cracks in composites by embedding in the composites hollow fibers that release repair chemicals when a crack propagates. No significant progress was made for this type of composite until 2001, when S. R. White and coworkers reported a polymeric composite with an epoxy resin matrix that contained a catalyst and encapsulated add-monomer (healing agent).

Figure 1 shows the autonomic healing concept. Dicyclopentadiene (DCPD) is microencapsulated as the healing agent, and a catalytic chemical trigger, called Grubbs' (ruthenium-based) catalyst, is embedded within the epoxy matrix. A propagating crack will rupture the embedded microcapsules, releasing the healing agent into the crack plane through capillary action. Polymerization of the dicyclopentadiene is then triggered by contact with the embedded catalyst to bond the crack faces. This damage-induced triggering mechanism provides site-specific repair in polymer composites, with a reported typical healing efficiency of 75% and an average healing efficiency of 60%.

While autonomic healing was clearly established by White and coworkers, they are currently addressing a few remaining problems with this approach, such as crack-healing kinetics and the stability of the catalyst to environmental conditions.

Multi-re-mendable polymer matrices. Polymers, especially highly cross-linked polymers, are widely used as matrices for the majority of modern structural composite materials. Carbon and glass fibers and other fillers are embedded in the polymer to obtain better mechanical properties. In all these composite materials, the polymer matrices are usually the weakest component and tend to produce microcracks after long-term use. A thermally re-mendable polymer matrix could provide a much longer service time and dramatically increased reliability.

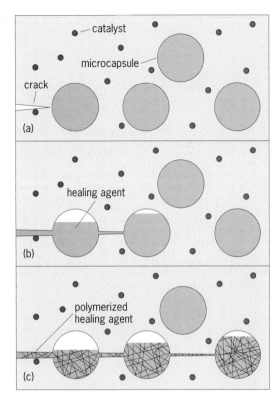

Fig. 1. Self-healing composite concept. Microcapsules containing DCPD (the healing agent) are embedded in an epoxy resin matrix containing a catalyst capable of polymerizing the healing agent. (*a*) A crack forms in the matrix wherever damage occurs. (*b*) The crack ruptures the microcapsules to release the healing agent into the crack plane through capillary action. (*c*) The healing agent contacts the catalyst to trigger polymerization that bonds the crack faces closed. (*Reproduced with permission from Nature, 409:794–797; copyright 2001 Nature Publishing Group.* http://www.nature.com)

During the past 20 years, thermally reversible reactions have been pioneered and studied for linear and cross-linked polymers, particularly the Diels-Alder reaction—the cycloaddition (unsaturated molecules combine to form a ring) of a diene and an alkene (dienophile). In 2002, X. Chen and coworkers reported a macromolecular network formed in its entirety by reversible cross-linking covalent bonds. The polymer exhibits multiple cycles of autogenous crack mending with simple and uncatalyzed thermal treatments and without additional ingredients. A special feature of the material is that a series of covalent bonds, which are chemically and physically the same as the original material, are formed at the interface of the mended areas. In hotplate welding and crack healing of thermoplastics, by contrast, only intermolecular, noncovalent interactions (for example, chain entanglements) are responsible for the mend and no new covalent bonds are formed between the mended parts.

As shown in the reaction below, a thermally reversible Diels-Alder cycloaddition of a multidiene (multifuran, nF) and multidienophile (multimaleimide, mM) was used to prepare a polymeric material. Monomer 1 (4F) contains four furan moieties on each molecule, and monomer 2 (3M)

Mechanical properties of multi-re-mendable polymers

	Properties	3M4F	2MEP4F	2ME4F
Ultrasonic testing (room temperature)	Poisson ratio	0.32	0.36	0.37
	Young's modulus (GPa)	4.70	4.41	3.6*
	Shear modulus (GPa)	1.78	1.62	
	Bulk modulus (GPa)	4.31	5.41	
	Density (g/cm^3)	1.34	1.31	1.33
	Specific Young's modulus	3.51	3.37	
Hopkinson bar compression testing (strain rate ~2000/s; room temperature)	Young's modulus (GPa)	4.55	4.14	3.4*
	Ultimate tensile (MPa)	241	234	
	Strain to failure (%)	25	24	
	Specific Young's modulus (GPa)	3.40	3.16	
	Specific strength (MPa)	180	179	116*

*Data from Instron® floor model (TT) testing system.

includes three maleimide moieties on each molecule. A highly cross-linked network (polymer 3M4F) can be formed via the Diels-Alder reaction of furan and maleimide moieties, while thermal reversibility can be accomplished by the retro Diels-Alder reaction.

nF + mM ⇌ mMnF

polymerization/ cross-linking or mending

heat or crack

(a = weakest link)

The mechanical properties of 3M4F and 2M4F were found to be well within the range of widely used engineering materials such as epoxy resins and unsaturated polyesters (see **table**).

Upon heating, the retro Diels-Alder reaction is preferred over (nonreversible) bond-breaking degradation reactions in the polymer network. The thermal reversibility of cross-linking was studied, and it was found that the Diels-Alder connections can be disconnected at 120°C (248°F) and above. About 30% of the cross-linking bonds are observed to disconnect when the material is heated to 150°C (302°F), and upon cooling to room temperature a new cross-linked network forms.

Because the bond strength between the diene and dienophile of the Diels-Alder adduct is much weaker than all the other covalent bonds, the retro Diels-Alder reaction should be the major reason for crack propagation. In principle, when the sample is reheated and cooled, the furan and maleimide moieties connect again, and the cracks heal and fractures mend.

Using scanning electron microscopy, Chen and coworkers showed that the fracture healed almost completely to produce a homogeneous material with a few minor defects, suggesting a remarkable mending efficiency.

This mending/healing efficiency has been tested using compact tension specimens, as shown in **Fig. 2**. Fracture tests were used to quantify the healing efficiency. Representative load-displacement curves for a polymer specimen are plotted in Fig. 2,

showing recovery of about 57% of the original fracture load. The second mending efficiency was determined to be about 45% of the original load, which strongly suggests that the material can be healed multiple times.

Because of its more flexible polymer chains, polymer 2M4F shows much better healing/mending efficiency, up to 81%. The second mending efficiency achieved was, on average, 78% of the original load, which is much higher than that of 3M4F, and clearly shows that the material can be effectively healed multiple times without any additional ingredients.

Polymers consisting of cross-linked furan-maleimide have also been developed by J. R. McElhanon and coworkers for solvent-removable epoxy resins and other applications. Although healing properties were not reported, in principle, these materials might heal cracks. Besides furan-maleimide linkage, other reversible linkages reviewed by L. P. Engle and K. B. Wagener might be used to prepare self-healing polymers. In addition, noncovalent reversible linkages, such as hydrogen bonding, may also be applied to prepare self-healing polymers.

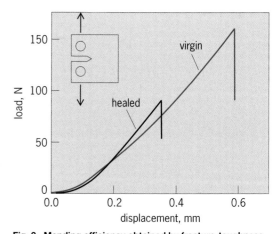

Fig. 2. Mending efficiency obtained by fracture-toughness testing of compact tension test specimens. The original and healed fracture toughness were determined by the propagation of the starter crack along the middle plane of the specimen at the critical load.

The multi-re-mendable polymers are being used as matrices for composite structures by various research groups. Potential applications involve, but are not limited to, transparent (from visible light to microwave) structural materials, optical lenses, and plastic insulators.

For background information *see* DIELS-ALDER REACTION; EPOXIDE; POLYETHER RESINS; POLYMER; POLYMERIC COMPOSITE; POLYMERIZATION in the McGraw-Hill Encyclopedia of Science & Technology.

Xiangxu Chen; Fred Wudl

Bibliography. X. Chen et al., A thermally remendable cross-linked polymeric material, *Science*, 295:1698, 2002; X. Chen et al., New thermally remendable highly cross-linked polymeric materials, *Macromolecules*, 36:1802, 2003; L. P. Engle and K. B. Wagener, A review of thermally controlled covalent bond formation in polymer chemistry, *J. Macromol. Sci.—Rev. Macromol. Chem. Phys.*, C33(3):239, 1993; J. R. McElhanon et al., Removable foams based on an epoxy resin incorporating reversible Diels-Alder adducts, *J. Appl. Polym. Sci.*, 85:1496, 2002; R. P. Sijbesma et al., Reversible polymers formed from self-complementary monomers using quadruple hydrogen bonding, *Science*, 278:1601, 1997; S. R. White et al., Autonomic healing of polymer composites, *Nature*, 409:794, 2001.

Sex differences in stress response

Male and female mammals, including men and women, respond to stress in largely similar but not identical ways. In both sexes, stress triggers a cascade of neuroendocrine responses including, for at least some stressors, the release of oxytocin and vasopressin and of the stress hormones: the catecholamines (epinephrine and norepinephrine) and corticosteroids (such as cortisol). Direct neural activation of the adrenal medulla triggers the release of norepinephrine and epinephrine, with concomitant responses of the sympathetic nervous system. Corticosteroids are released via the hypothalamo-pituitary-adrenal (HPA) axis. Stress triggers neurons in the hypothalamus to release corticotropin releasing hormone (CRF), which stimulates the release of adrenocorticotropin hormone (ACTH) from the anterior pituitary, which in turn stimulates the adrenal cortex to release corticosteroids. Both sexes exhibit these responses, with men showing somewhat stronger vascular responses and women somewhat stronger heart rate responses. Through these mechanisms, both males and females are mobilized to meet the short-term demands presented by stress. This stress response has been characterized as fight-or-flight. However, the range of responses to stress has proven to be more diverse than initially believed.

Fight-or-flight. The behavioral component of the fight-or-flight response is thought to depend upon the nature of the stressor. If the organism sizes up a threat and determines that it has a realistic chance of overcoming it, it is likely to attack. When the threat is more formidable, it is more likely to flee or withdraw. Contemporary manifestations of the fight response in humans assume the form of aggressive responses to stress, whereas flight responses in contemporary life are commonly manifested as social withdrawal or substance abuse, as through alcohol or drugs.

Research on the fight-or-flight response has been disproportionately based on males, and recent research suggests that females do not exhibit the same pattern of responses with the same frequency. For example, physical aggression is much more prevalent in male responses to stress, suggesting that the fight component is more descriptive of males than females. The flight component, substance abuse and social withdrawal in response to stress, is also more prevalent among males than females.

Social responses: tend-and-befriend. Humans and many animal species also demonstrate social responses to stress. Social affiliation during stressful times has long been known to protect against adverse changes in mental and physical health associated with stress. Among humans, social support from a partner, relatives, friends, or coworkers, and social and community ties reliably reduce cardiovascular and neuroendocrine stress responses. Findings such as these may account for the robust relations between social support and a reduced likelihood of illness, faster recovery, and greater longevity, and for the observation that social isolation has been consistently tied to a heightened risk of mortality in both animals and humans.

Just as the fight-or-flight response is more descriptive of males, social responses to stress are somewhat more descriptive of females. Women are reliably more likely to seek social support, gain social support, and show beneficial health effects of social support than are men. Women are also disproportionately the source of social support for other women, for men, and for children. Although the differences in men's and women's use of social support for managing stress are only moderate in size, they are extremely robust. For example, in a meta-analysis that examined 26 stress studies, 25 studies showed that women sought social support more than men for managing stress, with one study finding no difference.

Social responses to stress have been termed tend-and-befriend. Tending involves quieting and caring for offspring during stressful times, and befriending involves engaging the social network for help in responding to stress. This pattern is argued to have particular survival benefits for females and their offspring. Human studies document that whereas men are more likely to withdraw from or be combative with family members in response to extrafamilial stress, women are more likely to exhibit nurturant behavior toward offspring during stressful times. Females typically have primary responsibility for the care of offspring, and consequently female responses to stress have evolved in light of those pressures. Whereas fight-or-flight can leave offspring unattended with potentially fatal consequences, tending to offspring in times of stress may be highly adaptive and promote the survival of both mother and offspring. In addition, because protection of self and offspring is a difficult task under threatening circumstances, those who make use of the social group will be more successful against many threats than will those who do not. Thus, females' selective tendency to affiliate (befriend) in response to stress may maximize the likelihood that multiple group members will protect them and their offspring.

Biology of tending responses. In order for a female to tend in response to stress, she would need to regulate the activity of her sympathetic and HPA axis in order to quiet and calm offspring. Biological evidence consistent with this idea is provided by research on oxytocin and endogenous opioid peptides (or endorphins). Oxytocin is secreted by both sexes in response to stress. Evidence from a broad array of animal studies shows that oxytocin reduces HPA axis activity and sympathetic nervous system arousal in response to stress. In both animal and human studies, oxytocin is also associated with reduced anxiety and has mild sedative properties. These effects of oxytocin are enhanced by estrogen, which is why oxytocin's effects appear to be stronger in females than males. Oxytocin also promotes maternal behavior. For example, exogenous administration of oxytocin stimulates extensive grooming and touching of offspring. Thus, evidence for the role of oxytocin in female tending responses to stress is based on oxytocin's (1) secretion in response to stress, (2) enhancement by estrogen, (3) demonstrated role in down-regulating stress responses, and (4) relation to maternal behavior. Endogenous opioid peptides also appear to be implicated in these processes. In rhesus monkeys, for example, administration of naloxone, an opioid antagonist (inhibitor), is associated with less caregiving and protective behavior toward infants. Similarly, administration of naltrexone, another opioid antagonist, inhibits maternal behavior in sheep.

Evidence for maternal tending under stress is plentiful, especially in animal studies. Moreover, research indicates that maternal nurturance is important not only for short-term protection against stress but also for shaping the developing offspring's stress systems, especially the HPA axis, in ways that can be protective over the lifespan. Maternal care by rat dams permanently affects the expression of social, affective, and endocrine responses to stress in their offspring. For example, in response to laboratory stressors such as separation and reunification, maternal behavior reduces HPA axis activity and sympathetic arousal. Long-term effects of maternal nurturance on offspring include a better-regulated HPA axis response to stress, fewer behavioral signs of fear in novel situations, and slower age-related onset of HPA axis dysregulation, such as age-related cognitive deficits. Reinforcing these animal studies, the human literature implicates nurturant contact (and its absence) in a broad array of health and mental health outcomes in offspring, both in childhood and in adulthood.

Biology of befriending responses. The same neurocircuitry that may underlie tending responses to stress may also underlie befriending responses. Animal studies of affiliative behavior indicate that exogenous administration of oxytocin prompts an increase in social contact and grooming in several species. Studies have also reported that animals increase their affiliation with conspecifics if they have experienced high oxytocin and endogenous opioid levels in their company in the past. Administration of opioid antagonists has been shown to block affiliative behavior in both primates and humans, disproportionately so in females. As such, oxytocin and endogenous opioid peptides may be implicated in befriending responses as well as tending responses.

Future directions. For the most part, males and females, including men and women, show fairly similar responses to stress. Both sexes experience increases in sympathetic arousal and HPA axis activity in response to stress. Both males and females are capable of mounting aggressive responses to threats or withdrawing from threatening circumstances (the fight-or-flight response). Nonetheless, at the behavioral level the fight-or-flight response is more descriptive of male stress responses than of female stress responses. Tending to offspring and affiliation in response to stress (the tend-and-befriend pattern) is more characteristic of female behavioral responses to stress.

The exploration of sex differences in stress responses is a relatively new endeavor, in part because historically animal and human studies of stress responses have involved primarily males. As females have been included in stress study protocols, the fact that their responses to stress are more social than males' stress responses has come into focus. Nevertheless, men's tendencies to use social support for coping with stress, especially support from their female partners, are well documented. Moreover, animal studies suggest social responses to stress in males (such as guarding and protection of territory, mate, and offspring) that may implicate vasopressin, a hormone that is highly similar in molecular structure to oxytocin. These findings indicate that sex differences in stress responses are relative rather than exclusive in nature. A biosocial model that explores men's social responses to stress appears propitious. Further advances in the scientific understanding of sex differences in stress responses are anticipated.

For background information *see* ADRENAL CORTEX HORMONE; ENDOCRINE MECHANISMS; ENDORPHINS; HORMONE; STRESS (PSYCHOLOGY) in the McGraw-Hill Encyclopedia of Science & Technology.

Shelley E. Taylor

Bibliography. C. S. Carter, Neuroendocrine perspectives on social attachment and love, *Psychoneuroendocrinology*, 23:779–818, 1998; S. B. Hrdy, *Mother Nature: A History of Mothers, Infants, and Natural Selection*, Pantheon Books, New York, 1999; R. L. Repetti, S. E. Taylor, and T. E. Seeman, Risky families: Family social environments and the mental and physical health of offspring, *Psychol. Bull.*, 128:330–366, 2002; L. Tamres, D. Janicki, and V. S.

Helgeson, Sex differences in coping behavior: A meta-analytic review, *Personality Social Psychol. Rev.*, 6:2–30, 2002; S. E. Taylor, *The Tending Instinct: How Nurturing Is Essential to Who We Are and How We Live*, Holt, New York, 2002.

Small tree utilization (forestry)

In the western interior of the United States, more than 73 million acres of national forests are at risk from insect attack and catastrophic fire because of the major change over the years in their vegetative structure and composition. Some of this change has been caused by 50 years of successful fire suppression that has created forests with significantly overstocked stands of small-diameter trees. This dense understory in the forest creates a ladder-type fuel (allowing fire to grow from ground level into the canopy fueled by vegetation at different heights) that can lead to high-intensity crown fires, which spread fast and can cause catastrophic alterations to forested landscapes and watersheds. The western wildfires of 2000 demonstrated the devastating impact these dense, overstocked stands can have on forests.

To restore the open, parklike, more fire-resistant settings that existed in presettlement times, overstocked stands need thinning and, possibly, prescribed burning. Most small-diameter trees must be mechanically thinned (removed) before prescribed fire can be safely introduced. The cost to thin the forest, ranging $300–1000/acre, is usually more than the value of the thinnings. (The average cost for thinning is approximately $70/dry ton; traditional markets for thinned material, namely the energy and chip markets, pay approximately $25–35/dry ton, respectively.) However, in most cases the cost of thin-ning is less than the cost of fighting a forest fire. Prescribed fire costs less than mechanical removal. However, it is estimated that only about 10% of the overstocked forests can use prescribed fire without some type of prior mechanical thinning. The air pollution and smoke associated with prescribed burning are also of concern.

Economic factors. If cost-effective and value-added uses for thinned small-diameter trees could be found, forest management costs could be offset, economic opportunities could be created for rural forest-dependent communities, and catastrophic wildfires could be prevented or minimized. In addition to these benefits, finding value-added uses for small-diameter trees could improve stand species and quality mix, increase forest resiliency, reduce the risk of damage from insects and disease, provide healthier wildlife habitat, protect and improve watersheds, and expand the supply of wood and fiber to industry and individuals.

For the past 5–7 years, there has been tremendous activity within universities, federal research institutions, nonprofit groups, rural communities, and other organizations to explore and evaluate the potential of small-diameter trees, for both traditional lumber and value-added uses.

Lumber industry impact. The perception that wood from small-diameter trees is inferior in quality is not necessarily true. Since small-diameter trees are a nontraditional resource, appropriate uses cannot be determined until the wood properties and characteristics are understood. Grade yield and recovery studies are being conducted to determine the best lumber grades for small-diameter timber based on a particular species. Studies indicate that high-quality material can be found within small-diameter timber (**Fig. 1**). For example, preliminary research results

Fig. 1.   Small-diameter logs stacked and ready for processing into stud lumber. (*Photo by Jean Livingston, Forest Products Laboratory, USDA Forest Service, Madison, WI*)

show that much of the Douglas-fir that occurs as understory is suppressed growth and contains a large number of rings per inch and has small, tight knots, which improves the value of the wood.

Efficient and economical sawmill operation are necessary to maximize the use of small-diameter trees. A 1999 study of some of the economic issues associated with processing small-diameter trees in conventional sawmills demonstrated that small changes in diameter could drastically change their economics, with smaller diameters decreasing profitability and larger diameters increasing profitability.

Another barrier to developing uses for small-diameter trees is that not all such trees can be utilized for high-value applications. To overcome this obstacle, these trees must be sorted to isolate those of the best and highest-value use, a practice that usually takes place at a location where the logs are collected and sorted (that is, a log landing). Logs are sorted based on their physical quality, such as number of rings per inch, size of knots (lumber scars), and straightness. Estimates indicate that creating higher-value uses for 20–30% of the thinned material could improve the economics of forest management operations.

Structural roundwood. One potential use of small-diameter trees, particularly those in the range of 4–6-in. diameter at breast height, is structural applications such as roundwood trusses, beam-column elements for post-and-frame building systems, pile foundations for residential structures, space frame building systems, and a variety of other structures (**Fig. 2**). When small-diameter trees are left in the round, they are less susceptible to warping, can be more dimensionally stable, and take full advantage of the strength characteristics of the log; the natural taper makes them suitable for use in column applications; and processing costs are decreased. Research is addressing some technical issues about the structural use of roundwood, such as the problem of joining roundwood sections. Results have shown that several types of connectors (flitch plates, dowel nut connectors, and powder-driven nails) can be used for section joining. Another application for roundwood is in highway structures such as guardrails, posts, and poles. Survey results suggest a favorable market potential for roundwood used in these structures.

Engineered wood products. During the past decade, engineered wood products have experienced rapid growth and acceptance in the marketplace. Small-diameter trees from the right species can be ideal for engineered wood products such as laminated veneer lumber (a large panel formed by gluing together layers of thin sheets of wood), wood I-joists (wood beams resembling the letter I), oriented strand board (structural panel made of layers of glued strands of fiber oriented at right angles to one another), and glued-laminated timbers (large beams made from small pieces of lumber glued together). However, the capital costs of an engineered wood products manufacturing facility are high, and unstable timber supplies in the West have resulted in a shift to reliance on resources from industrial and nonindustrial private lands in the East and South.

Veneers and laminates. Some studies have focused on the quality of veneer produced from Douglas-fir,

Fig. 2. Structure built using small-diameter roundwoood; St. Marie's City Park, Idaho. (*Forest Products Laboratory, USDA Forest Service, Madison, WI*)

western larch, white fir, and ponderosa pine logs, ranging from 8 to 14 in. in diameter. Researchers concluded that veneer recovery for these small-diameter logs was comparable with that obtained from trees that regenerate after previously being harvested.

High-quality lumber is required only for the outer laminations of glued-laminated (glulam) beams, and lower-quality lumber can be used in the core laminations when the stock is E-rated, a method of nondestructively determining stiffness. This method is used extensively for Douglas-fir. Currently, design properties for glulam beams made of ponderosa pine use a more conservative approach than is used with Douglas-fir, which results in ponderosa pine being graded as lower quality. However, if ponderosa pine beams can be graded using the more efficient E-rating method, there is potential for improving the utilization of ponderosa pine in glulam beams.

Composites. Wood composite technology is easily adaptable to a changing resource base. Composite products can utilize a variety of wood and wood-based raw materials, including fibers, particles, flakes, and strands. Many species can be used in composites such as particleboard (nonstructural panel made of ground wood particles bonded together with a synthetic resin), medium-density fiberboard (nonstructural panel made of wood fibers bonded together with a synthetic resin), and oriented strand board and lumber. Current wood composite technology is focusing on wood fiber/plastic composites that utilize low-quality species such as juniper, pinyon pine, and insect-killed white fir.

Other uses. Other uses for small-diameter trees include variations on the traditional uses of large-diameter trees and their by-products, as well as some new and creative solutions to contemporary environmental concerns.

Pulp chips. Even with new value-added uses, pulping chips will remain an outlet for small-diameter and primary manufacturing residues. There have been concerns about using the chips from such trees because of problems related to wood fiber angle and length, and the cost of harvesting trees and delivering wood chips to the pulpmill. However, recent studies on the pulp quality from small-diameter Douglas-fir, western larch, and lodgepole pine have shown that, in general, with kraft pulping procedures (chemical degradation of wood chips to form paper), these species behaved similarly to kraft pulp made from traditional sawmill residue chips. However, some problems may occur if certain species are mixed with traditional sawmill residues. For example, cedar does not pulp well because it contains extractives (substances in addition to the major components of wood), and some species have too much bark.

Compost and mulch. Even when other uses for small trees become commonplace, a certain portion will end up as woody residue. Traditionally, this material has been burned or allowed to accumulate in huge piles. However, tighter environmental regulations designed to reduce the runoff of organic residues into the ground water have necessitated the development of different disposal systems. Composting transforms this material into a value-added product that increases soil fertility and can be sold to consumers. The primary technical problem in composting is increasing the efficiency.

Wood fiber mats. Wood fiber filters made from small-diameter trees are used to remove water pollutants. These filters absorb pesticides, herbicides, toxic heavy metals, oil and grease, phosphorus, and toxic organic compounds. Wood fiber mats can also be used for erosion control. *See* WATER DECONTAMINATION.

Energy. The requirements for using wood as fuel are less demanding than the requirements for other uses. Some residues of small-diameter trees and wastes generated in manufacturing various products are suitable for fuel with minimal processing. Fuels can be further refined through drying, pelletizing, or manufacturing charcoal. Possible additional pathways for using wood for energy include power-generating plants, industrial applications of cogenerating systems to produce heat and electricity, institutional heating facilities, and home heating.

For background information *see* WOOD ENGINEERING DESIGN; WOOD PROCESSING; WOOD PRODUCTS; WOOD PROPERTIES in the McGraw-Hill Encyclopedia of Science & Technology.   Susan L. LeVan

Bibliography. R. J. Barbour, Diameter and gross product value for small trees, *Proceedings of the 27th Annual Wood Technology Clinic and Show*, pp. 40–46, Miller Freeman Pub., San Francisco, 1999; J. R. Dramm, G. L. Jackson, and J. Wong, *Review of Log Sort Yards*, FPL-GTR-132, USDA Forest Service, Forest Products Laboratory, Madison, WI, 2002; *Forest Products Laboratory Research Program on Small-Diameter Material*, FPL-GTR-110 (rev.), USDA Forest Service, Forest Products Laboratory, Madison, WI, 2000; L. M. Guss, Engineered wood products: The future is bright, *Forest Prod. J.*, 45(7/8):17–24, 1995; R. Hernandez et al., Improved utilization of small-diameter ponderosa pine in glued-laminated timber, FPL-RP-XXX, USDA Forest Service, Forest Products Laboratory, Madison, WI, 2003; R. Wolfe and C. Moseley, Small-diameter log evaluation for value-added structural applications, *Forest Prod. J.*, 50(10):48–58, 2000.

Smart labels

Radio-frequency identification (RFID) is the use of electromagnetic radiation at radio frequencies, or thereabouts, to interrogate data stored in labels or tags on portable devices. The function has hitherto been fulfilled by visible barcoding in some cases, but RFID offers much greater opportunity for goods and information management. The term "smart label" encompasses all responsive, contactless, electronic tickets, laminates, or adhesive labels but, for the foreseeable future, the greatest commercial opportunities lie with RFID versions. Smart labels for

RFID are cheap enough to be disposable or fitted for life, and have few problems of obscuration or misorientation. Today they typically cost 10 cents to $1 each. Unlike barcode readers, the electronic reader employed to interrogate the smart label in RFID contains no moving parts and is highly reliable.

Objectives. One objective is to replace 10^{13} barcodes printed yearly with RFID smart labels, which will be more reliable and versatile. For example, smart labels can confer unique rather than generic identification, be hidden from sight, and deter counterfeiters. However, to be viable on, say, everything in a supermarket, they must cost no more than 1 cent each.

A data capacity of 96 bits and a range of a few centimeters may suffice for such a tag, though most proponents seek a tag with a capacity of 256 bits and a range of 1 m (3.3 ft), as well as a means of killing the tag, that is, of electronically rendering it permanently inoperative. Requiring a tag with the latter capabilities would make it harder to achieve the target price. It may be necessary to print the electronic circuit directly onto products. Unique item identification is sought in order to improve most supply-chain parameters by a factor of 10 such as labor cost, time from buying parts to having the finished product on sale in the shop, stocks, stockouts (sales that are lost because customers leave a store before they have found the item they want), recall integrity, theft, and counterfeit detection. To allay concerns about privacy, if the smart label is killed at the point of purchase of the product, then unfortunately it cannot be used to improve warranty control, repairs, recalls, and so forth.

While a major investment will be required to achieve item tracking by unique radio identification, the planned payback is several times larger for consumer-packaged goods, postal services, and other industries.

Electronic Product Code. The Auto-ID Centers, headquartered at the Massachusetts Institute of Technology, have developed the Electronic Product Code (ePC) and some enabling software. The codes are governed by their potential end users. The Electronic Product Code provides more comprehensive numbering than barcodes do.

Perhaps 10^{12} (10%) of the world's new barcodes will be replaced or supplemented by ePC smart labels by 2015. More immediately, even RFID smart labels costing $1 are economic when applied to vehicles, conveyances, and even multipacks (packages containing more than one item) by saving cost and providing other benefits.

Applications. New safety legislation worldwide is likely to lead to at least 6×10^8 RFID tags in the stomachs of cows (for disease traceability, preventing fraudulent claims for grants, auto feeding, and deterring rustling) and a similar number in car tires, in the latter case usually linked to the automatic monitoring of pressure. In health care, RFID smart labels can reduce errors and costs in drug dispensing, double-blind testing, blood sample tracking, trans-

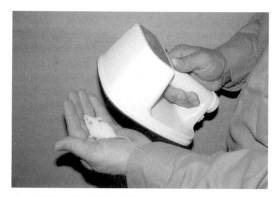

Fig. 1. Mouse in which an identity tag has been implanted is scanned with a hand-held reader that activates the tag to transmit a unique number. (*Avid Identification Systems Inc.*)

port and storage of food and drugs, and so forth (**Figs. 1** and **2**). Some of these labels can also record temperature, time, humidity, and so forth. Widespread use of smart labels is expected in postal services, the military, and elsewhere.

Silicon chips versus chipless labels. The smart label may have prerecorded digital data on an acoustomagnetic or electromagnetic microwire or film, or it may consist of inductor-capacitor pairs. None of these options employs a silicon chip, just as the 6×10^9 antitheft electronic article surveillance radio tags sold yearly function without such chips.

At present, 97.5% of RFID smart labels contain a simple silicon chip (usually with no more than 10,000 transistors) and an antenna, because these store and handle data better than current chipless versions. Indeed, some can be electronically rewritten at a distance. Unfortunately they are more expensive. The cost of the least expensive silicon chips has remained at 5–10 cents for decades. However, ultrasmall chips, 0.1–0.25 mm (0.004–0.01 in.) across, have been developed, and these can be assembled into smart labels at millions an hour using new automation. Such labels cost 10 cents and will cost 5 cents in a few years. If trillions of barcodes are to be replaced yearly, chips only 0.1 mm (0.004 in.) across are expected to be cost-effective, the total tag costing only 1 cent at these volumes. The smallness of the chip makes it cheaper, less likely to be damaged, and less susceptible to chip famines when, every 4 years or so, the world's factories cannot meet chip demand.

By 2010, 30% of RFID smart labels may be chipless. New types of inexpensive chipless smart label are expected to be widely used, notably surface-acoustic-wave (SAW) labels and printed thin-film transistor circuits on flexible polymer or paper substrates.

SAW labels. SAW RFID smart labels are already used in nonstop electronic highway toll collection and in car factories and, at 10 cents each (initially), they are being assessed for ePC item tracking. Unlike silicon chips, SAW tags operate at very low power, with no threshold voltage. They can retain 256 bits of data, work at a range of 10 m (33 ft), and are simple to

(a)   (b)   (c)

Fig. 2. Health-care applications of smart labels. (*a*) Smart label wrapped around a pill bottle intended for a blind or partially sighted person. The label is programmed to store prescription information, which is sensed by (*b*) a reader that is held near the label and speaks out the information on it using speech-synthesis technology (*En-Vision America*). (*c*) Very small test tube of the type used in the millions for drug trials, with a chip-sized tag embedded in its base (*Hitachi High Technologies*).

manufacture. They tolerate an unusually wide temperature range, and some can even sense temperature or pressure without being connected to extra sensors. Versions having as high a working frequency as 2.45 GHz are available, but none are as thin as alternative chipless smart-label technologies. Currently, a protected SAW smart label is at least 1 mm (0.04 in.) thick.

Thin-film labels. Thin-film transistor circuits are expected to appear from 2005 onward, and RFID labels incorporating them promise to cost a few cents or less. Four thin-film technologies now under development for the semiconductors in field-effect transistors incorporated in smart labels are polycrystalline thin-film silicon (at fairly low potential cost); amorphous thin-film silicon (low cost); insoluble organic molecules, usually oligomers (low cost); and soluble organic molecules, usually polymers (very low cost). The successful processes will be particularly useful in the simultaneous manufacture of microphones, loudspeakers, humidity sensors, batteries, and so forth in smart labels.

Since most of these technologies are capable in the laboratory of creating both *n*- and *p*-type transistors, complementary metal-oxide semiconductor (CMOS) devices can be manufactured. Most can achieve a satisfactory 10^6 ratio of on-to-off current. However, at present 13.56 MHz is the highest RFID frequency envisaged with these technologies, and development of labels that can operate at higher frequencies presents a challenge. Although organic versions can be usefully transparent, organic conducting layers are relatively resistive and, as yet, this degrades performance and power consumption.

Higher operating frequencies. Most RFID tags have employed the license-free frequency band of 125–135 kHz. At these low frequencies the H (magnetic) field component of the addressing electromagnetic radiation can be detected by an antenna coil in the tag replying to a similar coil in the interrogator, and this H field floods an area of many square meters so that the tag can even be seen around corners. At higher frequencies, communication is via the E (electric) field component of electromagnetic radiation, which must be emitted as a beam, and the tag antenna has to be a dipole and reply to an equivalent dipole on the interrogator.

However, the flooding property of the H-field component at low frequencies also means that the reader can confuse different tags. Most of the major new capabilities envisioned for smart labels require them to operate at higher frequencies. One such feature is the ability to read tags at high speed, which may be needed because the labels are on, for example, a railway train or because up to 1000 must be read as a group. For another feature, the remote location of labels, the necessary electromagnetic communicating E-field beams are possible only above 50 MHz. A battery in a smart label—which permits long range, signaling from the tag, and toleration of high ambient noise—is drained less at higher operating frequencies. On the other hand, for operation without the expense of a battery, and at power levels within the legal limits that are deemed safe for humans in the vicinity of the devices and guard against electromagnetic interference, achieving maximum range also requires high-frequency operation. Finally, operation at high frequencies allows the use of a label that is thinner (because higher frequencies avoid the need for many turns of antenna wire) and less expensive (because high-frequency smart labels use the least materials).

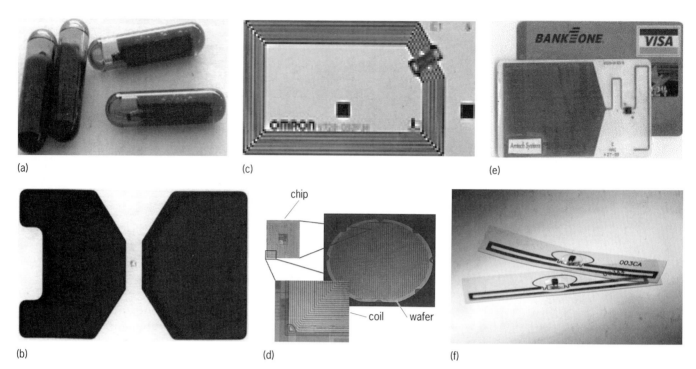

Fig. 3. Smart labels at various working frequencies. Labels at around 125 kHz: (*a*) in the form of glass bullets and (*b*) thin and credit-card size. Labels at 13.56 MHz: (*c*) credit-card size but thinner and (*d*) with the antenna coil deposited on a semiconductor chip, showing the chip itself (2.3 × 2.3 mm) and the slice of silicon (wafer) from which it is cut. (*e*) Label at ultrahigh frequency (around 900 MHz), credit-card size but thinner. (*f*) Label at 2.45 GHz, the size of a hairpin.

At low frequencies, tags typically take the form of a glass grain of rice or bullet (**Fig. 3***a*), although a low-frequency tag has been developed which is thin and credit-card size (Fig. 3*b*). Tags at 13.56 MHz (Fig. 3*c*) and at ultrahigh frequency (UHF, around 900 MHz; Fig. 3*e*) are typically credit-card size but thinner. At 13.56 MHz, when short range is tolerable, they can even be the size of the chip itself because the antenna coil can be deposited on the chip (Fig. 3*d*). (This is not possible for antennas at higher frequencies, which detect the E field.) At 2.45 GHz, a typical tag is the size of a hairpin (Fig. 3*f*), for example, 52 mm (2.0 in.) long and 60 mm (2.4 in.) across, and again very thin (0.2 mm, or 0.008 in.).

Most RFID smart labels in use in 2010 are expected to operate at 13.56 MHz and at ultrahigh frequency. Multifrequency, multiprotocol readers are becoming available to read several different types of smart label, but it will not be economic to have readers that accommodate all types. Electronic Product Code readers deployed in large numbers may have to filter data to avoid overload of the computer system behind them and wasted expense in transmitting and processing unwanted data downstream.

Varying parameters. To meet disparate needs, RFID smart labels are specified with various operating and other parameters whose values can differ by up to 11 orders of magnitude from one label to another (see **table**). As RFID smart labels are increasingly used to save cost, reduce theft, find lost items, improve safety, and so on in many disparate locations, the technologies will need to continue improving. Because the needs are so varied, no overall winner is expected in the technology employed, any more than there will be any sole frequency for all applications.

For background information *see* COMPUTER; COMPUTER SECURITY; DIFFRACTION; INTEGRATED CIRCUITS; PRINTED CIRCUIT; RADIO-SPECTRUM ALLOCATIONS; REAL-TIME SYSTEMS; SURFACE-ACOUSTIC-WAVE DEVICES in the McGraw-Hill Encyclopedia of Science & Technology.   Peter Harrop

Choices of parameters for RFID devices*

Parameter	Magnitudes of choice	Typical values
Number ordered: 1–5 × 10¹¹ units	>11	50–5,000,000 units (recent order for 500,000,000 units)
Transmission frequency: 3 Hz–30 GHz	10	125–135 kHz, 13.56 MHz, 433 MHz, UHF (around 900 MHz), 2.45 GHz, 5.6 GHz
Data transfer rate: 10–10⁷ bits/s	6	100–10⁶ bits/s
Size: 0.01–10⁶ mm³	8	Smart label 0.2 mm (0.008 in.) thick, the size of a postage stamp or credit card
Range: 40 µm–4 km (2.5 mi)	8	8 mm–10 m (0.3 in.–33 ft)
Price: 0.1 cent–$1000	6	Smart labels: 1 cent–$1.00

SOURCE: IDTechEx.
*Expected to be needed in the period 2003–2013.

Bibliography. R. Das, *The Smart Label Revolution*, IDTechEx, 2003; K. Finkenzeller, *RFID Handbook*, 2d ed., Wiley, 2003; P. J. Harrop, *The Future of Intelligent Packaging*, PIRA International, 2002; P. J. Harrop et al., *Total Asset Visibility*, IDTechEx, 2003; P. J. Harrop and R. Das, *The Future of Chipless Smart Labels*, IDTechEx, 2003.

Soft lithography

Soft lithography refers to a suite of techniques for replicating patterns of organic molecules or other materials (for example, ceramics or metals) on both planar (flat) and nonplanar (curved) substrates. It is applicable to structures ranging in size from tens of nanometers to centimeters. For most applications, soft lithography uses mechanical processes to transfer organic material by physical contact between a topographically patterned stamp or mold and a substrate. The mechanisms for pattern transfer (molding, embossing, and printing) are more similar to methods used for bulk manufacturing (for example, plastic parts and newspapers) than they are to those used commonly in fabricating microelectronic devices (for example, photolithography or electron-beam lithography, where beams of light or beams of electrons write patterns in polymeric materials). The term "soft" originally came from physics usage where organic materials are known as soft matter. Soft lithography initially referred to the rubbery, organic stamps used to transfer patterns. It now generally refers to both the system used for pattern transfer and to the organic or organometallic materials patterned, regardless of whether a rubber stamp or a hard stamp (usually fabricated of quartz or glass) is used, and has applications in electronics, optics, and biology.

Master versus replica. Most fabrication processes are divided into two steps: the fabrication of a master, and the replication of that master. To fabricate a master for use in soft lithography, a high-precision technique such as photolithography is used to form a three-dimensional pattern in a photosensitive polymer supported on a rigid substrate (**Fig. 1**). Then a liquid precursor for an elastomeric polymer, such as polydimethylsiloxane, is poured onto the topographically patterned master and cured to form a rubbery (elastomeric) solid. The polymer forms a negative replica of the original pattern with high fidelity. The topographically patterned, polymeric block acts as a stamp or mold for depositing or transferring the pattern from the master into organic structures supported on another substrate. Polydimethylsiloxane (which is a common material, often found in bathtub caulking and rubber toys) is used for the stamp or mold element in soft lithography because it is inexpensive, commercially available, nontoxic, elastomeric, and optically transparent. Relief features on the stamp define the shape, size, and distance of separation between the printed or molded regions. The flexibility of the polydimethylsiloxane stamp al-

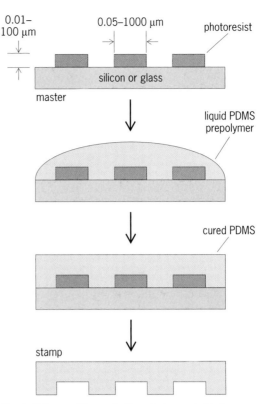

Fig. 1. Scheme of the replication of a topographically patterned substrate. The liquid precursor for polydimethylsiloxane (PMDS) is cast on the master and thermally cured. The solid, rubbery stamp is removed from the substrate; the molded surface is patterned with the inverse topography of the substrate.

lows molecular-level or conformal contact between the stamp and the surface. This feature is important for preparing uniform and reproducible patterns of molecules on rigid substrates.

Techniques. The two most common mechanisms for transferring patterns by soft-lithographic techniques are molding and printing (**Fig. 2**). A polydimethylsiloxane mold, pressed into contact with a liquid prepolymer, produces replicas with features that are inverted from the mold. Variations of this method, using rigid stamps, are used in the production of compact discs and holograms. Alternatively, a low-viscosity precursor to a polymer can fill the mold by capillary action. The molded precursor is cured by exposure to heat or to ultraviolet light or by evaporation of a solvent. The elastic properties and low surface energy of the polydimethylsiloxane make it easy to release the mold from the cured polymer. These procedures enable the replication of three-dimensional topologies in polymers, ceramics, glassy carbon, colloids, and salts in a single step.

Another soft-lithographic technique, microcontact printing, is analogous to a rubber stamp printing on paper; that is, a chemical "ink" is transferred upon contact between the stamp and a surface. The ink often is composed of organic molecules that spontaneously organize into an ordered array only one molecule thick on the surface. These self-assembled

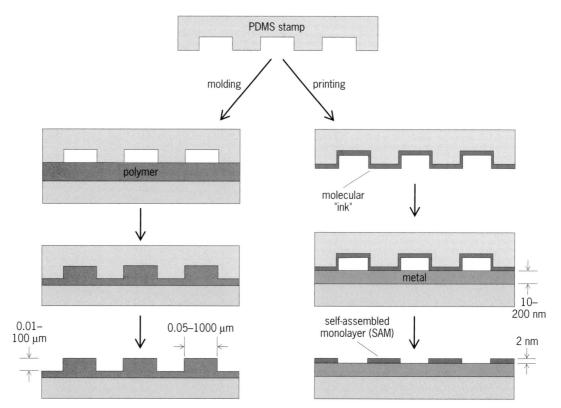

Fig. 2. Two methods common in soft lithography using a polydimethylsiloxane (PMDS) stamp.

monolayers present specific chemical functionalities on the surface that can modify its physical properties, for example, wettability or adhesion. Alkanethiols [$CH_3(CH_2)_nSH$] are used commonly to form self-assembled monolayers on coinage metals such as gold, silver, palladium, and copper. The sulfur in the thiol moiety binds to the surface of the metal, and the hydrocarbon chains arrange and form an ordered, hydrophobic surface.

Uses and applications. Soft lithography is particularly useful in rapidly prototyping micro- and nanostructures for applications in electronics, optics, and biology. The appropriate choice of soft-lithographic techniques can reduce the duration of the design cycle from an initial idea to a working prototype in less than 24 h. This feature is attractive at both the research and production levels.

Surface patterning and biology. The capability to control the macro- and microscopic properties of surfaces is useful (especially in biology) whenever nonspecific adsorption of proteins to surfaces is a significant problem. Soft lithography can generate patterns of biomolecules on a surface that direct proteins, cells, or other biologically relevant molecules to interact, spread, or avoid specific locations on a surface. This ability has enabled the development of applications such as cell-based assays for the study of cancer cells and their response to small molecules (**Fig. 3a**).

Micro/nanofabrication. Soft-lithographic techniques can pattern metallic and polymeric features over large areas in a single fabrication process for use in electronic and optical devices. The stamps are used as masks in a photolithographic technique, called phase-shifting lithography, which transfers the pattern of the edges of the features defined in the stamp to the photoresist material because the intensity of the light at the edges of the features is lower than that passing through the other regions of the stamp. Bandpass filters and polarizers are examples of optical elements fabricated by this technique. Microelectronic devices such as transistors and sensors (for H_2) are easily fabricated by techniques such as microcontact printing (Fig. 3b).

Microfluidics. Polymer-based microfluidic devices have two significant advantages over those produced in glass by etching: (1) the materials and the processes used to fabricate the polymer-based devices are inexpensive, and (2) the polymer-based channels are mechanically stable. The polydimethylsiloxane-based, two-dimensional channels can be sealed to rigid or flexible substrates and used to form complicated, three-dimensional fluidic networks. These channels are useful for biological assays (solution gradients and serial dilutors), for optical devices (waveguides), and for particle alignment (nanowires and colloidal crystals) [Fig. 3c]. *See* MICROFLUIDICS; MINIATURIZED ANALYSIS SYSTEMS.

Advantages and disadvantages. Four areas of science and technology where soft lithography offers significant advantages over other lithographic methods are (1) the fabrication of microstructures on curved surfaces, (2) the fabrication of nanostructures, (3) the development of microsystems for cell biology, and (4) the fabrication of micro- (and

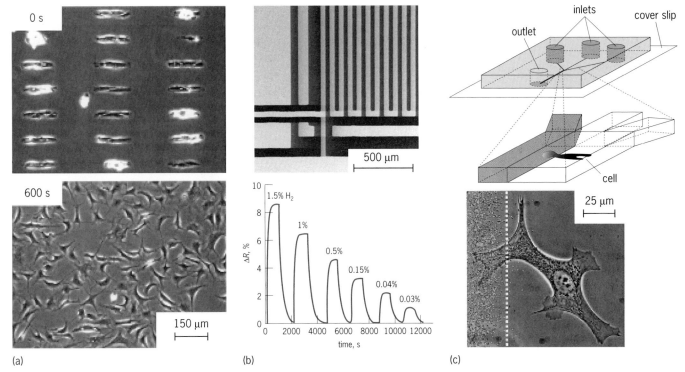

Fig. 3. Applications of soft lithography. (*a*) Cell biology. Adhesion sites for bovine capillary endothelial cells patterned by microcontact printing localize the cells to specific regions and geometries on a surface. The optical micrographs show an initial configuration of confined cells and an image of the system 10 min after a mild electrochemical pulse was applied to release the barriers on the surface that prevented the cells from spreading. This type of assay is useful for studying the spreading behavior of normal and cancerous cells. (*b*) Microfabrication. A palladium-based microsensor fabricated by microcontact printing and selective etching responds electrically to the presence of hydrogen (H_2) gas. The image is a scanning electron micrograph that shows one section of the palladium sensor. The graph shows the change in resistance of the sensor as a function of the H_2 concentration. (*c*) Microfluidics. Microfluidic channels can address different parts of a single cell by controlling the flow inside the channels. The scheme shows a channel that is placed over a cell. There are three inlets to the channel. Fluid streams introduced in each inlet do not mix by turbulence because the dimensions of the channel are small (<100-μm diameter). This physical phenomenon (known as laminar flow) allows the delivery of reagents to specific parts of a cell. In this case, the optical micrograph shows the mitochondria on the left side of the cell that have been stained with a fluorescent dye.

perhaps nano-) structures over large areas at low cost. Photolithography uses rigid, planar materials (for example, silicon wafers) as substrates and physical masks in microfabrication. The rigid masks cannot conform to curved substrates (for example, optical lenses), and optical systems require highly specialized lenses to project patterns in focus onto curved surfaces. Soft-lithographic methods, however, can generate microstructures on nonplanar surfaces (for example, fiber optics) because the rubber stamps easily conform to the surface.

Soft lithography also can fabricate simple arrays of nanostructures over large areas. State-of-the-art systems for photolithography can produce features down to 25 nm in the laboratory and features of 70-100 nm in mass production. Diffraction of the light used in photolithography ($\sim$150–200-nm wavelengths) physically limits the minimum resolution of photolithographic processes but, in principle, only the size of the molecules in the replica ($\sim$1 nm) limits the resolution of soft-lithographic techniques based on molding. Another limitation of photolithography is that the tools are expensive ($\sim$\$10 million). This cost may not deter microelectronics manufacturers, but it can be prohibitive to researchers in areas of bi-

ology, chemistry, physics, and materials science who wish to study systems with critical dimensions below 100 micrometers (and especially below 100 nm). The initial cost of defining a master with nanostructures for soft lithography is offset by the ease and low cost of producing polydimethylsiloxane replicas for further experiments to fabricate nanostructures.

Soft lithography also is useful for developing microtools to study living cells, as many biologists and chemists interested in studying cell behaviors are not familiar with, or do not have access to, the tools used in microelectronic manufacturing. Techniques for rapidly prototyping microstructures use transparency masks generated on high-resolution printers to define the patterns in a photoresist, and do not require special facilities such as cleanrooms. It is, therefore, straightforward to design and implement systems for studying cells on surfaces using polydimethylsiloxane molds.

The disadvantages of soft lithography using polydimethylsiloxane replicas are (1) the softness of polydimethylsiloxane can cause distortions in printed or molded structures; (2) polydimethylsiloxane is not stable to many organic solvents or at high temperatures; (3) polydimethylsiloxane is not a good choice when it is necessary to align or register multiple patterns on a substrate; (4) polydimethylsiloxane has a high coefficient of thermal expansion, that is, the size of the features in the mold varies significantly with small changes in temperature.

Outlook. Soft lithography allows research scientists in many disciplines to develop new tools for exploring scientific problems that require micrometer- and nanometer-scale structures in a simple, inexpensive manner. The discoveries from these studies could lead to new technologies for commercial markets. Soft lithography is currently well suited for producing systems with one level of patterning. However, further improvements in soft lithography should allow the fabrication of complex patterns of molecules that are composed of aligned or registered single-layer patterns.

For background information *see* INTEGRATED CIRCUITS; MONOMOLECULAR FILM; NANOSTRUCTURE; POLYMER; SURFACE AND INTERFACIAL CHEMISTRY; SURFACE TENSION in the McGraw-Hill Encyclopedia of Science & Technology.

J. Christopher Love; Daniel B. Wolfe;
Byron D. Gates; George M. Whitesides

Bibliography. S. Brittain et al., Soft lithography and microfabrication, *Phys. World*, 11:31, 1998; G. M. Whitesides et al., Soft lithography in biology and biochemistry, *Ann. Rev. Biomed. Eng.*, 3:335, 2001; G. M. Whitesides and J. C. Love, The art of building small, *Sci. Amer.*, 285:32, 2001; G. M. Whitesides and A. D. Stroock, Flexible methods for microfluidics, *Phys. Today*, 54:42, 2001; Y. Xia et al., Unconventional methods for fabricating and patterning nanostructures, *Chem. Rev.*, 99:1823, 1999; Y. Xia and G. M. Whitesides, Soft lithography, *Annu. Rev. Mat. Sci.*, 28:153, 1998.

Solid-state NMR spectroscopy

The processing and formulation (dosage form) of a drug may affect important pharmaceutical properties, such as form purity (whether the drug contains polymorphic or amorphous content, which might make it less therapeutically active), dissolution characteristics, chemical and physical stability, and bioavailability. Although it currently is not widely used for the characterization of pharmaceutical formulations, solid-state nuclear magnetic resonance (NMR) spectroscopy is a powerful technique that is useful for studying pharmaceutical materials. Because the spectrum for each polymorphic/amorphous form is often unique, the NMR method makes it possible to identify the state of the drug as well as the inactive ingredients used in drugs (excipients). Peaks from excipients are usually in a different region (chemical shift or frequency) of the NMR spectrum than are the peaks from drugs.

Solid-state NMR spectroscopy is also relatively insensitive to particle size and surface effects, eliminating some of the problems often associated with x-ray diffraction techniques, such as the preferred orienta-

tion and lengthy sample preparation. In many cases, samples for x-ray analysis are ground up, thereby increasing the possibility for phase transformations due to mechanical strain imposed on the crystal. With NMR spectroscopy, an intact formulated tablet can be analyzed directly. A further advantage of solid-state NMR spectroscopy is that quantitation of mixtures of polymorphic forms can be performed without the need to prepare standard curves.

Solid-state analysis. Nuclear magnetic resonance is a phenomenon exhibited when atomic nuclei in a static magnetic field absorb energy from a radio-frequency field of certain characteristic frequencies. The NMR spectrum contains peaks referred to as resonances. The intensity of the resonance is directly proportional to the number of nuclei that produce the signal. Nuclear magnetic resonance spectroscopy is routinely performed on solutions because rapid molecular motion results in a spectrum with narrow lines (high resolution). In the solid state, however, the absence of molecular motion leads to broad lines (low resolution). The two main sources of line broadening for nuclei that have nuclear spin (unpaired protons or neutrons) values of 1/2 (for example, ^{13}C, ^{15}N, and ^{1}H) are dipolar coupling and chemical shift anisotropy (CSA).

Dipolar coupling and decoupling. Because nuclei have magnetic moments, one nucleus can affect the magnetic field of a neighboring nucleus if they are close together. Such interactions between two (or more) spin-1/2 nuclei are known as dipolar couplings. The magnitude of the dipolar coupling has a $1/r^3$ dependence and is proportional to the magnetogyric ratios of the coupled nuclei. Either homonuclear (for example, between two ^{13}C nuclei) or heteronuclear (for example, between ^{1}H and ^{13}C nuclei) coupling is possible. The low natural abundance of ^{13}C (1.1%) makes the probability of two ^{13}C nuclei being in proximity very small, so in the case of solid-state ^{13}C NMR spectroscopy, only heteronuclear dipolar interactions between ^{1}H and ^{13}C are significant. High-power ^{1}H decoupling is used to eliminate or reduce ^{1}H and ^{13}C dipolar interactions. During the acquisition of the resonance signal, called the free induction decay, a continuous, high-power, radio-frequency pulse at the ^{1}H Larmor (angular) frequency is applied to the sample. The ^{1}H spins can be thought of as rapidly flipping, so the ^{13}C nucleus sees only the average dipolar interaction (which is zero).

Chemical shift anisotropy and magic-angle spinning. The other main broadening factor in solid-state NMR spectra is chemical shift anisotropy; that is, the chemical shift of a nucleus is highly dependent upon the orientation of the molecule with respect to the static magnetic field. When a polycrystalline sample is placed in the magnetic field, each crystallite has a different orientation in the static magnetic field. This results in a range of chemical shifts observed in the spectrum, which is typically called a powder pattern. Powder patterns can be ~200 parts per million (ppm) broad for a single sp^2 hybridized carbon (carbonyl, aromatic, and so on). When multiple carbon sites are present (as is typical for drugs), the result is a broad, featureless NMR spectrum. Magic-angle spinning (MAS) is used to average the chemical shift anisotropy. When a sample is spun rapidly at 54.74° (the magic angle) relative to the static magnetic field, the anisotropic (angular) component of the chemical shift is either partially or completely removed. The result is a high-resolution solid-state NMR spectrum.

Cross-polarization and combination techniques. Cross polarization (CP) provides a fourfold enhancement in signal intensity for ^{13}C experiments. In a cross-polarization experiment, the magnetization is transferred from the abundant spin (normally ^{1}H) to the dilute spin (for pharmaceuticals it is normally ^{13}C). It also leads to shorter acquisition times, because the recycle delay is determined by the ^{1}H relaxation rate, which is often much shorter than that of ^{13}C. The combination of these three techniques (CP, MAS, and high-power decoupling) was described by E. O. Stejskal and J. Schaefer.

The high-resolution ^{13}C CPMAS NMR spectra obtained for ibuprofen are shown in **Fig. 1**. In Fig. 1*a*, the spectrum was acquired using typical solution NMR conditions (magnetic field of 7.05 tesla). The peaks are very broad due to both chemical shift anisotropy and dipolar coupling present in the

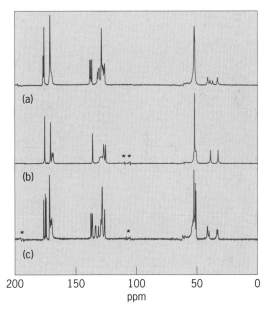

Fig. 2. ^{13}C CPMAS NMR spectra of three polymorphic forms of aspartame: (*a*) hemihydrate, (*b*) hemihydrate produced by ball milling form shown in *a*, (*c*) dihemihydrate, prepared by exposure of form in *a* to a high relative humidity environment (>98%) for 5 days.

sample. In Fig. 1*b*, magic-angle spinning was used to increase resolution by averaging out the chemical shift anisotropy. Even with spinning at 4000 Hz (240,000 rpm), the resulting NMR spectrum is still quite broad due to the dipolar coupling. The spectrum in Fig. 1*c* was acquired using high-power ^{1}H decoupling and magic-angle spinning. A dramatic improvement in resolution is apparent. The spectrum in Fig. 1*d* was acquired with cross polarization, magic-angle spinning, and high-power ^{1}H decoupling. The difference in signal-to-noise ratio between Fig. 1*c* and *d* is significant, as is the fact that the increase is also achieved in a much shorter time, about 1/30 of the time required for the normal single-pulse experiment with magic-angle spinning and decoupling. This is because the time between successive acquisitions in a cross-polarization experiment is only 1 s (determined by the proton relaxation), whereas 30 s is required for the ^{13}C relaxation in a single-pulse experiment.

Figure 2 shows the difference between polymorphs and solvates of the artificial sweetener aspartame. The spectra in Fig. 2*a* and *b* are hemihydrates (0.5 mole of water for each mole of aspartame), and the spectrum in Fig. 2*c* is a dihemihydrate (2.5 water:1 aspartame). The differences in these spectra are due to the differences in the orientation, packing, and intra- and intermolecular forces that are present in the crystal lattice.

Quantitation of drug forms. Nuclear magnetic resonance spectroscopy is inherently a quantitative technique. Under ideal circumstances, this should be true of an NMR spectrum in the solid state too. In most analytical techniques, a standard curve must be prepared to quantify the amount of each drug form present. However, ^{13}C solid-state NMR spectra often

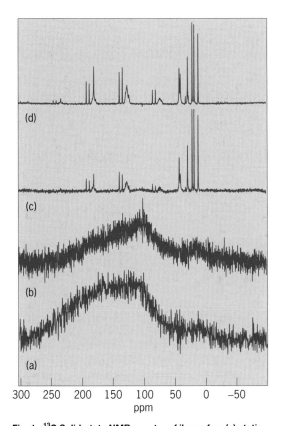

Fig. 1. ^{13}C Solid-state NMR spectra of ibuprofen: (*a*) static sample, 128 acquisitions, 30-s pulse delay, 64-minute acquisition time. (*b*) Single pulse MAS, 128 acquisitions, 30-s pulse delay, 4000 Hz spinning rate, 64-min acquisition time. (*c*) Single-pulse MAS with ^{1}H decoupling, 128 acquisitions, 30-s pulse delay, 4000 Hz spinning rate, 64-min acquisition time. (*d*) Cross polarization with MAS and ^{1}H decoupling, 128 acquisitions, 1-s pulse delay, 4000 Hz spinning rate, 2.1-min acquisition time.

are not quantitative because they are usually acquired using cross polarization, where the individual peak areas are determined by the amount of a particular form present in the sample and by the magnetization transfer dynamics from the abundant spin (normally ^{1}H) to the dilute (normally ^{13}C) spin. These dynamics must be taken into account to achieve accurate results.

Delavirdine mesylate. The use of solid-state NMR spectroscopy to quantify polymorphic mixtures of the human immunodeficiency virus (HIV) drug delavirdine mesylate was demonstrated by P. Gao. Binary mixtures of a minor polymorphic or pseudopolymorphic form were mixed with a major form, and quantitation was performed. Since the cross-polarization dynamics had to be considered, multiple contact times were used to measure the change in magnetization transfer with respect to time. The initial increase in magnetization was due to the rate of transfer of magnetization from ^{1}H to [13]C. This increase was described by the heteronuclear cross-polarization rate constant T_{CH}. The decrease of magnetization over time was due to the proton relaxation rate, $T_{1\rho}$. It was determined that the peaks used for quantitation had the same $T_{1\rho}$ rates for all of the forms. Thus, quantitation could be directly done based on the peak area to give a relative measure of the amount of one form with respect to the other. The peak areas corresponding to each polymorphic form were measured, and the experimental weight percent was determined for each polymorphic form. The experimental results shown in **Table 1** agree to within 1% of the actual amount present. The overall accuracy of the method was confirmed by plotting the actual weight percent versus the experimental weight percent determined by NMR spectroscopy. This work is beneficial to the pharmaceutical industry because it provides a nondestructive method for both the observation and quantitation of polymorphs or pseudopolymorphs in a pharmaceutical solid.

Neotame. The quantitation of mixtures of crystalline polymorphs and of crystalline and amorphous forms of the artificial sweetener neotame has been studied by T. J. Offerdahl and coworkers. Each crystalline polymorph had a different proton relaxation rate ($T_{1\rho}$), so direct integration could not be performed for accurate quantitation. To obtain quantitative data for each polymorph, the peak areas at various contact

TABLE 2. Quantitation of the crystalline polymorphs for neotame

Actual weight percent	NMR-determined weight percent
12.60	13.7
21.20	22.0
29.70	29.4
39.10	37.9
49.73	50.3
61.56	62.3
70.29	71.4
80.88	81.8
88.41	89.0

times were determined. A plot of natural logarithm (ln) of peak areas versus contact time was used to determine the total amount of each form present in the sample. The error in the amount of each form present was within 1% for the crystalline polymorphic mixtures as shown in **Table 2**. When the crystalline and amorphous forms were analyzed, a large deviation (~15%) in the results was noticed with increasing amorphous content. This was caused by the presence of amorphous material in the crystalline standards. The amorphous material in the standards was determined using the above protocol, and when the presence of amorphous material was accounted for in the standards, the error was ~1%.

This type of quantitative analysis is beneficial because a sample can be analyzed directly without any preparation. In other techniques, the sample needs to be dissolved, ground up, and so on prior to analysis. This can lead to the formation of other polymorphs or even degrade the sample prior to analysis. This research is also beneficial because direct quantitation can be done, eliminating the need for standard curves. In this study, it was found that the two crystalline polymorphs were not pure. Instead, each had an amorphous impurity of varying degrees. If these polymorphs had been used for the preparation of a standard curve, all the quantitative results would be based on impure standards, leading to incorrect results.

For background information *see* ANALYTICAL CHEMISTRY; MAGNETIC RESONANCE; NUCLEAR MAGNETIC RESONANCE (NMR); PHARMACEUTICALS TESTING in the McGraw-Hill Encyclopedia of Science & Technology.   Thomas J. Offerdahl; Eric J. Munson

Bibliography. C. A. Fyfe, *Solid State NMR for Chemists*, C.F.C. Press, Guelph, 1983; P. Gao, Determination of the composition of delavirdine mesylate polymorph and pseudopolymorph mixtures using ^{13}C CP/MAS NMR, *Pharm. Res.*, 13:7, 1095–1104, 1996; C. Gustafsson et al., Comparison of solid-state NMR and isothermal microcalorimetry in the assessment of the amorphous component of lactose, *Int. J. Pharmaceutics*, 174:1–2, 243–252, 1998; T. J. Offerdahl et al., Quantitation of crystalline and amorphous forms of Neotame using ^{13}C CPMAS NMR spectroscopy, in press, 2003; B. R. Rohrs, Tablet dissolution affected by a moisture mediated solid-state

TABLE 1. Quantitation of polymorphs for delavirdine mesylate

Actual weight percent	NMR-determined weight percent
2.00	2.7
3.00	2.8
5.00	5.4
10.00	9.4
15.00	14.5
19.90	20.2
25.00	24.0
50.00	48.7

interaction between drug and disintegrant, *Pharm. Res.*, 16:12, 1850–1856, 1999; E. O. Stejskal and J. Schaefer, Magic-angle spinning and polarization transfer in proton-enhanced NMR, *J. Magnetic Resonance*, 28:105–112, 1977; R. Suryanarayanan and T. S. Wiedmann, Quantitation of the relative amounts of anhydrous carbamazepine ($C_{15}H_{12}N_2O$) and carbamazepine dihydrate ($C_{15}H_{12}N_2O \cdot 2H_2O$) in a mixture by solid-state nuclear magnetic resonance (NMR), *Pharm. Res*, 7:2, 184–187, 1990; Y. Tozuka et al., Characterization and quantification of clarithromycin polymorphs by powder x-ray diffractometry and solid-state NMR spectroscopy, *Chem. Pharm. Bull.*, 50:8, 1128–1130, 2002.

Space flight

Space flight in 2002 featured a number of highlights both in human missions and in automated space exploration and commercial utilization. While the United States space budget remained relatively stable, international space activities continued their trends of reduced public spending and increased pressure on the private industrial sector for more entrepreneurship, particularly in developing and providing launch services. After reaching their lowest point since 1963 in 2001, launch activities in 2002 showed some recovery. A total of 61 successful launches placed about 81 payloads into orbit, compared to 57 flights in 2001 (81 in 2000). There also were four launch failures (up from two in 2001) [**Table 1**].

In its traditional leadership role among world space organizations, the United States National Aeronautics and Space Administration (NASA) racked up remarkable accomplishments in the advancing buildup and operation of the *International Space Station* (*ISS*); the successful launches of science missions such as the *HESSI* (*High-Energy Solar Spectroscopic Imager*), later renamed *RHESSI*; the *Aqua* satellite of the Earth Observing System (EOS); the *POES-M*; and the new upgrade of the Hubble Space Telescope. Interplanetary milestones included the start of the science mission of the *Mars Odyssey* spacecraft in Mars orbit after its arrival at the planet on October 24, 2001, and the last flyby of the tiny Jupiter moon Amalthea by the spacecraft *Galileo* before its final plunge into the giant planet, scheduled for September 21, 2003.

The commercial space market showed surprising recovery from the dramatic decline of previous years. Out of the 61 successful launches worldwide, 28 (46%) were commercial launches (carrying 43 commercial payloads), compared to 22 (38%) in 2001. Civil science satellite launches totaled 10 worldwide, the same number as in 2001.

TABLE 1. Successful launches in 2002 (Earth orbit and beyond)

Country	Number of launches (and attempts)
United States (NASA/DOD/ Commercial)	18 (18)
Russia	23 (25)
Europe (ESA/Arianespace)	11 (12)
People's Republic of China	4 (5)
Japan	3 (3)
India	1 (1)
Israel	1 (1)
TOTAL	61 (65)

TABLE 2. Some significant space events in 2002

Designation	Date	Country	Event
RHESSI	February 5	United States	Successful launch of NASA's *Ramaty High-Energy Solar Spectroscopic Imager* into orbit for basic physics research of elementary particles
ENVISAT	March 1	Europe	Successful Ariane 5 launch of ESA's highly advanced environmental satellite, with 10 instruments, gathering information to survey and protect the planet
STS 109 (*Columbia*)	March 1	United States	4th "service call" to the Hubble Space Telescope, with five spacewalks to upgrade the observatory with new systems and the Advanced Camera for Surveys (ACS)
Grace 1 and *2*	March 17	United States/ Europe/Russia	Successful launch on a Russian Rockot rocket of twin satellites for highly accurate mapping of the Earth's gravity field
Shenzhou 3	March 25	P.R. of China	3d crewless test flight (and recovery after several orbits) of an experimental manned capsule for future use by human crews on a Long March 2F rocket
STS 110 (*Atlantis*)	April 8	United States	*ISS* mission 8A, with the S-Zero (S0) truss segment, the Canadian-built Mobile Transporter (MT), and a crew of seven, four spacewalks to install S0 and MT
Soyuz TM-34/ISS-4S	April 25	Russia	Launch of 3d "taxi flight" to *ISS*, bringing a fresh Soyuz crew return vehicle (CRV) and carrying the 2d "space tourist," South Africa's Mark Shuttleworth, return on TM-33
Aqua	May 4	United States	NASA's *EOS-PM* satellite, launched on a Delta 2, to observe Earth's water cycle: evaporation, water vapor, clouds, precipitation, soil moisture, ice, and snow
STS 111 (*Endeavour*)	June 5	United States	*ISS* mission UF-2, carrying Multi-Purpose Logistics Module *Leonardo* (3d flight), Canada's SSRMS Mobile Base System, meteoroid protection shields, and resupply
POES-M (*NOAA-M*)	June 24	United States	Launch of polar operational weather satellite *POES-M* for NOAA, to provide, with *GOES-M*, global data for weather forecasting and meteorological research
STS 112 (*Atlantis*)	October 7	United States	*ISS* mission 9A, carrying the 2d truss element, the 14.5-ton S1, which was attached to S0, mission featured 3 spacewalks, bringing total EVAs for *ISS* to 46
Soyuz TMA-1/ISS-5S	October 30	Russia	Launch of 4th taxi flight to *ISS*, delivering a fresh Soyuz CRV; three-person crew included ESA/Belgium guest cosmonaut Frank De Winne, crew returned on *Soyuz TM-34*
STS 113 (*Endeavour*)	November 23	United States	*ISS* mission 11A, with 5th *ISS* replacement crew, the 4th truss element P1 (of 11 total), and other equipment and resupply, Expedition 4 crew return on shuttle
Shenzhou 4	December 29	P.R. of China	4th crewless test flight (and recovery after several orbits) of another experimental "Divine Vessel" for future use by human crews on a Long March 2F rocket

Russia's space program, despite chronic shortage of state funding, showed continued dependable participation in the buildup of the *International Space Station*. Europe's space activities in 2002 rose above the previous year's but suffered a crushing blow at year's end in the failure of the fourteenth Ariane 5 heavy-lift launch vehicle.

Seven crewed flights from the two major spacefaring nations (down from eight in 2001) carried 40 humans into space (2001: 44), including 4 women (2001: 5), bringing the total number of people launched into space since 1958 (counting repeaters) to 955, including 98 women, or 431 individuals (37 females). Some significant space events in 2002 are listed in **Table 2**.

International Space Station

Goals of the *ISS* are to establish a permanent habitable residence and laboratory for science and research and to maintain and support a human crew at this facility. The completed station will have a mass of about 1,040,000 lb (470 metric tons). It will measure 356 ft (109 m) across and 290 ft (88 m) long, with almost an acre (0.4 hectare) of solar panels to provide up to 110 kilowatts power to six laboratories. Led by the United States, the *ISS* draws upon the

scientific and technological resources of 16 nations, also including Canada, Japan, Russia, the 11 nations of the European Space Agency (ESA), and Brazil.

Operations and assembly. One of the partnership issues that emerged in 2002 is the provision of assured crew return capability after the Russian obligation to supply Soyuz lifeboats to the station expires in April 2006. In NASA's new space transportation planning, a U.S. crew rescue capability (other than via space shuttle) will be available only in 2010. Efforts continue by the partnership to work out a solution for dealing with the 4-year gap.

Following the recommendations of an independent advisory panel of biological and physical research scientists called Remap (Research Maximization and Prioritization), NASA in 2002 established the formal position of a Science Officer for one crew member aboard the *ISS*, responsible for expanding scientific endeavors on the station. Expedition 5 flight engineer Peggy Whitson became NASA's first Science Officer.

The initial milestones for the *ISS* program since the beginning of orbital assembly in 1998 included the arrival of the first long-duration station crew in November 2000 and the installation of the first set of U.S. solar array wings in December 2000. During

---

TABLE 3. *International Space Station (ISS) assembly progress through 2002*

Number	Launch date	Mission designation*	Flight	Primary cargo	Purpose
1	Nov. 20, 1998	1 A/R	Proton K	Control Module FGB	Assembly
2	Dec. 4, 1998	2A	Shuttle/STS 88	Node 1, pressurized mating adapters 1, 2	Assembly
3	May 27, 1999	2A.1	Shuttle/STS 96	Spacehab Double Module	Outfitting
4	May 19, 2000	2A.2a	Shuttle/STS 101	Spacehab Double Module	Outfitting
5	July 12, 2000	1R	Proton K	Service Module (SM)	Assembly
6	Aug. 6, 2000	1P	*Progress M1-3*	Consumables, spares, props	Logistics
7	Sept. 8, 2000	2A.2b	Shuttle/STS 106	Spacehab Double Module	Outfitting
8	Oct. 11, 2000	3A	Shuttle/STS 92	Z1 truss, 4 control moment gyroscopes, pressurized mating adapter 3	Assembly
9	Oct. 31, 2000	2R/1S	*Soyuz TM-31*	Expedition 1 crew	1st crew
10	Nov. 15, 2000	2P	*Progress M1-4*	Consumables, spares, props	Logistics
11	Nov. 30, 2000	4A	Shuttle/STS 97	P6 module, photovoltaics array	Assembly
12	Feb. 7, 2001	5A	Shuttle/STS 98	U.S. lab module, racks	Assembly
13	Feb. 26, 2001	3P	*Progress M-44*	Consumables, spares, props	Logistics
14	Mar. 8, 2001	5A.1	Shuttle/STS 102	Multi-Purpose Logistics Module *Leonardo*	Lab outfit
15	Apr. 19, 2001	6A	Shuttle/STS 100	Space Station Remote Manipulator System, Multi-Purpose Logistics Module *Raffaello*	Outfitting
16	Apr. 28, 2001	2S	*Soyuz TM-32*	1st "taxi" crew (with Tito)	New crew return vehicle
17	May 20, 2001	4P	*Progress M1-6*	Consumables, spares, props	Logistics
18	July 12, 2001	7A	Shuttle/STS 104	U.S. Airlock, high-pressure O_2/N_2 gas	Assembly
19	Aug. 10, 2001	7A.1	Shuttle/STS 105	Multi-Purpose Logistics Module *Leonardo*	Outfitting
20	Aug. 21, 2001	5P	*Progress M-245*	Consumables, spares, props	Logistics
21	Sept. 14, 2001	4R	"Progress 301"	Docking Compartment 1	Assembly
22	Oct. 21, 2001	3S	*Soyuz TM-33*	2d "taxi" crew	New crew return vehicle
23	Nov. 26, 2001	6P	*Progress M-256*	Consumables, spares, props	Logistics
24	Dec. 5, 2001	UF-1	Shuttle/STS 108	Multi-Purpose Logistics Module *Raffaello*	Utilization
25	Mar. 21, 2002	7P	*Progress M1-8* (257)	Consumables	Logistics
26	Apr. 8, 2002	8A	Shuttle/STS 110	S0 truss segment	Assembly
27	Apr. 25, 2002	4S	*Soyuz TM-34*	3d "taxi" (with Shuttleworth)	New crew return vehicle
28	June 5, 2002	UF-2	Shuttle/STS 111	Mobile Base System, Multi-Purpose Logistics Module *Leonardo*	Utilization
29	June 26, 2002	8P	*Progress M-24* (246)	Consumables, spares, props	Logistics
30	Sept. 25, 2002	9P	*Progress M1-9* (258)	Consumables, spares, props	Logistics
31	Oct. 7, 2002	9A	Shuttle/STS 112	S1 truss segment	Assembly
32	Oct. 30, 2002	5S	*Soyuz TMA-1* (211)	4th "taxi" (with De Winne)	New crew return vehicle
33	Nov. 23, 2002	11A	Shuttle/STS 113	P1 truss segment	Assembly

*A = American; R = Russian; S = Soyuz; P = Progress.

2001, astronauts and cosmonauts added U.S. Laboratory module *Destiny*, the Canada-supplied Space Station Remote Manipulator System (SSRMS) Canadarm2, the U.S. Airlock module *Quest*, and the Russian Docking Compartment (DC-1) *Pirs*, In April 2002, the first of eleven truss elements, S0 (Starboard Zero), was attached on top of *Destiny*, becoming the centerpiece of the 109-m-long (356-ft) truss for carrying the solar cell arrays of the station. In June, the Expedition 4 crew of Russian Commander Yuri Onufrienko and U.S. Flight Engineers Carl Walz and Dan Bursch was rotated. The new station crew (Expedition 5) of Russian Commander Valery Korzun, U.S. Flight Engineer/Science Officer Whitson, and Russian Flight Engineer Sergey Treschev delivered cargo, including the Mobile Base System (MBS), to provide mobility for the SSRMS, and the Italian-built Multi-Purpose Logistics Module (MPLM) *Leonardo* for cargo and equipment transport. The second truss segment, S1, arrived in October (**Fig. 1***a*) and was attached to S0 on the starboard side. Its counterpart on port, P1, followed in November and was also successfully mounted (Fig. 1*b*). The same shuttle mission brought the replacement crew (Expedition 6) of U.S. Commander Kenneth Bowersox, Russian Flight Engineer Nikolay Budarin, and U.S. Flight Engineer/Science Officer Donald Pettit, and returned the Expedition 5 crew to Earth. By end-2002, 33 carriers had been launched to the *ISS*: 16 shuttles, 2 heavy Protons (FGB/*Zarya*, SM/*Zvezda*), and 15 Soyuz rockets (9 uncrewed Progress cargo ships, the DC-1 docking module, and 5 crewed Soyuz spaceships). *ISS* assembly progress through 2002 is summarized in **Table 3**.

STS 110. The April 8–19 mission of *Atlantis*, designated *ISS* Mission 8A, carried the S0 truss, the Mobile Transporter (MT), and a crew of seven. Four spacewalks (EVAs) were conducted to install the S0 and the MT. On April 15, the MT railcart traversed a distance of 72 ft (22 m) during this first operation of a "railroad in space." Mission 8A was the first flight during which the SSRMS robot arm of the *ISS* was used to maneuver spacewalkers around the station, and the first where all EVAs of a shuttle crew were performed from the station's own airlock, *Quest*.

Soyuz TM-34. The April 25–May 4 Russian mission *Soyuz TM-34* (number 208), *ISS* mission 4S, was the third "taxi" flight to deliver a fresh crew return vehicle (CRV) to *ISS*. Its crew of three, under the command of Yuri Gidzenko, included an Italian flight engineer, Roberto Vittori, and the second "space tourist," Mark Shuttleworth of the Republic of South Africa. A science program, sponsored by Italy and South Africa, included four experiments in the Italian Marco Polo program, five payloads in the Shuttleworth program, and two new Russian experiments. Gidzenko, Vittori, and Shuttleworth returned in *Soyuz TM-33*, which had been docked to the *ISS* since October 2001.

STS 111. The June 5–19 flight of *Endeavour*, *ISS* Mission UF-2, carried a seven-member crew, including the fifth *ISS* resident crew (Korzun, Whitson, and

(a)

(b)

Fig. 1. Truss elements for the *International Space Station*. (*a*) Element S1 (Starboard One), following installation during shuttle mission STS 112 in October 2002. The Space Station Remote Manipulator System (SSRMS), Canadarm2, is visible below the truss. (*b*) Astronauts Michael E. Lopez-Alegria (left) and John B. Herrington working on element P1 (Port One), following installation during shuttle mission STS 113 in November 2002. (*NASA*)

Treschev). Its cargo comprised MPLM *Leonardo* on its third flight, loaded with 5600 lb (2500 kg) of equipment and consumables, the Mobile Base System (MBS), a replacement wrist roll (WR) joint for the SSRMS, meteoroid/orbital debris protection shields (ODPs) for the *Zvezda* Service Module, and new science payloads. During three spacewalks, astronauts installed the MBS on the MT (mobile transporter), temporarily stowed the ODPs, and replaced the

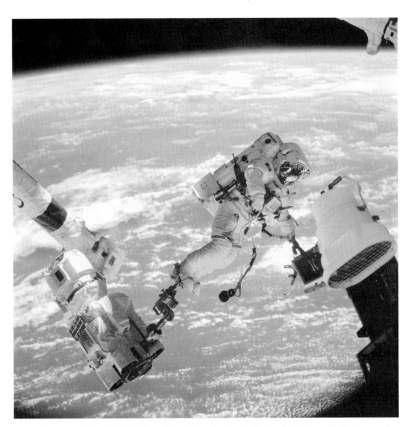

Fig. 2. Astronaut David A. Wolf, anchored to a foot restraint on the Space Station Remote Manipulator System (SSRMS), Canadarm2, carries the outboard nadir external camera for installation on the end of the S1 truss following the truss element's installation during shuttle mission STS 112 in October 2002. (*NASA*)

SSRMS WR joint. *Endeavour* returned to Earth with the fourth station crew (Onufrienko, Bursch, and Walz).

STS 112. The October 7–18 flight of *Atlantis, ISS* Mission 9A, carried a six-member crew. On October 10, its main payload, the 14.5-ton (13.1-metric-ton) S1 truss element, was transferred with the

Fig. 3. *International Space Station* photographed from the space shuttle *Endeavour* (mission STS 113), following the undocking of the two spacecraft on December 2, 2002. (*NASA*)

Canadarm2 from the shuttle cargo bay to the *ISS* and firmly attached to the S0 center truss segment atop the Lab module (Fig. 1*a*). The installation featured the SSRMS operating from its new MBS. The crew also transferred about 1000 lb (450 kg) of supplies, equipment, and other cargo to the station. Three spacewalks were performed to connect the S1 and S0 trusses with electrical, fiber-optical, data, and coolant umbilicals and to install external systems (**Fig. 2**).

Soyuz TMA-1. *Soyuz TMA-1* (number 211), *ISS* Mission 5S (October 30–November 10), was the fourth taxi flight to the *ISS*, to replace the previous CRV, *Soyuz TM-34/4S*, which reached the end of its certified lifetime on November 11. *TMA-1* was the first of a new version of the venerable Soyuz, with modifications financed by NASA for improved safety and widened crew-member size range. The crew consisted of Sergei Zalyotin, Yuri Lonchakov, and an ESA guest cosmonaut from Belgium, Frank De Winne. A science program, sponsored by Belgium and the Russian space agency Rosaviakosmos (RSA), included 19 experiments in the ESA Odissea program and four Russian experiments. The crew returned in the "old" *Soyuz TM-34*.

STS 113. Mission *ISS*-11A on *Endeavour* (November 23–December 7) carried a crew of seven, including the sixth *ISS* crew (Bowersox, Budarin, and Pettit), the fourth truss element, P1 (of 11 total), and other equipment and resupply. The 14.5-ton (13.1-metric-ton) P1 was transferred and installed on the S0 truss portside, and three spacewalks were conducted to connect the P1 and install other gear (Fig. 1*b*). After *Endeavour* undocked on December 2 (**Fig. 3**), it launched two tethered minisatellites called MEPSI (microelectromechanical systems–based Picosat Inspector), and returned with the Expedition 5 crew of Korzun, Whitson, and Treschev.

In addition to the above missions, Russia launched three crewless Progress logistics/resupply missions to the ISS: M1-8 (7P) on March 21, M-24 (8P) on June 26, and M1-9 (9P) on October 7.

United States Space Activities

Launch activities in the United States in 2002 showed a downturn from the already low level of the previous year. There were 18 NASA, Department of Defense, and commercial launch attempts, all of them successful (2001: 23 out of 24 attempts).

Space shuttle. Five shuttle mission were conducted during 2002, one less than in 2001: four of them for supply, logistics, outfitting, crew rotation, and orbital assembly of the *ISS*, and one for repairing and upgrading the Hubble Space Telescope. However, the human space flight program suffered a serious setback on February 1, 2003, with the loss of the shuttle *Columbia* and its crew of seven.

STS 109. The fourth "service call" to the Hubble Space Telescope, March 1–12, was performed by *Columbia* after undergoing more than 100 improvements since its last flight (STS 93), including a weight reduction of over 1000 lb (450 kg). Hubble received new, more durable solar arrays (SA) with their diode boxes (DBAs), a large gyroscopic reaction wheel assembly (RWA) for pointing, a new power-control unit (PCU), a new advanced camera for surveys (ACS), and an experimental cryocooler to restore the dormant Near Infrared Camera and Multi-Object Spectrometer (NICMOS) to operation. The crew conducted five spacewalks: EVA-1 to install the starboard SA and its DBA; EVA-2 for the port SA and DBA plus the RWA; EVA-3 to replace the PCU; EVA-4 for the ACS and its electronics box; and EVA-5 to install the NICMOS cryocooler. After reboost on March 8 by about 4 mi (6.4 km), Hubble was deployed from *Columbia* on March 9.

STS 107. Columbia, on its 28th flight, lifted off on January 16, 2003, on a research mission that gave more than 70 international scientists access to the microgravity environment in a sophisticated laboratory staffed by highly trained researchers working 32 payloads with 59 separate investigations. With its crew of seven, Commander Rick D. Husband, Pilot William C. McCool, and Mission Specialists David M. Brown, Kalpana Chawla, Michael P. Anderson, Laurel B. Clark, and Ilan Ramon, *Columbia* circled Earth for nearly 16 days. Both in the shuttle middeck and in the SPACEHAB Research Double Module (RDM), on its first flight in the cargo bay, the crew worked on a mixed complement of competitively selected and commercially sponsored research in the space, life, and physical sciences. Experiments were performed in the areas of astronaut health and safety, advanced technology development, and Earth and space science disciplines. Besides the RDM, payloads in the shuttle cargo bay included the FREESTAR (fast reaction experiments enabling science, technology, applications and, research) with six payloads, the MSTRS (miniature satellite threat reporting system), and the STARNAV (star navigation).

Columbia was lost with its crew during reentry on February 1, 16 min before the planned landing at Kennedy Space Center, when it violently disintegrated over Texas at 9 a.m. EST. Debris from the Orbiter fell in Texas, Louisiana, Arizona, and perhaps even California. The Columbia Accident Investigation Board (CAIB), chaired by Admiral (ret.) Harold W. Gehman, Jr., later concluded that, with crew and ground personnel unaware, one of the left wing's leading-edge reinforced carbon-carbon (RCC) elements or an associated filler strip of the heat shield had been compromised during ascent to orbit by a piece of debris, possibly a suitcase-sized chunk of foam insulation blown off the External Tank by the supersonic air stream, hitting the leading edge and rendering the wing unable to withstand reentry heating longer than about 8 min after entry interface. Further shuttle operations were halted for the duration of the CAIB investigation.

Advanced transportation systems activities. NASA's 5-year Space Launch Initiative (SLI) project, announced in 2001, is continuing, with 22 contracts awarded to industry in 2001 for developing the technologies that will be used to build an operational reusable space launch vehicle (RLV) before 2015. *See* SCRAMJET TECHNOLOGY.

As a cost-saving measure, NASA discontinued previously started work on a crew return vehicle, which had been intended to eventually take over the emergency lifeboat function for the Space Station from the currently chosen Russian Soyuz three-seater capsules by providing a shirtsleeve environment for seven crew members. The planned full-scale flight testing of the X-38 prototype test vehicle was terminated in 2002.

Space sciences and astronomy. In 2002, the United States launched three civil science spacecraft, down from six in the previous year: *RHESSI*, *CONTOUR*, and *Aqua*.

RHESSI. RHESSI (Reuven Ramaty High-Energy Solar Spectroscopic Imager, in honor of the late NASA scientist who pioneered the fields of solar-flare physics, gamma-ray astronomy, and cosmic-ray research), the latest member of NASA's Small Explorer class spacecraft, was launched on February 5, 2002, on a Pegasus XL vehicle dropped from an L1011 aircraft, and successfully reached its orbit (365 × 373 mi or 587 × 600 km, 38° inclination). *RHESSI*'s primary mission is to explore the basic physics of particle acceleration and explosive energy release in solar flares, using advanced imaging and spectroscopy instruments. On April 21, a spectacular X-class (extremely large) solar flare on the western limb of the Sun was observed by *RHESSI* and the *TRACE* (*Transition Region and Coronal Explorer*) spacecraft, as well as many other spacecraft and ground-based observatories. The combined *RHESSI* and *TRACE* data yielded beautiful time-lapse movies and revealed important new physics. Other *RHESSI* accomplishments include imaging in narrow gamma-ray lines, high-resolution x-ray and gamma-ray spectra of cosmic sources, and hard x-ray images of the Crab Nebula with 2-arcsecond resolution.

CONTOUR. NASA's *Comet Nucleus Tour* (*CONTOUR*) spacecraft, launched July 3, 2002, on a Delta 2 vehicle, was a disappointment. *CONTOUR* was to investigate the nuclei of comets, with a visit to Comet Encke in November 2003 and an intercept of Comet Schwassmann-Wachmann 3 in June 2006. After orbiting the Earth for more than a month, the probe was to ignite its solid-propellant rocket injection stage, necessary for escaping Earth's gravity pull and heading out for its comet encounters. At the time of the critical ignition of the 50-s burn, however, the spacecraft fell silent, and it became apparent later that it had exploded at about 140 mi (225 km) above the Indian Ocean.

Hubble Space Telescope. Twelve years after it was placed in orbit, the Hubble Space Telescope continued to probe far beyond the solar system. During the repair, maintenance, and upgrade mission by STS 109, Hubble received the new Advanced Camera for Surveys (ACS), which provides spectacular new glimpses of galaxies and other distant objects.

Astronomers using the Hubble Space Telescope measured the mass of the extrasolar planet Gliese

876b, in orbit around the star Gliese 876. The Hubble results for the star's transverse motion, combined with spectroscopic measurements of its radial motion, show that the planet is 1.89–2.4 times as massive as Jupiter. Gliese 876b is only the second planet outside the solar system for which astronomers have determined a precise mass.

Hubble also measured the diameter of Quaoar, the largest object discovered in the solar system since the discovery of Pluto in 1930. Approximately half the size of Pluto, the icy world is about 4.0×10^9 mi (6.4×10^9 km) away, more than 1.0×10^9 mi (1.6×10^9 km) farther than Pluto at present. Like Pluto, it dwells in the Kuiper belt.

Multiple observations made over several months with Hubble and NASA's *Chandra X-ray Observatory* captured the spectacle of matter and antimatter propelled to near the speed of light by the Crab pulsar, a rapidly rotating neutron star the size of Manhattan. *See* CRAB NEBULA.

Chandra Observatory. Launched on July 23, 1999, the massive (12,930-lb or 5870-kg) *Chandra X-ray Observatory* continued to provide views of the high-energy universe. Astronomers tracked the life cycle of x-ray jets from a black hole. A series of images from *Chandra* revealed that as the jets evolved they traveled at nearly the speed of light for several years before slowing down and fading. *Chandra* also provided the best x-ray image yet of two Milky Way–like galaxies in the midst of a head-on collision, in the galaxy Arp 220, which probably triggered the formation of huge numbers of new stars, sent shock waves through intergalactic space, and could likely lead to the formation of a supermassive black hole in the center of the new conglomerate galaxy. Since

all galaxies—including the Milky Way—may have undergone mergers, this provides insight into how the universe came to look as it does today.

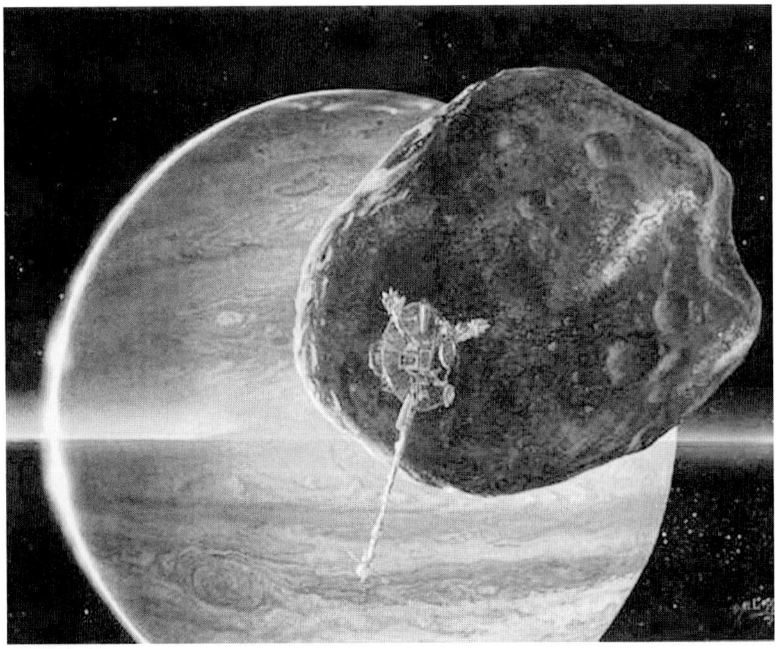

Galileo. *Galileo* (launched 1989) continued to return unprecedented data on Jupiter and its satellites. On November 4, 2002, conducting its last flyby of a Jupiter moon, *Galileo* reached its closest point to the volcanic satellite Io. The spacecraft passed by Io's orbit at about 6 Jupiter radii (267,000 mi or 429,000 km) from the planet on its way in to the inner system, and the radiation become intense enough that its star scanner and attitude control software could not properly determine the orientation of the spacecraft. Thus, the software was placed in hibernation for 9 h, to ignore the signals from the star scanner and remember its last calculated orientation and spin rate (which remained unchanged). A few hours later, *Galileo* reached its closest point to the tiny moon Amalthea, an irregularly shaped satellite approximately 168 mi (270 km) across at its longest dimension (**Fig. 4**). The closest distance to the surface of the body was 99 mi (160 km), and Galileo flew by Amalthea at a relative speed of 41,160 mi/h (18.4 km/s), taking less than 15 s to pass by. During the flyby, Jupiter's radiation caused a failure in computer circuitry that handles timing of the events on the spacecraft, shutting down operations. The problem was with gallium-arsenide light-emitting diodes, which were later repaired by "annealing" their crystal structure with an electric current. Stored science data from the flyby were then transmitted to Earth. *Galileo* was now nearly out of the hydrazine propellant needed to keep its antenna pointed toward Earth and do maneuvers. After a last loop away from Jupiter, the probe plunged into Jupiter's atmosphere on September 21, 2003, to ensure that there was no chance the spacecraft might hit (and possibly contaminate) Europa.

Cassini. NASA's 6-ton (5.4-metric-ton) spacecraft *Cassini* continued its 6.7-year, 2-billion-mile (3.2-billion-km) journey to the planet Saturn. The spacecraft will enter orbit around the ringed planet on July 1, 2004, and is scheduled to release its piggybacked European-built *Huygens* probe for descent through the thick atmosphere of the moon Titan on January 14, 2005. On April 3, 2002, *Cassini* successfully completed a course correction, firing its main engine for 9.8 s. The engine burn, the 13th since launch and the first since February 2001, was planned both for correcting *Cassini*'s trajectory and for maintenance of the propulsion system: To keep fuel lines flowing freely, the spacecraft is not allowed to go much longer than 1 year between engine firings.

WMAP. NASA's *Microwave Anisotropy Probe* (*MAP*), renamed the *Wilkinson Microwave Anisotropy Probe*, had been launched on June 30, 2001, on a Delta 2 rocket. Its differential radiometers measure the temperature fluctuations of the cosmic microwave background radiation (CMBR) with unprecedented accuracy. During the 12 months of 2002, scientists produced the first version of a full-sky map of

Fig. 4. Artist's concept of *Galileo* spacecraft passing near Jupiter's small inner satellite Amalthea on November 5, 2002. (*NASA/JPL/Caltech; image by Michael Carroll*)

the faint anisotropy or variations in the CMBR's temperature. One big surprise revealed in the data is that the first generation of stars to shine in the universe first ignited only 200 million years after the big bang, much earlier than many scientists had expected. See WILKINSON MICROWAVE ANISOTROPY PROBE.

Genesis. The solar probe *Genesis*, launched on August 8, 2001, on a Delta 2 rocket, had gone into orbit about the first Earth-Sun Lagrangian libration point, L1, on November 16, 2001. [Joseph Lagrange showed that, in a system of three orbiting bodies such as the Sun, the Earth, and a spacecraft (where the mass of the spacecraft is much smaller than that of the Earth or the Sun), there are five points at which the gravitational and centrifugal forces on the spacecraft are in equilibrium, so that the spacecraft remains at fixed distances from both the Sun and the Earth in the absence of perturbing forces. The point L1 is located on the line between the Sun and the Earth, about 0.9×10^6 mi (1.5×10^6 km) from the Earth and 92.0×10^6 mi (148.1×10^6 km) from the Sun.] After an unconventional orbit insertion, *Genesis* began the first of five loops around L1. Collection of samples of solar wind material expelled from the Sun started on December 3, 2001. On December 10, 2002, the spacecraft orbit around L1 was fine-tuned. In 2004, after 29 months in orbit, the sample collectors are to be restowed and returned to Earth, where the sample return capsule is to be recovered in midair by helicopter over the Utah desert in September 2004.

ACE. The *Advanced Composition Explorer (ACE)*, launched on August 25, 1997, is positioned in an orbit around L1. *ACE* continued to observe, determine, and compare the isotopic and elemental composition of several distinct samples of matter, including the solar corona, the interplanetary medium, the local interstellar medium, and galactic matter. The spacecraft has enough propellant on board to maintain an orbit about L1 until around 2019.

Wind. Launched on November 1, 1994, as part of the International Solar-Terrestrial Project (ISTP), *Wind* was first placed in a sunward, multiple double-lunar swingby orbit with a maximum apogee of 350 Earth radii, followed by an orbit at the L1 point. The spacecraft carries an array of scientific instruments for measuring the charged particles and electric and magnetic fields that characterize the interplanetary medium (or solar wind), a plasma environment. Nearly continuous plasma measurements made by *Wind* are being used to investigate the disturbances and changes in the solar wind that drive important geomagnetic phenomena in the near-Earth geospace (such as aurorae and magnetic storms), as detected by other satellites and ground-based instruments.

Stardust. NASA's comet probe *Stardust* was launched on February 3, 1999, on a Delta 2 rocket to begin its mission to intercept Comet Wild 2. The mission objective is to return closeup imagery to Earth, and to collect comet dust and interstellar dust particles during the comet encounter and return them to Earth for analysis. *Stardust*'s trajectory is mak-

ing three loops around the Sun before its closest approach to the comet. After one solar orbit, an Earth flyby was used to boost the spacecraft orbit on January 15, 2001, and there was a second period of interstellar dust collection from July to December 2002. On November 2, 2002, *Stardust* passed within 1900 mi (3000 km) of asteroid 5535 Anne Frank, at 4 mi/s (7 km/s) relative velocity. A second orbit of the Sun was completed in mid-2003, and the comet encounter will take place on January 2, 2004, with a closest approach of about 90 mi (150 km) at a relative velocity of about 3.8 mi/s (6.1 km/s). The sample collector is to be deployed in late December 2003, and to be retracted, stowed, and sealed in the sample vault of the sample reentry capsule after the flyby. Images of the comet nucleus should have a resolution of 100 ft (30 m) or better, with predicted coverage of the entire sunlit side. On January 15, 2006, the capsule is to separate from the main craft and return to Earth. A parachute is to be deployed, and the descending capsule recovered by a chase aircraft over the Utah desert.

Ulysses. The joint European/NASA solar polar mission *Ulysses*, launched in 1990, continued, with all spacecraft systems and the nine sets of scientific instruments remaining in excellent health. *Ulysses* arrived over the Sun's south polar regions for the second time in November 2000, followed by a rapid transit from maximum southern to maximum northern helio-latitudes that was completed in October 2001. Solar activity reached its maximum in 2000, so that *Ulysses* experienced a very different high-latitude environment from the one it encountered during the first high-latitude passes. The spacecraft is now heading away from the Sun toward aphelion at the end of June 2004.

Pioneer 10. Launched from Cape Kennedy in 1972 aboard an Atlas/Centaur rocket on a mission to Jupiter planned for only 2 years, the 570-lb (258-kg) *Pioneer 10* has become the Earth's longest-lived interplanetary explorer. By the end of 2002, at a distance of 7.52×10^9 mi (12.10×10^9 km) from Earth and 81.86 astronomical units from the Sun (more than twice the mean distance of Pluto), *Pioneer 10* was passing through the transitional region between the farthest traces of the Sun's atmosphere, the heliosphere, still noticeable at that distance, and free intergalactic space. Signals transmitted by the spacecraft need more than 11 h to reach Earth. On December 5, 2002, the Deep Space Station near Madrid found the weak signal but could not lock onto the receiver, and so no telemetry was received. Project Phoenix also picked up the signal from *Pioneer 10* at Arecibo in Puerto Rico.

Mars exploration. After the failures of two Mars probes in 1999, NASA's Mars exploration program rebounded in 2001 and 2002. The *Mars Odyssey* probe, launched April 7, 2001, reached Mars on October 24, 2001, after a 6-month journey. Entering a highly elliptical orbit over the poles of the Red Planet, it began to change orbit parameters by aerobraking, reducing its ellipticity to a circular orbit at 250 mi (400 km) by end of January 2002. The orbiter is circling Mars for at least 3 years, with the objective of conducting a detailed mineralogical analysis of the planet's surface and measuring the radiation environment.

Since its antenna was successfully deployed in March 1999, *Mars Global Surveyor (MGS)* has returned more data about Mars than all other missions combined, transmitting a steady stream of high-resolution images which show a world constantly being reshaped by forces of nature, including shifting sand dunes, monster dust devils, wind storms, frosts, and polar ice caps that grow and retreat with the seasons. *See* MARS GLOBAL SURVEYOR.

Earth science. The *Aqua, POES-M,* and *GRACE* satellites were launched in 2002.

Aqua. NASA launched *Aqua* on May 4. The 3858-lb (1750-kg) satellite carries six instruments weighing 2385 lb (1082 kg) and designed to collect information on water-related activities worldwide. It was placed in a polar, Sun-synchronous orbit of 438 mi (705 km) altitude by a Delta 2 rocket from Vandenberg Air Force Base in California. During its 6-year mission, *Aqua* will observe changes in ocean circulation and study how clouds and surface water processes affect the Earth's climate. The mission is collecting huge amounts of information about the Earth's water cycle, including evaporation from the oceans, water vapor in the atmosphere, clouds, precipitation, soil moisture, sea ice, land ice, and snow cover. Additional variables being measured include radiative energy fluxes, aerosols, vegetation cover on the land, phytoplankton and dissolved organic matter in the oceans, and air, land, and water temperatures. This information will help scientists better understand how global ecosystems are changing, and how they respond to and affect global environmental change. The *Aqua* mission, formerly named EOS PM (signifying its afternoon equatorial crossing time), is a part of the NASA-centered international Earth Observing System (EOS). *Aqua* joins *Terra*, launched in 1999, and is to be followed by *Aura* in 2004.

POES-M. On June 24, the operational weather satellite *POES-M (Polar-orbiting Operational Environmental Satellite-M)* was launched from Vandenberg Air Force Base on a Titan 2 rocket. The satellite, now called *NOAA-M,* is part of the POES program, a cooperative effort between NASA and the National Oceanic and Atmospheric Administration (NOAA), the United Kingdom, and France. It joined the *GOES-M,* launched in July 2001. Both satellites, operated by NOAA, provide global coverage of numerous atmospheric and surface parameters for weather forecasting and meteorological research.

GRACE. Launched on March 17 on a Russian Rockot launcher, the twin *GRACE (Gravity Recovery and Climate Experiment)* satellites are able to map the Earth's gravity field by making accurate measurements of the distance between the two satellites, using the Global Positioning System (GPS) and a microwave ranging system. The unprecedented accuracy of this mapping is yielding crucial information about the distribution and flow of mass within the Earth and its surroundings. *See* GRACE (GRAVITY RECOVERY AND CLIMATE EXPERIMENT).

Department of Defense space activities. United States military space organizations continued their efforts to make space a routine part of military operations across all service lines. One focus concerns plans for shifting the advanced technology base toward space in order to build a new technology foundation for more integrated air and space operations, as space is increasingly dominant in military reconnaissance, communications, warning, navigation, missile defense, and weather-related areas. Many of the recommendations from the Space Commission Report, also known as the Rumsfeld Commission Report, were implemented in 2002. The use of space systems within military operations reached a new level in 2002 for the war on terrorism and operations in Afghanistan. The increased use of satellites for communications, observations, and—through the GPS—navigation and high-precision weapons targeting was of decisive importance for the military command structure. *See* COUNTERTERRORISM.

In 2002, there was one military space launch: the fifth Milstar FLT satellite on a Titan-4B/Centaur vehicle from Cape Canaveral Air Force Station, Florida, completing the ring of communications satellites around the Earth and providing ultrasecure, jam-resistant transmission virtually anywhere on the planet.

Commercial space activities. In 2002, commercial space activities in the United States exhibited slow recovery from the 2001 crisis in the communications space market caused by failures of satellite constellations for mobile telephony. Iridium Satellite LLC, having bought the assets of bankrupt Iridium LLC in 2000, launched seven satellites in 2002 which ensured the life span of the mobile-phone satellite constellation to at least mid-2010 (five on a U.S. Delta 2, two on the new Russian commercial Rockot/ Briz-KM). However, commercial ventures continued to play a relatively minor role in U.S. space activities, even less than in 2001 (50%), amounting in 2002 to about 26% of commercial satellites and associated launch services worldwide.

Of the 18 total launch attempts by the United States in 2002 (versus 24 in 2001), 8 carried commercial payloads (NASA, 9; military, 1). In the launch services area, Boeing launched three Delta 2 vehicles, and Lockheed Martin launched three Atlas 2A. Both companies also had successful first launches of their next-generation EELV (evolved expendable launch vehicle) rockets, Lockheed Martin with the Atlas 5 (French *Hot Bird 6* comsat), and Boeing with the Delta 4 (*Eutelsat W5* comsat). Orbital Science Corp. had a successful launch of NASA's *RHESSI* with a Pegasus XL airplane-launched rocket; and the partnership of Boeing, RSC-Energia (Russia), NPO Yushnoye (Ukraine), and Kvaerner Group (Norway) successfully launched a Russian Zenit-3SL rocket, carrying PanAmSat's *Galaxy 3C* geosynchronous comsat from the *Odyssey* sea launch platform floating at the Equator. *See* SATELLITE LAUNCH VEHICLES.

Russian Space Activities

Marginally supported by a slowly improving national economy, Russia in 2002 showed what may be a slight rebound in space operations from 2001. Its total of 23 successful launches (out of 25 attempts, excluding one Zenit-3SL launch by an international partnership) equaled the previous year's 23 (out of 23 attempts): four Soyuz-U (one crewed), two Soyuz-FG (one crewed), eight Protons, two Rockot (first launched in 1994), two Molniya, four Kosmos-3M, and one Dnjepr-1 (first launched in 1999). The upgraded Soyuz-FG rocket flew twice; its new fuel injection system provides a 5% increase in thrust over the Soyuz-U, enhancing its lift capability by 440 lb (200 kg) and enabling it to carry the new Soyuz-TMA spacecraft, which is heavier than the Soyuz-TM ship that had been used to ferry crews to the *ISS*. Soyuz-TMA was flown for the first time on October 30, 2002, as *ISS* mission 5S.

There were two major losses: A Soyuz-U, carrying the 13th *Foton-M1* spacecraft, with 44 microgravity experiments from Russia, the ESA, the United States, and Japan in its recovery capsule, failed on October 15 due to automatic engine shutdown 29 s after liftoff; and a Proton-K, on November 25, stranded Europe's *Astra 1K*, the world's largest communications satellite, in a useless orbit when its Block DM-3 upper stage failed to reignite for the second of three planned firings. The DM failure was the fourth in as many years for this stage, which is built by RSC Energia. However, on December 30, the Canadian *Nimiq-2* comsat, with 32 Ku-band relay transponders for direct broadcast to Canadian customers, was launched successfully on a Proton-K, equipped with the new Briz-M (Breeze-M) upper stage from Khrunichev.

The Russian space program's major push to enter into the world's commercial arena by promoting its space products on the external market, driven by the need to survive in an era of severe reductions of public financing, continued with slow increase in 2002. First launched in July 1965, the Proton heavy lifter, originally intended as a ballistic missile (UR500), by end-2002 had flown 218 times since 1980, with 14 failures (reliability, 0.936). Its launch rate in recent years has been as high as 13 per year. Of the nine Protons launched in 2002 (2001: six), five were launched for commercial customers (*Intelsat 903, DirecTV-5, EchoStar 8, Astra 1K, Nimiq-2*), and the others launched for the state, including military. From 1985 to 2002, 166 Proton and 388 Soyuz rockets were launched, with 10 failures of the Proton and 10 of the Soyuz, giving a combined reliability index of 0.964. Until its launch failure on October 15, 2002, the Soyuz rocket had flown 74 consecutive successful missions, including 12 with human crews on board.

European Space Activities

Europe's efforts to reinvigorate space activities, after their long decline since the mid-1990s, in 2002 remained modest compared to astronautics activities of NASA, the U.S. Department of Defense, and Russia. Work was underway by the European Union on an emerging new strategy for the ESA to achieve an autonomous Europe in space, under Europe's new constitution that makes space and defense an EU responsibility. On orders from European countries' government ministers unhappy with the cost and competitiveness of the launch-vehicle sector, ESA was charged with readying this sector for a potentially sweeping reorganization. This task was given particular emphasis in 2002 by another failure of the Ariane 5 heavy launcher.

The year 2002 did not bring the much-needed breakthrough of Europe's commercial space industry in its faltering attempts at recovery, despite continued successful launches of Arianespace's Ariane 4 workhorse. The heavy-lift Ariane 5 was launched four times, bringing its total to 14, but the failure of the upgraded Ariane 5 EC-A on December 11, carrying the European high-value comsats *Hot Bird* 7 and *STENTOR*, dealt a severe blow to European space flight. The new EC-A version of the Ariane 5, designed to lift 10 tons (9 metric tons) to geostationary transfer orbit, enough for two big communications satellites at once, uses a new cryogenic upper stage, an improved Vulcain 2 main-stage engine, and solid boosters loaded with 10% more propellant. The rocket failed shortly after liftoff when the Vulcain 2 caused the vehicle to lose control and self-destruct after 456 s. For the Ariane 4, 2002 saw eight successful launches, up two from 2001 (when there was a problem with satellite delivery delays), carrying ten (2001: seven) commercial satellites for customers such as India, the United Kingdom, and the United States. Among its payloads were two Intelsats, the *Spot 5* imaging satellite, and two satellites of the New Skies Satellites (NSS) system, a promising emerging global satcom operator. At year's end, the Ariane 4 had flown 144 times, with 7 failures (95.1% reliability) from its Kourou/French Guyana spaceport. The 3 successful Ariane 5 launches carried five satellites, including the Italian *Atlantic Bird-1* comsat and *Envisat*, Europe's advanced environmental satellite.

The most significant space event for Europe in 2002 was the unanimous decision, on March 26, by the transport ministers of 15 European countries, after more than 3 years of difficult negotiations, to launch the Galileo navigation and global positioning system. Starting in 2008, this will enable Europe to be independent of the U.S. GPS system, an area where major strategic and commercial stakes are at play. By the end of 2002, the project was already oversubscribed, exceeding 125%. *See* GALILEO NAVIGATION SATELLITE.

In the human space flight area, while the *ISS* remains ESA's biggest ongoing program and its only engagement in the human space flight endeavor, the European *ISS* share (8.6%) remains unchanged due to a top-level agreement signed by previous governments of the participating nations. France has a relatively large and active national space program. The Italian Space Agency, ASI, participates in the *ISS* program through ESA but also has a protocol with NASA for the delivery of multipurpose logistics modules (MPLM) for the *ISS*. Germany is the second major ESA contributor after France, but it has only a very small national space program.

Envisat. ESA's operational environmental satellite *Envisat* was launched on March 1 on an Ariane 5. The 18,100-lb (8200-kg) satellite reached its polar orbit at 500 mi (800 km) altitude with great precision, completing a revolution of Earth every 100 min. Because of its polar Sun-synchronous orbit, it flies over and examines the same region of the Earth every 35 days under identical conditions of lighting. It is 82 ft (25 m) long and 33 ft (10 m) wide, about the size of a bus, and is equipped with ten advanced instruments (seven from ESA, the others from France, the United Kingdom, and Germany/Netherlands), including an Advanced Synthetic Aperture Radar (ASAR), a Medium Resolution Imaging Spectrometer (MERIS), an Advanced Along Track Scanning Radiometer (AATSR), a Radio Altimeter (RA-2), a Global Ozone Monitoring by Occultation of Stars (GOMOS) instrument, a Michelson Interferometer for Passive Atmosphere Sounding (MIPAS), and a Scanning Imaging Absorption Spectrometer for Atmospheric Cartography (SCIAMACHY).

Spot 5. Launched on May 4 by an Ariane 4, the fifth imaging satellite of the commercial Spot Image Company arrived in its polar Sun-synchronous orbit of 505 mi (813 km) altitude. The 6680-lb (3030-kg) satellite is similar to its predecessor *Spot 4*, launched in 1998, but carries more advanced instruments with improved image quality and rate of delivery. With two special cameras it can obtain three-dimensional images with a resolution of 33 ft (10 m) over a 75-mi (120-km) footprint, photographing 48,600 mi^2 (126,000 km^2) every 24 h, convertible into stereo images after processing.

INTEGRAL. ESA's *INTEGRAL* (*International Gamma-Ray Astrophysics Laboratory*), a cooperative project with Russia and the United States, was launched on October 17 on a Russian Proton rocket into a 72-h orbit with an inclination of 51.6°, a perigee height of 5600 mi (9000 km), and an apogee height of 96,000 mi (155,000 km). The most sensitive gamma-ray observatory ever launched, *INTEGRAL* provides new insights into the most violent and exotic objects of the universe, such as black holes, neutron stars, active galactic nuclei, and supernovae. It also helps scientists to understand processes such as the formation of new chemical elements and the mysterious gamma-ray bursts, the most energetic phenomena in the universe. This is made possible by *INTEGRAL*'s combination of fine spectroscopy and imaging of gamma-ray emissions in the energy range from 15 keV to 10 MeV.

Newton XMM. The *XMM* (*X-ray Multi Mirror*) *Newton* observatory, launched on December 10, 1999, on an Ariane 5, is the largest European science research satellite ever built. Operating in an orbit of 71,216 × 4375 mi (113,946 × 7000 km), inclined at 40° to the Equator, the telescope has a length of nearly 36 ft (11 m), with a mass of almost 8800 lb (4 metric tons). Using its three x-ray detecting instruments—a photon imaging camera, reflection grating spectrometer, and optical telescope—it obtained the first reliable measurement of the mass-to-radius ratio of a neutron star (EXO 0748-676). These objects are believed to be among the densest in the universe. The investigation indicated that the neutron star probably contains normal matter and not the exotic plasma of dissolved matter that scientists hypothesized might occur if the great gravitational forces present caused elementary particles like protons and neutrons to fuse. *XMM-Newton* also detected unexpectedly large amounts of iron in the spectrum of a distant quasar (APM 08279 + 5255), which could mean that (1) the universe was much older at the time of the quasar than the 1.5 billion years currently believed, or (2) iron can be produced in a highly efficient manner that is at present totally unknown.

Asian Space Activities

China, India, Japan, and Israel have space programs capable of launch and satellite development and operations.

China. China's space program continued strong activities in 2002, with four successful launches (2001: one) of its Long March rocket in two versions and one launch failure, on September 15, of the first new all-solid four-stage launch vehicle Kaituozhe 1 (Explorer 1). Since China separated its military and civil space programs in 1998, the China National Space Administration (CNSA) has been responsible for planning and development of space activities. Its top priority today is the development of piloted space flight, followed by applications satellites.

After the launch and recovery, on November 21, 1999, of its first inhabitable (but still crewless) capsule *Shenzhou* (Divine Vessel), a 16,000-lb (7200-kg) modified version of the Russian Soyuz vehicle, China successfully launched and recovered *Shenzhou 2* on January 9, 2001. In 2002, the third *Shenzhou* spaceship was placed in orbit on March 25, followed by *Shenzhou 4* on December 29. Both were recovered after a parachute landing in central Inner Mongolia. Reportedly, *Shenzhou 4* was completely equipped to carry a crew, including food, medicine, and sleeping bags, and it was also upgraded from the three earlier missions in its control systems. After the launch, Beijing indicated that the first crew would fly in the latter half of 2003.

The launch vehicle of the *Shenzhou* spaceships is the human-rated Long March 2F rocket. China's Long March (Chang Zheng, or CZ) series of launch vehicles consists of 12 versions which by the end of 2002 had made 69 flights, sending 78 payloads (satellites and spacecraft) into space, with 90% success rate.

India. The Indian Space Research Organization (ISRO) has continued its development programs for satellites and launch vehicles. Main satellite programs are the INSAT (Indian National Satellite) telecommunications system, the IRS (Indian Remote Sensing) satellites for earth resources, the METSAT weather satellites, and the new GSat series of large (up to 2.5 tons or 2.3 metric tons) experimental geostationary comsats. India's main launchers are the PSLV (Polar Space Launch Vehicle) and the Delta 2–class GSLV (Geostationary Space Launch Vehicle). In 2002, India augmented its weather forecasting ability by launching a dedicated METSAT with a PSLV rocket into a highly elliptical orbit for later maneuvering into its geosynchronous (stationary) orbital slot using on-board propulsion. The launch was the seventh flight of the four-stage PSLV, in a modified version, and the first to place a satellite into geosynchronous transfer orbit.

Japan. Japan's central space development and operations organization, the National Space Development Agency (NASDA), has developed the launchers N1, N2, H1, and H2. In 1999, it was decided to focus efforts on the H2-A vehicle, an uprated and more cost-effective version of the costly H-2. The H2-A had its maiden flight on August 29, 2001, and in 2002 it had three missions, all successful, launching eight satellites, including one for communications, two for remote sensing, one for microgravity research, and several for technology development. The four successful H2-A launches in a row have greatly restored commercial confidence in Japan's launch vehicle technology, even if more verification and some improvement of the vehicle's availability are needed to make it commercially fully competitive.

ISS program. The *ISS* program continues to be an area of great promise for Japan, whose contributions are the Japanese Experiment Module (JEM), called *Kibo* (Hope), along with its ancillary remote manipulator system and porchlike exposed facility; and the H-2 transfer vehicle, which will carry about 6.6 tons (6 metric tons) of provisions to the *ISS* once or twice a year, launched on an H2-A.

Nozomi. During 2002, Japan's Mars mission *Nozomi* (Planet-B) proceeded along its rocky path toward the planet after executing a series of course-correction maneuvers in early September. The probe was launched in July 1998, but an engine problem that December forced controllers to reroute the spacecraft, delaying its arrival at Mars from October 1999 to January 2004. In April 2002, while approaching Earth for a gravity-assist maneuver scheduled for December, *Nozomi* was hit by solar flare radiation that damaged its on-board communications and power systems. The probe completed the first Earth flyby as scheduled; a second was slated for June 2003. It remained doubtful whether *Nozomi* could survive until its Mars arrival and maneuver itself into Mars orbit in January 2004.

Israel. The Israeli Space Agency successfully launched the Earth-observing satellite *Ofeq-5* on a three-stage solid-propellant Shavit rocket on May 28. The 660-lb (300-kg) satellite, 7.5 ft (2.3 m) high and 4 ft (1.2 m) in diameter, carries an imaging reconnaissance camera.

For background information *see* COMET; COMMUNICATIONS SATELLITE; COSMIC BACKGROUND RADIATION; GAMMA-RAY ASTRONOMY; JUPITER; MARS; MILTARY SATELLITES; REMOTE SENSING; SATELLITE ASTRONOMY; SATURN; SPACE FLIGHT; SPACE PROBE; SPACE SHUTTLE; SPACE STATION; SPACE TECHNOLOGY; SPACE TELESCOPE, HUBBLE; SUN; X-RAY ASTRONOMY in the McGraw-Hill Encyclopedia of Science & Technology. Jesco von Puttkamer

Bibliography. *Aerospace Daily*; *AIAA AEROSPACE AMERICA*, December 2002; *Aviation Week & Space Technology*; European Space Agency, press releases; NASA Public Affairs Office, news releases; *SPACE NEWS*.

Stone tool origins

The human lineage is unique in the animal world for having an adaptation that is based upon the complex use of tools and technology. The archeological record bears testament to the origins and development of technology through time in tandem with major changes in biological evolution. By definition, the archeological record starts as soon as we can identify intentionally or unintentionally modified materials (artifacts) made by humans or protohumans (early hominids) in the prehistoric record. Over the last century and a half, the dates for the origins of the earliest stone tools have been pushed further and further back, as fieldwork intensified in the Old World and dating techniques became more refined.

Tool behavior. In the animal world, a number of species are known tool-users; for example, the mudwasp uses a pebble to tamp mud in nest building, the Galapagos finch uses a cactus spine to probe for insects, the Egyptian vulture cracks open bird eggs by dropping a rock, and the California sea otter uses a pebble as a hammer or anvil to crack open shellfish. These behaviors probably have a strong instinctive component.

Among the nonhuman primates, the chimpanzee in particular shows a range of tool uses, including shaping and using sticks and grass stems to "fish" for termites and ants, using stones or wood as hammers and anvils to crack nuts, and using wads of chewed leaves as sponges to dip for drinking water or to clean itself. Moreover, chimpanzee tool use sometimes includes the careful shaping of implements. These tool-making and tool-using skills appear to be learned and are indications of a cultural system of shared behaviors. The chimpanzee provides one model of what early hominid tool behavior might have been like before the advent of flaked stone tools (tools made from thin pieces of stone, or flakes, chipped off a larger parent stone).

Early hominids could have learned to flake stone as a by-product of other activities, such as rock throwing in hunting or defense, nut-cracking, or bone-breaking. A stone missile, hammer, or anvil may have accidentally broken during its use for these activities, producing unnaturally sharp edges. This observation, which may have occurred countless times during the course of human evolution, could have led to the intentional fracture of stone to produce sharp cutting and chopping edges.

Oldowan industrial tradition. The earliest recognizable stone artifacts (and the earliest fossil hominids) are found at archeological sites on the African continent. These sites are normally assigned to the

Oldowan (a term coined by Louis and Mary Leakey from their work at Olduvai Gorge in Tanzania) industrial tradition. Oldowan sites date back between approximately 2.5 and 1.5 million years and include Gona, Omo Valley, and Fejej in Ethiopia; West Turkana and East Turkana (Koobi Fora) in Kenya; Olduvai Gorge in Tanzania; Ain Hanech in Algeria; and Sterkfontein and Swartkrans in South Africa. This was a time of cooling and drying on much of the African continent, during which the extinction of numerous animal forms as well as the evolutionary emergence of many new species took place. The East and North African archeological sites tend to be found along streams, lake margins, and deltas, whereas the South African sites are found in cave breccias (rock composed of angular or sharp fragments). Environmental evidence indicates that these sites were commonly located in grasslands and woodlands with close access to water sources. These ancient landscapes would have teemed with wildlife that included prehistoric forms of elephants, hippos, crocodiles, wild cattle, wild pigs, antelopes, giraffes, carnivores (such as large cats, hyenas, and hunting dogs), monkeys, birds, smaller reptiles, insects, and hominids.

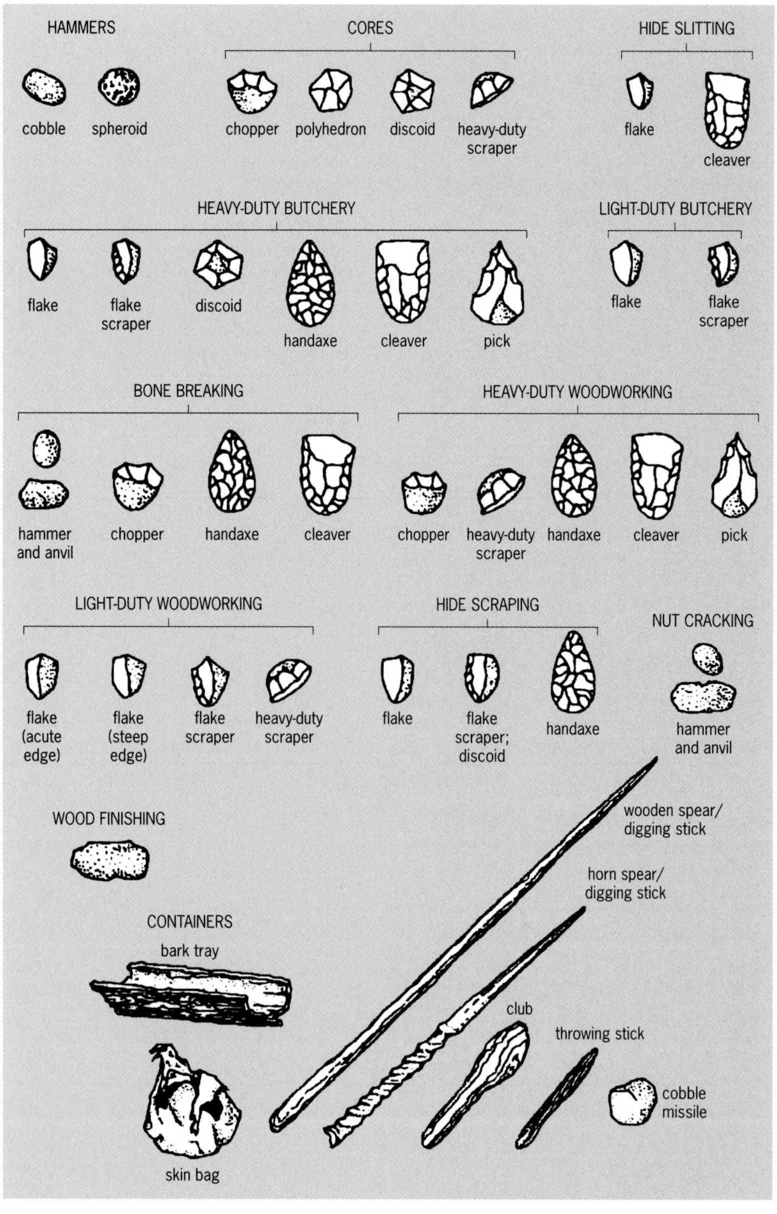

Tool types. The Oldowan artifacts consists of battered hammerstones and spheroids, flaked cores (the parent piece of rocks from which flakes are struck), sharp-edged flakes and fragments, and sometimes retouched flakes exhibiting chipping along one or more edges. Cores are typically classified as choppers, discoids, polyhedrons, and heavy-duty scrapers; and retouched forms are categorized as scrapers and awls (small pointed tools; **Figs. 1** and **2**). The Oldowan stone artifacts were usually made from cobbles or chunks of lava, quartz, quartzite, ignimbrite ("welded tuff"), and chert, sometimes transported several kilometers from their geological sources and found in the hundreds or thousands at particular places on the ancient landscapes, perhaps in proximity to water and food resources and safety. It is likely that a wide range of tools were also made from organic materials such as bone, wood, hide, horn, and shell. Much of the variation seen in these Early Stone Age sites can probably be explained by differences in and proximity to available raw materials.

Production techniques. Experiments have shown that these Oldowan forms are relatively easy for modern humans to produce. Modern African apes such as bonobos (pygmy chimpanzees) can also learn to make simple Oldowan cores and flakes, although their skill level in flaking stone does not seem as high as that seen at prehistoric Oldowan sites. Oldowan artifacts can be produced by the direct percussion technique (striking a stone hammer against a core), anvil technique (striking a core against a stone anvil on the ground), and bipolar technique (setting a core on an anvil and hitting it with a hammer).

Functions. Experiments with early stone tools have shown that these implements could have been effectively used for a wide range of activities. Unmodified cobbles and battered spheroids have been used

Fig. 1. Potential functions of Oldowan and early Acheulean stone tools, based upon feasibility experiments. Such tools could have expanded the range of foods that could potentially be exploited, decreased the amount of time required to process them, and increased the amount of food that could be carried by early Stone Age hominids—all important adaptive strategies. (*From K. D. Schick and N. Toth, Making Silent Stones Speak, p. 175, Simon and Schuster, New York, 1993*)

as hammerstones to strike cores and produce flakes, break open bones to obtain marrow and brains, crack nuts, and be thrown as missiles in hunting or defense. Cores are clearly a source of flake production, and larger sharp-edged cores such as choppers can be used for chopping and shaping wooden tools such as digging sticks or spears. Sharp-edged flakes make excellent butchery tools for cutting skin, dismembering, and defleshing. Flakes can be retouched to resharpen an edge for cutting activities or can be used to scrape or saw wood and scrape the fat and hair off animal hides. Some tools may have functioned as clubs or containers (Fig. 1).

The actual archeological evidence for early tool use includes cobbles that show signs of battering and chipping from their use as hammers or pitting from their use as hammers or anvils. Cut marks on fossil animal bones show that sharp-edged flakes were used to skin, dismember, and deflesh animal carcasses. Fracture patterns and striations in long bones (referring to the larger bones of the arms and legs) also show that these bones were broken with stone hammers to obtain marrow. Interestingly, many of these signs of modification are found on large mammals that had weighed hundreds of pounds; whether these carcasses were obtained through hunting, chasing predators from a kill, or scavenging is hotly debated by anthropologists at present. Microscopic analysis of a sample of Oldowan stone artifacts from East Turkana have also shown polishes and wear patterns consistent with animal butchery, woodworking, and cutting of soft plant matter such as grasses.

Hominid evolution. The human paleontological record shows that hominids had become upright walkers several million years before the first recognizable tools, suggesting that bipedalism emerged for some other reason, perhaps as a feeding adaptation. Between 2.5 and 1.5 million years ago, the hominids that were contemporaneous with the Oldowan sites included members of the smaller-brained and larger-toothed genus *Australopithecus* [in East Africa, *A. garhi*, *A. (Paranthropus) aethiopicus*, and *A. (Paranthropus) boisei*; in South Africa, *A. africanus* and *A. (Paranthropus) robustus*] and the larger-brained genus *Homo* (*H. habilis*, *H. rudolfensis*, and *H. ergaster/erectus*). Although it is possible that both *Australopithecus* and *Homo* made and used stone tools, many anthropologists believe that the increased brain expansion and tooth reduction in *Homo* suggests that they were more technological creatures, with material culture beginning to supplement biology in a more significant way than in modern apes or earlier hominids. By 1 million years ago *Australopithecus* was extinct, *Homo* continued to evolve (ultimately leading to modern humans), and stone tool technologies continued to evolve in tandem with biological evolution.

It is thought by many paleoanthropologists that the emergence of stone tools indicates a dietary shift that included more animal meat and fat through hunting or scavenging and an overall increase in diet

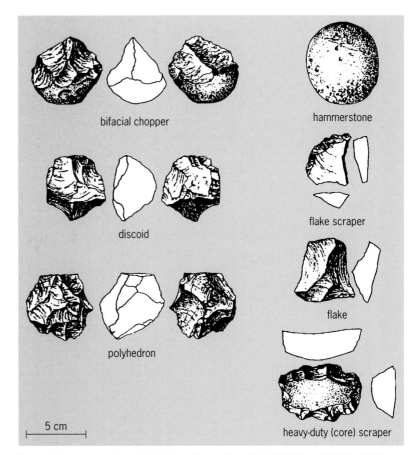

bifacial chopper

discoid

polyhedron

hammerstone

flake scraper

flake

heavy-duty (core) scraper

5 cm

Fig. 2. Oldowan forms from Koobi Fora, Kenya. (*From K. D. Schick and N. Toth, Making Silent Stones Speak, p. 113, Simon and Schuster, New York, 1993*)

breadth through the use of tools and technology. A shift to a higher-quality diet could have decreased the amount of energy required for the digestive tract and thus increased the amount of energy that could be devoted to the brain, ultimately leading (via natural selection) to hominids with larger cranial capacities (from 400–550 cm^3 in the early forms of *Australopithecus* to 600–850 cm^3 in early *Homo*, with modern humans averaging about 1350 cm^3) and higher orders of intelligence for more complex social, communicative, and adaptive behaviors.

The earliest stages of stone tool technology show a relatively slow rate of change over time and relatively little diversification across large geographical ranges. This is in contrast to much more rapid technological change and geographical variation seen in later periods of the Stone Age, particularly after anatomically modern humans emerged and became widespread within the past 100,000 years.

For background information *see* AUSTRALO-PITHECINE; EARLY MODERN HUMANS; FOSSIL HUMANS; PALEOLITHIC; PREHISTORIC TECHNOLOGY in the McGraw-Hill Encyclopedia of Science & Technology.

Nicholas Toth; Kathy Schick

Bibliography. G. Li. Isaac, *The Archaeology of Human Origins: Papers by Glynn Isaac*, ed. by B. Isaac, Cambridge University Press, New York, 1989; R. G. Klein, *The Human Career*, University of

Chicago, 1999; K. D. Schick and N. Toth, *Making Silent Stones Speak: Human Evolution and the Dawn of Technology*, Simon and Schuster, New York, 1993; K. Schick and N. Toth, Palaeoanthropology at the Millennium, in D. Feinman and G. Price (eds.), *Archaeology at the Millennium: A Sourcebook*, pp. 39–108, Kluwer Academic/Plenum, New York, 2001; N. Toth, The Oldowan reassessed: A close look at Early Stone artifacts, *J. Archaeol. Sci.*, 12:101–120, 1985; N. Toth and K. D. Schick, The first million years: The archaeology of protohuman culture, *Adv. Archaeol. Method Theory*, 9:1–96, 1986.

Supernova Remnant 1987A

Supernova 1987A was discovered on February 23, 1987. The explosive death of a massive star, it quickly rose to become the brightest supernova in 383 years. Located 170,000 light-years away in the Large Magellanic Cloud, a small companion galaxy to the Milky Way Galaxy, it is the closest supernova in modern times. Because Supernova 1987A occurred in a well-surveyed region of the Large Magellanic Cloud, it is one of a handful for which the preexisting star, the progenitor, has been clearly identified. The progenitor was a blue supergiant star, born with a mass of roughly 20 solar masses. After a lifetime of around 10 million years, the star had fused all of its hydrogen fuel into an unstable iron core that rapidly collapsed into a neutron star, releasing of the order of 10^{46} joules (10^{53} ergs) of energy as a pulse of neutrinos. Collision of the infalling surrounding layers with the rigid neutron star, aided by poorly understood heating from the neutrino pulse, generated a powerful outgoing shock wave that expelled the atmosphere of the star into space. Approximately 10 solar masses of gas, called ejecta, were launched into space with a total kinetic energy of around 10^{44} joules (10^{51} ergs) and observed speeds as high as 30,000 km/s (19,000 mi/s).

Short-lived massive stars typically experience periods of heavy mass loss in the form of strong winds, so their ejecta are destined to collide with denser, slow-moving circumstellar matter. When this happens two shock waves result: A forward blast wave is driven into the circumstellar gas, racing ahead of the ejecta, while a reverse shock is driven back into the expanding ejecta (**Fig. 1**). Gas crossing these shocks is heated to temperatures in the range 10^6–10^8 K and becomes a strong source of thermal x-ray emission. Electrons from ionized atoms can be accelerated to relativistic energies (that is, energies much greater than their rest energies, with speeds close to that of light) by the shocks, resulting in strong radio synchrotron emission. In cases of higher gas density and lower shock velocity, the shocked gas can cool rapidly to temperatures near 10^4 K, where it radiates strong atomic emission lines at ultraviolet, optical, and infrared wavelengths. As the shocks convert the kinetic energy of the ejecta into photons, a glowing nebula called a supernova remnant is born.

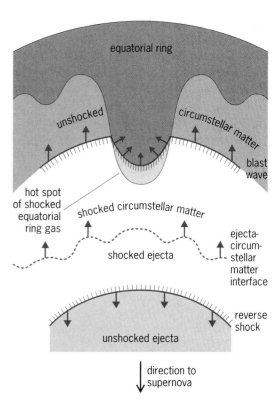

Fig. 1. Shocks in Supernova Remnant 1987A. As the ejecta expand and sweep up the surrounding circumstellar matter at the interface, a reverse shock is driven backward into the ejecta and a blast wave races out ahead into the circumstellar matter. As the blast wave encounters a protrusion on the equatorial ring, it drives a forward shock into this denser gas, creating a "hot spot." Hot gas between the reverse shock and blast wave radiates strong x-ray and radio emission, while the denser gas in the hot spot radiates ultraviolet, optical, infrared, and low-energy x-ray photons.

Circumstellar rings. The presence of large amounts of circumstellar matter in some form of shell-like structure around the progenitor of Supernova 1987A was first deduced from ultraviolet spectroscopy during the years immediately following the original detection. Once the Hubble Space Telescope was launched in 1991, high-resolution imaging revealed the circumstellar matter to have a surprisingly complex distribution (**Fig. 2**). Ionized by x-ray and ultraviolet radiation from the supernova, it appears concentrated into three fading ringlike structures. The rings contain roughly 0.07 solar mass of photoionized gas in inclined circular bands lining the walls of a large, hourglass-shaped bipolar cavity inside a much larger nebula thought to contain several solar masses of invisible, neutral gas. The bright, innermost equatorial ring is a roughly circular structure with a radius of 0.65 light-year, inclined by $44°$ and expanding at 10 km/s (6 mi/s). The formation of the bipolar nebula and concentration of gas into the three rings is not well understood, but may have resulted from the merger of a close binary star system roughly 20,000 years prior to the explosion.

Radio observations. Supernova 1987A was a temporary source of radio emission for 7 months following the explosion, an unusually short period

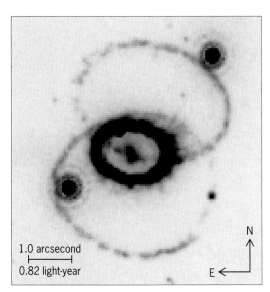

Fig. 2. Hubble Space Telescope image of SN 1987A taken in red light (655–658 nm). The densest glowing central regions of the ejecta (central spot) are surrounded by three rings of gas. The rings are thought to line the interior of an hourglass-shaped bipolar cavity within a much larger nebula of circumstellar matter. Coincidentally, two bright stars (upper right and lower left) lie in projection on the outer rings. (*B. Sugerman, Space Telescope Science Institute; NASA*)

compared to that of other radio supernovae. This brief phase was interpreted as the ejecta sweeping up the most concentrated inner region of the low-density wind blown off by the blue supergiant progenitor. But the radio emission resumed in 1990 and has been increasing ever since, announcing the birth of Supernova Remnant 1987A. Since late 1992, the Australia Compact Telescope Array has produced interferometric radio images of the supernova remnant. The radio emission comes from an annular shell that was 0.80 times the radius of the optical equatorial ring in late 1992. The size of this radio shell is increasing; the initial radius and expansion rate indicate that the blast wave expanded at an average rate greater than 27,000 km/s (17,000 mi/s) prior to 1992, but at roughly 3000 km/s (1900 mi/s) since. The simultaneous reappearance of radio emission and strong deceleration of the expansion resulted when the ejecta encountered a denser layer of photoionized gas lining the interior of the equatorial ring and the bipolar cavity. The radio images provided the first evidence that the newborn supernova remnant was not cylindrically symmetric, with eastern and western lobes of brighter emission aligned with the major axis of the equatorial ring.

X-ray observations. The supernova remnant became a detectable source of x-rays roughly 6 months after the radio emission began, and the x-ray emission is increasing at an accelerating pace. Resolved images of the supernova remnant from the *Chandra X-ray Observatory*, launched in 1999, show a remarkable overall similarity with the radio synchrotron emission: The x-ray emission is distributed in a ringlike shell interior to the equatorial ring with a strong east-west asymmetry. The correlation with the radio

is best at higher x-ray energies, indicating that both originate in the hottest gas contained between the reverse shock and the blast wave. At lower x-ray energies the remnant correlates better with shock features on the equatorial ring seen in optical and ultraviolet images from the Hubble Space Telescope (see below), indicating that the blast wave is already contacting the innermost edge of the equatorial ring. X-ray spectra suggest blast wave velocities near 3000 km/s (1900 mi/s), consistent with the radio expansion results.

Optical and ultraviolet observations. After the radio and x-ray detections, the next development in the birth of Supernova Remnant 1987A was discovered with the Space Telescope Imaging Spectrograph (STIS) on the Hubble Space Telescope in 1997. The first STIS spectra of the supernova remnant revealed that a brightened "hot spot" on the inner edge of the fading equatorial ring contained gas with velocities reaching more than 250 km/s (160 mi/s) and displaced from the slow expansion of the ring by 80 km/s (50 mi/s). A second hot spot was discovered in ground-based infrared images in early 2000, followed immediately by four more in optical Hubble Space Telescope images (**Fig. 3**). Subsequent spots have appeared rapidly; as of October 2002 there were 20 hot spots scattered around the entire equatorial ring circumference. The first spots to appear were predominantly in the eastern half of the equatorial ring. This timing, along with the radio and x-ray asymmetries, suggests either that the eastern ejecta are denser and have been decelerated less or that the progenitor was not centered within the equatorial ring.

The hot spots are interpreted as sites where the blast wave has swept past an inward-pointing protrusion on the clumpy inner edge of the equatorial ring (Fig. 1). The forward tip of the protrusion likely has a strong 500-km/s (310-mi/s) shock driven into

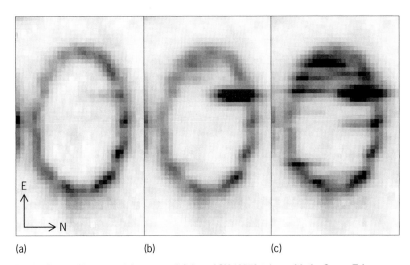

(a)   (b)   (c)

Fig. 3. Spectral images of the equatorial ring of SN 1987A taken with the Space Telescope Imaging Spectrograph on the Hubble Space Telescope, showing evolution in the red emission line of atomic hydrogen (656.3 nm). Images show the rapid onset and brightening of successive hot spots, which appear as horizontal smears due to Doppler broadening of high-velocity gas within their internal shocks. (*a*) April 1997. (*b*) May 2000. (*c*) April 2001. (*Stephen Lawrence, Space Telescope Science Institute; NASA*)

it, creating very hot gas with long cooling times; such nonradiative shocks presumably are responsible for the soft x-rays seen in *Chandra* images. As the blast sweeps down the length of the protrusion, slower oblique shocks are driven inward from the walls of the protrusion. Gas behind these slower radiative shocks cools rapidly, generating the hot spots in optical and ultraviolet images.

Evolution of the individual hot spots varies: Their optical fluxes double on time scales ranging from 100 to 500 days, while the first spot appears to have leveled off at a temporary maximum. These differences are likely due to variations in the geometries and densities of the shocked protrusions. Precise measurements of the centers of the brightest spots reveal radial motions outward from the supernova. These motions are interpreted as shifts in the average position of the light emission as the blast wave sweeps farther down the protrusions and adds more radiating shocked gas at greater distances from the supernova. These apparent motions indicate a blast wave speed of at least 2000 km/s (1200 mi/s), consistent with results from radio and x-ray data.

In addition to the forward blast wave, high-resolution ultraviolet and optical spectral imaging with the Hubble Space Telescope traces out the location of the reverse shock in exquisite detail. At these wavelengths, the reverse shock forms an annulus roughly 0.8 times the radius of the equatorial ring. Asymmetry is also seen here, with greater amounts of shocked ejecta to the east.

Future developments. Only the very initial stages have been observed in the birth of this supernova remnant. In mid-2003, the forward blast wave had reached only the most inward protrusions of the equatorial ring; as it sweeps into the main volume of gas, the numerous hot spots will merge into a continuous ring of bright emission. The reverse shock will continue to move "upstream" into ejecta from deeper within the progenitor, eventually generating emission from gas enriched in heavy elements by explosive nucleosynthesis. The increasing volumes of shocked gas from these two fronts will cause Supernova Remnant 1987A to brighten at x-ray, ultraviolet, optical, infrared, and radio wavelengths by factors of 100–1000 over the next decade, and to remain bright for centuries. This increasing x-ray and ultraviolet flux will photoionize a greater volume of the bipolar nebula, progressively illuminating the mass-loss history of the progenitor. Supernova Remnant 1987A will be an object of intense study for many decades to come.

For background information *see* ASTRONOMICAL SPECTROSCOPY; INTERSTELLAR MATTER; NEBULA; RADIO ASTRONOMY; SHOCK WAVE; SPACE TELESCOPE, HUBBLE; SUPERNOVA; ULTRAVIOLET ASTRONOMY; X-RAY ASTRONOMY in the McGraw-Hill Encyclopedia of Science & Technology.   Stephen S. Lawrence

Bibliography. W. D. Arnett et al., Supernova 1987A, *Annu. Rev. Astron. Astrophys.*, 27:629–700, 1989; L. A. Marschall, *The Supernova Story*, Princeton University Press, 1994; R. McCray, Supernova 1987A, in K. Weiler (ed.), *Supernovae and Gamma-ray Bursters*, pp. 219–242, Springer, Heidelberg, 2003; R. McCray, Supernova 1987A revisited, *Annu. Rev. Astron. Astrophys.*, 31:175–216, 1993; P. Murdin, *End in Fire: The Supernova in the Large Magellanic Cloud*, Cambridge University Press, 1990; J. C. Wheeler, *Cosmic Catastrophes*, Cambridge University Press, 2000.

System families

Large systems are often formed from a variety of component systems: custom systems, which are newly engineered from the "ground up;" existing commercial-off-the-shelf (COTS) systems that are subsequently tailored for a particular application; and existing or legacy systems. Such related terms as systems of systems (SOS), federations of systems (FOS), federated systems of systems (F-SOS), and coalitions of systems (COS) are often used to characterize these systems. These appellations capture important realities brought about by the fact that modern systems are not monolithic. Rather, they have five characteristics, initially summarized by Mark Maier, that make one of the system family designations appropriate:

1. *Operational independence of the individual systems.* A system of systems is composed of systems that are independent and useful in their own right. If a system of systems is disassembled, the constituent systems are capable of independently performing useful operations by themselves and independently of one another.

2. *Managerial independence of the systems.* The component systems generally operate independently in order to achieve the technological, human, and organizational purposes of the individual unit that operates the system. These component systems are generally individually acquired, serve an independently useful purpose, and often maintain a continuing operational existence that is quite independent of the larger system of systems.

3. *Geographic distribution.* Geographic dispersion of the constituent systems in a system of systems is often quite large. Often, these constituent systems can readily exchange only information and knowledge with one another, and not substantial quantities of physical mass or energy.

4. *Emergent behavior.* The system of systems performs functions and carries out purposes that do not reside uniquely in any of the constituent systems. These behaviors arise as a consequence of the formation of the entire system of systems and are not the behavior of any constituent system. The principal purposes supporting engineering of these individual systems and the composite system of systems are fulfilled by these emergent behaviors.

5. *Evolutionary and adaptive development.* A system of systems is never fully formed or complete. Development of these systems is evolutionary and adaptive over time, and structures, functions, and

purposes are added, removed, and modified as experience of the community with the individual systems and the composite system grows and evolves.

Unfortunately, there is no universally accepted definition of these system families, and at present there are no definitive criteria that distinguish a system of systems from other systems. In a formal sense, almost anything could be regarded as a system of systems. A personal computer is a system. However, the monitor, microprocessor, disk drive, and random-access memory are also systems. Thus, should a personal or a mainframe computer be called a system of systems? Is the Internet a system of systems? Instinctively, they are quite different. The big difference in these two examples is that the former system may be massive in size but monolithic in purpose, while the latter is capable of supporting the myriad communications and commerce purposes of utilizing organizations and humans. There is also a distinction based on operational and managerial independence of the system components and in evolution and emergence possibilities.

Federations and coalitions of systems.
Often, appropriate missions exist for relatively large systems of systems in which there is a very limited amount of centralized command-and-control authority. Instead, a coalition of partners has decentralized power and authority and potentially differing perspectives of situations. It is useful to term such a system a "federation of systems" and sometimes a "coalition of systems." The participation of the federation or coalition of partners is based upon collaboration and coordination to meet the needs of the federation or coalition.

Innovation and change.
Support for innovation and change of all types is a desirable characteristic of these system families. Innovation includes both technological innovation and organizational and human conceptual innovation. Accomplishing this requires continuous learning, a reasonable tolerance for errors, and experimental processes to accomplish both the needed learning and the needed change. The systems fielded in order to obtain these capabilities will not be monolithic structures in terms of either operations or acquisition. Rather, they will be systems of systems, coalitions of systems, or federations of systems that are integrated in accordance with appropriate architectural constructs in order to achieve the evolutionary, adaptive, and emergent cooperative effects that will be required to achieve human and organizational purposes and to take advantage of rapid changes in technology. They can potentially accommodate system life-cycle change, in which the life cycle associated with use of a system family evolves over time; system purpose change, in which the focus in use of the system emerges and evolves over time; and environment change, in terms of alterations in the external context supporting differing organizational and human information and knowledge needs, as well as in the technological products that comprise constituent systems.

Autonomy, heterogeneity, and dispersion.
One can contrast system families and identify relationships between conventional systems, systems of systems, and federations or coalitions of systems with regard to three characteristics, as initially noted by A. J. Krygiel: autonomy, heterogeneity, and dispersion. A federation of systems, federated system of systems, or coalition of systems will generally have greater values of these characteristics than a (nonfederated) system of systems.

Many conventional systems are built for special purposes, as a mixture of commercial-off-the-shelf systems and custom developments of hardware and software. These constituents are generally provided by multiple contractors who are used to supporting a specific customer base and working under the leadership of a single vertical program management structure. For best operation, these systems should be managed as a system of systems, federation of systems, or coalition of systems.

A system of systems generally has achieved integration of the constituent systems across communities of contractors, and sometimes across multiple customer bases, and is generally managed by more horizontally organized program management structures, such as integrated product and process development (IPPD) teams. When the IPPD team effort is well coordinated, the team is generally well able to deal with conflict issues that arise due to business, political, and other potentially competing interests.

Federations of systems face the same dilemmas identified for systems of systems but are generally much more heterogeneous along transcultural and transnational socio-political dimensions, are often managed in an autonomous manner without great central authority and direction such that they satisfy the objectives and purpose of an individual unit in the federation, and often accommodate a much greater geographic dispersion of organizational units and systems. Thus, the delimiters between systems of systems and federations of systems or coalitions of systems, while generally subjective, are nonetheless principled.

The notions of autonomy, heterogeneity, and dispersion are not independent of one another. Increasing geographic dispersion will usually lead to greater autonomy and consequently will also increase heterogeneity. The Internet is perhaps the best example of a system that began under the aegis of a single sponsor, the U.S. Department of Defense, and has grown to become a federation of systems.

Federalism.
This approach, initially studied by C. Handy as "new federalism," addresses the necessary considerations for the structuring of loosely coupled organizations in order to help them adapt to changes in the information and knowledge age. The application of federalist political principles to the management of systems of systems and federations of systems is an appealing way of obtaining a systems engineering ecology—in other words, a sustainable systems engineering approach that possesses adaptation, evolution, and emergence characteristics

analogous to those in a natural ecology. This is so because many contemporary organizational alliances, including those for engineering and development, take the form of virtual organizations or virtual teams rather than the classical physical organizations and teams. The concept of federalism is particularly appropriate since it offers a well-recognized way to deal with the systems engineering management paradoxes of power and control such that the desired systems ecological balance is obtained. Generally, this is accomplished by making things big by keeping them small, a goal which, in turn, is accomplished by expanding the domain of an enterprise by instantiating multiple quasiautonomous units as opposed to acquiring mass by aggregation around a centralized command-and-control authority base; encouraging autonomy but within appropriate bounds set by process and architecture standards; and combining variety with shared purpose, and individuality with partnerships at national and global levels.

Federalism is based on five principles: subsidiarity, interdependence, a uniform and standardized way of doing business, separation of powers, and dual citizenship. Subsidiarity, the most important principle, suggests that power belongs to the lowest possible point within the engineering team of a federation of systems. Interdependence, or pluralism, requires that the autonomous development units or teams of a development federation of a federation of systems stick together because they need one another as much as they need management leadership and leadership authority. Having a uniform and standardized way of doing business ensures interdependence within federated engineering organizations of a system of systems or federation of systems through agreement on basic rules of conduct, common traditions of communicating, and common units of measurement of progress and quality. Separation of powers requires that management, monitoring, and governing aspects of engineering programs and projects of federations of systems be viewed as separate functions to be accomplished by separate bodies whose membership may overlap. Finally, dual citizenship requires that every individual is a "citizen" in two communities: the local development group, professional group, or union, and the overall program of the federation of systems.

Systems engineering approaches. These systems-of-systems and federations-of-systems concepts have numerous implications for systems engineering and management.

Grand design approach. Contemporary organizations often treat the engineering of systems of systems or federations of systems with systems engineering protocols that are, at best, suitable only for monolithic systems. The archetype of such ill-advised protocols is the "grand design" life cycle, which is based on the waterfall model that came into prominence around 1970. A large number of problems have been encountered with grand design efforts to engineer a system. Today, the classic waterfall approach is suggested only in those rare cases where user and system-level requirements are crystal clear and unlikely to change at all during or after engineering the system, and where funding for the grand design is essentially guaranteed. This is rarely the case for major systems, especially those that are software-intensive, and would be the rarest of all cases for a system of systems or federation of systems. Changing user and organizational needs and changing technologies virtually guarantee that major systems cannot be developed using the grand design approach.

Incremental and evolutionary approaches. Two leading alternatives to the grand design approach for the engineering of systems were initially termed incremental and evolutionary, although the term "evolutionary" is now generally used to characterize both of these. In incremental development, the system is delivered in preplanned phases or increments, in which each delivered module is functionally useful. The overall system capability improves with the addition of successive modules. The desired system capability is planned to change from the beginning, as the result of "build N" being augmented and enhanced through the phased increment of "build $N+1$." This approach enables a well-functioning implementation to be delivered and fielded within a relatively short time and augmented through additional builds. It also allows time for system users to thoroughly implement and evaluate an initial system with limited functionality compared to the ultimately desired system. Generally, the notion of preplanning of future builds is strong in incremental development. As experience with the system at build N is gained, requirements changes for module $N+1$ may be more easily incorporated into this, and subsequent, builds.

Evolutionary life-cycle development is similar in approach to its incremental complement; however, future changes are not necessarily preplanned. This approach recognizes that it is impossible to initially predict and set forth engineering plans for the exact nature of these changes. The system is engineered at build $N+1$ through reengineering the system that existed at build N. Thus, a new functional system is delivered at each build, rather than obtaining build $N+1$ from build N by adding a new module. The enhancements to be made to obtain a future system are not determined in advance, as in the case of incremental builds. Evolutionary development approaches can be very effective in cases where user requirements are expected to shift dramatically over time, and where emerging and innovative technologies allow for major future improvements. They are especially useful for the engineering of unprecedented systems that involve substantial risk and allow potentially enhanced risk management. Evolutionary development may help program managers adjust to changing requirements and funding priority shifts over time, since new functionality introductions can be advanced or delayed in order to accommodate user requirements and funding changes. Open, flexible, and adaptable system architecture is central to the notion of evolutionary and emergent development.

These are major elements in the contemporary U.S. Department of Defense Initiatives in evolutionary acquisition.

For background information *see* DISTRIBUTED SYSTEMS (CONTROL SYSTEMS); INTERNET; LARGE SYSTEMS CONTROL THEORY; REENGINEERING; RISK ASSESSMENT AND MANAGEMENT; SYSTEMS ARCHITECTURE; SYSTEMS ENGINEERING; SYSTEMS INTEGRATION in the McGraw-Hill Encyclopedia of Science & Technology.   Andrew P. Sage

Bibliography. P. G. Carlock and R. E. Fenton, Systems of systems (SoS) enterprise systems engineering for information-intensive organizations, *Sys. Eng.*, 4(4):242–261, 2001; P. Chen and J. Clothier, Advancing systems engineering for systems-of-systems challenges, *Sys. Eng.*, 6(3):170–183, 2003; C. Handy, Balancing corporate power: A new Federalist Paper, *Harvard Bus. Rev.*, 70(6):59–72, November-December 1992; C. Handy, Trust and the virtual organization, *Harvard Bus. Rev.*, 73(3):8–15, May/June 1995; A. J. Krygiel, *Behind the Wizard's Curtain: An Integration Environment for a System of Systems*, CCRP Publication Series, Vienna, VA, 1999; M. W. Maier, Architecting principles for systems-of-systems, *Sys. Eng.*, 1(4):267–284, 1998; J. Morganwalp and A. P. Sage, A system of systems focused enterprise architecture framework and an associated architecture development process, *Inform. Knowl. Sys. Manag.*, 3(4):87–105, 2003; A. P. Sage and C. D. Cuppan, On the systems engineering and management of systems of systems and federations of systems, *Inform. Knowl. Sys. Manag.*, 2(4):325–345, 2001; A. P. Sage and W. B. Rouse (eds.), *Handbook of Systems Engineering and Management*, Wiley, New York, 1999.

Telecommunications network security

A telecommunications network is a collection of communication devices interconnected in some fashion. In general, a network includes the terminal equipment (such as telephone handsets, Ethernet access devices, and 802.11 Wi-Fi cards) and the network routing equipment and control infrastructure (such as switches, routers, and billing servers). The logical structure and the physical equipment of a network have a profound effect on security, either lending to greater security or creating specific vulnerabilities. *See* WIRELESS FIDELITY TECHNOLOGY.

Telecommunications has come to refer to any large, interconnected structure of communications equipment that provides voice or data transmissions over a distance. Examples of telecommunications networks include cell phone networks (for example, Global System for Telecommunications, or GSM, coverage throughout Europe) and television and Internet networks (for example, access to broadcast television and the Internet from a fixed location). Network services normally include a service contract that specifies a certain level of network service

in exchange for a contract fee. The service contract is important because it defines the minimum level of network services that will be supplied to the network user and also specifies the proper behavior of the network user while accessing the network.

Conflicting objectives. Network security includes the mechanisms and practices that protect a network so as to ensure the proper operation of the network and the realization of all network services for each participant of the network. Participants of the network include users of the network as well as network operators. The desired operation of the network may differ for different participants, and at times network security may interfere with a participant's view of proper operation of the network.

For example, a typical user may want instant access to the network, while the network operator may wish to maximize operating profits by ensuring that all users of the network pay for their access. This network security requirement (that is, authenticating all users to ensure that they are paying customers) may cause an initial delay when users access the network, frustrating the desires of the network user. Similarly, a network operator may wish to maximize the number of users on the network, but users may require extra bandwidth on the network in order to encrypt or authenticate their data to achieve data privacy. The extra bandwidth required by encryption may reduce the number of users that can be effectively serviced by the network.

General security concerns. Telecommunications network security, from a network operator's perspective, is concerned with (1) authentication, whereby only authorized users are allowed to access the network; (2) access control, whereby users are allowed to employ only those services that they are assigned; (3) authorization, whereby only approved service provider personnel are allowed access to protected network infrastructure and services; and (4) availability, whereby network users are prevented from exceeding their service contract limits to ensure that all network users receive adequate service.

Network users are generally concerned with (1) protecting the secrecy of their personal data, which is referred to as data privacy; (2) preventing malicious modification of these data, which is achieved through data integrity; and (3) preventing other users from gaining access to their network access devices without permission—this is another perspective of authorization.

Operator concerns. The most common form of authentication is a username and password pair. Network operators utilize protocols such as RADIUS and DIAMETER to exchange the usernames and passwords in a secure manner.

Most network service providers use a firewall to partition off public network resources from private network resources (the network infrastructure). Firewalls are usually installed with filtering routers to restrict the types of traffic that can pass through to protected areas of the network.

While some networks, such as the Internet, are normally offered with just access services, many networks provide additional services. For example, a cell phone network usually provides voice mail, short message service (SMS), multimedia message service (MMS), and call waiting. Most networks that provide these additional services protect them using an authorization mechanism. In GSM cell phone networks, authorization is linked with authentication through the use of a subscriber identity module (SIM) that is implemented on a smartcard or integrated circuit card (ICC).

Availability on a network is typically provided through some type of ingress filtering. Ingress filtering prevents network users from sending more traffic than their service contract allows, ensuring that all network users gain fair access to the network. Normally, if all users are operating within their ingress filtering parameters, there will not be any egress (output) issues.

User concerns. Data privacy and data integrity may be standard services of the network itself (as in the case of GSM cell phone network protocols, which provide encryption over the wireless link) or may be provided by an additional element. Data privacy on the Internet is normally assured through the application that is handling the data. For Web browsing, a type of protocol that encrypts data packets is used, such as Secure Sockets Layer (SSL) or Transport Layer Security (TLS). For electronic mail, different encryption software is commonly available, such as Pretty Good Privacy (PGP) or Secure Multipart Internet Message Extensions (S-MIME).

An alternate method is to provide encryption for all data that are transmitted or received over a network connection. This logically creates a private network since no one, other than the intended receiver, can understand the packets. Called a virtual private network (VPN), this encrypted communication tunnel allows two parties to communicate securely over a public network.

The VPN is created using a special protocol stack that encrypts all sent packets and decrypts all received packets (**Fig. 1**). Two parties communicating over one of the encrypted tunnels have their data protected from everyone else. No other party can insert or modify the packets sent between them without detection. The technology allows multiple tunnels to be set up to different parties with different encryption keys, logically creating multiple secure tunnels. Most VPNs utilize a set of protocols from the International Engineering Task Force (IETF) called the Internet Protocol Security (IPSec) suite; however, it is also possible to use special VPN clients that utilize SSL and TLS.

Attackers use two primary mechanisms to gain access to other users' network access devices without permission: viruses and network protocol attacks. A virus is an executable program or script that performs some task of which the user was not aware and did not intend. Network protocol attacks take advantage of defects or anomalous effects in the software or hardware that implement the network protocols in network access devices. The effects of these network protocol attacks range from inconveniences (such as shutting down a device) to major catastrophes (such as allowing the attacker to have full control of the device to perform any task of a user or service provider). Most users understand the threat that a virus poses and utilize protections against them (such as virus scanners, and proper care and restraint when downloading programs and executing content from e-mails), but they may have difficulty understanding network protocol attacks and protecting against them. Protections against network protocol attacks include personal firewall software and maintaining the software on devices so that it has the most updated fixes for known network protocol vulnerabilities.

Classification of attacks and threats. When a security concern is violated, network security has been breached. A breach in network security is the result of an attack. An attack is classified according to the threat that carries out the attack. A threat may be natural (physical), intentional (malicious), or unintentional (benign). Physical threats include earthquakes, fires, hurricanes, and power outages. A benign threat is caused by an honest mistake by a user or an operator. An example of a benign threat is a user accidentally hitting the "delete" key and removing a file, or a construction worker accidentally cutting a fiber-optic network cable with a backhoe. Malicious threats are conscious acts by persons committed for enjoyment or personal gain, often with the intention of causing harm to the network or its users or providers.

General network vulnerabilities. Networks vulnerabilities are characteristics of the network system that make the network susceptible to network protocol attacks. Each type of network has vulnerabilities that are unique to it, but all networks share some common vulnerabilities. These include (1) intrusion, an

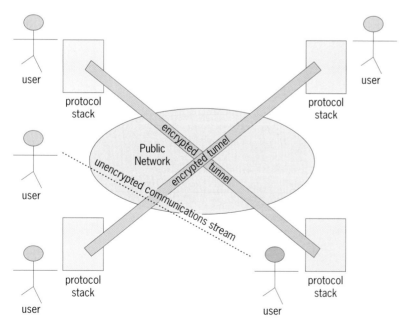

Fig. 1. Using a virtual private network (VPN).

active attack whereby the attacker gains access to another user's communication device (for example, his or her computer attached to the Internet) or a component of the service provider's network infrastructure; (2) interception, a passive attack whereby the attacker merely listens to (captures) data on the network; (3) misrouting, an active attack whereby the attacker changes the address or handling instructions that accompany network data, causing data to be sent to the wrong network station; (4) forgery (insertion), an active attack whereby the attacker adds data to a network channel or data stream, or constructs data messages that are incorrect or misleading; (5) masquerade, an active attack whereby the attacker claims to be another (authorized) entity in order to gain access to protected resources or services; and (6) denial of service, an active attack whereby the attacker purposely consumes a large portion of network resources in order to prevent other (valid) users from gaining access to or utilizing the network.

Motivations for an attack. It is extremely important to understand the purpose behind an attack in order to understand the methods the attacker may use. There are many motivations and combinations of them; a few important ones will be discussed.

Destruction and disruption. An attacker who is angry at a particular entity may wish to destroy or disrupt that entity's operations. In this case, merely destroying data or machines or interrupting the flow of communications is sufficient. A common method for this type of attack is a distributed denial of service (DDoS) attack to bring down parts of the Internet and specific Web sites. One type of DDoS attack uses a mechanism called a SYN Flood. For example, a DDos attack on February 7, 2000, used SYN Flooding to overwhelm the capacity of networks that service companies that maintain major Web sites.

Normal communication is initiated between a computer (the client) and a server on the Internet with a start-up handshake that includes a SYN packet, a SYN-ACK packet, and an ACK packet (**Fig. 2**). The SYN packet is sent from the client to the server to request that a communications port be opened from the server back to the client. The SYN packet contains the address of the client so the server knows to whom to send the response. The server automatically responds with a SYN-ACK packet that contains the address of the port the server is opening for the client. The client is supposed to respond with an ACK packet transmitted to the specified port. Because a server has a limited amount of memory, it can support only a certain number of ports (typically in the tens of thousands), so that it can support only a certain number of clients simultaneously.

Certain procedures are specified if a client does not respond to a SYN-ACK packet, or if a server does not respond to a SYN packet. In the former case, the server resends the SYN-ACK packet several times, and if no response if received after a minute or two, it frees the resources and ignores the SYN packet.

In a SYN Flood attack, a hacker sends a SYN packet but uses a crafted address (that is, an address that does not correspond to any client on the Internet but is still valid). The attacking client can send thousands of these packets a second to the same server. The server cannot tell the difference between valid SYN packets and crafted SYN packets, so it must respond as if the latter were valid. This requires the server to reserve resources (for example, a port) for the potential communication with a client, making the server capable of handling fewer legitimate clients. If the server receives enough of these crafted SYN packets, it will crash. This SYN Flood attack is an example of a DoS attack. Other types of packets can be used to mount DoS attacks, for example, an ICMP Ping Flood.

While it is possible for a few clients, or possibly one client, to mount a SYN Flood attack, it is relatively easy for a network operator to trace the source of the crafted packets and shut off their flow using ingress filters on a router or firewall. In a DDoS attack, a hacker compromises hundreds or thousands of computers on the Internet using viruses or worms which carry a program that is referred to as a zombie. Once a zombie infects a computer, it runs in the background whenever the computer is on, in a nearly undetectable way, and waits for a command. The hacker uses a program such as Internet Relay Chat (IRC) to communicate with the zombies and tell them which server to attack. Each zombie sends out only a few hundred crafted SYN packets, but since there are thousands of zombies, the server receives hundreds of thousands of packets all at once. If the zombies are distributed all around the Internet, the attack seems to be coming from everyone, distributed over the whole world, so that it cannot be quenched with ingress filtering and is nearly impossible to shut off.

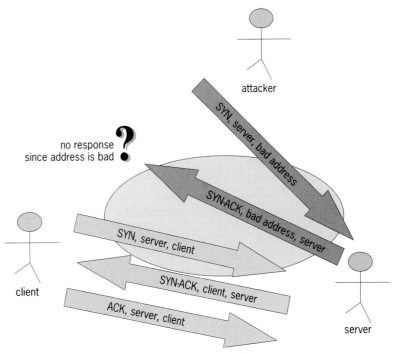

Fig. 2. Normal communication on the Internet and a SYN Flood attack.

Notoriety. In this case, the attacker wants as much publicity as possible, and therefore would not want to destroy or interrupt communications, but might in fact wish to increase or expand them. The attacker would leave some type of message or indication of his or her exploits in order to make a credible claim to his or her accomplishments. There are sites on the Internet that track these types of Web defacements, adding to the notoriety of these actions. Notoriety is normally attributed in some form of pseudonym, which the attacker believes to be untraceable to himself or herself.

Financial motivations. These motivations may be direct or indirect. A direct financial motivation includes stealing credit card and personal information for the purposes of fraud. An indirect financial motivation includes publishing damaging information on a company in order to affect stock prices, and thereby make financial gains by different stock market schemes. In these attacks, the perpetrators prefer that they remain anonymous and also that their attacks go unnoticed. The longer the attacks go unreported, the better chance the attackers will remain unpunished for their crimes.

Internet mitigation techniques. Most attacks on the Internet are thwarted using a combination of network firewalls, filtering routers, VPNs, personal firewalls, and authentication. Good system administration practices, including keeping the network software and operating system updated with the latest patches and fixes, are also necessary. These techniques will handle most intrusion, interception, insertion, and masquerade attacks.

Only careful monitoring and filtering can respond to and mitigate the most sophisticated DoS and DDoS attacks. No way is currently known to prevent DDoS attacks on the Internet as it exists today.

For background information *see* COMPUTER SECURITY; INTERNET; MOBILE RADIO; WORLD WIDE WEB in the McGraw-Hill Encyclopedia of Science & Technology.   David M. Wheeler

Bibliography. W. R. Cheswick, S. M. Bellovin, and A. D. Rubin, *Firewalls and Internet Security*, 2d ed., Addison-Wesley, Reading, MA, 2003; R. L. Freeman, *Fundamentals of Telecommunications*, Wiley, New York, 1999; R. L. Freeman, *Practical Data Communications*, 2d ed., Wiley, New York, 2001; S. Northcutt, *Network Intrusion Detection: An Analyst's Handbook*, New Riders Publishing, Indianapolis, 1999; A. S. Tanenbaum, *Computer Networks*, 4th ed., Prentice Hall, Upper Saddle River, NJ, 2002; E. D. Zwicky, S. Cooper, and B. D. Chapman, *Building Internet Firewalls*, 2d ed., O'Reilly & Associates, Sebastopol, CA, 2000.

Temperature measurement

A common definition of temperature, as stated by one dictionary, is: "Degree of hotness or coldness measured on a definite scale based on some physical phenomenon (as the expansion of mercury in a thermometer)." However, to rigorously define the concept of temperature, one can use the idea of operationalism, which asserts that a concept is synonymous with a corresponding set of operations. Thus, "temperature" can be defined by the five operations needed to construct a temperature scale. There are many arbitrary decisions in doing this.

Temperature scales. The five operations, with appropriate arbitrary choices, are:

1. The selection of a standard substance: helium gas.

2. The selection of a physical property of this standard substance that changes with "degree of hotness": volume at constant pressure and amount of substance.

3. The selection of an interpolation formula, that is, an equation connecting changes in the physical property, V_t, with degree of hotness, t; Eq. (1).

$$V_t = V_0(1 + \alpha t) \qquad (1)$$

4. The selection of a sufficient number of fixed points, that is, conditions in nature with an invariant degree of hotness. (These are typically melting and boiling points at constant pressure, or a triple point, where three physical phases are in equilibrium.) Two suitable fixed points are the melting and boiling points of water at 1 atmosphere pressure.

5. The assignment of numerical values to these fixed points: mp (H_2O) = 0 and bp (H_2O) = 100.

The numerical values of the two constants in the interpolation formula, V_0 and α, can be determined by measuring the volume of helium gas (at constant pressure and mass), V_t, at the melting point and boiling point of water. This results in a centigrade scale since the range between defined points is 100. For an ideal gas (helium is very close to ideal in behavior), or for any gas as the pressure approaches zero, $\alpha = 1/273.15$.

Absolute temperature. This temperature scale may be converted to an "absolute" one using the following transformations. First, t^* is set equal to $1/\alpha$. Then, V_t is given by Eq. (2).

$$V_t = V_0(1 + t/t^*) = V_0(t^* + t)/t^* \qquad (2)$$

Then the absolute temperature is defined by $T = t + t^*$ to get Eq. (3),

$$V_t = V_0(T/t^*) = (V_0/t^*)T = aT \qquad (3)$$

since V_0 and α are constants and their product can be set equal to another constant a. This absolute temperature scale requires only one fixed point (for one constant). By international agreement, this is the triple-point of water, where solid, liquid, and vapor are in equilibrium; and this point is assigned the value of 273.16 K (kelvin). Absolute scales have zero as the lower limit.

The fundamental temperature scale is the thermodynamic temperature scale, which is based on the Carnot cycle, a kind of idealized engine, whose properties are independent of the substance used in

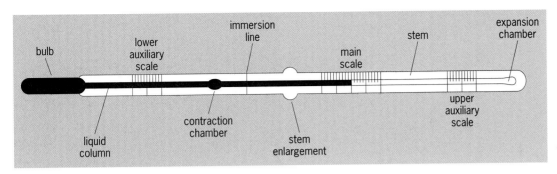

Fig. 1. Mercury-in-glass thermometer.

the cycle. In practice, an ideal gas thermometer can be defined to be exactly the same as this theoretical scale, but using the same single fixed reference point. Temperature is then what is measured using this temperature scale.

International Temperature Scale of 1990. This scale (abbreviated ITS-90) realizes temperature measurement in the range from 14 to 1235 K (−259 to 962°C) via a standard platinum resistance thermometer (SPRT) and a number of fixed points. The electrical resistance of the SPRT is measured at each of these fixed points with temperatures determined by a constant-volume ideal gas thermometer whose pressure can be measured very accurately. Elaborate equations connect electrical resistance to the ideal gas thermometer temperatures. In this way, laboratories throughout the world can measure temperature on the same basis. These temperatures can be typically known accurately to a few thousandths of a kelvin. The scale is defined in terms of the vapor pressure of helium-3 (^{3}He) and helium-4 (^{4}He) between 0.65 and 5.0 K, and by interpolating constant-volume gas thermometry between 3.0 and 25 K. (In the overlapping temperature ranges, the different definitions have equal status.) Above 1235 K, the scale is defined by radiation thermometry, with fixed points up to 1358 K (1085°C), the freezing point of copper.

The international temperature scale has been extended into the millikelvin region down to 0.9 mK. Temperatures in this range are now defined by three phase transitions and a minimum in the melting pressure of helium-3 (^{3}He), with interpolation through specification of this pressure as a polynomial function of the temperature.

Thermometers. There are many ways to measure temperature. The most common instrument is the liquid-in-glass thermometer (**Fig. 1**). Mercury is a good thermometric medium since it is easily puri-

fied, opaque, and a good thermal conductor, and its expansion coefficient is well known. Mercury thermometers are useful in the range from −35 to 510°C. Low-temperature liquid-in-glass thermometers are either filled with ethanol (−80 to 60°C) or *n*-pentane (−200 to 30°C). Typically, they can be calibrated to be reliable to a few tenths of a degree. There are many specialty versions, like the Beckmann thermometer, which can determine changes of a few degrees to within a thousandth of a degree.

Resistance thermometers. Electrical resistance is useful for temperature measurement, the resistance of metals increasing with temperature. Copper, nickel, and nickel alloys are used for inexpensive devices; and platinum is used for the more stable expensive thermometers. **Figure 2** shows the working end of an SPRT, where a fine coil of platinum wire is wound strain-free on a mica cross support in a quartz tube, which is hermetically sealed with an inert gas. With calibration and care, such SPRTs are accurate to a few thousandths of a degree. A variety of high-temperature platinum resistance thermometers with resistances between 0.25 and 2.5 ohms have been used to extend their coverage of ITS-90 up to 1064°C.

Recent advances in the use of SPRTs and other resistance thermometers have been in the development of automatic bridges with digital readouts (versus manual bridges using standard resistors and galvanometers). One such bridge has an accuracy of 1.0 mK. Modern electronics has made these devices easy to use, stable, and encompassing a wide range of temperatures. For example, a glass-capsule SPRT has a range from 13 to 373 K (−260 to 100°C). The National Institute of Standards and Technology (NIST) will calibrate probes to as low as 0.65 K.

The popular and inexpensive thermistor is made of a sintered mixture of metal oxides, and can be quite small. Thermistors show a large decrease in

Fig. 2. Detail of the sensor end of a standard platinum resistance thermometer (SPRT).

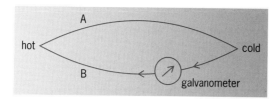

Fig. 3. Basic thermocouple, composed of dissimilar metals, A and B.

resistance as the temperature increases, and are useful in the range of −200 to 1000°C. An inexpensive indoor/outdoor thermometer uses two thermistors and has a liquid-crystal display (LCD) reading to 0.1°C (accuracy about 2°C).

Standard bead-in-glass thermistors using a digital bridge have been tested which are accurate to 0.005°C (5 mK) in the range from 0 to 100°C, and with a resolution of 0.0001°C (0.1 mK). These new, highly stable thermistors are a practical, and less expensive, alternative to platinum resistance thermometers. One type of thermistor is sensitive to 0.05 mK.

Bimetallic thermometers. The inexpensive bimetallic thermometers consist of two strips of dissimilar metals bonded together, normally in a coil. When heated, the coil twists, and a needle mounted at one end is used as an indicator. Large-display dial thermometers use bimetallic sensors.

Thermocouples. T. J. Seebeck discovered the phenomenon of thermoelectricity in 1821. A thermocouple is made up of two dissimilar metals whose ends are joined (mechanically or welded) as in **Fig. 3**. An electric current flows when the junctions are at different temperatures, and the voltage generated is a function of the temperature difference of the junctions. One junction is used as a reference junction and the other as the indicator. Relatively inexpensive readout devices using thermocouple probes are available. The signals are small—of the order of 4 mV/100°C for a copper/constantan thermocouple. Devices called thermopiles have thousands of junctions and are quite sensitive. Depending on the thermocouple, temperatures may be measured from −190 to 2300°C.

Infrared thermometers. Infrared thermometry measures temperature by using the radiant energy emitted from a heat source. Planck's law is an equation relating radiant energy emitted by an object, as detected by various sensors, to the temperature of the object. These devices are used for measuring temperatures well above 750°C, such as those found in blast furnaces, and they read temperature from a distance. Fiber-optic thermometers use a detector at the end of an optical fiber; they are useful in the range from 100 to 4000°C.

Fluorescence thermometers. One modern thermometer uses the fluorescence of a special film as a temperature sensor. This film is mounted at the end of an optical fiber to detect and convert the optical signals to a temperature reading. These thermometers can

measure temperatures from as low as −200° to over 700°C.

Quartz thermometers. The mineral quartz has different frequencies of vibration depending on how it is cut. One cut yields frequencies independent of temperature, and is used to maintain time in quartz watches. A special cut produces a crystal where the frequency is a linear function of the temperature. A quartz thermometer (all digital) is useful in the range from −80 to 250°C, and needs only one calibration point. Using two probes allows differential measurements to as low as a few millionths of a degree.

Thermal noise thermometers. These have been used to measure temperatures between 0.09 and 0.4 K. An R-SQUID (resistively-shunted quantum interference device) noise thermometer, believed to be accurate to better than 0.3%, spans the range from 6.3 to 650 mK.

Nuclear resonance thermometers. Nuclear quadrupole resonance spectroscopy has been used for ultralow-temperature thermometry. In particular, self-calibrating millikelvin and microkelvin thermometers have been developed that are based on intensity ratio measurements of magnetic resonance lines. This technique provides a direct measure of the nuclear spin temperature and yields reliable first-principles thermometry at ultralow temperatures.

For background information *see* CARNOT CYCLE; ELECTRICAL NOISE; FIBER-OPTIC SENSOR; FLUORESCENCE; GAS THERMOMETRY; HEAT RADIATION; LOW-TEMPERATURE THERMOMETRY; NUCLEAR QUADRUPOLE RESONANCE; PYROMETER; SQUID; TEMPERATURE; TEMPERATURE MEASUREMENT; THERMISTOR; THERMOCOUPLE; THERMOMETER in the McGraw-Hill Encyclopedia of Science & Technology.   Rubin Battino

Bibliography. P. R. N. Childs, *Practical Temperature Measurement*, Butterworth-Heinemann, Oxford, 2001; K. T. V. Grattan and Z. Y. Zhang, *Fiber Optics Thermometry*, Chapman & Hall, London, 1995; I. M. Klotz and R. M. Rosenberg, *Chemical Thermodynamics: Basic Theory and Methods*, 6th ed., Wiley, New York, 2000; W. E. K. Middleton, *A History of the Thermometer and Its Uses in Thermometry*, Johns Hopkins University Press, Baltimore, 1966; H. Preston-Thomas, The international temperature scale of 1990 (ITS-90), *Metrologia*, 27:3–10, 1990; J. F. Schooley (ed.), *Temperature: Its Measurement and Control in Science and Industry*, vol. 6, pts. 1 and 2, American Institute of Physics, 1992.

Therapeutic protein synthesis

Transgenic animals can be used to make large quantities of complex human therapeutic proteins. These animals are genetically engineered by the addition of a gene designed to direct the synthesis of a foreign protein without affecting the health of the animal. Some common applications for these medicinal proteins include the treatment of congenital deficiencies or augmentation of the naturally occurring protein during life-threatening disease states such as sepsis (blood poisoning) or trauma from accidents and surgery. For example, transgenic animals may become a major source of therapeutics used to treat the common forms of hemophilia.

Only certain human and animal cells are capable of producing functional recombinant (genetically engineered) versions of complex proteins. Plant, insect, and taxonomically lower-level cells, such as yeast and bacteria, do not possess sufficiently sophisticated biochemical machinery to make many human proteins that are important therapeutics. Furthermore, after extensive research only a few candidate mammalian cells have been identified as having the specialized characteristics needed for both synthesis and secretion of the recombinant protein. Mammals most often used to produce recombinant proteins are rabbits, pigs, sheep, goats, and cows.

Traditional production methods. Traditionally, therapeutic proteins have been collected from human tissues in which they are naturally found, such as blood plasma, or produced by genetically engineered animal cells grown in stainless steel bioreactors. One restriction of this method is that the mammalian cells that are suited to be grown at production scale must be capable of cancerlike growth to sufficiently populate the bioreactor under culture conditions. However, even with immortalized growth characteristics, an increase to commercial production levels is difficult because the technology does not yet exist to grow these cells in concentrations high enough to produce large amounts of harvestable protein. Thus, production capacity for recombinant human proteins has been limited and difficult to expand.

Current production methods. Currently, however, the most productive method for making a pharmaceutical protein is to engineer an animal to secrete the protein into its milk. Many of the same molecular biological and biochemical techniques used to make recombinant proteins in animal cells grown in culture are used to make transgenic animals that make the protein in milk. In contrast to the cancerlike growth characteristics needed to produce large quantities of recombinant proteins in bioreactor systems, healthy normal animal tissues with very high cell densities, such as those found in the liver and mammary gland, are capable of making very high levels of harvestable protein. The mammary gland can make complex recombinant proteins at about 1 gram per liter per hour compared with about 0.005 gram per liter per hour or less for mammalian cell culture bioreactors.

Importantly, both liver and mammary cells have sophisticated internal biosynthetic machinery that allows them to make and secrete a diversity of proteins at high levels. The mammary gland has the additional advantages of exportation of the protein via milk, making exsanguination of the animal to harvest blood or tissue unnecessary. Thus, the production of a recombinant protein in the milk of a transgenic mammal combines the productivity and harvestability of high levels of protein in milk with the quality of animal cell biochemistry. Cows, goats, sheep,

pigs, and rabbits are all capable of producing sufficient quantities of milk to produce certain recombinant proteins at industrial scale. While the mammary gland of each species will have different biosynthetic capabilities, the choice of production animal is made by matching both the biochemical complexity and clinical market demand of the needed therapeutic protein to the specific animal's biochemical abilities and milk production volume.

Structure and function of milk genes. Harnessing transgenic livestock as sources for making therapeutic proteins requires precise genetic control to direct protein synthesis exclusively into milk. Over the past 15 years, the relationship between the structure and function of milk genes that is necessary for this task has been well characterized. Several examples of different deoxyribonucleic acid (DNA) control elements taken from natural milk genes have been used to precisely target gene activity to the mammary gland. Recombinant proteins have been produced in the milk of cows, goats, sheep, pigs, and rabbits using the control of sheep beta-lactaglobulin, mouse whey acidic protein, and cow alpha-S1-casein milk genes. Because milk gene structure and control is highly conserved among mammals, the control elements from these genes have been used to direct the synthesis of recombinant proteins into the milk of animals other than those from which the gene was derived. For example, the DNA control elements derived from the mouse whey acidic protein gene have been used to very efficiently direct the production of recombinant human protein C (a clotting inhibitor) and Factor IX (a clotting promoter) into the milk of transgenic pigs. (This makes genetic sense, since a milk protein similar to the mouse whey acidic protein has been recently discovered in the pig.) Thus, using the control elements of the foreign gene, the pig expresses the recombinant proteins in the mammary gland and only during lactation.

Transgenesis methods. Several methods can be used to make transgenic animals.

Microinjection. The most common method, microinjection, directly introduces the transgene (foreign DNA) into the nucleus of an embryo at the one-cell or two-cell stage of development. The embryos, which are microinjected with about 500–1000 copies of the transgene, are then transferred into a surrogate mother that is hormonally synchronized to become pregnant. About 1–10 copies of the transgene are usually successfully integrated into the recipient animal's chromosomes. Sometimes the integration can occur on more than one chromosome. It is believed that the natural mechanisms for chromosomal repair and maintenance are responsible for integration of the transgene into the hereditary makeup of the animal. In most cases, the integration of the transgene into the embryo's chromosomes at the one-celled stage of development results in transmission of the transgene to both somatic and germ tissues of the animal.

Making transgenic animals by microinjection requires the harvesting of large amounts of embryos,

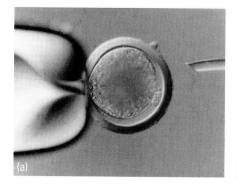
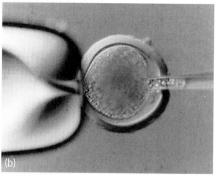
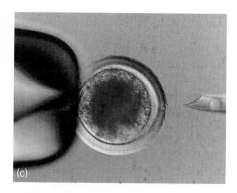
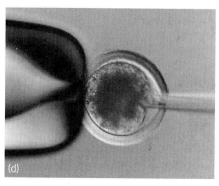
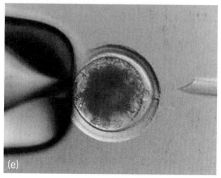

Making a transgenic animal using nuclear transfer. (*a*) A nontransgenic, one-celled pig embryo about 110 microns in diameter is immoblized on a suction pipette to prepare for enucleation. (*b*) The nucleus and a small amount of cytoplasm are removed using gentle suction by a micropipette. (*c*) A micropipette containing a nucleus aspirated from a transgenic donor fetal cell is positioned prior to injection into the enucleated pig embryo. The donor cell was a fetal cell made transgenic by introducing engineered DNA into a multitude of host cells propagated by tissue culture. (*d*) The microinjection pipette is inserted into the enucleated embryo and the nucleus is transferred. (*e*) The microinjection pipette is removed, and the reconstituted embryo is further cultured along with others in order to assess viability. The viable embryos are then transferred into a surrogate pig mother that is hormonally synchronized to become pregnant. (*Photomicrographs courtesy of Revivicor Inc.*)

since less than about 10–20% of the animals born are transgenic. Thus, microinjection is better suited for making transgenic pigs, mice, and rabbits because these species can be hormonally synchronized and induced to produce many ova (superovulation) that are suitable to produce many viable embryos.

Nuclear transfer of transgenic cells. Bulk methods of creating transgenic animals utilize nuclear transfer into embryonic cells from transgenic donor cells. Transgenesis of the donor cells usually is staged by bulk delivery of DNA into the tissue culture media while the cells are being propagated or sustained in mass. For example, the cultured cells can be bathed in a mixture of foreign DNA and chemicals that encourage uptake of the DNA by the cells. Alternatively, this DNA bath can be combined with electric pulses that allow the DNA to pass into the cell by momentarily disorganizing the cell membrane. The cultured cells can be embryonic cells or cells excised from fetal or somatic tissues. Since select cells taken from embryos or from fetal animal tissue can be cultured and expanded for up to tens of cell generations, this method provides many cells for the transgenesis procedure, increasing the probability of obtaining a transgenic cell. Fluorescent in-situ hybridization (FISH) can be used to identify the transgenic cells by exposing the cells to fluorescently labeled DNA that selectively hybridizes with the integrated transgenes. Next, the nucleus of a nontransgenic embryo

or ovum is removed (enucleation) and replaced with the transgenic cell nucleus (nuclear transfer) [see **illus.**]. The transgenic embryo or ovum is then transferred into a surrogate mother.

Posttranslational modifications and circulation half-life. Therapeutic proteins are diverse in their structural requirements needed for biological activity. Their biological functions vary from enzymatic activity to inhibition of enzymatic activity. The structural requirements most critical to imparting functionality are often produced by posttranslational modifications (PTMs) of certain amino acids contained within the polypeptide backbone of the protein. PTMs of amino acids, such as the glycosylation of asparagine and serine, gamma-carboxylation of glutamic acid, phosphorylation of serine, and sulfation of tyrosine, are important for proper functioning of proteins such as Factor IX, which is used to treat type B hemophilia.

With the exception proteins like fibrinogen, which is used to make the fibrin glue of blood clots, most proteins must circulate in the bloodstream in order to reach their target tissues or to continually perform their function. Thus, circulation half-life is an important part of protein functionality. A chief determinant of circulation half-life is the glycosylation structure attached to amino acids such as asparagine and serine. Glycosylation structures consist of sugar molecules that have been polymerized. In particular, the presence of *N*-acetylneuraminic acid (NANA) in the terminal positions of the branched chains of the glycosylation structure is important for interactions with receptors within the human liver. These receptors act to recognize, capture, and remove proteins with NANA deficiency from circulation. Thus, NANA content can be a central determinant of the circulation half-life of blood plasma proteins.

However, the glycosylation structures made by a given host animal are species-specific. For example, the mammary glands of ruminants tend to add less complex glycosylation structures, lacking NANA, that are called high oligomannosidic glycosylation structures, because they contain mostly mannose and are deficient in other sugars such as fructose and galactose. Proteins with oligomannosidic structures are thought to have inherently shorter circulation half-lives. In contrast, the pig mammary gland tends to synthesize proteins with more complex structures containing NANA that will likely have longer circulation half-lives. Although cows and sheep tend to produce less complex glycosylation patterns, they can be very desirable sources of proteins that do not need to circulate in the blood to perform their therapeutic function, such as fibrinogen (which is used to control bleeding during surgery) or alpha1-antitrypsin (which is inhaled as a localized inhibitor of elastase, an enzyme that destroys the protein necessary for lung elasticity).

The quality of the PTMs occurring in transgenic mammary cells is ultimately judged by the efficacy of the protein in human clinical trials.

Human clinical trials. The production of protein therapeutics in the milk of transgenic animals has advanced to the level of evaluation in human clinical trials. There are three proteins currently in clinical trials that are produced in the milk of transgenic animals (see **table**). These are alpha1-antitrypsin made in the milk of transgenic sheep, used to reduce lung elasticity degradation resulting from emphysema-induced inflammation; antithrombin III made in the milk of transgenic goats, used to reduce clotting complications during cardiac bypass surgery; and alpha-glucosidase made in the milk of transgenic rabbits, used to treat congenital deficiency of glycogen storage occurring in Pompe's disease. Several other clinical trials are planned for the near future to evaluate

the efficacy of Factor IX made in the milk of transgenic pigs and fibrinogen made in the milk of transgenic cows.

For background information *see* BIOTECHNOLOGY; GENE; GENETIC ENGINEERING; PROTEIN; SOMATIC CELL GENETICS in the McGraw-Hill Encyclopedia of Science & Technology.

William H. Velander; Kevin Van Cott

Bibliography. A. G. Bijvoet et al., Human acid alpha-glucosidase from rabbit milk has therapeutic effect in mice with glycogen storage disease type II, *Hum. Mol. Genet.*, 8:2145–2153, Nov. 8, 1999; A. Colman, Dolly, Polly and other 'ollys': Likely impact of cloning technology on biomedical uses of livestock, *Genet. Anal.*, 15:167–172, 1999; J. D. Pollack et al., Transgenic milk as a method for the production of recombinant antibodies, *J. Immunol. Meth.*, 231:147–157, 2000; W. H. Velander, Shepherding new medicines from biotechnology: Using the milk of transgenic livestock to treat hemophilia, *Virginia Tech Scholarly Rev.*, no. 1, March 2003; W. H. Velander, H. Lubon, and W. Drohan, transgenic livestock as drug factories, *Sci. Amer.*, 276(1):54–58, January 1997; W. Velander and K. Van Cott, Protein expression using transgenic animals, in V. A. Vinci and S. R. Parekh (eds.), *Handbook of Industrial Cell Culture: Mammalian, Microbial, and Plant Cells*, Human Press, Totawa, NJ, 2002.

Time-reversed signal processing

In applications of remote sensing—including radar, sonar, biomedical imaging, and nondestructive evaluation—the main intent is the detection, localization, and identification of distant objects or features that either scatter or generate electromagnetic or acoustic waves. Although the disparity in objectives between military-radar and ultrasonic-imaging systems may be considerable, they have in common that random wave scattering and diffraction tend to limit the accuracy and confidence with which such systems can be used and the distances over which such systems can operate. For radar and sonar, turbulence and wave motions in the ocean or atmosphere, or rough terrestrial or ocean surfaces, may cause such random wave scattering and diffraction. In ultrasonic remote sensing, the random scattering and diffraction may be caused by grain structure within metals or by variations between and within tissues. Time-reversed signal processing is emerging as means to enhance the performance of a wide variety of remote sensing systems through compensation for the deleterious effects of random wave scattering and diffraction.

Applications. Exploitation of time reversal in wave propagation problems dates back to the early 1960s, when similar concepts arose in the fields of geoacoustics, radio waves, and underwater sound propagation. In 1965, Antares Parvulescu and Clarence Clay published a study in which acoustic signals were recorded with a single transducer 20 nautical miles

Therapeutic proteins made in the milk of transgenic animals

Recombinant protein	Transgenic animal source	Native human source
Alpha1-antitrypsin*	Sheep	Plasma
Alpha-glucosidase*	Rabbit	Intracellular (that is, muscle)
Antithrombin III*	Goats	Plasma
Fibrinogen	Cows	Plasma
Factor IX	Pigs	Plasma
Monoclonal antibodies	Goats	Hybridoma cell culture

*Currently in human clinical trials.

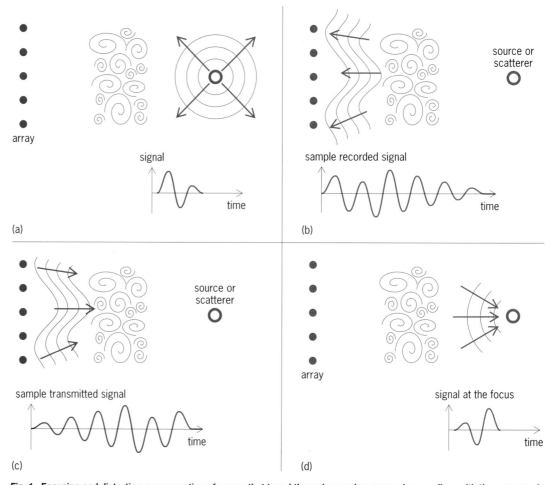

Fig. 1. Focusing and distortion compensation of waves that travel through an unknown random medium with time-reversed signal processing. (*a*) Step 1: signal generation. (*b*) Step 2: ordinary recording. (*c*) Step 3: time-reverse playback. (*d*) Step 4: focus formation.

(37 km) from a sound source in water that was approximately 1 nautical mile (1.85 km) deep. The recorded signals were time-reversed, and retransmitted through the ocean. The signal received after the second transmission was much clearer and far less distorted than the signal received after the first transmission. Since that time, the sophistication of transmitters, receivers, and signal processing algorithms has allowed time reversal concepts to permeate nearly every application of remote sensing that relies on wave propagation to or from an object or feature of interest. Components for aircraft engines have been inspected with time-reversing ultrasonic arrays. The U.S. Navy is considering time-reversal concepts for new active sonar systems and underwater communication links. Time-reversal concepts are now being applied in biomedical diagnostic ultrasound to increase the clarity of images and in therapeutic ultrasound to better target tumors and kidney stones. Feasibility studies are getting underway for radar systems that incorporate time reversal to enhance operating distances and detection of targets under foliage, and the future of time-reversed signal processing is likely to include structural monitoring of buildings and machines.

Time-reversal process. The basic process of time reversal can be described by four steps (**Fig. 1**). First, waves are generated or scattered by an object or feature of interest and travel forward through the environment to an antenna or array of transducers. The paths that waves follow between the object and the array may be complicated and unknown. Second, the array records the signal in the usual manner. These recordings will include signal distortion from echoes, scattering, and diffraction in the environment. Third, the signal recorded by each element in the array is retransmitted from that element with the direction of time inverted; the end of the signal is transmitted first, and the start of the signal is transmitted last. In the final step, these array-transmitted time-reversed waves travel backward through the environment, retracing their paths to converge at the location where they originated. Although specific applications of time reversal typically involve more steps and greater processing of the array-received signals, all are based, directly or indirectly, on these simple steps.

When time reversal is working properly, the waves that return to the location of their origin focus tightly and are undistorted, even though they may have

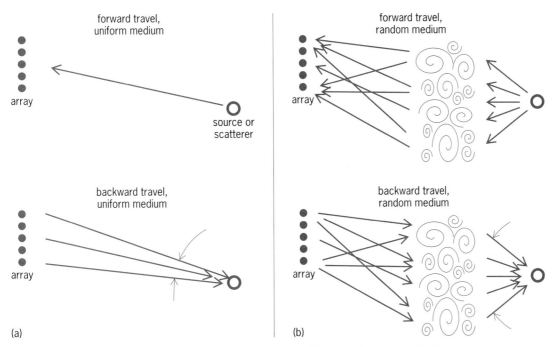

Fig. 2. **Time-reversed focusing.** (*a*) **Without random wave scattering.** (*b*) **With random wave scattering.**

passed through a complicated environment that generates echoes and causes random scattering and diffraction. In fact, the size of the focal region may be much smaller when the environment is complicated than when it is uniform in all directions. Both the tight focusing and distortion-removal capabilities of time reversal are of interest in remote sensing.

Distortion removal and focusing. The time-reversed waves are able to remove distortion and focus well when the transducer array is many wavelengths long, the absorption of wave energy by the environment is weak, background noise is low, and the environment changes little (or not at all) between forward and backward wave travel. The process by which time reversal accomplishes distortion removal may be understood by considering an environment containing three travel paths—fast, middle, and slow—with different travel times between the object and the transducer array. In this case, a signal that starts at the object will be received at the array as three possibly overlapping signals. This phenomenon is called multipath distortion, and it occurs when there are well-defined wave-travel paths and when there is wave scattering from particles, turbulence, or other fluctuations between the object and the transducer array. In the three-path environment, the time-reversed broadcast will launch the signal on the slow path first, the middle path second, and the fast path last. Here, the timing of the signal launches will exactly undo the multipath distortion, and the signal will arrive undistorted back at its place of origin.

The superior focusing characteristics of time-reversed waves come from their ability to exploit reflection, scattering, and diffraction within the environment. The range of angles through which waves converge determines how tightly the converging

waves focus. If all the time-reversed waves come from the same direction, the focus is larger than if the waves converge from above, from below, and from either side. When the environment is uniform, there is only one travel path between the object (a source or scatterer) and each element of the transducer array, so the focus size of the time-reversed waves is determined by the array's angular aperture. When the environment produces echoes, random scattering, and diffraction, there may be many travel paths between the object and the transducer array, and some of these paths can be used by the time-reversed waves to increase the range of convergence angles occurring during backward travel to decrease the focus size compared to what would occur in a uniform environment (**Fig. 2**). However, when the environment changes between forward and backward wave travel, the advantages of time reversal may be degraded or even lost.

Techniques. There are several signal processing techniques that have been used to convert time reversal from an alluring oddity of wave physics into a useful remote sensing tool. These techniques generally fall into two categories: those that require an active transducer array that can transmit and receive waves, and those that merely require a passive receiving array.

Active techniques. The most popular of the active techniques, referred to by the French acronym DORT (décomposition de l'operateur de retournement temporel; decomposition of the time-reversal operator), was developed by Claire Prada and Mathias Fink for nondestructive evaluation and biomedical ultrasound applications. This method can be used to detect and separately illuminate distinct scatterers. In its simplest implementation, DORT requires the measurement of the object-scattered signal at every array element when the object is separately illuminated by each array element. This carefully constructed measurement matrix can then be mathematically analyzed to determine the number or scattering objects, and the directions toward the objects in an environment with some random scattering and diffraction but no strong echoes. In such an environment, an image of the scattering objects may be formed using DORT.

A second active technique may be combined with DORT and involves identifying and selectively time-reversing segments of the received signal that may contain scattered wave energy from the object or feature of interest. When the correct signal segment is time-reversed and transmitted, the waves that return to the transducer array will concentrate or peak at one particular time when an item of interest is present. When the rebroadcast signal segment corresponds to an ordinary section of the medium, the returning waves will not concentrate at any particular time. This technique has been used to detect the presence of internal flaws in titanium, an important aerospace material.

Other active array techniques have been proposed for underwater acoustic communication, security barriers, and reverberation reduction in active sonar systems.

Passive techniques. The passive array signal processing techniques based on time reversal are also important. The best known of these is matched-field processing (MFP). Here, the array-received signals are delivered to a computer program that can simulate the backward wave travel. The computed location where the backward traveling waves converge is the presumed location of the source or scatterer. Thus, this technique can be used by the computer operator to locate remote sources or scatterers based on the array reception alone; a transmission by the array is unnecessary. Although MFP can theoretically locate or image objects with subwavelength accuracy, this technique cannot be implemented successfully without enough environmental information to ensure the accuracy of the computer model.

There are also passive time-reversed signal processing techniques that do not require any knowledge of the environment. One of these, passive phase conjugation (PPC), may be used for underwater communication between a cooperating remote source and a receiving array. If the source wishes to send a coded information stream to the array, it first sends a single pulse to characterize the multipath distortion of the environment. When the source starts its coded-information broadcast, the receiving array uses the measurements of the distorted single-pulse signal to unravel the distortion put into the coded message by the environment. Similar methods are used by modems to correct for variations in telephone lines. Another promising passive array technique, artificial time reversal, is similar to PPC but does not require the initial single-pulse broadcast.

For background information *see* ACOUSTIC SIGNAL PROCESSING; BIOMEDICAL ULTRASONICS; MATCHED-FIELD PROCESSING; NONDESTRUCTIVE EVALUATION; RADAR; REMOTE SENSING; SONAR; ULTRASONICS; UNDERWATER SOUND in the McGraw-Hill Encyclopedia of Science & Technology.　　David R. Dowling

Bibliography. J. Berryman et al., Statistically stable ultrasonic imaging in random media, *J. Acous. Soc. Amer.*, 112:1509–1522, 2002; D. R. Jackson and D. R. Dowling, Phase-conjugation in underwater acoustics, *J. Acous. Soc. Amer.*, 89:171–181, 1991; E. Kerbrat et al., Imaging in the presence of grain noise using the decomposition of the time reversal operator, *J. Acous. Soc. Amer.*, 113:1230–1240, 2003; W. A. Kuperman et al., Phase-conjugation in the ocean: Experimental demonstration of an acoustic time reversal mirror, *J. Acous. Soc. Amer.*, 103: 25–40, 1998; N. Mordant, C. Prada, and M. Fink, Highly resolved detection and selective focusing in a waveguide using the D.O.R.T. method, *J. Acous. Soc. Amer.*, 105:2634–2642, 1999; A. Parvulescu and C. S. Clay, Reproducibility of signal transmissions in the ocean, *Radio Electr. Eng.*, 29:223–228, 1965.

Tissue engineering

Tissue and organ transplantation has saved countless lives and significantly changed medical practice. New technologies aim to improve clinical treatment, addressing the difficulties of donor shortages and the effects of lifetime immunosuppressive drug therapy required for transplants. Biomaterials, engineered for implantation in biological systems, are able to replace or repair tissue function in some applications but cannot fully mimic normal tissue behavior. Tissue engineering is a field that has evolved over the past 10 years as a method for creating normally functioning tissues and organs to replace those lost due to disease, congenital abnormalities, or traumatic injury. Pioneers in the field bridged engineering and medicine, creating an interdisciplinary field requiring expertise in materials science and polymer chemistry, cell and molecular biology, biomedical engineering, and medicine. Today, practically every tissue or organ system in the body has been "engineered" at least to the proof-of-concept level. A few tissue products have been approved by the U.S. Food and Drug Administration (FDA) and have reached clinical practice in the area of skin and cartilage repair, with potentially many more to follow. While originating in the United States, the field has globalized to a large degree with strong stem cell and biomaterial research in Asia and significant clinical experience of tissue engineering technologies in Japan.

The general strategy of tissue engineering is to place cells on a biomaterial scaffold, a three-dimensional support for the cells that is designed to provide the appropriate signals to promote desired cell function and tissue development. The cell-scaffold construct can be incubated in vitro in

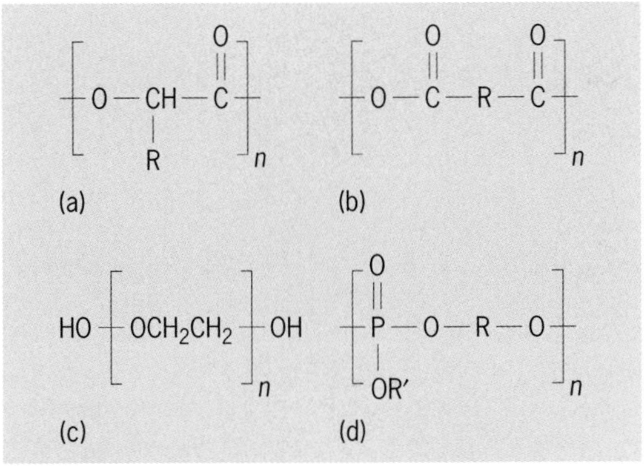

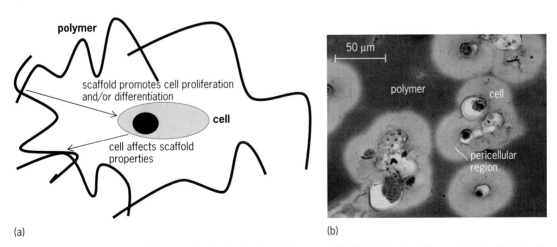

Fig. 1. Examples of commonly used polymers in tissue engineering: (a) poly(α-hydroxy esters), (b) poly-(anhydrides), (c) poly(ethylene glycol), and (d) poly-(phosphoesters).

bioreactors with a highly controlled environment or implanted directly in vivo to promote tissue formation.

Recent advances in polymer science and the discovery of both adult and embryonic stem cell capabilities have made a significant impact on tissue engineering and will further stimulate research in this field over the next few years. Moreover, as our ability to create more complex materials is combined with an increasing knowledge of the biological signals required to modulate stem cell phenotype and tissue morphogenesis, we will be able to rationally design tissue engineering systems that more closely mimic normal tissue properties and are clinically practical to use.

Intelligent materials. Polymers have enjoyed a long history in numerous biomedical applications ranging from artificial organs and limbs to medical devices. Polymers applied to the creation of tissue engineering scaffolds have unique requirements, including degradability (so that only functional tissue remains), compatibility with cells that need to proliferate or differentiate, and the ability to incorporate biological signals to control cell function. A large array of poly-

mers have been applied to tissue engineering, including poly(α-hydroxy esters), poly(phosphoesters), poly(ethylene glycol), and poly(anhydrides). In addition to the variable of polymer composition, scaffolds can be processed into structures such as highly porous foams, fibers, or gels to further manipulate tissue development. The chemical structures of polymers commonly used for tissue engineering applications are shown in **Fig. 1**.

Researchers are learning more about the mechanisms of cell-scaffold interactions and tissue development. This knowledge has provided the groundwork for synthesizing sophisticated scaffolds that are able to send complex biological signals to the developing tissue and respond specifically to the biological environment (**Fig. 2**). Understanding structure-function relationships between scaffold properties and cell behavior/tissue development is critical when designing tissue engineering systems.

Materials that can respond to cell behavior and their biological environment are described as intelligent or bioresponsive materials. The scaffold provides a protective three-dimensional environment for cells to develop and produce tissue. However, when tissue formation reaches a critical level, the scaffold can interfere with the new tissue's extracellular matrix organization and even reduce desired cell activities. Thus, a balance must be reached with tissue development and scaffold degradation rate. Cells from people of varying ages or disease states, or cells placed in the unique environments of an individual, will respond and develop tissue at different rates. Even implanting a cell-scaffold system underneath the skin versus the mechanically harsh environment of a joint will cause significant changes in the rate of tissue formation. If a scaffold degrades before the newly engineered tissue has reached a critical level of function, the construct will fail. Therefore, it would be ideal to create a scaffold system that degrades specifically in response to cell behavior and the rate of tissue development.

Fig. 2. The general principle of tissue engineering is to place cells on a scaffold that is designed to promote the desired cell phenotype and tissue development. (a) Intelligent scaffold materials can respond to their environment and tailor their properties based on cell behavior or the rate of tissue development. (b) Osteoblasts encapsulated in a bioresponsive material degrade the polymer (pericellular regions) as they secrete extracellular matrix.

An elegant work by M. P. Lutolf and colleagues used intelligent materials for bone regeneration. Specific peptide sequences that are responsive to degradative enzymes, adhesion peptide sequences that promote cell function, and growth factors that promote tissue formation were incorporated in hydrogels (polymer networks that swell in water). As cells migrate from the surrounding tissue and travel into the hydrogel, they secrete degradative enzymes that break down the gel, giving more space for migration and tissue deposition, and release growth factors to stimulate bone formation.

Bioresponsive materials for musculoskeletal tissue engineering have also been created using a simple polymer synthesis scheme. Hydrogels were created using biopolymers derived from materials normally found in the native cartilage extracellular matrix and from synthetic polymers with degradable phosphoester bonds, both of which are sensitive to enzymes naturally produced by bone- and cartilage-forming cells. The polymer degradation products also may be integrated into the matrix of the developing tissue. Figure 2b shows a hydrogel section containing bone marrow–derived mesenchymal stem cells. The cells are encapsulated in a hydrogel and then provided with the appropriate signals to become bone-forming cells. The region surrounding the cells is depleted of the degradable phosphate-containing poly(ethylene glycol), as seen by the loss of the violet background-staining characteristic of the polymer. Moreover, gene expression of typical bone markers increased in the presence of the phosphate-containing polymer.

Stem cells. The quality of the cell that is seeded on the tissue engineering scaffold is a critical component to the success of the system. The cell is responsible for secreting and organizing the matrix that forms the engineered tissue. Stem cells are a potentially powerful tool for tissue regeneration, as they can proliferate many times while still maintaining their ability to differentiate and produce tissue. Adult stem cells have been found in numerous places, including the bone marrow, muscle, and fat. Our knowledge of adult stem cell capabilities is continually increasing. For example, bone marrow–derived mesenchymal stem cells (MSCs) were originally discovered to differentiate into musculoskeletal-related tissues, such as bone, cartilage, tendon, and fat. Researchers have now shown that these adult stem cells can also differentiate into neural, liver, and skin cells, further expanding the potential for MSCs in tissue engineering.

Cartilage tissue lines the surface of bones in joints, providing a smooth gliding surface. Cartilage loss leads to significant pain during movement, and often joint replacement is required. MSCs have been encapsulated into photopolymerizing hydrogels for applications in cartilage and bone tissue engineering. When given the appropriate signals, the MSCs can be pushed toward a chondrocyte (cartilage cell) phenotype in the photopolymerizing hydrogels and secrete a matrix found in normal cartilage (**Fig. 3**). In the photopolymerization process used to encapsulate or trap the cells in the scaffold, light converts a liquid solution to a hydrogel. Using light to form hydrogels allows significant temporal and spatial control over the gelation process, while providing mild conditions amenable to cell encapsulation. Enough light is transmitted through tissues such that photopolymerization can occur through skin to create hydrogels in the subcutaneous space, or a light may be introduced through an arthroscope to reach the joint space in a minimally invasive manner.

While our understanding of how to control embryonic stem cells is still very limited, we are starting to see how they may be applied to tissue engineering. S. Levenberg and colleagues placed embryonic stem cells on poly(glycolic acid) meshes with Matrigel® (extracellular matrix extract from the Engelbroth-Holmes-Swarm mouse sarcoma) to create vascular networks. Others have devised differentiation schemes to push stem cells toward a neural phenotype before implantation, without using a

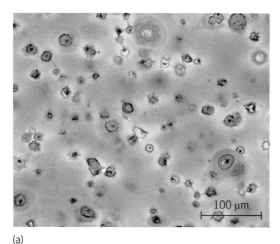
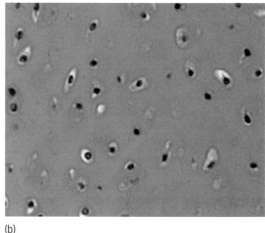

100 µm

(a)   (b)

Fig. 3. Bone marrow–derived mesenchymal stem cells encapsulated in a hydrogel that is formed using light. The cell-laden hydrogel is then incubated in conditions that promote cartilage differentiation. (a) The embedded cells secrete extracellular matrix similar to (b) native cartilage.

biomaterial scaffold. As stem cell research progresses the impact on tissue engineering could be substantial, from clarifying the basic mechanisms of stem cell biology to developing useful clinical technologies.

Tissue integration. As tissue engineering systems move from the laboratory to animal models and people, numerous challenges emerge. For cartilage, integration of engineered tissue with the surrounding native tissue is critical for implant function and long-term survival. This integration is impeded by the native structure of cartilage, which has a dense extracellular matrix without a blood supply, limiting the cell migration needed to heal the interface. The maturity of the engineered tissue is a factor that determines its integration potential. Moreover, strategies can also be engineered to integrate biomaterials into tissues.

Cartilage extracellular matrix is composed primarily of crosslinked collagen fibers with embedded, negatively charged proteoglycans (protein-polysaccharide complex). The collagen fibers are responsible for the tensile strength of the tissue and would be an ideal target for immobilization of a polymer scaffold matrix. Basic protein chemists at the National Institutes of Health have shown that when collagen is oxidized a tyrosyl radical is formed. This phenomenon has been applied to polymerizing and integrating a hydrogel scaffold. Collagen fibers are exposed on the cartilage surface by chondroitinase, which removes the proteoglycans. The surface can then be treated with a mild oxidative agent (2% hydrogen peroxide) to create tyrosyl radicals on the tyrosine residues located at the N-terminus of the protein. **Figure** 4 shows a hydrogel polymerized on the surface of a cadaveric cartilage using photopolymerization on the articular surface of a cadaver leg. The hydrogel fills the defect and is bound to the surrounding tissue through covalent linkages of the synthetic polymer to the tyrosine residues in collagen. This example of research demonstrates how tissue engineering integrates basic science, engineering, and clinical research to solve the problem of cartilage tissue repair and regeneration.

Outlook. Tissue loss in disease processes such as arthritis is progressive and relentless. Since there are currently no available means to replace the tissue, a person suffers until finally receiving an artificial implant to reduce pain and improve joint function and mobility. Novel tissue engineering technologies using sophisticated biomaterials and stem cells will allow for earlier treatment of tissue loss in simple, minimally invasive procedures. While numerous issues remain to be solved, it is clear that tissue engineering has sparked the interest of researchers in diverse fields and will ultimately change clinical practice.

For background information *see* ACQUIRED IMMUNOLOGICAL TOLERANCE; BIOMEDICAL CHEMICAL ENGINEERING; BIOMEDICAL ENGINEERING; BIOPOLYMER; CARTILAGE; CELL (BIOLOGY); COLLAGEN; POLYMER; STEM CELLS; TRANSPLANTATION BIOLOGY in the McGraw-Hill Encyclopedia of Science & Technology.
Jennifer Elisseeff

Bibliography. K. S. Anseth et al., Photopolymerizable degradable polyanhydrides with osteocompatibility, *Nat. Biotech.*, 17(2):156–159, 1999; J. Elisseeff et al., Transdermal photopolymerization for minimally invasive implantation, *Proc. Nat. Acad. Sci. USA*, 96:3104–3107, 1999; Y. Jiang et al., Pluripotency of mesenchymal stem cells derived from adult marrow, *Nature*, 418(6893):41–49, 2002; R. Langer and J. Vacanti, Tissue engineering, *Science*, 260:920–926, 1993; R. S. Langer and J. P. Vacanti, Tissue engineering: The challenges ahead, *Sci. Amer.*, 280(4):86–89, 1999; E. Lavik et al., Seeding neural stem cells on scaffolds of PGA, PLA, and their copolymers, *Meth. Mol. Biol.*, 198:89–97, 2002; S. Levenberg et al., Endothelial cells derived from human embryonic stem cells, *Proc. Nat. Acad. Sci. USA*, 99(7):4391–4396, 2002; M. P. Lutolf et al., Repair of bone defects using synthetic mimetics of collagenous extracellular matrices, *Nat. Biotech.*, 21(5):513–518, 2003; D. J. Mooney and A. G. Mikos, Growing new organs, *Sci. Amer.*, 280(4):60–65, 1999; B. E. Reubinoff et al., Neural progenitors from human embryonic stem cells, *Nat. Biotech.*, 19(12):1134-1140, 2001; R. C. Thomson et al., Polymer scaffold processing, in R. P. Lanza et al. (eds.), *Principles of Tissue Engineering*, pp. 263–272, R. G. Landes, Austin, 1997; X. Xu et al., Peripheral nerve regeneration with sustained release of poly(phosphoester) microencapsulated nerve growth factor within nerve guide conduits, *Biomaterials*, 24(13):2405–2412, 2003.

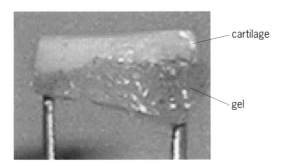

Fig. 4. Manipulating the tissue-biomaterial interface allows integration of biomaterials to the extracellular matrix of tissues. Cartilage can be pretreated so that the collagen fibers in the matrix covalently bind to a hydrogel biomaterial.

Tree of Life

Assembling the universal Tree of Life is an ambitious project that intends to provide a framework for what scientists interpret as the genealogy of life on planet Earth. The branch of biology involved in this tree-building processs is called phylogenetics, and it combines efforts from all areas of organismal biology together with mathematics and computer science. Phylogeneticists reconstruct phylogenetic trees (hypothetical genealogies of organisms) which are primarily derived from analyses of morphological, anatomical, physiological, developmental, behavioral, and molecular attributes assembled in data matrices. Molecular attributes include primarily deoxyribonucleic acid (DNA) and ribonucleic acid (RNA) sequence data, but also include the structure of proteins or genomes. Although molecular sequence data are at the forefront of assembling the Tree of Life, more traditional morphological data are still important, especially for fossils and extinct or rare species, as these are the only data available.

The Tree of Life project is a multidisciplinary international initiative that brings together previously independent laboratories that were addressing more narrowly focused phylogenetic questions. Such a large collective effort allows broad evolutionary questions to be addressed that would simply be beyond the scope of individuals or small teams to answer. An example of such collaborative effort was spearheaded by plant scientists, and the main objective now is to expand such broad collaboration to the remainder of the branches of the Tree of Life. The series of important biological achievements at the end of the twentieth century, led by the rapid DNA sequencing and annotation of the human genome, have set the foundation for the scientific community to start projects of even larger magnitude such as the Tree of Life. The main challenge is to integrate data collection on a massive scale, filling in major gaps in taxonomy for certain groups of organisms, such as microbes, and producing detailed data for a broad range of applications for science and society.

A Tree of Life is essential for a predictive understanding of biodiversity and tightly linked to successful expansion of postgenomic knowledge. Biodiversity affects our lives at a variety of levels. For example, a comprehensive Tree of Life can be used to predict functional diversity within ecosystems, set priorities for conservation, target biological control of invasive species, bioprospect for pharmaceutical and agrochemical products, or even track the origin and spread of diseases and their vectors. Since traditional systematic projects are insufficient to attain goals of the scale of the Tree of Life, a major international integrated effort to obtain data and tools needed for this world-leading resource is now being organized, so that the Tree of Life will also become the ultimate taxonomic reference.

General methodology. Generally, the approach to building such a large phylogenetic tree involves comparison of molecular attributes such as DNA sequences. Homologous (same) genes are sequenced for the largest possible sample of species, and a set of relationships between the species is hypothesized using either a maximum likelihood or a maximum parsimony method of computational phylogenetic inference. The maximum likelihood method starts with potential phylogenetic trees and a prespecified model of DNA sequence evolution, and assesses the probability that each tree would give rise to the observed data given the specified model of sequence evolution; the maximum parsimony method creates the simplest possible tree (that is, the one that assumes the fewest evolutionary events). Since these methods are extremely computer time intensive for very large trees, simple distances (that is, percentage similarity in DNA sequences) can be calculated, and species are linked in the tree in proportion to their evolutionary relatedness. Nonmolecular data can be used to build trees as well, using the same sort of similarity indices.

Tree of Plant Life. One of the first large-scale phylogenetic trees was based on ribosomal RNA and encompassed all major lineages of living organisms (**Fig. 1**); it provided a framework for much more detailed studies. For example, during the last decade botanists have produced several thousand phylogenetic trees based on molecular data, with particular emphasis on sequencing *rbcL*, the plastid gene coding for the large subunit of ribulose-1,5-biphosphate carboxylase/oxygenase (RuBisCO), the most abundant enzyme on Earth, which fixes carbon dioxide during photosynthesis. However, many other genes from the three plant genomes (plastid, nuclear, and mitochondrial) have also been used, and phylogenetic trees retrieved from these three genomes have been highly congruent, that is, in agreement with each other.

As a result, the general picture of land plants has changed. Flowering plants (angiosperms) are sister to gymnosperms (conifers, cycads, ginkgo and allies), both groups being sister to horsetails and ferns; altogether they form the leafy plants group (Euphyllophytina). Related to leafy plants are club mosses, mosses, hornworts, and liverworts (**Fig. 2**).

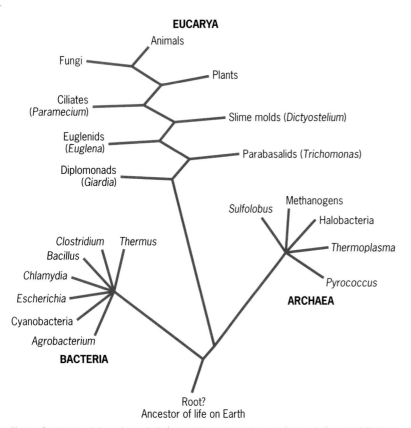

Fig. 1. Summary of the universal phylogenetic tree based on analyses of ribosomal RNA sequences by C. R. Woese and coworkers.

More specifically for flowering plants, the Angiosperm Phylogeny Group, an international botanical network of more than 30 laboratories, has aimed to objectively interpret published DNA-based phylogenetic trees and compile them into a hierarchical system at and above the level of family. The broad picture has changed, with the first split among flowering plants not being that of monocots versus dicots, as found in most botany textbooks, but between plants with one opening in their pollen grains versus three openings with the latter type now termed eudicots (Fig. 2). Ideas regarding angiosperm relationships have also changed. For example, the sacred lotus (*Nelumbo*) was always considered a close relative of the water lilies based on their morphological and habitat similarity, whereas based on DNA sequence data the lotus is in fact the closest relative of the plane tree and the Southern Hemisphere Protea family.

Tree of Animal Life. In zoology, the revolution of molecular phylogenetics together with modern morphological and anatomical studies using electron and confocal laser microscopy, and even more sophisticated x-radiographic computed axial tomography, has led scientists to reevaluate much of the classical animal phylogenetic work. The new era of comparative developmental biology, especially the integration of studies on cell lineage fate through embryonic development and gene expression data with other more traditional sources of information, has been extremely fruitful, and the application of rigorous phylogenetic methods has led to the current view of multicellular animal (metazoan) evolution (Fig. 2).

Unlike the plant scientists, zoologists have yet to integrate the work from multiple laboratories in a collective effort. Perhaps the lack of coordination is a reflection of the controversies that still remain about the relationships of animal phyla. Horizontal transfer of genes (the exchange of genetic material between very distantly related organisms) at this deep level of phylogenetic relationships makes precise links between lineages and rooting of this tree extremely difficult to assess, thus fueling the dissension. However, some consensus has been achieved. Sponges are recognized as the sister group to the remainder of the metazoans, followed by cnidarians (for example, jellyfish) and ctenophores (for example, comb jellies). It seems clear that animals with three distinct body layers (triploblasts) share a common ancestor, perhaps similar to living acoels, a group of worms previously classified with the flatworms. Within triploblastic animals, a major division is recognized between protostomes and deuterostomes. Deuterostomes include echinoderms (for example, sea urchins), hemichordates, and chordates (including vertebrates). Protostomes include most other animal phyla, such as wormlike animals, mollusks, and arthropods.

The protostome phylogeny has been challenged with the analysis of mostly ribosomal DNA sequences, and later corroborated by other genes. Since the time of the French naturalist G. Cuvier, most zoologists believed that a group of segmented

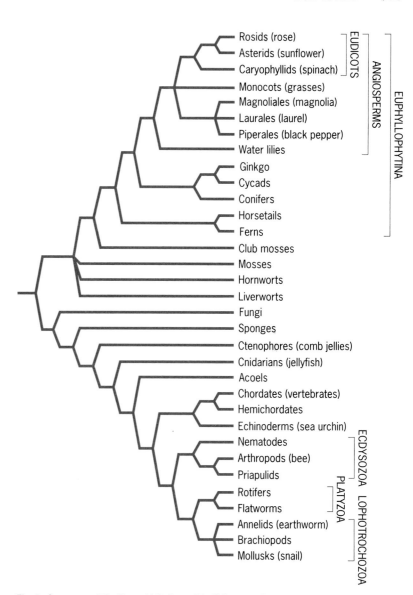

Fig. 2. Summary of the Tree of Life for multicellular organisms.

animals that included annelids (segmented worms, for example, earthworms) and arthropods belonged to the Articulata group. However, molecular data show that the segmented arthropods are not directly related to the annelids; instead they are closely related to nematodes (roundworms), priapulans, and the like. Tardigrades (microscopic, bilaterally symmetrical invertebrates commonly known as water bears) and kinorhynchs (microscopic free-living marine invertebrates) seem to play an important role in the reinterpretation of the morphological evidence in support of this new group known as Ecdysozoa. Widely accepted by the scientific community, Ecdysozoa includes all animal phyla that molt their cuticles at least once during their life cycle (for example, nematodes and arthropods).

Project completion. Several large continental initiatives have emerged, all aiming to make significant contributions toward assembling the Tree of Life in the next few decades. For example, the National Science Foundation in the United States has set up a

15-year program on the Tree of Life, awarding over $15 million in 2002. In Europe, an integrated project on similar topics has received the attention of the European Commission, and delegates of this initiative include virtually all phylogeneticists in Europe. Likewise, Australia and Canada are forming their own Tree of Life programs.

Assembling the Tree of Life is a difficult but essential task for scientific enterprise. The number of branches that need to be assembled can be counted in millions. Difficulties in assembling information of all species is evident; resolving the relationships among them is hard to envision due to the number of possible trees for these species. The number of possible trees connecting a mere 1 million species (much less than the known diversity on Earth) is $5 \times 10^{68,667,340}$. Fortunately, several algorithms allow us to concentrate on the most likely fraction of possible phylogenetic trees; however, the Tree of Life project will require new bioinformatic developments, on both the algorithmic and computation sides. It is hoped that biological and computer sciences will reach a level of interdependence rarely seen in other branches of science so that the Tree of Life will be assembled before the end of the century (but probably not as fast as was previously hoped).

For background information see ANIMAL SYSTEMATICS; CHEMOTAXONOMY; PHYLOGENY; PLANT PHYLOGENY; PLANT TAXONOMY; TAXONOMY in the McGraw-Hill Encyclopedia of Science & Technology.

Vincent Savolainen; Gonzalo Giribet

Bibliography. A. M. A. Aguinaldo et al., Evidence for a clade of nematodes, arthropods and other moulting animals, *Nature*, 387:489–493, 1997; Angiosperm Phylogeny Group, An update of the Angiosperm Phylogeny Group classification for the orders and families of flowering plants: APG II, *Bot. J. Linnean Soc. Lond.*, 141:399–436, 2003; G. Giribet, Current advances in the phylogenetic reconstruction of metazoan evolution: A new paradigm for the Cambrian explosion?, *Mol. Phylogenet. Evol.*, 24:345–357, 2002; A. H. Knoll, Life on a young planet: The first three billion years of evolution on Earth, Princeton University Press, 2003; D. R. Maddison, *The Tree of Life Web Project* ; K. M. Pryer et al., Horsetails and ferns are a monophyletic group and the closest living relatives to the seed plants, *Nature*, 409:545–648, 2001; A. M. Sugden et al., Charting the evolutionary history of life, *Science*, 300:1691, 2003; C. R. Woese, Interpretating the universal phylogenetic tree, *Proceedings of the National Academy of Sciences USA*, 97:8392–8396, 2000.

Ultrasonic trapping of gases

Acoustic levitation has been applied as a technique for suspending liquid and solid samples of micrometer-to-millimeter range in the pressure nodes of a stationary ultrasonic field (SUSF). This effect was discovered in the 1930s and mainly developed in the 1970s and 1980s by the National Aeronautics and Space Administration (NASA) and the European Space Agency (ESA), who used this technique to fix small samples for various zero-gravity experiments in their space laboratories. However, in the last two decades acoustic levitation has also been applied as a useful and powerful tool for handling small samples under terrestrial conditions, that is, with gravity. Several investigations based on acoustic levitation have been carried out in different research areas, such as fluid dynamics, materials, atmospheric sciences, and analytical chemistry.

Levitation of liquid and solid samples. Stationary ultrasonic fields can easily be arranged between an ultrasonic transducer and a concave reflector by emitting a sound wave along the central levitation axis perpendicular to the surface of the transducer. If the distance between the transducer and reflector is a multiple of half of the sound wavelength, resonance conditions are met and a series of nodes and antinodes of sound pressure and sound velocity occurs along the levitation axis. At high sound pressure levels in the range 150–180 dB, second-order forces in the acoustic field can be used to suspend small liquid and solid samples at each pressure node of the field (**Fig. 1**). This single-axis geometry of an acoustic levitator is convenient and provides easy access to the suspended samples, while a three-axis geometry offers stronger acoustic forces.

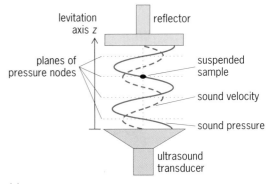

(a)

(b)

Fig. 1. Single-axis acoustic levitator. (*a*) Stationary waves of sound pressure and sound velocity between the ultrasound transducer and reflector as a function of distance along the levitation axis *z*. (*b*) Polystyrene balls of a diameter of approximately 5 mm (0.2 in.) suspended in the pressure nodes of the stationary ultrasonic field of a 20-kHz levitator.

Acoustic levitation is limited by the gravity of the sample, which has to be balanced by the acoustic levitation forces. However, W. J. Xie and B. Wie have recently demonstrated with a single-axis levitator that acoustic levitation is possible for very high density materials such as tungsten (specific density 18.92 g/cm³).

The size of the suspended samples is also limited by the stationary ultrasonic field and its geometry. The maximum diameter of the samples has to be smaller than half of the wavelength used for the ultrasonic field. For example, with an ultrasonic frequency of 20 kHz and at standard conditions of pressure and temperature of the gaseous environment (for example, 1013 kPa and 20°C, or 29.92 in. of mercury and 68°F) only samples with a diameter smaller than 8.5 mm (0.33 in.) can be levitated. The lower limit of sample diameter for acoustic levitation is given mainly by the viscosity of the gaseous environment and the sound frequency. Under standard conditions, this limit is of the order of a few tenths of a micrometer for typical frequencies of 20–100 kHz.

Trapping of heavy gases. In the late 1990s, it was shown that under suitable experimental conditions samples of cold gases, such as ice aerosol, are sucked into the stationary ultrasonic field, and gather in temporarily stable, rotational ellipsoidal systems around the pressure nodes of the field (**Fig. 2**). These systems are well separated in temperature from the ambient gaseous environment by a strong temperature gradient at the borders of the systems. Therefore, they can be called cold-gas traps.

The trapping of cold gases has been studied in more detail, using sample gases with various mass densities (such as octafluorocyclobutan, krypton, bromine, carbondioxid, nitrogen, and helium) and at various temperatures in the range from approximately −100°C (−148°F) to 100°C (212°F). In these experiments it was demonstrated that the effect is primarily controlled not by the difference in temperature but by the difference in density of the sample gas and the surrounding gaseous environment. Sample gases with higher mass densities than the gaseous environment are always sucked into the stationary ultrasonic field, while sample gases with lower mass densities are always displaced from this field. There is no evidence for a direct temperature effect in this process. Therefore, not only cold gases but also heavy gases gather in ellipsoidal systems around the pressure nodes in the stationary ultrasonic field, so that this phenomenon can be described more generally as heavy-gas trapping in stationary ultrasonic fields.

A first theoretical explanation of the effect was based on a model of acoustic levitation of liquid and solid samples. The difference in the density of the gases in the stationary ultrasonic field causes a levitation potential at the pressure nodes, which is attractive if the sample gas is heavier than the surrounding gas and repulsive in the opposite case. Besides the difference in densities, the effect depends mainly on the supply of the (heavy) sample gas and the sound pressure level of the stationary ultrasonic field. Both effects have been experimentally demonstrated. The effect is increased by a more rapid supply of sample gas into the stationary ultrasonic field, in the range 5–25 mL/s, and disappears within a second after the supply of sample gas is stopped. This abrupt termination is caused by gas convection and diffusion within the stationary ultrasonic field. Ultrasonic trapping of heavy gases has been detected using frequencies of 4, 20, and 58 kHz, and at sound pressure levels between approximately 120

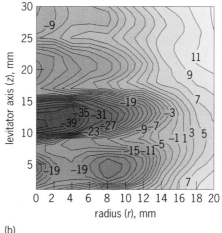

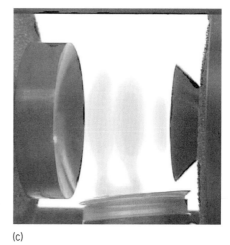

(a)   (b)   (c)

Fig. 2. Trapping of heavy gases. (*a*) Rotational ellipsoidal zones of cold-gas traps visualized by ice aerosol in the stationary ultrasonic field (SUSF) of a 20-kHz levitator. The ice aerosol, which was generated from ambient air using a metal rod precooled with liquid nitrogen, is sucked into the SUSF in the planes of the pressure nodes. There is a sharp boundary between the cold-gas traps and the ambient host medium. (*b*) Typical temperature field of cold-gas traps. The temperature in °C is plotted as a function of the radius *r* and the height *z* in the levitator axis of the SUSF. Gaseous nitrogen at approximately −100°C (−148°F) and a flow rate of 10 mL/s was used as the cold-gas source. (*c*) Trapped gaseous bromine in an SUSF of a horizontally positioned 20-kHz levitator. The heavy bromine at homogeneous temperature is sucked into the SUSF against gravity and forms rotational ellipsoidal zones. (*Reprinted from R. Tuckermann et al., Trapping of heavy gases in stationary ultrasonic fields, Chem. Phys. Lett., 363:349–354, 2002, with permission from Elsevier*)

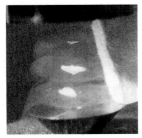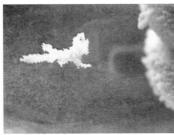

Fig. 3. Ice particles generated from primary ice aerosol in the stationary ultrasonic field of a 20-kHz levitator. The ice particles formed vary in size and shape as a result of different mechanisms and physico-chemical parameters involved in the particle formation process.

and 180 dB, with a maximum between 150 and 160 dB. The decrease of the trapping effect at higher sound pressure levels may be caused by an increase of acoustic streaming and convection inside the stationary ultrasonic field.

Ice particle formation. When cold ice aerosols are trapped in stationary ultrasonic fields in an ambient environment, the formation of larger ice particles from the primary aerosol particles can be observed. These secondary ice particles generated in the stationary ultrasonic field show a great variety of sizes (ranging from tenths of a micrometer to a few millimeters), shapes, and structures (**Fig. 3**). Four different mechanisms and effects are responsible for the ice particle formation in stationary ultrasonic fields:

1. *Acoustic agglomeration of primary particles.* This effect is well known for liquid aerosols, such as liquid water and ethanol fog, and can also be used for cleaning industrial exhaust fumes. The sound generates relative motion of the primary particles inside the stationary ultrasonic field depending on their size and shape and the acoustic field. The relative motion of the particles leads to a higher collision rate with other particles, a precondition of agglomeration.

2. *Quasi-liquid layers (QLLs) on the surfaces of the primary aerosol particles.* Due to acoustic agglomeration, the primary particles stick together by adhesive forces. The adhesion between the primary particles is caused by liquid layers on the surfaces of the particles (or is even better if the particles are completely liquid). Ice particles down to temperatures of approximately $-25°C$ ($-13°F$) are covered by such a liquid layer, so that agglomeration occurs. (The suggestion that ice crystals are covered by a thin layer of liquid water is attributed in part to Michael Faraday, and that idea has been developed during the past few decades.) At lower temperatures of the ice particles and also for dry aerosol particles, like flour dust or icing sugar, particle formation due to the stationary ultrasonic field has not been observed. This is due to the absence of the liquid layer which produces adhesive forces between the primary particles.

3. *Acoustic levitation.* Particles with sizes larger than a few of tenths of a micrometer are gathered and suspended in the pressure nodes of the stationary ultrasonic field because of the acoustic levitation forces. There ice particles with sizes up to a few

millimeters can be generated. The fixed location of particle formation in the pressure nodes of the stationary ultrasonic field allows these processes to be studied easily with physical and chemical methods.

4. *Dendritic growth of ice particles.* The ice particles can degrade or grow by sublimation or desublimation of water molecules from the gas phase due to a typical dendritic crystallization structure on the surface of the ice particles.

Scientific and technical potential. Acoustic levitation is a useful tool for the containerless handling of liquid and solid samples and for physical and chemical research. The trapping of cold and heavy gases in stationary ultrasonic fields offers the possibility of adjusting gaseous environments within the field with regard to the composition, species, and temperature of the gases and aerosols trapped around the pressure nodes of the field and the suspended liquid and solid samples. This can be useful not only in various technical applications but also for basic research in such areas as material science and atmospheric science. Acoustic levitation using the ice-particle-formation process in stationary ultrasonic fields seems to be a promising tool for physical and chemical research on ice particles and snowflakes. The trapping effect for cold and heavy gases can be used to develop miniaturized laboratories for gas or particle-gas reactions without any wall contact. But it is questionable whether the trapping effect is strong enough to be useful in gas separation processes.

For background information *see* ACOUSTIC LEVITATION; AEROSOL; CRYSTAL GROWTH; NONLINEAR ACOUSTICS; PARTICULATES; ULTRASONICS in the McGraw-Hill Encyclopedia of Science & Technology.

Rudolf Tuckermann; Sigurd Bauerecker

Bibliography. M. Barmatz and P. Collas, Acoustic radiation potential on a sphere in plane, cylindrical, and spherical wave fields, *J. Acous. Soc.*, 77:926–945, 1985; S. Bauerecker and B. Neidhart, Cold gas traps for ice particle formation, *Science*, 282:2211–2212, 1998; S. Bauerecker and B. Neidhart, Formation and growth of ice particles in stationary ultrasonic fields, *J. Chem. Phys.*, 109(10):3709–3712, 1998; W. J. Xie and B. Wie, Parametric study of single-axis acoustic levitation, *Appl. Phys. Lett.*, 79(6):881–883, 2001; R. Tuckermann et al., Trapping of heavy gases in stationary ultrasonic fields, *Chem. Phys. Lett.*, 363:349–354, 2002.

Urban geology

Urban geology is a branch of geology that emerged in the mid-1960s with the growing concern about the impact of urban sprawl on the environment. What differentiated urban geology from prior geological studies in urban areas was the involvement of diverse branches of earth sciences, such as engineering geology, geomorphology, hydrogeology, and geochemistry, to study the processes related to the equilibrium of the geological environment and how human activities can upset this equilibrium.

Information components. A major revolution in urban geology studies occurred in the mid-1980s with the arrival of desktop computers and geographic information systems (GIS) capable of processing large georeferenced data sets. In parallel, many producers of geoscience data were releasing digital versions of geological maps and databases, along with digital topographic maps (digital elevation models, or DEM). The availability of geoscience information in digital format permitted the integration of these documents to produce derived geological maps and models required for regional planning.

Major sources of information for the production of geological models are surficial and bedrock geology maps, DEMs, and borehole stratigraphic logs with accompanying regional studies. The role of the urban geologist is to translate documents that were produced for various purposes into a coherent system that can be understood and used by nongeoscientists. Data derived from geoscience source documents are reconfigured into thematic maps, cross sections, and three-dimensional models. Thematic maps provide information on various aspects of the regional geology, such as areas of potential natural hazards (such as landslides), suitability for urban development, and identification of natural resources. Cross sections provide a vertical view of the stratigraphy of geological formations between two points on a map; the **illustration** shows how stratigraphic sections are derived from source documents. The third type of data derived from source documents are three-dimensional models, which provide a global view of the regional geology.

Issues in urban development. The purpose of urban geology studies is to provide engineers, planners, decision-makers (various levels of government), and the public with the geoscience information required for sound planning of densely populated areas. Sound planning is making best use of the land, including preventing pollution of soils and ground water, mitigating natural hazards, preserving natural resources, and defining the best practices in regional infrastructure development. Urban geology studies must include the cities and the surrounding areas, as urban development can create an impact on the environment at the regional scale.

Pollution. Discharge of harmful substances into the environment can contaminate soils and ground water, making the land unsuitable for urban development and ground water unfit for use. Pollution can have many sources; some are accidental and hard to prevent, whereas others result from inappropriate regulations or practices. Accidental pollution, such as chemical spills due to road accidents or from leakage from holding tanks, reservoirs, or pipelines due to natural hazards, are impossible to predict. The role of urban geology in these cases is to mitigate the damages by assembling all the information (such as the nature of soils, drainage, stratigraphy, and ground-water flow) that is required to monitor where the spill can propagate, and to determine measures that can be taken to contain and eventually clean the polluted areas.

Accepted (nonaccidental) pollution results from a compromise between socioeconomic issues and protection of the environment. Examples of accepted pollution are burying of domestic and industrial waste; the use of chemical herbicides, pesticides, or fertilizers; and intensive livestock farming close to urban development. Urban geology studies can help in the planning of underground waste disposal by identifying lands underlain by relatively impermeable soils or bedrock formations. These impermeable layers will limit the penetration of contaminants into the local ground-water system. However, even with the best sanitary landfill practices, some pollution occurs because geological formations are never completely impermeable. These sites are of limited use in urban development because contaminated soils endanger public safety, and because methane gas, generated through decomposition of organic matter, will escape into the atmosphere for long periods unless extensive remedial actions are taken to capture it. Pollution caused by the use of chemicals and intensive livestock farming is more difficult to control as they cover large areas and rainwater runoff can carry organic and chemical contaminants into the local drainage system and ground water, making the water unsafe for human consumption.

Urban planning for accidental or controlled pollution relies on urban geology studies to define the best practices to avoid soil and ground-water contamination within the city and surrounding rural areas.

Natural hazards. Natural hazards are the conditions or processes (biological, geological, meteorological) in the natural environment which may cause disaster. Examples of natural hazards are earthquakes, floods, and landslides. Natural hazards can also be triggered by human intervention (anthropogenic) when the natural equilibrium of the geological environment is upset by human activities. Urban geology studies provide the required geoscience information to prevent anthropogenic geological hazards and to mitigate natural geological hazards. Planning for natural hazards in urban areas is of prime importance because of the possible impact on the safety of large populations.

Anthropogenic geological hazards can be avoided or reduced by identifying the impact of urban development on the original geological environment. Zoning regulations can specify the type of development that a region can sustain, and can flag possible

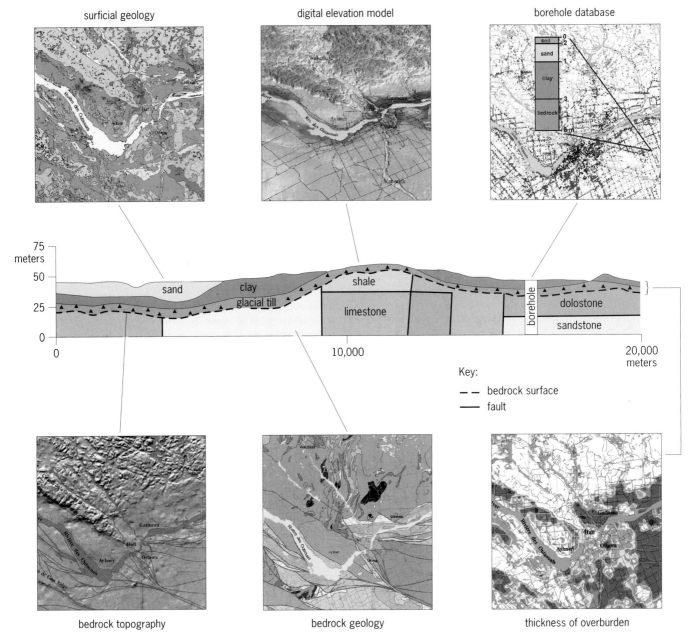

surficial geology

digital elevation model

borehole database

bedrock topography

bedrock geology

thickness of overburden

Construction of a stratigraphic cross section from source documents.

geotechnical problems. For example, areas suscepti-ble to slope failure (landslides, mudflows, rock falls, and rockslides) can be secured by remediation en-gineering work, zoned as recreational, or preserved in their original state. In areas flagged as susceptible to land subsidence, special measures can be taken to avoid lowering of the water table, and the proper design of underground tunneling or mining can be enforced to prevent the collapse of the overlying geological formations. Geological maps that include geotechnical and geochemical characteristics of sur-ficial and bedrock formations provide valuable infor-mation to planners and engineers for avoiding pos-sible biological or chemical reactions (such as the swelling of rocks) with building materials (such as concrete), or structural damage due to differential

settlement of the ground (such as buildings having foundations on two types of material that have dif-ferent physical properties).

Natural hazards, such as floods, earthquakes, or slope failure caused by severe earthquakes, usually cannot be predicted. The effects of these natural hazards on urban development can be mitigated by identifying the possible hazards that can occur in an area and by taking precautionary measures to reduce the loss of property and human life. Areas identi-fied as susceptible to flooding should not be zoned for urban development unless appropriate measures are taken to circumvent the problem. In earthquake areas, building codes should include regulations to have buildings and infrastructure (bridges, under-ground and aboveground transportation, energy corridors, and water and sewage systems) withstand ground disturbance. Areas of potential slope failure should be zoned as inappropriate for urban development unless prevention measures are taken. Another factor when zoning for seismic hazards is the sensitivity of the ground to disturbance. Buildings located on solid bedrock will require different structural design from those located on easily disturbed soils, such as sensitive clays.

Natural resources. The rapid increase in population and urbanization throughout the world has generated major concern about the sustainable development of natural resources. Natural resources can be nonrenewable such as minerals and fossil fuels, or renewable such as water, biodiversity, and agricultural resources. Urban geology provides information on the nature and location of underground resources such as aggregate materials (sand or stones used for fill or in concrete), ground water, and minerals, but regional planning relies on other sources of information for resources of nongeological origin, such as forestry or agriculture. Planning for natural resources implies the identification of all the natural resources in the urban and surrounding area, and development of plans and regulations to preserve, protect, and promote careful use of these resources.

Value of urban geology. The benefits of urban geology go beyond regulating regional planning and development since it has a net positive value to society, including government, the private sector, the academic sector, and the public. The prime users of earth sciences in the urban environment are engineering and environmental consulting firms that use the information in preliminary project planning and design work. Since site-specific investigations are expensive, urban geology maps and models are used to provide estimates of the work to be done and the problems to be anticipated, resulting in considerable savings to the private sector. Substantial savings are also reported by governments in planning highways, facilities, land use, and public servicing in the most appropriate locations based on the nature of soils and bedrock. Mitigation of natural hazards can also result in substantial monetary savings and prevent the loss of human lives. Educational and research bodies, especially universities, often use geoscience maps and models provided by urban geology for teaching or training purposes. The public is also a major user of regional geoscience information for safety and knowledge of the local environment.

For background information *see* CONSERVATION OF RESOURCES; ECOLOGY, APPLIED; ENGINEERING GEOLOGY; ENVIRONMENTAL MANAGEMENT; GEOGRAPHIC INFORMATION SYSTEMS; GEOMORPHOLOGY; LAND-USE CLASSES in the McGraw-Hill Encyclopedia of Science & Technology.   Robert Bélanger

Bibliography. P. F. Karrow and O. L. White (eds.), *Urban Geology of Canadian Cities,* Geol. Ass. Can. Spec. Pap. 42, 1998; R. F. Legget, *Cities and Geology,* McGraw-Hill, New York, 1973; G. J. H. McCall, E. F. J. De Mulder, and B. R. Marker (eds.), *Urban Geoscience,* AGID Spec. Pub. Ser., no. 20, 1996.

Visible Human Project

For centuries, the study and findings of anatomy have traditionally been recorded in anatomical atlases—books of two-dimensional pictures representing three-dimensional structures. The early atlases contained idealized illustrations of anatomical features to give the viewer some sense of their three-dimensionality. Many modern atlases use photographs of actual dissections. These often contain artist renderings in order to make the photographs more understandable. Still, all atlases rely on two-dimensional art forms to represent complicated three-dimensional structures.

Realizing the limitations of the printed page, David L. Bassett from 1948 to 1962 used a three-dimensional photographic technique, popularly known as View-Master® stereo pair transparency technology, to photograph his dissections. In this way, he was able to capture the depth and spatial orientation of the anatomical structures. However, the stereo pairs still limited the user to the orientation of the photographer at the time the picture was taken. There was no way to change one's point of view of the object, to view it from the side instead of the front, or to move in closer to see detail.

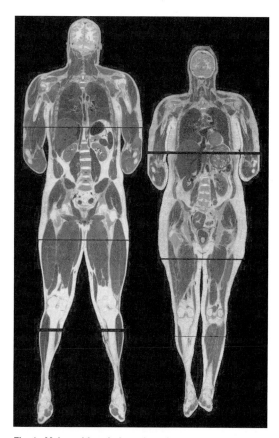

Fig. 1. Male and female frontal sections created by loading the Visible Human anatomy cross-section data into computer memory, aligning the sections in the memory, and then viewing the data as if a cut were made though the middle of the body from head to toe.

The Visible Human Project® of the U.S. National Library of Medicine (NLM) was designed to provide the sample-based data needed to reproduce in three dimensions any part of human anatomy down to 1 mm in detail (**Fig. 1**). To obtain the images that make up the Visible Human data sets, a male and a female cadaver were magnetic resonance imaging (MRI) scanned, then computerized tomography (CT) scanned, and finally frozen. Each frozen cadaver was filed down using a mill to produce a series of cross-sectional views of the body. Digital photographs were taken of the cadaver each time 1 mm was removed from the male and each time 0.33 mm was removed from the female to produce cross-sectional anatomy images.

The complete male data set contains 1871 cross-sectional anatomy images and is 15 gigabytes in size. The complete female data set contains 5189 cross-sectional anatomy images and is 39 gigabytes in size. Each anatomy image is made up of an array of dots (2048 dots wide and 1216 dots high), where each dot is one of 16,777,216 colors. Each CT image is 512 × 512 dots, where each dot is one of 4096 gray tones, and each MR image is 256 × 256 dots, where each dot is one of 4096 gray tones.

The new data acquired by the Visible Human Project, in conjunction with developing data manipulation technologies, have greatly increased our visual understanding of human anatomy.

Digital imaging. During the 1990s, it became clear that the digital computer could provide the missing technology needed to acquire, store, display, and interact with three-dimensional images. These computer images are reproduced by two fundamentally different methods: object-based and sample-based.

Object-based method. In this method, the picture is broken down into its fundamental geometric objects, such as a collection of polygons, whose shapes and surface qualities can be calculated by the computer to produce a likeness of a real-world object. The underlying representation is based on mathematical formulas for the geometric shapes that are combined to make up the complete object. The mathematics allows the computer to control groups of individual geometric components making up objects within the picture.

Sample-based method. This method is based on the acquisition, storage, and display of image samples taken of an original object. The pictorial data can be thought of as individual sample points (pixels) in a

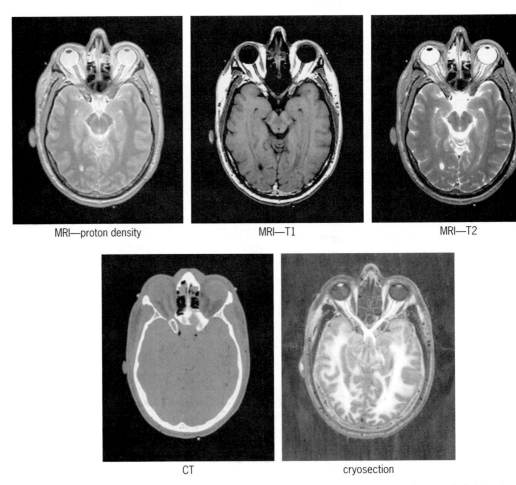

MRI—proton density   MRI—T1   MRI—T2

CT   cryosection

Fig. 2. Corresponding cross sections from the Visible Human MRI (various), CT, and anatomy (cryosection) data sets showing the nose, nasal sinuses, eyes, optic nerves, and brain.

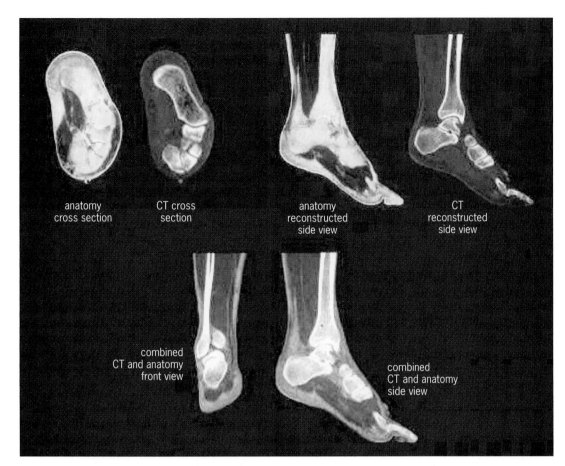

anatomy
cross section

CT cross
section

anatomy
reconstructed
side view

CT
reconstructed
side view

combined
CT and anatomy
front view

combined
CT and anatomy
side view

Fig. 3. Reconstructions of the Visible Human male foot, as seen from the front and side, made from the CT and anatomy data sets to demonstrate the absolute coincidence between them.

large two-dimensional field which, when displayed together, create an object that is a representation of the original object. There are no mathematical formulas in this scheme, so the computer cannot control any of the objects within the image.

Image acquisition. The distinction between the object- and sample-based methods in terms of acquiring, storing, and displaying a digital picture is crucial to computer-based biomedical imaging. Most clinical images acquired with techniques such as magnetic resonance imaging and computerized tomography use the sample-based method. Thus, varying black, gray, and white patterns are displayed such that a physician can interpret them as various anatomical structures. But to the computer, the components of the image are not manipulable as independent structures (for example, lung, heart, or bone), except to the extent that some tissues share a unique gray-tone level (for example, bone). For perception of each element within a complex medical image, the identification of objects and object boundaries within the images is essential.

The need for anatomical, CT and MR images is based on the fact that physicians view radiological images but are expected to interpret and treat anatomy. Anyone who has seen an x-ray photograph will immediately understand that the image seen on an x-ray is very different from the anatomy itself. The same is true for anatomy seen on CT or MRI photographs. In fact, anatomy viewed on a CT photograph looks different from the same anatomy viewed on an MRI photograph.

Computerized tomography. Computerized tomography is an x-ray-based technique. Tomography is a technique by which only a single selected cross section through a three-dimensional object is photographed. Using a CT scanner, the radiologist obtains consecutive and equally spaced x-ray cross sections through a region of a patient's anatomy. Typical spacing between the cross sections is 3 mm for clinical use. The cross sections are then computer-assembled to create a three-dimensional scene.

Unexposed film is transparent, so it looks white. When film is exposed to light or x-rays, it becomes opaque and looks black. When x-rays pass through the body, they are reflected by hard tissue (for example, as bone) but are transmitted through soft tissue (for example, muscle). Therefore, on an x-ray photograph or a CT image, bone appears gray and white, and muscles generally appear black. There is no color, just shades of gray from black to white.

Magnetic resonance imaging. The basis of this technique is that water molecules have very weak but detectable magnetic properties. If living tissue is put into a very strong magnetic field, sensors will detect the interference to the magnetic field caused by the water molecules in the tissue. Using sophisticated computer techniques, this effect can be interpreted as sequential cross-sectional pictures of a patient's anatomy. Typical spacing between the cross sections is 5 mm for clinical use. As for CT images, these cross sections can also be reassembled by a computer to create a three-dimensional scene.

To the untrained eye, MRI photographs look similar to CT photographs. This is illustrated in **Fig. 2**. In an MRI, soft tissue (for example, muscle) contains more water than hard tissue (for example, bone), so soft tissue causes more interference to the magnetic field. In the MRI photograph, this effect is represented as a lighter area. In a CT, hard tissue is denser than soft tissue and therefore reflects x-rays with greater efficiency. In the CT photograph, this effect is represented as a lighter area.

MRI technology is more sensitive to differences in soft tissue, while CT technology is more sensitive to difference in hard tissue. Clinically, CT is best for broken bones, and MRI for pulled muscles. In CT photographs, bones appear from gray to white depending on the bone density, while in MRI photographs bones generally appear black. In MRI photographs muscles appear from gray to white depending on the muscle mass, while in CT photographs muscles generally appear black. The physician must know how to interpret each of these imaging formats in order to relate them to a specific patient's anatomy.

Teaching aid. The Visible Human data sets contain coincidental CT, MRI, and anatomical images of the same cross section. These can be overlaid by the computer so that one can see the relationships between MRI and CT photographs and interpret their respective anatomical data (**Fig. 3**).

The consecutive cross-sectional nature of the Visible Human data sets also allows any part of the human body to be reconstructed by computer. The reconstructed organ can be rotated up or down and right or left, zoom in for a closer look at details and, given the proper software, even zoom inside. One can construct a reusable virtual cadaver that can be used indefinitely for the study and appreciation of anatomical complexity.

Outlook. In the near future, it will be possible to apply the techniques learned in building this virtual cadaver to build a personal virtual cadaver for each of us based on our own CT and MRI data. The National Library of Medicine, though a continuation of its Visible Human Project, will be working on computer-based software tools and Web-based methods to bring this futuristic dream into reality.

[This article was written by the author in his private capacity. No official support or endorsement by the National Library of Medicine is intended or should be inferred.]

For background information *see* ANATOMY, REGIONAL; COMPUTER GRAPHICS; COMPUTERIZED TOMOGRAPHY; MAGNETIC RESONANCE; MEDICAL IMAG-

ING; RADIOLOGY in the McGraw-Hill Encyclopedia of Science & Technology.   Michael J. Ackerman

Bibliography. M. J. Ackerman, The Visible Human Project, *Proc. IEEE*, 86(3):504–510, March 1998; A. R. Smith, Geometry and imaging: Clarifying the major distinctions between the two domains of graphics, *Computer Graphics World*, vol. 11, November 1988; V. Spitzer et al., The Visible Human Male: A technical report, *J. Amer. Med. Informatics Assoc.*, 3(4):118–130, 1996.

Water decontamination

For 1.5 to 2.5 billion people in the world, lack of clean water is a critical issue. It is estimated that by the year 2025 there will be an additional 2.5 billion people who will live in regions already lacking sufficient clean water. In the United States today, it is estimated that 90% of citizens live within 10 mi of a body of contaminated water. Large numbers of point (single, identifiable) and nonpoint sources having low flow volume [50 gal (190 L) per minute or less] contribute significantly to these water contamination problems. These sites pose a major unsolved problem because they also can be intermittent, reducing the cost effectiveness of many current mitigation technologies. The northeastern United States—with its large population, concentrated residential areas, industrial sites, livestock confinement operations, and the like—has many such sites where low-volume-flow water runoff and discharges need to be treated. In addition, it is estimated that there are approximately 500,000 abandoned hard-rock mine sites in the United States, many of them located in or near watersheds where acid mine drainage may release heavy metals into thousands of public drinking-water systems.

Decontamination systems. There are many technologies used today to remove contaminants from water, including reverse osmosis, synthetic resins, activated carbon, sand filtration, and inorganic substrates. Several of these technologies are very effective but can be expensive. Low-cost filtration systems are needed to remove an array of pollutants such as heavy metals, pesticides, herbicides, and other toxic chemicals; bacteria; particulates; nutrients; phosphorus; oil and grease; and nitrogen. Research has shown that lignocellulosic (plant-derived) resources such as wood and agricultural residues (for example, stalks, nut shells, and grasses) have ion-exchange capacity and general sorptive characteristics derived from their constituent polymers and structure. The polymers include extractives (those chemicals removed by solvent extraction), cellulose, hemicelluloses, pectin, lignin, and protein, which are adsorbents for a wide range of solutes, particularly divalent metal cations.

Lignocellulosic materials. Lignocellulosic materials are very porous and have a very high free-surface volume that allows accessibility of aqueous solutions to the cell wall components. One cubic inch ($16.4 \, cm^3$) of a lignocellulosic material, for example, with a specific gravity of 0.4, has a surface area of $15 \, ft^2$ ($1.4 \, m^2$). Even when the lignocellulosic material is ground, the adsorptive surface increases only slightly. Thus, the sorption of heavy-metal ions by lignocellulosic materials does not depend on particle size. Lignocellulosics are both hygroscopic (have the ability to absorb water) and hydrophilic (have an affinity for water). Water is able to permeate the noncrystalline portion of cellulose and all of the hemicellulose and lignin. Thus, through a combination of absorption (sorption into a three-dimensional matrix) and adsorption (sorption onto a two-dimensional surface), aqueous solutions come into contact with a very large surface area of different cell wall components. The cell wall polymers of lignocellulosics contain acid, phenolic hydroxyl, and other hydroxyl groups that can act as ion-exchange sites. Thus, sorption of heavy-metal ions by lignocellulosics can be accomplished by ion exchange, complexation, and precipitation. Removal of heavy metals from solution using lignocellulosic resources is dependent on temperature, pH, sorption time, and metal concentration. Generally, the optimum temperatures are 15 to $30°C$ (59 to $86°F$), pH range is 4 to 6, sorption time is as long is practicable, and metal concentration is low.

Many different types of lignocellulosics have been studied to remove heavy-metal ions from aqueous solution. Several types of sawdust have been used to remove cadmium and nickel, and several types of barks have been used to remove cadmium, copper, lead, zinc, nickel, and cobalt from solution. The fibrous remains of other plant products, such as sugarcane bagasse, corncobs, kenaf, cotton, coconut coir, tea leaves, sugarbeet pulp, and various types of straws and ground nut shells, have also been used to remove heavy metals. One of the best lignocellulosic materials found to date to remove heavy metals from solution is base-extracted fiber from juniper trees, which cover up to 124 million acres in the southwestern United States; the reason this particular fiber seems to work best is presently under investigation. Isolated cellulose as well as kraft and organosolv lignin (by-products from pulp and paper production) has also been used to remove copper and cadmium from aqueous solutions.

Filtration methods. The lignocellulosic fibers can be used as filters in several ways. Filtration containers can be packed with fiber, but this can result in restricted flow due to uneven distribution of the fiber bed. The fibers can be made into webs using several technologies: fibers can be carded into a uniform web, formed into a web using nonwoven needling equipment, or combined with a thermoplastic fiber and bonded thermally. Flow rate through the webs is then a function of web density and thickness.

The exact mechanism of heavy-metal removal using lignocellulosic materials is open to debate. Lignin content, type and amount of extractives, free acid functional groups, and hydroxyl content have been suggested, but no direct correlations have been found. Cell wall structure and surface area may also influence heavy-metal sorption.

Heavy-metal sorption capacity of lignocellulosic materials can be increased in several ways. Solvent extractions, alkali treatment, sulfonation, acetylation, reactions with multifunctional carboxylic acids, and surface functionalization have been used to increase the sorption capacity of heavy metals.

Future research. This technology is presently being tested in two National Forests by the U.S. Forest Service to remove toxic heavy metals from abandoned coal and hard-rock metal mines. These sites, where the heavy-metal ions concentration is high and the water flow is low, are ideal for this technology. Many other locations where acid mine drainage is a problem are presently under consideration for possible sites for this technology.

For background information *see* FILTRATION; ION EXCHANGE; WATER POLLUTION; WATER TREATMENT; WOOD PRODUCTS in the McGraw-Hill Encyclopedia of Science & Technology.   Roger Rowell

Bibliography. M. C. Basso, E. G. Cerrella, and A. L. Cukierman, Lignocellulosic materials as potential biosorbents of trace toxic metals from wastewater, *Ind. Eng. Chem. Res.*, pp. 3580–3585, 2002; P. Kumar and S. S. Dara, Modified barks for scavenging toxic heavy metal ions, *Indian J. Environ. Health*, 22:196, 1980; J. A. Laszlo and F. R. Dintzis, Crop residues as ion-exchange materials: Treatment of soybean hull and sugar beet fiber (pulp) with epichlorohydrin to improve cation-exchange capacity and physical stability, *J. Appl. Polymer Sci.*, 52:521–528, 1994; J. E. Phalman and J. E. Khalafalla, *Use of Ligochemicals and Humic Acids to Remove Heavy Metals from Process Waste Streams*, U.S. Department of Interior, Bureau of Mines, RI 9200, 1988; J. M. Randall et al., Use of bark to remove heavy metal ions from waste solutions, *Forest Prod. J.*, 24(9):80–84, 1974; J. S. Sedell et al., *Water and the Forest Service*, USDA, Forest Service Washington Office, ES 660, January 2000; T. Vaughan, C. W. Seo, and W. E. Marshall, Removal of selected metal ions from aqueous solution using modified corncobs, *Bioresource Tech.*, 78:133–139, 2001.

West Nile virus (equine)

West Nile (WN) virus is one of a group of disease-causing viruses known as the flaviviruses, which are spread by insects (usually mosquitoes) and infect a variety of animals, including humans (**Fig. 1**). WN virus is an emerging infectious disease in the United States; in 1999–2002 it infected increasing numbers of people each year, and has had a similar effect on horses. West Nile virus was first found in Africa in the 1930s, and disease from WN virus has become more common in Europe, Asia, and the South Pacific over the last 30–50 years. In 1999, it was discovered to be the cause of brain infection and death in birds, humans, and horses in the New York City area. Several new trends have recently been noted with the WN

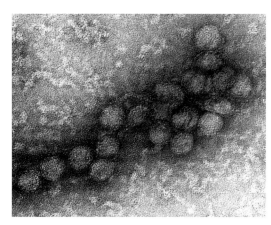

Fig. 1. Electron micrograph of the West Nile virus. (*Photo courtesy of the CDC/Cynthia Goldsmith*)

virus, including (1) movement into new geographic areas, (2) rapid development of new types of the virus in these areas, (3) increase in the frequency of human and horse outbreaks, (4) increase in severity of human infection, and (5) incredibly high avian losses associated with these outbreaks.

Epizootiology in horses. The 1999-2002 North American epizootic, or epidemic, of equine WN virus is notable for having infected the highest number of horses ever reported. West Nile virus encephalomyelitis (WNE), the spinal and brain infection associated with the virus, first occurred in United States horses in the initial North American encroachment of this virus in 1999. During that year, 25 horses were diagnosed with clinical disease in Suffolk and Nassau counties in New York. Clinical cases occurred from late August through late October. During the 2000 mosquito season, WN virus in horses spread slightly south and west to 7 other states and 25 counties, affecting at least 60 horses. In 2001, 738 horses were reported to have clinical WNE, involving 20 states and 130 counties; the majority of clinical cases for that year occurred in southern Georgia and northern Florida. With movement of the virus into a subtropical climate, its active season increased to extend from July to December 2001 and, in Florida in 2002, year-round. Infection of horses in 2002 brought a phenomenal westward expansion of the virus, with WNE reported in 14,717 horses from 40 states. The yearly mortality rate has ranged 33–38% in clinically affected horses, resulting in the loss of more than 5000 horses since 1999. In addition, WN virus exposure or actual clinical disease has been reported in cats, dogs, mountain goats, llamas, alpacas, sheep, alligators, big brown bats, chipmunks, raccoons, striped skunks, gray squirrel, and the domestic rabbit.

A case control study performed by the U.S. Department of Agriculture (USDA) in 2000 demonstrated that blackbird roosts and waterfowl congregations were usually present within a $\frac{1}{2}$-mi radius of a WN virus–positive location. Horses likely to develop WNE were primarily used for pleasure and less likely to be stalled at night. Ecological factors which in-

creased the likelihood of cases included the presence of broadleaf forests and clay soils. These are associated with slow water drainage from an environment, resulting in a tendency toward permanent or recurrent development of marsh- and swamplands (and increased mosquito populations). The study also determined that amplification (increased infection, spread, and virus levels) in birds likely preceded infection in horses. Evaluation of horses from the 2001 outbreak demonstrated an increased susceptibility in older horses.

Transmission. West Nile virus is primarily transmitted to horses by mosquitoes. *Culex pipiens* mosquitoes have demonstrated the highest infection rate throughout the eastern United States, with a high degree of vector competency (that is, ability to become infected by an avian or mammalian host and spread the virus to high numbers of other hosts). However, the exact species of mosquito responsible for the high degree of equine infectivity is unknown at this time. *Aedes albopictus* has been used to experimentally infect horses with the New York strain of WN virus. These horses developed viremia (low levels of virus in the blood) but few signs of encephalomyelitis. *Culex nigripalpus* and *C. salinarius* are likely important in subtropical locations such as Florida, while *C. tarsalis* is an important vector in the Midwest and West.

Susceptibility and clinical disease. There are two lineages of WN virus: type I, which includes all strains from northern Africa, Europe, Israel, and the United States, and type II, which includes strains from west, central, and east Africa and Madagascar. Horses have demonstrated susceptibility to neurally invasive lineage type I WN virus, but appear to be resistant to most African lineage type II viruses. A mild low-grade fever, feed refusal, and depression are common systemic signs. Unlike most viral encephalitides in the horse, where seizure and coma are most common, WNE horses primarily exhibit spinal cord disease and moderate mental aberrations. Neurologic disease is frequently sudden and progressive, and characterized by problems in maintaining balance (ataxia) and strength (paresis) [**Fig. 2**]. Over 90% of affected horses develop some degree of spinal cord disease, while 60% develop behavior changes. Many horses have periods of hyperexcitability, apprehension, and/or sleepiness. Fine and coarse tremors of the face and neck muscles are very common, and paralysis of the nerves of the head (cranial nerves) can occur. A weakness of the tongue, muzzle deviation, and head tilt are the most common manifestations. The exact course of disease in any one animal is extremely difficult to predict. After initial signs abate, about one-third of WNE horses experience an increase in severity of clinical signs within the first 7–10 days of onset. Some horses progress to complete paralysis of one or more limbs. Most of these are euthanized for humane reasons or die spontaneously.

Treatment and response. There is no specific antiviral therapy for WNE in horses. The focus of therapy is

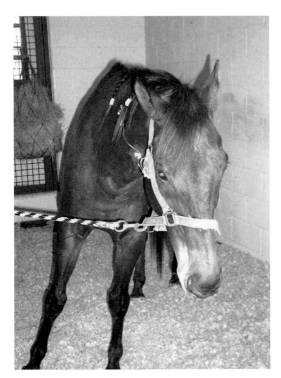

Fig. 2. Horse with clinical signs of West Nile virus that has head tilt and inability to maintain balance.

to reduce brain inflammation, treat fever, and provide supportive care, which may require 1–4 weeks of intermittent therapy. Common anti-inflammatory medications are used, and sometimes fluid therapy is administered to animals that are not able to drink or eat. Forced oral or intravenous feeding may also be necessary. For recumbent horses, slinging is necessary to assist standing, and head and leg protection are also frequently needed. In general, many horses will improve within 3–7 days of displaying clinical signs. However, horses that are recovering or stable may exhibit a sudden recurrence of clinical signs after 3–5 days. This may be of short duration, or the horses may suddenly become recumbent and either die or require prolonged treatment. Horses that become recumbent often require a return to aggressive treatment. Once significant improvement has been demonstrated, full recovery within 1–6 months can be expected in 90% of the horses. Residual weakness and ataxia appear to be the main problems. Some horses have experienced long-term inflammation of the meninges (membranes in the brain and spinal cord), and others have experienced long-term loss of the use of one or more limbs. Mild-to-moderate persistent fatigue upon exercise has also been noted.

Research. Induction of clinical signs after experimental infection has been unrewarding. In general, for every 12 horses that are infected by either mosquito feeding or intravenous inoculation, only one develops clinical signs. Therefore, similar to human infection, it is likely that many more horses are infected than become clinically ill. Upon experimental infection, horses develop a low-grade viremia within 3–5 days. This level of virus is not high enough

to allow transmission back to mosquitoes; thus, the horse is a dead-end host. West Nile virus causes a polioencephalomyelitis (infection of gray matter) in horses that is similar to that seen in humans. A mononuclear meningitis (inflammation of the meninges), perivascular inflammation (inflammation around blood vessels), and neuronal necrosis (nerve death) is a frequent finding in the midbrain (**Fig. 3**). These changes progress through the hindbrain, increasing in severity caudally (toward the tail) throughout the spinal cord.

Laboratory diagnosis and prevention. Confirmatory diagnosis in horses that demonstrate clinical signs is by detection of antibodies in the blood. Horses develop a sharp rise and fall in immunoglobulin M (IgM) during the first 6–8 weeks after exposure to WN virus, and their neutralizing antibody titers (primarily immunoglobulin G, or IgG) develop slowly during this time and stay elevated for several months. Thus, the IgM capture enzyme-linked immunosorbent assay (MAC) is the test of choice for detection of recent exposure to the virus. In most states, a veterinarian submits samples for the test, and any horse demonstrating clinical symptoms of encephalitis must be reported to the state agriculture service. Tracking and reporting of encephalitic disease in the United States is important for surveillance of diseases that can affect humans. Unfortunately, testing for IgG or neutralizing antibody is confounded by vaccination in the United States, and not reliable for diagnosis on a single sample.

The first WN virus vaccine marketed for United States horses was released conditionally in Fall 2001, and full licensure was granted in March 2003. The optimal time for vaccination is before the mosquito season. The manufacturer recommends a once-yearly vaccination, but veterinarians in the United States have recommended vaccination up to two to three times per year in epidemic or endemic areas. No information is available regarding long-term immunity; however, independent testing has demonstrated a definitive waning of antibody by 6 months postvaccination. Horses undergoing initial vaccination

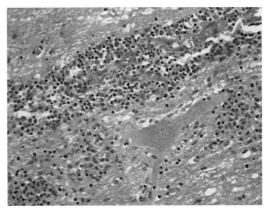

Fig. 3. Photomicrograph of a lesion in the brain of a horse with West Nile virus, showing several layers of inflammatory cells around a blood vessel. (*Photo by Carol Detrisac*)

must have the full primary series, consisting of an initial injection followed in 3–6 weeks by a second injection. Other control methods are equally important. Horses should be sprayed with repellents frequently during mosquito season, and the repellents should be reapplied after rain. Stable and farm conditions can also decrease mosquito activity. In stables, air movement through the use of fans is crucial to reduction of mosquito activity. At night, artificial light should be reduced and stalling increased to minimize exposure. If stock tanks or ponds are used for watering, placement of organic larvicides and mosquito-feeding fish in the water will decrease insect activity. Since mosquitoes breed in wet organic debris, removal of brush piles, litter, and manure will inhibit development of significant mosquito habitat on farm premises.

For background information *see* ANIMAL VIRUS; ARBOVIRAL ENCEPHALITIDES; INFECTIOUS DISEASE; VIRUS in the McGraw-Hill Encyclopedia of Science & Technology.   Maureen Long

Bibliography. M. L. Bunning et al., Experimental infection of horses with West Nile virus and their potential to infect mosquitoes and serve as amplifying hosts, *Ann. N. Y. Acad. Sci.*, 951:338–339, 2001; E. N. Ostlund et al., Equine West Nile encephalitis, United States, *Emerg. Infect. Dis.*, 7:665–669, 2001; M. B. Porter et al., West Nile virus encephalomyelitis in horses: 46 cases (2001), *J. Amer. Vet. Med. Assoc.*, 222:1241–1247, 2003; H. L. Wamsley et al., Findings in cerebrospinal fluid of horses infected with West Nile virus: 30 cases (2001), *J. Amer. Vet. Med. Assoc.*, 221:1303–1305, 2002.

Whale evolution

Cetaceans, or whales (a group which includes the animals commonly known as dolphins and porpoises), are interesting from an evolutionary perspective because they are so different from their ancestors. The ancestors of whales were fur-covered land mammals that rarely entered the water, whereas whales today are fully aquatic and thrive in the oceans, seas, and—in some instances—great rivers of the world. The ancestors were four-footed and hoofed for running on land, whereas whales today have forelimbs modified as flippers, lack external hindlimbs, and move by pushing themselves through water using a powerfully muscled tail with a broad terminal water-foil called a fluke. The ancestors were omnivorous to herbivorous; in contrast, whales today are piscivorous (fish-eating) and planktivorous (plankton-eating). The ancestors perceived their surroundings using their eyes and noses; however, whales today "see" their surroundings with their ears.

The origin of whales involved a macroevolutionary transition from a terrestrial to an aquatic way of life. And, compared with the general trend of vertebrate evolution from sea to land, the evolution of whales

was backward, because they moved from land back to the sea. New fossil discoveries illuminate the origin and early evolution of whales.

Whale diversity and classification. Similarities and differences of whales are expressed in a classification (see **table**), which is also a reflection of their diversity. All whales are classified in the mammalian order Cetacea. Within Cetacea, there are two general kinds of whales living today: toothed whales in the suborder Odontoceti and baleen whales in the suborder Mysticeti. Odontocetes have simple teeth embedded in sockets of bone in the upper and lower jaws. This socketing is a general mammalian characteristic. Mysticetes have baleen, which is a keratinous tissue that frays into hairs. Hair, too, is a general characteristic of mammals. Living Odontoceti and Mysticeti go back some 35 million years in the fossil record to the Oligocene Epoch of geological time.

The third major group of whales is now extinct (or "pseudoextinct" in the sense that it includes ancestors of later whales, but has no representatives living today). This is the suborder Archaeoceti, known only from fossils in the Eocene Epoch, spanning an age range from about 53 to 35 million years before present. Archaeocetes have complex teeth with double-rooted molars, and this double-rooted condition is a diagnostic feature of mammals. Archaeoceti includes the earliest whales and can be shown to have given rise to later Odontoceti and Mysticeti through Oligocene intermediates. The transition from archaic to modern whales coincides with a major tectonic reorganization of ocean currents in the Oligocene that changed the distribution of heat on the surface of the Earth, leading to formation of ice caps at the Poles and causing cold nutrient-rich seawater to well up along the western coasts of continents. Whales responded by developing simplified teeth and high-frequency sonar to aid in feeding on fish (odontocetes) or by developing baleen to strain plankton from seawater directly, bypassing fish in the food chain altogether (mysticetes).

Fossil record. Whales have large bones, making them relatively easy to find as fossils. Whales are by definition water-dwelling animals, and watery environments are generally ideal for preserving bones as fossils. Whales thrive in shallow marine environments on the margins of continents, and the Cenozoic Era when whales lived is represented in the geological record by widespread shallow marine deposits. However, since whales reside at or near the top of the food chain (that is, they are rare compared with other animals), they have a good, but not yet great, fossil record. The fossil record is sufficiently dense and continuous, though, that broad lineages can be traced up and down through successive strata, moving forward and backward through earth history and evolutionary time.

The most interesting whales, in some respects, are those that document the transition from land to sea, and here an archaic whale from the Eocene named *Rodhocetus*, classified in the family Protocetidae, is

Summary classification of whales (order Cetacea)*

Suborder	Family	Common name	Number of genera	Geological epoch
Odontoceti	Delphinidae	Dolphins	17	Miocene–Recent
	Phocoenidae	Porpoises	8	Miocene–Recent
	Monodontidae	Narwhals	4	Miocene–Recent
	Platanistidae	Indian river dolphins	3	Miocene–Recent
	Pontoporiidae	South American river dolphins	6	Miocene–Recent
	Iniidae	Amazon river dolphins	6	Miocene–Recent
	Lipotidae	Chinese river dolphins	2	Miocene–Recent
	Kogiidae	Pygmy sperm whales	5	Miocene–Recent
	Acrodelphinidae	Acrodelphinids	5	Miocene
	Physeteridae	Sperm whales	17	Oligocene–Recent
	Hyperoodontidae	Beaked whales	23	Oligocene–Recent
	Squalodontidae	Squalodontids	15	Oligocene–Miocene
	Rhabdosteidae	Rhabdosteids	6	Oligocene–Miocene
	Kendriodontidae	Kendriodontids	11	Oligocene–Miocene
	Agorophiidae	Agorophids	1	Oligocene
Mysticeti	Eschrichtiidae	Grey whales	1	Pleistocene–Recent
	Balaenopteridae	Rorquals	8	Miocene–Recent
	Balaenidae	Right whales	6	Miocene–Recent
	Cetotheriidae	Cetotheres	29	Oligocene–Pliocene
	Aetiocetidae	Aetiocetids	1	Oligocene
	Mammalodontidae	Mammalodontids	1	Oligocene
	Llanocetidae	Llanocetids	1	Eocene–Oligocene
Archaeoceti	Basilosauridae	Basilosaurids	8	Middle–late Eocene
	Protocetidae	Protocetids	12	Middle Eocene
	Remingtonocetidae	Remingtonocetids	4	Middle Eocene
	Ambulocetidae	Ambulocetids	2	Middle Eocene
	Pakicetidae	Pakicetids	4	Early–middle Eocene

*Numbers of genera are approximate. Fossil whales are from the Cenozoic Era and range from early Eocene in age, approximately 53 million years ago, to the present.

crucially important. *Rodhocetus* is similar in some respects to the later Eocene basilosaurid *Dorudon* (**Fig. 1**). Both are known from nearly complete skeletons. Similarities include long tapering skulls with pointed incisors and canine teeth, complex puncturing and shearing premolar and molar teeth, necks of medium length, a narrow and deep chest or thorax, and mobile forelimbs modified to some degree for swimming. *Dorudon* is similar, in turn, to primitive mysticete and odontocete whales known from the subsequent Oligocene Epoch. Comparison between *Rodhocetus* and *Dorudon* shows a reduction of the neural spines rising above the thoracic vertebrae, great elongation of the vertebral column with addition of lumbar vertebrae, and great reduction of the hindlimbs in *Dorudon*.

Skeletons of primitive archaeocete whales with associated forelimbs and hindlimbs were discovered in 2000. The most complete of these, *Rodhocetus*, is important in showing that primitive archaeocetes had morphological characteristics of land mammals in general, and of the anthracotheriid group of artiodactyls in particular. Artiodactyls are hoofed animals with even numbers of toes; anthracotheriids are a group of artiodactyls that lived in the Eocene and Oligocene and are thought to have given rise to living *Hippopotamus*. The most important similarities between *Rodhocetus* and the anthracotheriids are those of the ankle bones, including a double-pulley astragalus, notched cuboid, and convex fibular facet of the calcaneum found only in artiodactyls (**Fig. 2**). Both *Rodhocetus* and *Elomeryx*, the best-known skeleton of the family Anthracotheriidae, retain five fingers on the hand, both are mesaxonic (having the central axis of the hand in or near the middle finger), and both have small hooves on the terminal phalanges of the second, third, and fourth digits. Compared with *Elomeryx*, *Rodhocetus* has an elongated skull, a shorter neck and forelimb, enlarged lumbar and caudal vertebrae, an elongated tail, and hindfeet that are modified into elongated, webbed flippers.

Artiodactyls were first convincingly linked to whales on the basis of their elevated mutual immune reactions compared with those of other pairs of mammals. Deoxyribonucleic acid sequencing has corroborated this relationship and, somewhat controversially, identified the hippo as the living artiodactyl closest to whales. Similarity of *Rodhocetus* to *Elomeryx* does not prove this connection because there is a lack of evidence of *Elomeryx*-like artiodactyls at the beginning of the Eocene; however, the new fossil evidence makes such a relationship to hippos more plausible than previously thought.

Transition from land to sea. *Elomeryx* is Oligocene in age, so it cannot be ancestral to *Rodhocetus*, but it nevertheless makes a good model for the kind of terrestrial artiodactyl thought to have given rise to whales about 55 million years ago at the beginning of the Eocene. The proportions of *Elomeryx*, *Rodhocetus*, and *Dorudon* are compared in Fig. 1.

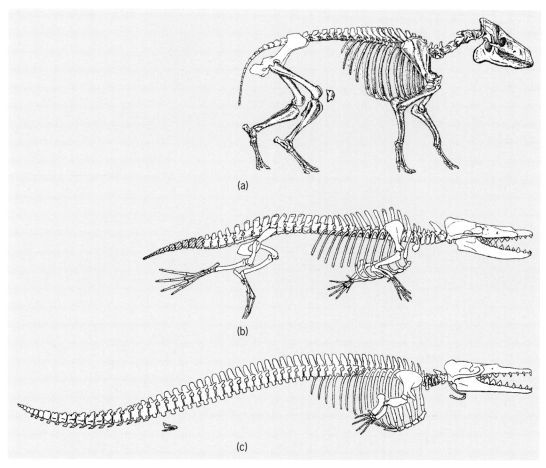

Fig. 1. Comparison of skeletons showing the morphology of (*a*) a model ancestral land mammal (*Elomeryx*; skeleton is about 2 m or 6 ft in length); (*b*) a semiaquatic middle Eocene protocetid (*Rodhocetus*; 3 m or 10 ft in length); and (*c*) a fully aquatic middle-to-late Eocene basilosaurid (*Dorudon*; 6 m or 18 ft in length). All are standardized to approximately the same head and thorax size. Note that the archaeocete whales *Rodhocetus* and *Dorudon* have longer skulls and shorter necks, progressively shorter forelimbs, and progressively longer tails and more reduced hindlimbs compared with land-dwelling *Elomeryx.*

Here the relative sizes of the skull, neck, and thorax (rib cage) are standardized, making it easy to visualize how archaeocetes differ from *Elomeryx*: the archaeocetes have longer skulls and shorter necks, progressively shorter forelimbs, progressively longer tails, and more reduced hindlimbs (so small as to be vestigial for locomotion in *Dorudon*).

There are more than 100 species of semiaquatic mammals living today. These range from the more terrestrial tapirs, bears, and hippos to the more aquatic seals and sea lions. Comparison of the overall trunk, limb, and hand and foot proportions of *Rodhocetus* with a large sample of living semiaquatic mammals shows *Rodhocetus* to have been a desman-like foot-powered swimmer, and the presence of feet modified into elongated, webbed flippers is consistent with this. Thus it appears that descendents of hoofed terrestrial mammals such as *Elomeryx* passed through a foot-powered swimming stage before more whalelike tail-powered swimmers such as *Dorudon* evolved.

The oldest fossil whale known to date, *Himalay-acetus*, is a pakicetid about 53 million years old that was found in marine strata of the lower Eocene of India. *Pakicetus* itself is a little younger and comes from riverine deposits of middle Eocene age in Pakistan. Skeletal remains of *Pakicetus* have been interpreted as being terrestrial, but these are much too fragmentary to interpret reliably. The bones that are complete enough to compare resemble those of semiaquatic Protocetidae, so it is doubtful that *Pakicetus* was terrestrial. Recovery of older *Himalay-acetus* from marine deposits is powerful evidence, too, that the earliest whales known to date were already semiaquatic.

Conclusion. What we infer of whale evolution begins with an as yet unidentified early Eocene terrestrial artiodactyl ancestor (here represented provisionally by *Elomeryx*). The first stage documented in the fossil record is an early-to-middle Eocene pakicetid-protocetid stage represented by the semiaquatic foot-powered swimmer *Rodhocetus*. This is followed by a middle-to-late Eocene basilosaurid stage represented by the fully aquatic tail-powered swimmer *Dorudon*. Modern baleen and toothed whales differentiated in the latest Eocene or early Oligocene, giving rise to the diversity known today. Molecular evidence from living animals is consistent with this historical perspective in recognizing a close relationship of whales and artiodactyls.

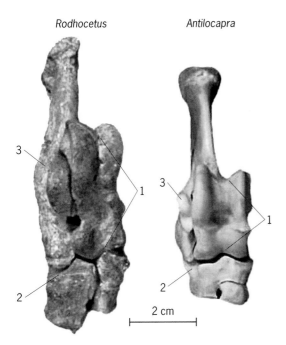

Rodhocetus   *Antilocapra*

2 cm

Fig. 2. Comparison of ankle bones of the middle Eocene protocetid *Rodhocetus* with those of an extant pronghorn *Antilocapra*. Note the presence of (1) a double-pulley astragalus bone; (2) a notch in the cuboid for insertion of the calcaneum; and (3) a large convex fibular facet on the calcaneum. This whole complex of characteristics unites early whales with artiodactyls, or even-toed hoofed mammals. Late Eocene basilosaurid whales have greatly simplified ankle bones, and later fossil and living whales lack these entirely.

For background information *see* ANIMAL EVOLUTION; ANIMAL SYSTEMATICS; ARTIODACTYLA; CETACEA; FOSSIL; MACROEVOLUTION in the McGraw-Hill Encyclopedia of Science & Technology.

Philip D. Gingerich

Bibliography. L. Bejder and B. K. Hall, Limbs in whales and limblessness in other vertebrates: Mechanisms of evolutionary and developmental transformation and loss, *Evol. Dev.*, 2002; F. E. Fish, in J.-M. Mazin and V. d. Buffrénil (eds.), *Secondary Adaptation of Tetrapods to Life in Water*, Friedrich Pfeil, Munich, 2001; R. E. Fordyce and C. de Muizon, in J.-M. Mazin and V. d. Buffrénil (eds.), *Secondary Adaptation of Tetrapods to Life in Water*, Friedrich Pfeil, Munich, 2001; J. E. Gatesy and M. A. O'Leary, Deciphering whale origins with molecules and fossils, *Trends Ecol. Evol.*, 2001; P. D. Gingerich et al., Origin of whales from early artiodactyls: Hands and feet of Eocene Protocetidae from Pakistan, *Science*, 2001; J. E. Heyning and G. M. Lento, in A. R. Hoelzel (ed.), *Marine Mammal Biology*, Blackwell Scientific, Oxford, 2002.

Wildlife forensic science

Wildlife is often killed because it is considered a nuisance. For example, strychnine has been used illegally to eliminate top predators such as wolves and coyotes, although it also kills birds of prey feeding on contaminated carcasses. The hunting of wolves and seals has been supported by government bounties and subsidies. At one time (between World Wars I and II) the Canadian government even used bombs to eradicate beluga whales, which eat economically valuable cod. Laws have been enacted to restrict such abuses, and wildlife forensic pathologists, like their medical counterparts, aim to link the crime scene, suspect, and "victim." To achieve this goal, they identify carcasses, document ballistic evidence or trauma, and detect poisons in carcasses. Over the last decades, illegal trading of wildlife products ($3.4 billion a year worldwide, the second largest type of illegal commerce after drug trafficking) and large-scale environmental contamination by industrial products and by-products have confronted wildlife forensic scientists with new challenges.

Species identification. To link the suspect and victim at the crime scene, the victim must be identified unambiguously. Often wildlife forensic scientists must determine the species from an otherwise unidentifiable sample to find out if a threatened species is involved. Such samples include, but are not limited to, skinned carcasses, canned meat, or products made from body parts (such as ivory, leather purses, wallets, powder, or fluids). Modern molecular techniques have been adapted for this purpose.

Polymerase chain reaction. The polymerase chain reaction (PCR) is a technique for synthesizing hundreds of thousands of copies of a selected region of deoxyribonucleic acid (DNA). This technique is now widely used by forensic wildlife scientists because it requires very little DNA and it can amplify even extensively damaged DNA such as DNA isolated from a blood spot, canned meat, or hair.

To exert some control on whale hunting, country members of the International Whaling Commission (IWC) agreed to forbid the hunting of most whale species in a 1986 global moratorium. Japan, a member of the IWC, was allowed to hunt the Minke whale, a more abundant species, maintaining it was for scientific purposes. However, many doubted that all the whale meat sold in Japan was from the Minke whale. Before the advent of PCR, canned whale meat could not be analyzed outside Japan because international conservation laws prohibited exporting samples of threatened animal species for any purpose. To get around this restriction, United States researchers went to Japan with a PCR instrument and the necessary reagents in their luggage. They bought canned whale meat and, in their hotel room, isolated and amplified DNA from the meat. The amplified DNA was brought back to the United States for analysis. (Since the new DNA had been biochemically synthesized, it was not subject to export regulations.) The analysis showed, in fact, that the cans contained numerous whale species other than Minke whales; many of these species were considered endangered.

PCR is also used to identify animal species ingested by predators. Wild felids (cats), such as cougars, are increasingly seen in urban areas where they may attack pets and people, often with fatal consequences.

Bears also occasionally attack joggers or cross-country skiers if surprised. The analysis of DNA isolated from the feces of suspect animals (captured or not) can reveal, unambiguously, if they are the perpetrators.

High-pressure liquid chromatography. DNA is not always in sufficient quantity or sufficiently well preserved to use PCR for identifying an animal. Other techniques for analyzing proteins or other types of molecules have been adapted for this purpose.

Bile, the fluid produced by liver cells and collected in the gallbladder, is highly prized in Asia for traditional medicine and as a general health tonic. In North America, an increasing number of bears are killed illegally for their gallbladders. However, it has been difficult to prove that suspected products contain bear bile. High-pressure liquid chromatography, a technique used to separate and identify molecules, has been used to identify a bile acid found only in bear bile and, as a result, to convict poachers.

Pollution. Large amounts of contaminants have been released into the environment by industrial processes either inadvertently [for example, polychlorinated biphenyls (PCBs) and dioxins] or as by-products [for example, polycyclic aromatic hydrocarbons (PAHs)]. Some contaminants, such as PCBs, are not found in nature, while others, such as PAHs, are normally present only in small amounts. Surprisingly, not much is known about the long-term effect of these pollutants on wildlife, even though thousands of tons of them have been released into the environment both locally and globally. This is partly due to lack of funding and to the absence of incentives for industrial manufacturers and governments to better define the toxicity of contaminants. There are strong legal and financial incentives not to document these effects. Animal studies to define the chronic toxicity of chemicals cost millions of dollars.

Even the negative effects of environmental exposure of people to contaminants, whose toxicity has been demonstrated experimentally and epidemiologically, have been disputed in court. For instance, the Minamata disaster, probably the most publicized case of environmental contamination, where an entire community was poisoned by mercury, still pended in court in 2001. Similarly, the effects of lead on the nervous system of children, and of asbestos on the lungs (tumors) of exposed workers have been the subjects of litigation until very recently, despite overwhelming epidemiological evidence. It is not surprising then that a causal relationship between environmental contamination and diseases is even harder to substantiate in wildlife.

The contamination by PAHs of a small population of beluga whales living in the St. Lawrence Estuary, near the mouth of the Saguenay River (Quebec, Canada), illustrates these difficulties (**Fig. 1**). This population suffers a high rate of cancer (**Fig. 2**). From 1937 to 1980, an estimated 43,000 tons of PAHs were released by the local aluminum smelters downwind into the air; and from 1956 to 1976, an esti-

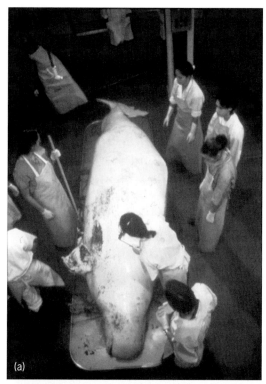

Fig. 1. Examination of beluga whales found stranded on the St. Lawrence River shoreline (Quebec, eastern Canada) at the College of Veterinary Medicine, Université de Montréal. (*a*) The carcass of an adult beluga whale is examined in the necropsy room by a team of veterinary pathologists and veterinary students. (*b*) A primary adenocarcinoma (cancer originating in the intestinal lining) distorts the normal architecture of the intestine (the cancer lesion is between the two arrows) and obstructs the intestinal lumen. The obstruction causes the dilation of the upstream intestine and the decreased diameter of the downstream intestine.

mated 1370 tons of PAHs were scrubbed and released directly into the river as liquid discharges. A total of 40,000 tons of PAHs is estimated to have accumulated in the Saguenay River and its watershed as the result of the net emissions by the aluminum smelters into the air and water between 1937 and 1980. It has also been clearly documented that these PAH emissions have contaminated the Saguenay River sediments and terrestrial and aquatic animals, including the beluga whales, living in that region. (Beluga whales are one of the rare toothed whales species feeding in significant amounts on animals living in

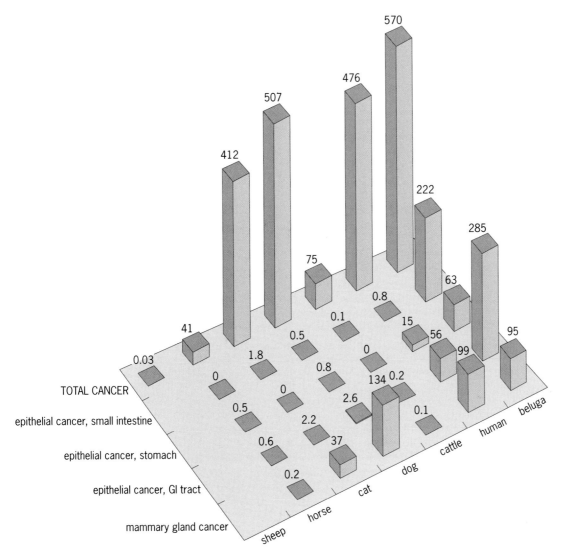

Fig. 2. Annual rates of cancer (number of cases per 100,000) in the St. Lawrence beluga whale population compared with that of humans and domestic animals. (*Reproduced with permission from D. Martineau et al., Environ. Health Perspect., 110(3):285–292, 2002*)

sediments.) In addition, local aluminum workers are affected by cancers that have been epidemiologically related to PAH exposure (for which affected workers are now compensated). Increased cancer rates have also been observed in the human population living in that area, where drinking water is taken mostly from surface waters, also contaminated by PAHs.

The aluminum companies claim that there is no proof that PAHs cause cancer in beluga whales. Yet PAHs became the first compounds whose carcinogenicity was empirically demonstrated when Percival Potts successfully recommended in 1775 that London chimney sweepers improve their personal hygiene to avoid scrotal cancer. Since then, many PAHs have been defined as carcinogens in mammals and humans according to assessments by the International Agency for Research on Cancer (IARC).

Other human impacts on wildlife. Human activities, such as fishing, can have severe impacts on animal species that are not the primary targets of these activities. Since the late 1940s, the tuna fishing industry has used a method, based on the fact that eastern tropical Pacific Ocean dolphins often swim along with tuna, whereby aircraft and speedboats pursue dolphins in order to catch tuna with a purse seine—a net used to encircle and entrap surface-schooling fish. It is estimated that at least 4 million dolphins died in those nets between 1959 and 1972. The Marine Mammal Protection Act, continued improvement of techniques and equipment, and increased attention of commercial fishers have decreased significantly the annual mortality. Because of concern that purse seine fishing may still be detrimental to the dolphin stock, additional studies were conducted to determine if this fishing method still had negative effects. Examination of dolphin carcasses found in seine nets revealed biochemical abnormalities suggesting a condition termed exertional myopathy, which is characterized by severe muscle damage and blood acidosis. Exertional myopathy results from strenuous exercise under severe stress. It is a frequent cause of mortality in wild mammals, especially in free-ranging ungulates (hoofed animals) that have been chased by vehicles.

Outlook. In March 1989, 38,000 tons of crude oil were spilled in an ecologically sensitive area of Alaska from the *Exxon Valdez* supertanker. The accident triggered a strong response from state authorities and from the population. The company was fined $900 million, which was placed in a trust to compensate the Alaskan government and population, to preserve and rehabilitate the habitat, and to monitor and fund research on wildlife biology and toxicology.

The *Exxon Valdez* accident is also the first example of a large-scale multidisciplinary approach, including toxicology, pathology, ecology, meteorology, and even biological and physical oceanography. For instance, veterinary pathologists and toxicologists determined that sea otters suffered severe interstitial pulmonary emphysema (the presence of air in lung tissue other than alveoli or airways) and that seals were affected by severe lesions of the nervous system, probably because both species inhaled the volatile components of crude oil. To estimate the total mortality of sea otters due to the spill (to account for the carcasses that were not recovered), a computer model was used to estimate the distribution and amount of oil over the ocean surface over time. This model was integrated with the distribution and abundance of sea otters in the region (before the spill) and with estimates of site-specific sea otter mortality obtained both by counting the carcasses and by calculating the mortality rate of captured oiled otters.

Wildlife forensic science will have to move increasingly to a multidisciplinary approach to address issues more complex (and more expensive) than those posed by poaching and poisoning.

For background information *see* CETACEA; DEOXYRIBONUCLEIC ACID (DNA); FORENSIC BIOLOGY; LIQUID CHROMATOGRAPHY; MOLECULAR PATHOLOGY; ONCOLOGY in the McGraw-Hill Encyclopedia of Science & Technology.   Daniel Martineau

Bibliography. Forensic investigational techniques for wildlife law enforcement investigations, in A. Fairbrother, L. N. Locke, and G. L. Hoff (eds.), *Noninfectious Diseases of Wildlife*, 1996; T. R. Loughlin (ed.), *Marine Mammals and the Exxon Valdez*, Academic Press, San Diego, 1994; D. Martineau et al., Cancer in wildlife: A case study—Beluga from the St. Lawrence Estuary, *Environ. Health Perspect.*, 110(3):285–292, 2002; J. N. Smith and E. M. Levy, Geochronology for polycyclic aromatic hydrocarbon contamination in sediments of the Saguenay Fjord, *Environ. Sci. Technol.*, 24:874–879, 1990; *Wildlife Forensic Field Manual*, Association of Midwest Fish and Game Law Enforcement Officers, 1992; G. A. Wobeser, Forensic (medicolegal) necropsy of wildlife, *J. Wild Dis.*, 32(2):240–249, 1996.

Wilkinson Microwave Anisotropy Probe

The *Wilkinson Microwave Anisotropy Probe* is a space-based instrument to measure the very small temperature fluctuations in the cosmic microwave background. The project to develop the *Microwave Anisotropy Probe* was started in 1994, based on a proposal by Charles L. Bennett. In 2003 it was renamed the *Wilkinson Microwave Anisotropy Probe* (*WMAP*) in honor of David T. Wilkinson, a pioneer in observing the cosmic microwave background, who played a major role in the design, construction, and early operation of *WMAP*. *WMAP* is the second medium-class Explorer mission to be launched by the National Aeronautics and Space Administration (NASA). Results from the first year of *WMAP* operation were released in February 2003.

Cosmic microwave background. The cosmic microwave background is a blackbody radiation field that is left over from the hot big bang. The mean temperature of the cosmic microwave background was shown to be 2.725 K by the *Cosmic Background Explorer* (*COBE*) satellite, launched in 1989. This temperature is very nearly isotropic, meaning the same in all directions, but one side of the sky is 3 mK hotter while the opposite side is 3 mK cooler due to the solar system's velocity of 370 km/s (230 mi/s) relative to the observable universe. *COBE* also found the first evidence for anisotropy other than the dipole pattern, specifically a ± 30-μK variation in $10°$ patches. This intrinsic anisotropy is due to primordial density perturbations. These primordial density perturbations also became the current large-scale structures in the universe: clusters of galaxies and superclusters of clusters. But the $7°$ beam of *COBE* corresponds to structures much larger than the observed clusters and superclusters.

Observations of the anisotropy with higher angular resolution can be used to measure the parameters that describe the large-scale structure of the universe. The mass-energy of the universe is believed to be composed of (1) observable matter, over 99.9% of whose mass resides in baryons, primarily protons and neutrons; (2) dark matter, which so far has not been directly observed, but whose existence is inferred from a number of observations including those of the motions of stars and galaxies; and (3) vacuum energy, an invisible form of energy that enters into the cosmological equations as a repulsive term, and is believed to be responsible for the acceleration in the expansion of the universe that was discovered in 1998 from observations of type Ia supernovae. The higher-resolution observations can reveal interference between fluctuations in the density of dark matter and the corresponding sound waves in the baryon-photon fluid, the hot plasma that filled the universe after the big bang. The amplitudes of the peaks in this interference pattern reveal the ratios of the baryon density and dark-matter density to the photon density, while the angular scale of the peaks (approximately $1°$) gives information about

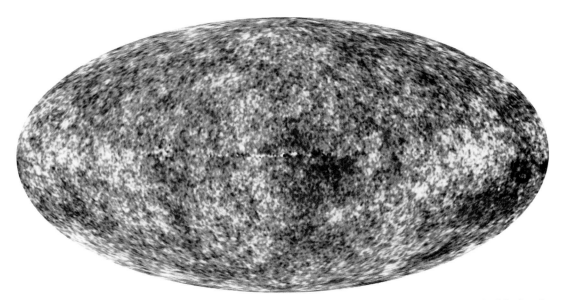

Fig. 1. Map of the universe based on cosmic microwave background temperature fluctuations. The whole celestial sphere is mapped into an oval by an equal-area Molleweide projection. The map is in galactic coordinates.

the vacuum energy density as well as the curvature of the universe.

First-year results. The data from the first year of *WMAP* observations were used to make the map shown in **Fig. 1**, which is an image of the universe as it was when the temperature of the cosmic microwave background had fallen to the point where electrons and protons could combine to form neutral hydrogen; according to *WMAP* data, this occurred 380,000 years after the big bang. The map is in galactic coordinates, so the plane of the Milky Way Galaxy runs horizontally across the oval, but the five frequency bands observed by *WMAP* have been combined in a way to cancel almost all of the foreground emission from the Milky Way Galaxy, as discussed below. The darkest areas on this map are about 0.5 mK colder than the lightest areas. This difference is much greater than the ± 30-μK variation observed by *COBE* for two reasons. First, the *COBE* $7°$ beam is not sensitive to the anisotropy between the 0.2 and $7°$ scales, and this region includes the main acoustic peak of the cosmic microwave background anisotropy angular power spectrum. As a result, *WMAP* sees a much larger root-mean-square anisotropy, $\sigma = 70$ μK. Second, the whole range of the map is about 7σ (the variation is $\pm 3.5\sigma$) due to the large number of beam cross sections in the sky.

The parameters derived from this map are consistent with results from the Hubble Space Telescope on the expansion rate of the universe, and with results from supernova observations indicating that the expansion of the universe is accelerating. When combined with these data, the universe is found to have very low spatial curvature. This agrees with the prediction of the inflationary scenario that the universe will be very nearly flat with nearly zero spatial curvature. The amount of fluctuation power is the same at all length scales, as predicted by the inflationary scenario.

Flat models consistent with the cosmic microwave background anisotropy data from *WMAP* have ages of 13.7 ± 0.2 billion years. The baryonic and dark-matter densities of the universe are found to be 0.42 and 2.1 yoctograms per cubic meter, respectively. (1 yoctogram is 10^{-24} gram; the proton mass is 1.67 yoctograms.) The universe is composed of approximately 4.4% baryons, 22% dark matter, and 73% vacuum energy. *WMAP* also found evidence from the polarization pattern of the cosmic microwave background that the universe was reionized by an early

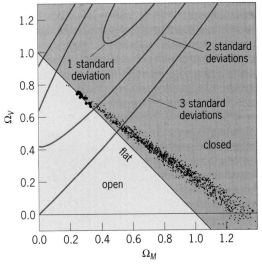

Fig. 2. Diagram of the vacuum energy density of the universe versus the matter density, both expressed as a fraction of the critical density ($\Omega_v = \rho_{vac}/\rho_{crit}$, $\Omega_M = \rho_{matter}/\rho_{crit}$), which divides open infinite models of the universe from closed finite models. Models consistent with cosmic microwave background data are shown as dots. Those models that are also consistent with the Hubble Space Telescope value of the Hubble constant are shown as larger dots. The contours show the 1-, 2- and 3-standard deviation confidence intervals based on supernova evidence.

generation of hot, ultraviolet-emitting stars about 200 million years after the big bang.

Figure 2 shows the values for two important parameters, the density of matter and the density of vacuum energy, that are given by cosmic microwave background, Hubble Space Telescope, and supernova observations, as well as the prediction of a flat universe. Both quantities, ρ_{matter} and ρ_{vac}, are expressed as a fraction of the critical density, ρ_{crit}; that is, they are expressed as the parameters $\Omega_M = \rho_{matter}/\rho_{crit}$ and $\Omega_V = \rho_{vac}/\rho_{crit}$. Here, the critical density is the total density of a flat universe, so that in such a universe $\rho_{matter} + \rho_{vac} = \rho_{crit}$, and $\Omega_M + \Omega_V = 1$. The latter is the equation of the diagonal line labeled "flat" in Fig. 2, and this line divides open infinite models from closed finite models. (The value of the critical density is given by $\rho_{crit} = 3H^2/8\pi G$, where G the gravitational constant, and the Hubble constant H is the expansion rate of the universe.) Models consistent with the *WMAP* and other cosmic microwave background data are shown as dots in Fig. 2. Those models that are also consistent with the Hubble Space Telescope value of the Hubble constant are shown as larger dots. The contours show the 1-, 2- and 3-standard-deviation confidence intervals based on the supernova evidence for an accelerating expansion rate. The large dots are consistent with all three data sets (cosmic microwave background, Hubble Space Telescope, and supernovae) and are also consistent with a flat universe.

Design. The design of *WMAP* was based on experience gained in the *COBE* project. The *COBE* satellite had demonstrated that the best contrast between the cosmic microwave background and the foreground emission from the Milky Way Galaxy was at a frequency of about 70 GHz. In order to separate the Milky Way foreground from the cosmic microwave background fluctuations, a large number of different frequencies were needed. The best technique available in 1995 to measure radiation temperatures at frequencies around 70 GHz used high-electron-mobility transistor (HEMT) amplifiers, which will operate at frequencies up to 100 GHz. HEMT amplifiers will work well at 90 K (−298°F), a temperature easily achieved in space by shielding an apparatus from the Sun and the Earth. Thus *MAP* has HEMT amplifiers—chilled to 90 K by passive radiative cooling—which measure the cosmic microwave background temperature differences in five frequency bands at 23, 33, 41, 61, and 94 GHz. The multiple frequencies allow the emission from foreground sources such as the Milky Way Galaxy to be separated from the cosmic

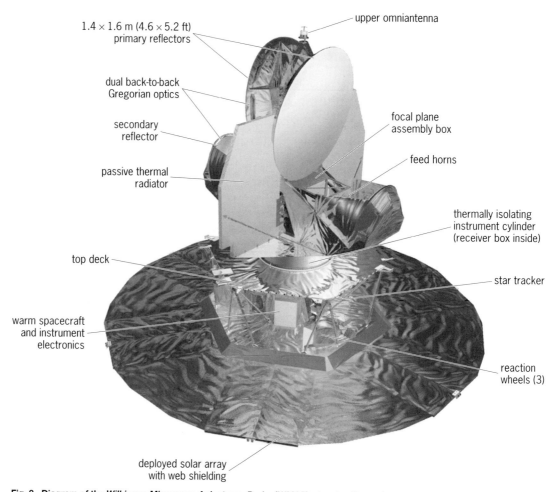

Fig. 3. Diagram of the *Wilkinson Microwave Anisotropy Probe* (**WMAP**), showing the major components of the instrument and spacecraft.

microwave background anisotropy. The optics consist of two 1.4×1.6 m (4.6×5.2 ft)–aperture off-axis Gregorian radio telescopes arranged back to back to allow precise differential measurements of cosmic microwave background temperature differences. **Figure 3** is a diagram of *WMAP*, showing how the two telescopes look out on opposite sides of the spin axis.

Angular resolution and sensitivity. The beam sizes scale with the wavelength, giving 0.93, 0.78, 0.58, 0.35, and 0.23° in the five bands. This gives *WMAP* an angular resolution that is 30 times better than the 7° resolution of the *COBE* differential microwave radiometer (DMR) instrument. In order to make useful measurements in these beams, which give 1000 times more beam-sized patches than the *COBE* DMR, the *WMAP* radiometers need to be capable of reaching a given sensitivity in 1000 times less integration time than the *COBE* DMR. Advances in microwave technology between the *COBE* launch in 1989 and the *WMAP* launch in 2001 actually yield a 2000-fold advantage in integration time, so 1 year of *WMAP* observations is equivalent to 2000 years with *COBE*. *WMAP* has one dual-polarization differential radiometer in each of the 23- and 33-GHz bands, but it has two radiometers in each of the 41- and 61-GHz bands, and four radiometers in the 94-GHz band. This increase in the number of detectors with frequency, along with a bandwidth proportional to the band center frequency, leads to a final temperature sensitivity that is almost the same for all five *WMAP* bands.

Scan pattern. The beam patterns from the two telescopes are centered 70.5° from the spacecraft spin axis but with a 180° difference in azimuth. The spacecraft spins around an axis 22.5° away from the antisolar direction with one rotation every 2.2 minutes, but this spin axis precesses around the antisolar direction once per hour. This complex scan pattern covers a large area of sky quite quickly: The "doughnut" from 87 to 132° away from the Sun that contains 36% of the sky is covered in 1 hour. Over the course of 6 months, this scan pattern covers the whole sky as the Earth moves around the Sun.

Orbit. *WMAP* is in a halo orbit around the second Lagrange point, L2, in the Earth-Sun system. This point is on the line from the Sun to the Earth but 1.5×10^6 km (9.4×10^5 mi) farther from the Sun than the Earth. As a result the bright microwave emissions from the Sun, the Earth, and the Moon are always in the same part of the sky, which is blocked by the sun shield on *WMAP*. The sun shield also carries solar panels to provide power to the spacecraft. The L2 orbit is only metastable, so orbit corrections are necessary every few months. The current fuel supply is adequate for decades of these routine orbit corrections, but *WMAP* was designed for a 2-year lifetime and in late 2003 was funded for up to 4 years of operations. *See* JAMES WEBB SPACE TELESCOPE.

For background information *see* BIG BANG THEORY; COSMIC BACKGROUND RADIATION; COSMOLOGY; GALAXY, EXTERNAL; HUBBLE CONSTANT; IN-FLATIONARY UNIVERSE COSMOLOGY; MICROWAVE SOLID-STATE DEVICES; QUANTIZED ELECTRONIC STRUCTURE (QUEST); SUPERNOVA; UNIVERSE in the McGraw-Hill Encyclopedia of Science & Technology.

Edward L. Wright

Bibliography. C. L. Bennett et al., First year Wilkinson Microwave Anisotropy Probe (WMAP) observations: Preliminary maps and basic results, *Astrophys. J. Suppl.*, 148, 1, 2003; C. L. Bennett et al., The Microwave Anisotropy Probe Mission, *Astrophys. J.*, 583:1–23, 2003; W. Freedman et al., Final results from the Hubble Space Telescope key project to measure the Hubble constant, *Astrophys. J.*, 553:47–72, 2001; J. L. Tonry et al., Cosmological results from high-z supernovae, *Astrophys. J.*, in press, 2003.

Wireless fidelity technology

The international standard 802.11 of the Institute of Electrical and Electronics Engineers (IEEE) governs wireless local-area network (LAN) technologies that operate on unlicensed portions of the radio-frequency spectrum. Such technologies enable computers to communicate wirelessly with each other or with a wired LAN over a distance of up to about 100 m (300 ft) in a normal office environment. Wireless fidelity (Wi-Fi®) refers to IEEE 802.11–based wireless LAN technologies that have passed interoperability tests designed by the Wi-Fi Alliance (formerly the Wireless Ethernet Compatibility Alliance), an industry organization of vendors of 802.11 wireless LAN products. Wi-Fi is extensively deployed in public areas, such as airport terminals, hotel lobbies, and coffee shops, so that customers can easily use their laptop computers to access the Internet and retrieve and send e-mails and other forms of communication.

The IEEE 802.11 standard consists of medium access control (MAC) specifications and physical layer (PHY) specifications. The MAC specifications define how a wireless LAN entity exchanges data with others using a shared wireless medium. The PHY specifications define the wireless signals that carry data between wireless LAN entities and the wireless channels over which the wireless signals are transmitted. IEEE 802.11 was introduced in 1997 with a MAC specification and three PHY specifications. It has been amended with many extensions, including three high-speed PHY specifications: 802.11b for 2.4 GHz (1999), 802.11a for 5 GHz (1999), and 802.11g for 2.4 GHz (2003). At present, Wi-Fi specifically means technologies based on the 802.11b PHY, 802.11a PHY, or 802.11g PHY, the 802.11 MAC, and Wi-Fi protected access (WPA).

IEEE 802.11 network architecture. IEEE 802.11 defines two networking components, a station and an access point (AP). In general, a station is a network adaptor or a network interface card that connects to the wireless medium. It has a MAC layer and a PHY layer. It is responsible for sending or receiving data to or from another station over the wireless medium, and does not relay data for any other station. An access point is a special station whose main task is to relay data for other stations. An access point is most likely attached to a wired LAN and relays data between stations associated with it and network components deployed on the wired LAN.

The data transmitted by the MAC layer are packed into frames. A frame consists of a header, a payload, and a tail. The header contains control information such as frame type, frame length, and MAC addresses. The payload can accommodate up to 2304 bytes of data. The tail is a 32-bit cyclical redundancy check (CRC) code that can tell a receiving MAC layer whether the received frame is corrupted.

IEEE 802.11 supports three networking modes: independent basic service set (IBSS), basic service set (BSS), and extended service set (ESS). An IBSS (**Fig. 1a**) is a wireless LAN consisting of stations only, and is usually set up on a temporary basis for a specific purpose. Every station competes for the wireless medium to send data frames to other stations directly, and there is no guarantee that any two stations in the same IBSS are in direct communication range. A BSS (Fig. 1b) is an infrastructure-based wireless LAN that consists of an access point and a number of stations associated with it. The access point may be attached to a wired LAN. Each station communicates only with the access point, and the access point relays data frames exchanged between these stations or between them and the wired LAN. An ESS (Fig. 1c) is a set of BSSs with all access points connected through a backbone network (a wired or wireless LAN). It provides mobility support for stations that move from one access point to another. The access point with which a station previously associated must ascertain the identity of the access point with which the station is currently associating, and must forward received inbound data frames for the station, if any, to that access point.

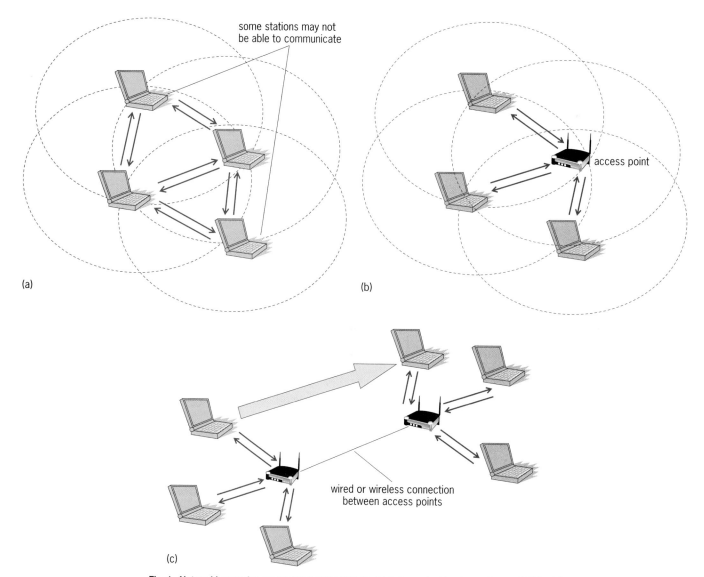

(a)

(b)

(c)

Fig. 1. Networking modes supported by IEEE 802.11. (*a*) Independent basic service set (IBSS). (*b*) Infrastructure-based basic service set (BSS). (*c*) Extended service set (ESS). The large arrow indicates that a station can move within a large area without loss of connectivity by associating with different access points.

IEEE 802.11 MAC. The IEEE 802.11 MAC specification provides three functions: fair access to the shared wireless medium with optional quality of service, reliable delivery of data frames over the wireless medium, and security protection for data frames.

Fair access. The IEEE 802.11 MAC employs carrier sense multiple access with collision avoidance (CSMA/CA), distributed coordination function (DCF) mode as the basic medium access mechanism to assure fair sharing of wireless medium by many stations. When a station has a data frame to transmit, the MAC needs to go through a CSMA/CA process illustrated by the flowchart in **Fig. 2.** (The figure shows only the logic of the procedure; the actual implementation uses a complicated event-driven approach for efficient use of the central processing units on the station.)

The IEEE 802.11 MAC also defines a point coordination function (PCF) mode of CSMA/CA. This mode can provide differentiated services for two classes of stations. It runs in conjunction with DCF and must be carried out by an access point. The access point maintains a list of stations that request higher quality of service. It periodically enters PCF mode from DCF mode by transmitting a contention-free polling frame, which announces the reserved time window for PCF-mode operation. No station may compete for the wireless medium during this period. The access point then polls stations one by one and relays traffic for them with a guaranteed bandwidth and delay bound in this contention-free period. To serve more complicated quality-of-service requirements, the 802.11e draft standard defines enhanced DCF (EDCF) and hybrid coordination function (HCF) modes. Both of them can support differentiated services for eight traffic categories for every station.

Reliable delivery. Since an 802.11 wireless LAN operates in unlicensed spectrum, the wireless medium could be very noisy due to the existence of other types of radio devices operating on the same band

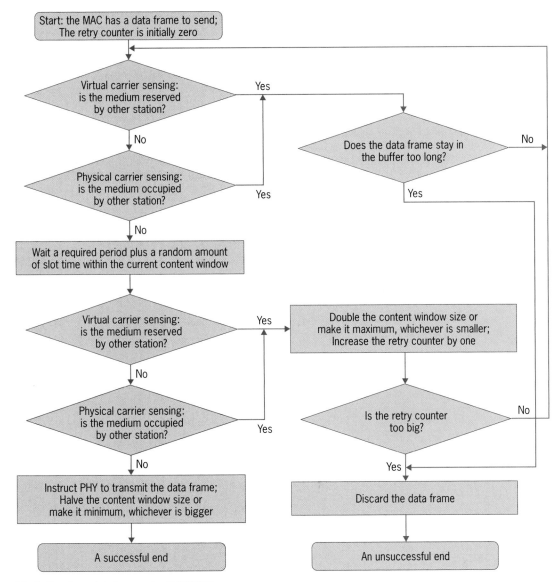

Fig. 2. Sending a data frame using CSMA/CD.

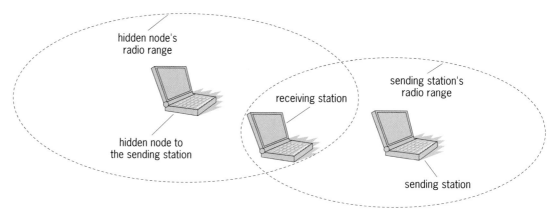

Fig. 3. Hidden-node problem for a wireless network.

and not conforming to CSMA/CA, and thus data frames can be frequently corrupted. In order to assure reliable delivery of data frames over the noisy wireless medium, the IEEE 802.11 MAC employs a data-frame exchange protocol for transmission of every data frame. It has a two-message form and a four-message form. The two-message form is suitable for short data-frame transmissions. It works as follows: (1) Station A sends a data frame to Station B; (2) if it is correctly received, Station B sends a short acknowledge (ACK) frame to Station A; and (3) if Station A does not receive the ACK frame on time, it tries to resend the data frame unless the MAC decides to drop the data frame after a sufficient number of retries. The four-message form is designed to overcome the hidden-node problem. A hidden node is a station that is outside the sending station's radio range but inside the receiving station's radio range (**Fig. 3**). Even if both the sending station and the hidden node run CSMA/CA, their data frames could collide at the receiving station, because neither of them can detect the other's signal or time window reservation announcement. The four-message data frame exchange protocol works as follows: (1) Station A sends a short request-to-send (RTS) frame to Station B, which tells the time window that must be reserved for the forthcoming data frame transmission; (2) Station B sends a short clear-to-send (CTS) frame to Station A, which announces the reserved time window to all stations in Station B's radio range, including hidden nodes to Station A, and prevents them from sending during that period; (3) Station A sends the data frame to Station B in the reserved time window; and (4) if it is correctly received, Station B sends a short ACK frame to Station A.

Security protection. Since anyone in the range of a wireless LAN can receive data frames transmitted over the wireless medium and can send data frames to stations or access points on the wireless LAN, authentication and data encryption must be implemented for access control and privacy. IEEE 802.11 MAC defines wired equivalent privacy (WEP) to serve these needs. WEP supports shared-key authentication and per-packet encryption, all based on a 40-bit or 104-bit WEP key. Unfortunately, WEP has a number of significant security flaws and completely fails

its missions. It allows authentication spoofing (falsification of a trusted identity). More seriously, an eavesdropper can break a 104-bit WEP key after collecting about 4 to 6 million encrypted data frames, which can be done in a time as short as a few hours. In order to fix WEP, the 802.11i draft standard has been developed. It specifies a new data security protocol, temporal key integrity protocol (TKIP), to replace the flawed WEP. It also introduces a 128-bit advanced encryption standard (AES) as an enhanced data security method. However, since the development of 802.11i has been slow, many prestandard wireless LAN security technologies have been rushed to the market. To respond to the market demand, the Wi-Fi Alliance has started interoperability tests for these technologies based on the 802.11i draft standard version 3.0, which is called WPA (Wi-Fi protected access). *See* TELECOMMUNICATION NETWORK SECURITY.

IEEE 802.11 PHY. Currently, three 802.11 PHY standards are in use. They are 802.11b, 802.11a, and 802.11g. Among them, 802.11b has the largest installed base.

The 802.11b standard defines a direct sequence spread spectrum (DSSS) radio PHY for the 2.4-GHz industrial, scientific, and medical (ISM) band. The detailed worldwide frequency plan is shown in **Table 1**. In North America, there are 11 channels. Each channel spans 22 MHz, and its center-frequency distance from that of adjacent channels is 5 MHz. This means every channel partially overlaps on several adjacent channels, and thus there are effectively only three nonoverlapping channels available. The 802.11b standard specifies four data rates for the DSSS radio: 1, 2, 5.5, and 11 megabits per second (Mbps). The major interfering sources for 802.11b-based wireless LAN are microwave ovens, 2.4-GHz cordless phones, and Bluetooth devices.

The 802.11a standard defines an orthogonal frequency-division multiplexing (OFDM) radio PHY for the 5-GHz unlicensed national information infrastructure (U-NII) band. The detailed worldwide frequency plan is shown in **Table 2**. In the United States, there are 12 nonoverlapping channels available in three groups with different limits on transmitter power. In order to satisfy regulatory requirements for operation in the 5-GHz band in Europe, some

TABLE 1. 2.4-GHz frequency plan and power regulation applied to standard 802.11b

Channel number	Channel center frequency, MHz	North America		ETSI* regulatory domain		Japan	
		Channel allocation	Power regulation (maximum output power)	Channel allocation	Power regulation (maximum effective isotropic radiation power)	Channel allocation	Power regulation (maximum output power)
1	2412	X	1000 mW	X	100 mW	X	220 mW†
2	2417	X		X		X	
3	2422	X		X		X	
4	2427	X		X		X	
5	2432	X		X		X	
6	2437	X		X		X	
7	2442	X		X		X	
8	2447	X		X		X	
9	2452	X		X		X	
10	2457	X		X		X	
11	2462	X		X		X	
12	2467			X		X	
13	2472			X		X	
14	2484					X	220 mW†

*ESTI = European Telecommunications Standards Institute.
† 10 mW/MHz.

spectrum and power management functions must be added to 802.11a, such as dynamic frequency selection for radar signal avoidance and uniform channel spreading. These functions are specified in the 802.11h draft standard. The 802.11a standard defines eight data rates for the OFDM radio: 6, 9, 12, 18, 24, 36, 48, and 54 Mbps. Compared with 802.11b, 802.11a offers higher speed and more nonoverlapping channels but in general has a smaller operating range since 5-GHz wireless signals are attenuated faster than 2.4-GHz ones. For these reasons, 802.11a is more suitable than 802.11b for dense, high-speed installations.

The 802.11g standard defines an OFDM radio PHY that supports the same high data rates offered by 802.11a but remains compatible with the popular 802.11b. Specifically, the compatibility means that (1) every 802.11g station supports 802.11b PHY, so

TABLE 2. 5-GHz frequency plan and power regulation applied to standard 802.11a

Channel number	Channel center frequency, MHz	United States		CEPT* regulatory domain	
		Channel allocation	Power regulation (maximum output power with up to 6 dBi antenna gain)	Channel allocation	Power regulation (maximum effective isotropic radiation power)
36	5180	X	40 mW	X	200 mW
40	5200	X		X	
44	5220	X		X	
48	5240	X		X	
52	5260	X	200 mW	X	200 mW
56	5280	X		X	
60	5300	X		X	
64	5320	X		X	
100	5500			X	1000 mW
104	5520			X	
108	5540			X	
112	5560			X	
116	5580			X	
120	5600			X	
124	5620			X	
128	5640			X	
132	5660			X	
136	5680			X	
140	5700			X	
149	5745	X	800 mW		
153	5765	X			
157	5785	X			
161	5805	X			

*CEPT = European Conference of Postal Telecommunications Administrations.

it can communicate with an 802.11b station using 802.11b; and (2) in a BSS consisting of an 802.11g access point and a mixture of 802.11g stations and 802.11b stations, the 802.11g stations can communicate with the 802.11g access point using 802.11g for high speed, while 802.11b stations can communicate with the 802.11g access point using 802.11b, with no interference between 802.11g stations and 802.11b stations.

For background information *see* LOCAL-AREA NETWORKS; MOBILE RADIO; MULTIPLEXING AND MULTIPLE ACCESS; SPREAD SPECTRUM COMMUNICATION; TELEPROCESSING in the McGraw-Hill Encyclopedia of Science & Technology.   Hui Luo

Bibliography. W. Arbaugh et al., Your 802.11 wireless network has no clothes, *IEEE Wireless Commun. Mag.*, pp. 44–51, December 2002; Daqing Gu and Jinyun Zhang, QoS Enhancement in IEEE 802.11 Wireless Local Area Networks, *IEEE Commun. Mag.*, pp. 120–124, June 2003; P. Henry and Hui Luo, Wi-Fi: what's next?, *IEEE Commun. Mag.*, pp. 66–72, December 2002; B. O'Hara and A. Petrick, *802.11 Handbook: A Designer's Companion*, IEEE Press, 1999; W. Pattara-Atikom and P. Krishnamurthy, Distributed mechanisms for quality of service in wireless LANs, *IEEE Wireless Commun. Mag.*, pp. 26–34, June 2003.

Wood anatomy

As one of the world's most abundant and accessible renewable resources, wood has been of immense utility and value to humans. Wood can be durable (resistant to decay and insect attack) and long-lasting, and has often survived as fossil or archeological material; it is therefore one of the major sources of information on past vegetation, climate, and civilization. When wood is burned to form charcoal, the carbonized product is completely inert, but retains most of the anatomical characteristics of the original wood and can similarly be studied. Charcoal with good preservation is known from at least as far back as the Carboniferous Period, when conifers first appeared in the fossil record.

Wood anatomy has long been of value in taxonomic and systematic studies. The very fact that it is usually possible to identify a piece of wood on the basis of anatomical characteristics means that these can be used to influence, confirm, or refute natural classifications (that is, those characteristics that are considered to reflect the evolutionary history of the group in question). With the advent of molecular phylogenies of flowering plants, it is now possible to see where discrepancies lie in the two forms of evidence, and to create firmer hypotheses on when and how often particular character traits evolved. This, of course, does not simply apply to wood anatomy but to many other morphological features such as leaves, pollen, and flowers.

Microscopic examination. Wood is usually examined in three planes: transverse section (or surface), tangential longitudinal (section), and radial longitudinal (section) [**Fig. 1**]. This provides a good three-dimensional view of anatomical structure. For examination with a light microscope, sections between 15 and 30 micrometers are cut in these three planes using a sliding microtome. The sections can be mounted unstained or stained. Charcoal cannot be sectioned because it is too brittle, but can be examined with an epi-illuminating microscope, which reflects light off the subject, or with a scanning electron microscope, which provides good three-dimensional images of wood but is an expensive technique.

Those studying developmental aspects of cambial activity, and therefore secondary xylem and phloem production, often use a transmission electron microscope. Wood anatomists often wish to examine individual cells, either to see fine structural features or to measure cell length. To do this, they macerate small samples, either using Jeffrey's solution (a mixture of nitric and chromic acids) or a mixture of hydrogen peroxide and glacial acetic acid. Another technique is to infiltrate the wood lumina with resin, dissolve the cell walls with acid, and examine the resulting resin casts and wall architecture with a scanning electron microscope. There are many techniques for examining wood, depending on the information sought.

Vascular tissue structure. Wood is composed of secondary xylem, which is produced by the vascular cambium, a region of secondary tissue growth in dicotyledonous angiosperms (hardwoods) and in gymnosperms (softwoods). The cambial cells divide outward to form the secondary phloem and inward to form the secondary xylem. The secondary xylem (and therefore wood) can be divided into an axial, or vertical (along the grain), system and a radial, or horizontal, system.

Angiosperms. Angiosperms have more complex wood than gymnosperms, and the appearance of different taxa is much more diverse. The axial tissue usually comprises a combination of vessels, fibers, axial parenchyma, and sometimes tracheids (Fig. 1). All axial tissue originates from fusiform cambial initials, which are spindle-shaped cells with tapered ends. The radial system consists of ray parenchyma cells (rows of cells involved in lateral conduction and storage of food, water, and other materials). Vessels are perforate (open-ended) tubes, dead at maturity, that transport water up the plant from the roots to the leaves, where water is lost by transpiration. Vessels vary in diameter from approximately 20 to 500 μm, and are often about the same length as the fusiform initials from which they developed. Fibers are also usually dead at maturity, have strongly lignified, often thick cell walls, and provide structural strength to the wood. They grow intrusively, so at maturity they are much longer than their originating fusiform cells. They can be 1 to 2 mm or longer. Axial parenchyma cells often remain alive for much longer, and are the repository for storage materials such as starch grains. They are rarely much taller than the fusiform initials from which they develop, but cross-walls are produced by cell division, and individual strands can be two-, four-, or eight-celled (or occasionally odd-numbered) or even longer than eight cells. Tracheids, when present, are imperforate tubes, generally narrower and longer than associated vessels but with similar wall pitting; since they have no perforation plates, water flows between tracheids through the pits.

The arrangement and proportions of these various cell types are usually strongly indicative of the taxonomic position of the species from which the wood comes. Some extreme examples are vesselless angiosperms, such as *Trochodendron, Tetracentron,* and Winteraceae, which look superficially like softwoods and *Hebe,* one of the few woody dicotyledons which lacks rays. Most dicotyledons have a combination of vessels, fibers, axial parenchyma, and rays. Angiosperm woods can be broadly divided into those that are ring-porous—that is, which have wide earlywood vessels (formed early in the growing season) and much narrower latewood vessels in a given growth ring, such as *Fraxinus,* most *Quercus, Ulmus, Robinia, Carya*—and are found entirely in temperate regions; and those that are diffuse-porous (with all vessels more or less the same diameter), and are found in temperate, subtropical, and tropical regions, which are the majority.

Gymnosperms. The wood of gymnosperms is generally less complex than that of dicotyledons, and is usually recognizable as such to the naked eye. Virtually all softwood comes from conifers, which have axially oriented tracheids (mostly with large circular bordered pits in the radial walls and no pits in the tangential walls), a varying amount of axial parenchyma, and radially arranged parenchymatous ray cells (**Fig. 2**). In some taxa, ray tracheids run along the top, bottom, or within a ray. The axially oriented tracheids are squarish or rectangular in cross section, and the wood is very uniform in appearance (apart from changes in shape across growth rings). Most softwoods grow in temperate regions and usually have well-defined growth rings. The rays are nearly always uniseriate (only one cell wide), except where they contain radial resin canals (Fig. 2). Resin canals transport resin and presumably have a protective function; they are found in both bark and wood, and will seal a wound when damage occurs.

Identifying softwoods to the family level is usually relatively straightforward. For example, Araucariaceae have distinctive alternate bordered pits in two or three rows in radial longitudinal section, whereas all other softwoods have larger, more circular uniseriate or occasionally biseriate rows of bordered pits. Pinaceae usually have radial and axial resin canals, and in combination with the appearance of the crossfield pits (pits between a ray and the adjacent axial tracheid), can usually be identified to genus level. Helical thickenings (outgrowths of the secondary cell wall) are found in the tracheids of few softwood genera, such as *Taxus,* or yew (which lacks resin

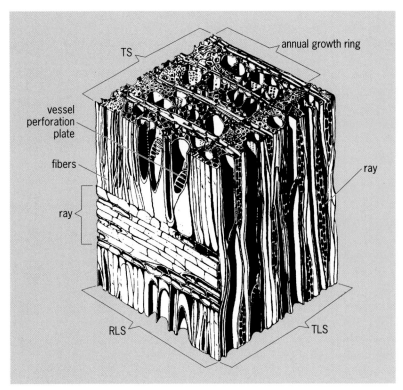

Fig. 1. Structure of a typical angiosperm, from the USDA. RLS = radial longitudinal section; TLS = tangential longitudinal section; TS = transverse section. (*Adapted from R. B. Miller, Wood anatomy, McGraw-Hill Encyclopedia of Science & Technology, 9th ed., 2002*)

canals), and *Pseudotsuga,* or Douglas-fir (which has resin canals).

Wood identification. Many professions, from archeology and paleontology to furniture restoration, require wood samples to be identified from time to

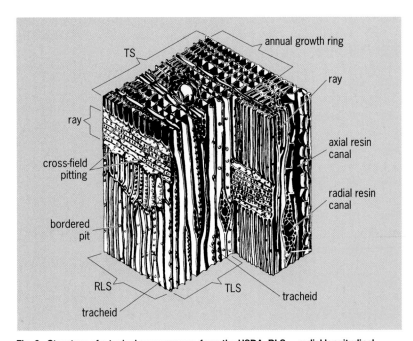

Fig. 2. Structure of a typical gymnosperm, from the USDA. RLS = radial longitudinal section; TLS = tangential longitudinal section; TS = transverse section. (*Adapted from R. B. Miller, Wood anatomy, McGraw-Hill Encyclopedia of Science & Technology, 9th ed., 2002*)

time. Wood identification is a specialized field which requires a knowledge of the various cell types that form woody tissue and how much variation is possible in the wood of a given species, genus, or family. A good reference collection of microscope slides is really essential, but books and computerized databases are very helpful initially. There is, unfortunately, no substitute for direct comparison with authentically named wood samples and microscope slides. The level at which an identification can be made depends on a number of factors. Very tiny samples may not allow sectioning in all three planes or may not possess all the features needed, and similar difficulties apply to degraded, waterlogged archeological material.

The geographical origin of a specimen may be unknown, or even sought by the identification. A common question is whether the wood used to make a piece of furniture is from Europe or North America. This can be impossible to answer, because identifications are generally safest to the genus level; oak, birch, elm, and pine, for example, commonly grow and are used in both places. Some families, such as Meliaceae (the mahoganies) are widespread in the tropics, and can be very difficult to tell apart macroscopically and microscopically. Knowing their geographical origin is often very helpful for determining their identity; for example, *Swietenia*, the true mahogany, is from the New World (although it is now plantation-grown elsewhere), *Entandrophragma* and *Khaya* are from Africa, and *Dysoxylum* is from Indomalesia. Many other reddish-brown woods are called mahogany, including some southeast Asian dipterocarps and even some eucalypts, so using botanical names is much more precise than using common names.

It is increasingly necessary to be able to distinguish between wood genera for ecological reasons. *Swietenia*, for instance, is now listed by the Convention on International Trade of Endangered Species (CITES). Deforestation of rainforests and overexploitation of certain timber species are controversial political issues, and a number of timber species, which can be identified by wood anatomists, are now controlled in international trade and covered by CITES regulations. Temperate, subtropical, and tropical forests are the sources of timbers with a variety of uses, including boat building, construction, and furniture, depending on their properties. Wood and charcoal are declining resources used as fuel by indigenous and local people, especially in the more arid parts of Africa, Asia, and South America, and this is a cause for concern in these areas.

Ecological anatomy and dendrochronology. Wood structure can provide information on the environmental conditions when the plant was growing. Growth rings are found most often in temperate plants, where the rings reflect annual periodicity. Growth rings can also occur in plants in the tropics, and often relate to variations in wet and dry seasons. Dendrochronologists can use variations in ring width to identify the specific years in which the

growth occurred and to produce chronologies over long periods, often into prehistory.

Some wood anatomists specialize in studying and interpreting the ecological and physiological implications of wood structure. Wood anatomy is often related to habit, habitat, or even the position of the sample in the tree. For example, lianas often have wider and longer vessels than trees, and this can be very apparent when studying members of a given taxonomic group (for example, Bignoniaceae, Malpighiaceae, and Leguminosae) with both growth forms. Wide vessels are much more efficient at water transport than narrow ones (if the diameter of a vessel is doubled, its conducting capacity increases by 16 times) but are at much greater risk of embolisms. Lianas also often have fewer fibers and a higher proportion of ray and axial parenchyma. These are all characteristics indicative of plants that do not need to support themselves but do need to transport water a long distance in a small cross-sectional area. Conversely, trees need to combine structural support with water transport.

For background information *see* DENDROCHRONOLOGY; MAGNOLIOPHYTA; PINOPHYTA; PLANT EVOLUTION; PLANT PHYLOGENY; WOOD ANATOMY; WOOD PROPERTIES; XYLEM in the McGraw-Hill Encyclopedia of Science & Technology.   Peter Gasson

Bibliography. S. Carlquist, *Comparative Wood Anatomy: Systematic, Ecological, and Evolutionary Aspects of Dicotyledon Wood*, Springer Series in Wood Science, 2001; N. J. Chaffey (ed.), *Wood Formation in Trees: Cell and Molecular Biology Techniques*, Taylor and Francis, New York, 2002; CITES (Convention on International Trade in Endangered Species); T. Fujii, Application of a resin casting method to wood anatomy of some Japanese Fagaceae species, *IAWA J.*, 14:273–288, 1993; IAWA Committee, IAWA list of microscopic features for hardwood identification, ed. by E. A. Wheeler, P. Baas, and P. E. Gasson, *IAWA Bull. 10*, pp. 219–332, 1989; *IAWA Journal* (*IAWA Bulletin* up to 1992), published quarterly by the International Association of Wood Anatomists; J. Ilic, *CSIRO Atlas of Hardwoods*, Springer-Verlag, 1991; A. Miles, *Photomicrographs of World Woods*, HMSO, London, 1978; J. Ohtani, *Wood Micromorphology: An Atlas of Scanning Electron Micrographs*, Hokkaido University Press, 2000; F. H. Schweingruber, *Trees and Wood in Dendrochronology*, Springer-Verlag, 1993; M. T. Tyree and M. H. Zimmermann, *Xylem Structure and the Ascent of Sap*, Springer Series in Wood Science, 2002.

Wood reuse and recycling

Although the general sources of discarded wood products are readily identified, the quantities discarded are difficult to document, and the proportion recycled or reused is even more difficult to quantify. Despite a general lack of data on the subject, it is evident that there is a growing emphasis on recycling and reuse of wood products that extends the useful life of wood, sometimes for many years.

Sources of wood discards. At the end of their useful life, wood products generally end up in the solid waste stream. They may be managed as a component of municipal solid waste or as a component of construction and demolition (C&D) debris.

Municipal solid waste (MSW). The U.S. Environmental Protection Agency (EPA) issues an annual report that characterizes the generation, recovery, and disposal of the various materials in municipal solid waste from residential homes, commercial establishments (such as office buildings and retail stores), institutional sources (such as schools, libraries, hospitals, and prisons), and industrial plants exclusive of process wastes (produced as a by-product of industrial processes). As of 2001, wood products' representation in municipal solid waste (in 10^3 tons) was as follows:

Overall MSW generation: 229,228
Overall MSW recovery: 67,989
Overall MSW percent recovery: 29.7
Wood waste generation: 13,188
Wood waste recovery: 1,250
Wood waste percent recovery: 9.5

In this listing, wood represents 5.7% of overall MSW generation and only 1.8% of total MSW recovery. Wood recovery includes wood used as a fuel. All data are from Franklin Associates, Ltd, *Working Papers for Wood Pallets and Wood Packaging.*

Wood products entering the waste stream in 2001 included wood framing in furniture and wood in other durable goods, such as cabinets for electronic products like television sets; however, the principal source of wood in municipal solid waste is wood pallets (platforms used for handling and transporting goods in large volumes) and other wood packaging (such as crates). Virtually all wood recovered from municipal solid waste for recycling consists of retired pallets (**Table 1**), mostly in the form of broken pallets or boards from pallet repairs. Pallets are usually shredded to produce mulch, animal bedding, or fuel. In 2001 the quantity of wood packaging converted to fuel was about 485,000 tons; used for mulch and animal bedding, 650,000 tons; and used for specialty products, 600,000 tons—for a total of 1,250,000 tons.

TABLE 1. Sources of wood wastes in MSW, 2001, in 10^3 tons

Category	Quantity	Recovery	Percent
Furniture	4,118	Negligible	0
Other durables	870	0	0
Pallets	8,170	1,250	15.3
Other packaging	30	Negligible	0
Total	13,188	1,250	9.5

Source: Franklin Associates, Ltd, *Working Papers for Wood Pallets and Wood Packaging.*

TABLE 2. Nonpallet sources of wood waste from MSW and C&D in 2001, in 10^6 tons

Source	Generation	Recovered, not usable
Municipal solid waste		
Wood Products	12.8	6.6
Wood yard trimmings	16.1	12.7
Total	28.9	19.4
Construction and demolition waste		
Construction	9.0	2.1
Demolition	27.4	16.2
Total	36.4	18.3

Source: D. McKeever, *BioCycle Magazine.*

Construction and demolition debris. Wood in various forms is a common component of C&D debris. Discarded wood products from the construction, renovation, and demolition of residential and commercial buildings include forming and framing lumber, siding, interior trim, window frames, doors, plywood, laminates, flooring, and scraps. In addition, trees, stumps, and branches enter the waste stream as a result of the clearing of land, storm damage cleanup, and normal maintenance such as tree trimming or removal of dead trees.

Table 2, based on recent USDA Forest Service Forest Products Laboratory research, shows a breakdown of some of the (nonpallet) sources of wood waste from both municipal solid waste and C&D for the year 2001, and the amount of each that was recovered but determined to be not usable.

Wood pallet recovery. For financial as much as environmental reasons, improved distribution systems are continuously pursued in the wood pallet industry, which in turn leads to more efficient management of wood resources. There are currently three types of pallet distribution systems: pallet pooling, pallet exchange, and expendable pallets.

Pallet pooling. Pallet pooling is a growing, highly efficient reuse system in which pallets are leased or rented to users. After use, the pallets are returned to a central location, where they are inspected, repaired as necessary, and put back into service. This closed system is designed to be a perpetual reuse system in which unusable wood components are recycled and essentially no solid waste is produced.

Pallet exchange. Pallet exchange involves the purchase of pallets, usually of standard dimensions (such as 48 by 40 in.) by the original shipper of goods. When the goods are delivered to a distribution center or to a retail or industrial plant, empty pallets are picked up in exchange for the fully loaded pallets. Usually the empty pallets are delivered to a processing center to be repaired or recycled, and repaired pallets are then sold to a new user. Repaired and refurbished pallets may make several trips before being retired or disposed of. Reuse is an important factor in this type of pallet distribution system, but disposal in landfills or by combustion is more common than it is in pallet pooling.

Expendable pallets. One-way (expendable) pallets are the lightest and least durable class of pallets. They are not intended for repeated use, although some may be reused or repaired and reused. However, by definition this type of pallet becomes solid waste after its initial use.

It is estimated that about 1 billion wood pallets were in circulation in 2001. About 487 million reusable pallets were taken out of service in 2001, of which about 338 million were refurbished and returned to service. (On average, reusable pallets make about five to ten trips in their useful life before being retired or disposed of; pooled pallets operating in a closed system have a significantly higher reuse rate.) The same year, 5 million tons of reusable pallets became unusable or were disposed of, 3 million tons of expendable pallets were discarded, and about 0.3 million tons of other wood packaging were disposed of.

Construction and deconstruction site recovery. Construction sites offer the best potential for recovery of materials, such as wood scraps and excess lumber, because there is minimal contamination of wood there. By contrast, demolition lumber is usually mixed with other building materials and may be difficult to separate.

The term "deconstruction," as used by the National Association of Home Builders, refers to dismantling or removal of materials from buildings before or instead of demolition. For many years it has been common practice to salvage building components, including wood (along with other building components such as plumbing), prior to demolition. Wood

siding, interior trim, window panes, doors, and flooring are examples of wood products that are routinely salvaged and reused. Interest in full deconstruction is growing, although the costs of the labor and methods required to rapidly deconstruct a building remain a major concern. Selling used lumber is one economic benefit of deconstruction. A major salvager of used lumber is an organization called ReStore, a part of Habitat for Humanity, which has about 60 facilities across the United States.

C&D recycling facilities that accept wood may convert it to mulch, animal bedding, compost bulking agent, or fuel. Forest products industries often use wood waste as a fuel. Activity in processing C&D material for reuse and/or recycling is growing. An initial report on C&D waste issued by the EPA in 1998 states that there were at least 1800 C&D recycling facilities in the United States in 1996, of which 500 were wood processing facilities; by 1998, the number of C&D recycling facilities in operation had grown to 3500.

Progress. Wood recycling and reuse is growing each year, as various industries are finding more cost-effective ways to recover and utilize end-of-life wood products rather than disposing of them in municipal or C&D landfills.

Recovered wood use in the pallet industry has been increasing steadily, according to research conducted by the Center for Forest Products Marketing and Management at Virginia Polytechnic Institute. In 1992, recovered wood accounted for 13% of the total board feet of lumber used in the pallet industry. In 1995, an estimated 171 million pallets were recovered, a 160% increase from 1992; of this recovered material, 87% was used again in pallets, raising the pallet industry's recovered wood use to 30%. Continued growth in pallet recovery and wood reuse increased recovered wood use in pallets to 42% by 1999.

Resource Recycling magazine carries out an annual survey of scrap wood processing facilities. Of the companies surveyed and reported in the March 2003 issue, the following breakdown was given for incoming materials: pallets and crates 22%; woody landscape and yard debris 26%; construction lumber 15%; the balance is from other sources such as construction and demolition lumber manufacturing scrap, sawmill residue, stumps, and painted wood. The principal products of the facilities were mulch, fuel, lumber, compost, and animal bedding.

William E. Franklin; Beverly J. Sauer

Recycled wood products. Discarded wood can be recycled into a wide range of products (see **illus.**). Almost all wood residuals can be recycled, with pressure-treated and painted, stained, or lacquered wood being the most problematic. Such recycled products can be broadly categorized as large-volume (commodity) or specialty products. The former encompasses mainly mulch and boiler fuel, whereas the latter consists of a wide variety of products.

Mulch. Mulch is a protective ground covering used, for example, to prevent soil drying, erosion, or weed

Value-added recycled wood products: (*a*) recycled wood-cement block; (*b*) compressed mulch form; (*c*) prefinished cherry flooring from discarded pallets; (*d*) fingerjointed, reassembled wood stud; (*e*) colored mulch; (*f*) threaded tubes from compressed sawdust; (*g*) wood fiber–plastic decking material with recycled wood flour; (*h*) cat litter from compressed wood residuals (this material is also used as boiler fuel).

growth. It can be made from all types of wood residuals, including yard debris; contaminated wood and plastic residuals, however, can render the final product useless. The wood is chipped to more or less uniform particles using a grinder. Colored mulch has gained popularity over the last few years and carries a price premium.

Boiler fuel. Boiler fuel can be created from wood residuals reduced to chips of a specified size. Sizes vary, with 2 in. (5 cm) being very common. Boiler fuel is most likely the lowest-margin product that is made from recycled wood. Depending on the boiler for which it is meant to be used, wooden boiler fuel can be contaminated with certain materials to some degree without ill effect. Due to its seasonality opposite that of mulch, the same processors often supply boiler fuel in the winter and mulch in spring and summer.

Specialty products. A wide variety of specialty products are made from recycled wood, such as compressed-wood fire logs, animal litter, and finger-jointed lumber. Some of these products have struggled to gain markets and become profitable. Successful examples include flooring made from discarded pallet deckboards, wooden concrete bricks (made from wood chips bonded together with cement), and wood fiber–plastic decking material (containing 60% finely ground wood flour and 40% recycled or virgin thermoplastic).

For background information *see* LUMBER; RECYCLING TECHNOLOGY; WOOD PRODUCTS in the McGraw-Hill Encyclopedia of Science & Technology.

Urs Buehlmann

Bibliography. R. J. Bush et al., Fewer pallets are reaching landfills, more are processed for recovery, *Pallet Enterprise*, May 2001; Down but not out: Habitat ReStore deconstructs home and buildings and then recycles the materials, *Kansas City Star*, May 31, 2002; EPA Office of Solid Waste, *Municipal Solid Waste in the United States: 2001 Facts and Figures*, July 2003; Franklin Associates, Ltd, *Working Papers for Wood Pallets and Wood Packaging*, 2001; D. McKeever, Taking inventory of woody residuals, *BioCycle*, pp. 31–35, July 2003; Mulch to be happy about, *Waste Age*, pp. 62–65, May 2003; Salvaged goods from deconstruction create community marketplace, *BioCycle*, pp. 19, March 2003; When recycling comes out of the woodwork, *Resource Recycling*, pp. 14–19, March 2003.

Xenotransplantation

Transplantation of solid organs or tissues has become a well-accepted clinical practice in the treatment of patients with end-stage organ failure. In particular, after the introduction of effective immunosuppressive drugs, prevention and treatment of rejection has become manageable and long-term survival has been achieved. However, due to this success there is an ever-increasing demand for donor organs, and the number of patients on the waiting list for an organ transplant far exceeds the number of available organs. In the search for alternative sources, organs from animals have been proposed. Xenotransplantation is defined as any procedure that involves the transplantation, implantation, or infusion into a human recipient of either (1) live cells, tissues, or organs from a nonhuman animal source or (2) human body fluids, cells, tissues, or organs that have had ex-vivo contact with live nonhuman animal cells, tissues, or organs. This article discusses the transplantation of cells, tissues, or organs.

Pig donors. The species that is generally proposed as a donor for clinical application is the pig. This choice is based on a number of facts. First, the size of its organs is compatible with that in humans, in particular when using miniature swine that grow to a maximum size of about 125 kg as compared with 450 kg in other pig strains. Second, organ physiology is similar in pigs and human beings. Third, pigs can be bred in large litters under conditions of minimal infectious risk (barrier facilities, using procedures such as early weaning and cesarian section). Fourth, ethical aspects of using pigs for xenotransplantation are generally considered acceptable.

There have been a number of clinical trials performed using porcine transplants in patients, in particular cell or tissue transplants (pancreatic islets of Langerhans in diabetic patients and neuronal cells in patients with Huntington or Parkinson's disease); the study of solid organs has been limited thus far to a few solitary explorations. A number of aspects are presently addressed in preclinical research before xenotransplantation can be developed to a clinical procedure.

Immune barrier. Since a transplant essentially is seen as foreign, it will elicit an immune response in the recipient, resulting in rejection (**Fig. 1**). In (clinical) allotransplantation (transplantation in which the donor and recipient are of the same species), this rejection reaction is mainly mediated by the cellular immune system (T-lymphocytes).

Immunosuppression. A spectrum of immunosuppressives drugs have been developed that particularly target this rejection reaction. Examples of currently used drugs are cyclosporine, tacrolimus, mycophenolate mofetil, rapamycin, and steroids; biologicals such as antithymocyte globulin or monoclonal antibodies to T-lymphocytes or activated T-lymphocytes are also used. These are often applied in combination to achieve the optimal balance between efficacious prevention of rejection and avoidance of potential side effects. However, side effects are not uncommon, and can be either direct toxicity of drugs or indirect consequences of a depressed immune system such as infections (in particular, viral infections) and tumors (in particular, of the lymphoid system; also called posttransplant lymphoproliferative disease).

Immunologic tolerance. Another approach to achieve long-term graft survival is the induction of immunologic tolerance. This is achieved by using certain drugs or by eradication of the recipient's immune system followed by bone marrow transplantation to

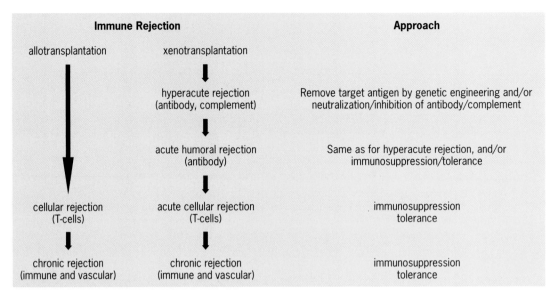

Fig. 1. Potential mechanisms involved in the rejection of an allograft and a porcine-to-human solid organ xenograft, and approaches to preventing graft rejection.

rebuild immunity so that it recognizes a graft as "self." Immunologic tolerance has been achieved only in small series of patients and has not yet reached the clinical status of the use of drugs for chronic immunosuppression.

Hyperacute rejection in xenotransplantation. In xenotransplantation the immune rejection is different and can include all branches of the immune system (Fig. 1). For a pig-to-human transplant, this at first involves the presence of naturally existing antiporcine antibodies that are specific for a sugar antigenic determinant (Gal epitope) found on glycoproteins and glycolipids on the surface of most pig cells; these antibodies emerge as a result of contact with intestinal flora when the gastrointestinal system becomes colonized after birth. This response is similar to the formation of blood group isohemagglutinins (red blood cell antibodies that induce clumping of red blood cells of other individuals of the same species) in the red blood AB0 system. Human and Old World nonhuman primates lack the enzyme Gal-transferase, which is involved in the synthesis of the Gal epitope, and thus recognize it as foreign; other species such as pigs have this enzyme and show a widespread tissue expression of the Gal epitope. A major part of the anti-Gal antibodies are formed in a T-cell-independent manner; therefore conventional immunosuppressants targeting T-lymphocytes are not effective in managing this antibody synthesis.

As a consequence of this natural antibody activity, solid organ xenografts undergo hyperacute rejection. Immediately after reperfusion of the graft (that is, return of blood flow to it), the antibodies stick to the endothelium lining the blood vessels, and this is followed by activation of the complement system, a powerful cascade of proteins that in sequence activate each other and finally result in destruction of the target. The complement system normally is important in host defense, for example, against bacterial

infections. As a result, the vasculature is damaged, bleeding starts throughout the organ, and the tissue parenchyma is destroyed within minutes to hours. There are indications that hyperacute rejection is less important for cell or tissue grafts that are not directly connected to the recipient's blood circulation upon transplantation. For instance, hyperacute rejection has not been observed for porcine islet cell transplants.

Prevention of hyperacute rejection. To avoid hyperacute rejection, a number of approaches have proven effective. The natural antibodies can be removed by injection of sugar-containing compounds or by perfusion of the recipient's blood through a column containing Gal molecules. Complement activity can be depleted or inhibited by using certain drugs. Modification of the donor has also been attempted. For example, pigs have been genetically engineered to express human regulators of complement activation, such as decay-accelerating factor and membrane cofactor protein. (The rationale for introducing these molecules by transgenesis is that the native-species molecules normally do not work across the xenogeneic barrier.) These molecules on the surface of, for example, endothelial cells prevent complement activation on the cell surface and thus cell destruction.

In a rather new development, pigs have been genetically altered to not express the main target, the Gal sugar moiety. This is done by gene targeting, which inactivates the gene encoding the enzyme Gal-transferase. Gene targeting by embryonic stem cell technology has long been possible in rodents (knockout mice), but became possible in other species just a few years ago using the nuclear transfer/embryo transfer technology. The first Gal-deficient pigs were generated in the second half of 2002, and the first transplantation studies using organs from these animals are ongoing. The generation of Gal-deficient pigs represents a major step forward, as the modification is directly related to the absence of the target (Gal) and is, therefore, proposed to be more biologically stable in controlling antibody-mediated rejection.

Long-term prevention of rejection. If hyperacute rejection can be avoided, rejection can still occur after sensitization of the immune system. This includes both antibody synthesis and the development of a cellular response (cytotoxic T-cells and mediator molecules synthesized by these cells). In animal transplantation models it has become clear that immunosuppressive drugs can yield graft survival, but the long-term xenogeneic immune response is much more difficult to suppress than a similar response to an allograft. For solid organs such as the heart and kidneys, it has also become apparent that antibody-mediated rejection is much more frequent than cell-mediated rejection. This can involve both natural antibodies that are stimulated further and the generation of antibodies with novel specificities.

In a number of studies using organs from donors with transgenic complement regulatory proteins, long-term survival has been achieved in individual animals: the best results reported have been a median survival time of about 3 months in a series of heart transplants into baboons. Due to the heavy chronic immunosuppression, side effects, mainly drug toxicity and/or infections, occur frequently. Although current survival rates are considered important progress, they are lower than those proposed by advisory committees in consideration of clinical trials (for example, for a life-supporting graft in a nonhuman primate, up to a half-year survival rate has been proposed). Thus, it is hypothesized that immunologic tolerance might be the ultimate goal in further developments in xenotransplantation of porcine donors. This includes rebuilding the immune system with a porcine thymus and/or bone marrow after whole-body irradiation and cytotoxic drugs. Initial explorations of this approach in nonhuman primate models have shown that it is feasible.

Microbiological safety. A major complication in allotransplantation (as well as blood transfusion) is the transmission of viruses such as human immunodeficiency virus (HIV), hepatitis viruses, and herpesviruses, in particular cytomegalovirus and γ-herpesvirus or Epstein-Barr virus (**Fig. 2**). These viruses can also be present in the recipient and become reactivated as a result of immunosuppression. In porcine xenotransplantation, concerns about microbiological safety include the possibility not only that infectious agents from a porcine donor will infect the recipient but also that the recipient may harbor a porcine-derived microorganism and become generally infectious. Viruses are of particular concern, exemplified by the presumed nonhuman primate origin of HIV in the acquired immune deficiency syndrome (AIDS) outbreak.

For porcine xenografts, potential infectious agents can be classified as exogenous or endogenous (Fig. 2).

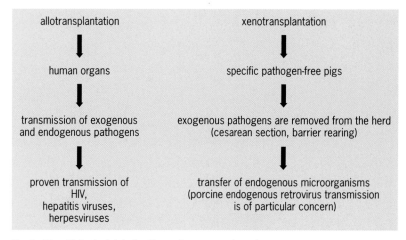

Fig. 2. Microbiological risks in allo- and xenotransplantation.

Exogenous bacteria and viruses. Exogenous microorganisms are acquired, for instance bacteria and viruses such as the herpesviruses. Such contaminants can be removed from the pig herd by using barrier facilities and cesarian section or early weaning in the breeding program. A large number of bacteria, fungi, protozoa, and viruses can so be removed in rendering pig donors "clean" for xenotransplantation purposes. Prion proteins (which cause diseases such as Creutzfeldt-Jakob in humans) have also been considered; however, they appear to be of less importance, as they have not been identified in the porcine species.

Endogenous viruses: PERV. The main issue in safety is, therefore, the presence of endogenous viruses. The identification of porcine endogenous retrovirus (PERV) a decade ago led to a number of studies on its characteristics, cross-species transmission, and induction of disease. Up to three subtypes (PERV-A, PERV-B, and PERV-C) have been identified. PERV is not related to retroviruses such as HIV, but bears resemblance to some leukemia viruses in rodents and cats. However, PERV has not yet been associated with any disease in pigs. In in-vitro transmission studies using co-cultures of human cells and PERV-containing porcine cells, a few human cell lines that are susceptible to transmission by PERV-A and PERV-B have been identified; although most cell lines studied were not susceptible, this observation indicates the potential of cross-species transmission.

A number of studies have been performed on humans that have been directly exposed to living porcine tissue, for example by extracorporeal blood perfusion through a porcine spleen or a device containing porcine liver cells, or by transplantation of porcine islets or fetal neuronal cells. Analysis using very sensitive molecular assays indicated that none of these patients actually showed the presence of PERV, suggesting that in-vivo cross-species transmission is an unlikely event. Cross-species transmission has also not been found in animal studies, including infectivity experiments in nonhuman primates. Thus, it appears that the current safety risk regarding

PERV is manageable, and in general, clinical trials are approved when there is a careful long-term monitoring system in place.

Organ physiology. It can be questioned whether cells, tissues, or organs from a pig will sustain life in a patient with end-stage organ failure, as at many levels incompatibilities may exist between species. (For example, regulators of complement activation at the endothelial cell surface do not function appropriately between the pig and human species, which is why pig donors were genetically engineered to express human versions of these molecules). However, one of the main reasons why pigs have been selected as donor species is their physiologic compatibility with humans at the organ level. This compatibility has been demonstrated in preclinical transplantation models, in which pig kidneys sustained life in bilaterally nephrectomized nonhuman primates, and a pig heart supported the blood circulation in nonhuman primates. The same results have been demonstrated for xenotransplantation of porcine islets in diabetic nonhuman primates, supporting the general use of xenogeneic insulin in diabetic patients. Although xenotransplantation of organs with metabolic function at multiple levels, such as the liver, might not be expected to achieve the same success, there are indications that a porcine liver supports life in a hepatectomized nonhuman primate, at least for a short time. It seems more likely, however, that liver xenotransplantation will be restricted so that the graft has to exert only part of its complex function, for example, the extracorporeal removal of toxic substances in acute liver failure. It is anticipated that aspects of physiology, in particular homeostasis in multifactorial processes such as coagulation, will be properly addressed once long-term organ survival has been achieved.

Conclusion. The potential of xenotransplantation as a viable solution for the growing shortage of donor organs for patients with end-stage organ failure was boosted 10 years ago, when pigs transgenic for complement regulatory proteins first became available. The progress of subsequent research in nonhuman primate transplantation has demonstrated that the immune barrier is quite large, requiring extensive immunosuppression or immune tolerance induction.

Although substantial survival has been achieved, more work is needed to reach long-term survival before development into clinical trials can be considered. Substantial progress is predicted from the genetic modification of pigs that lack a major target epitope for naturally existing antibodies. The most promising tissue and solid organ candidates for xenotransplantation include the pancreatic islets of Langerhans, to treat diabetes, and the kidneys and heart.

The identification of PERV and its potential transmission to some human cell lines in vitro has raised concern about the safety of xenotransplantation. However, the lack of PERV infection in humans exposed to living PERV-containing porcine tissue and the finding that PERV infection does not occur in living animal subjects, together with the absence of any disease association of PERV infection, make it likely that this concern can be managed, in particular, by stringent monitoring of patients in clinical trials.

For background information *see* ACQUIRED IMMUNOLOGICAL TOLERANCE; ANTIGEN-ANTIBODY REACTION; GENETIC ENGINEERING; IMMUNITY; IMMUNOSUPPRESSION; TRANSPLANTATION BIOLOGY in the McGraw-Hill Encyclopedia of Science & Technology.   Henk-Jan Schuurman; Julia L. Greenstein

Bibliography. A. Auchincloss and D. H. Sachs, Xenogeneic transplantation, *Annu. Rev. Immunol.*, 16: 433–470, 1998; J. Blusch, C. Patience, and U. Martin, Pig endogenous retroviruses and xenotransplantation, *Xenotransplantation*, 9:242–251, 2002; D. K. C. Cooper and R. P. Lanza, *Xeno, the Promise of Transplanting Animal Organs into Humans*, Oxford University Press, New York, 2000; Council of Europe, Working Party on Xenotransplantation, *Report on the State of the Art in the Field of Xenotransplantation*, Feb. 21, 2003; M. S. Sandrin, B. E. Loveland, and I. F. C. McKenzie, Genetic engineering for xenotransplantation, *J. Card. Surg.*, 16:448–457, 2001; U.S. Department of Health and Human Services, Food and Drug Administration, Center for Biologics Evaluation and Research, *Guidance for Industry: Source Animal, Product, Preclinical, and Clinical Issues Concerning the Use of Xenotransplantation Products in Humans: Final Guidance*, April 2003.

Contributors

Contributors

The affiliation of each Yearbook contributor is given, followed by the title of his or her article. An article title with the notation "coauthored" indicates that two or more authors jointly prepared an article or section.

A

Ackerman, Dr. Michael J. *Visible Human Project, National Library of Medicine, Bethesda, Maryland.* VISIBLE HUMAN PROJECT.

B

Balachandra, Prof. Ramaiya. *College of Business Administration, Northeastern University, Boston, Massachusetts.* EFFECTIVENESS OF HIGH-TECHNOLOGY RESEARCH.

Barrett, Prof. Spencer C. H. *Canada Research Chair, Department of Botany, University of Toronto, Ontario, Canada.* FLOWER DIVERSITY AND MATING STRATEGIES.

Battino, Prof. Rubin. *Department of Chemistry, Wright State University, Dayton, Ohio.* TEMPERATURE MEASUREMENT.

Bauerecker, Dr. Sigurd. *Institute of Physical and Theoretical Chemistry, Technical University of Braunschweig, Germany.* ULTRASONIC TRAPPING OF GASES—coauthored.

Beeson, Rob. *Senior Member of Technical Staff, Hewlett Packard Co., Corvallis, Oregon.* INKJET PRINTING.

Bélanger, Dr. J. Robert. *Geological Survey of Canada, Natural Resources Canada, Ottawa, Ontario.* URBAN GEOLOGY.

Bennett, Dr. Joan W. *Department of Cell and Molecular Biology, Tulane University, New Orleans, Louisiana.* FUNGAL GENOMICS.

Bergenti, Federico. *Researcher, Consorzio Nazionale Interuniversitario per le Telecomunicazioni, Parma, Italy.* COLLABORATIVE FIXED AND MOBILE NETWORKS—coauthored.

Beroza, Dr. Gregory C. *Department of Geophysics, Stanford University, Stanford, California.* PRECISE EARTHQUAKE LOCATION—coauthored.

Berry, Dr. Pam M. *Research Assistant, Environmental Change Institute, University of Oxford, United Kingdom.* PLANT VULNERABILITY TO CLIMATE CHANGE.

Blondel, Dr. Christophe. *Laboratoire Aime-Cotton, Centre National de la Recherche, Université Paris, France.* PHOTODETACHMENT MICROSCOPY.

Böltner, Dietmar. *Consejo Superior de Investigaciones Científicas, Estación Experimental del Zaidín, Granada, Spain.* BIODEGRADATION OF POLLUTANTS—coauthored.

Bonello, Prof. Pierluigi E. *Department of Plant Pathology, Ohio State University, Columbus.* PATHOGEN RESISTANCE IN PLANTS—coauthored.

Branscomb, Dr. Lewis M. *Harvard University, Cambridge, Massachusetts.* COUNTERTERRORISM.

Braiser, Dr. Martin D. *Professor of Palaeobiology, Department of Earth Sciences, University of Oxford, United Kingdom.* CYANOBACTERIAL FOSSILS—coauthored.

Breneman, Dr. Curt M. *Department of Chemistry, Rensselaer Polytechnic Institute, Troy, New York.* MOLECULAR MODELING—coauthored.

Brooks, Prof. Daniel R. *Department of Zoology, University of Toronto, Ontario, Canada.* HUMAN/PARASITE COEVOLUTION.

Buehlmann, Prof. Urs. *Assistant Professor and Extension Specialist, Department of Wood and Paper Science, North Carolina State University, Raleigh.* WOOD REUSE AND RECYCLING—coauthored.

Bullerman, Prof. Lloyd B. *Department of Food Science and Technology, University of Nebraska-Lincoln.* MYCOTOXINS IN FOOD.

Byrd, Dr. Robert. *Associate Professor in Clinical Pediatrics, University of California, Davis Medical Center, Sacramento.* AUTISM.

C

Caplan, Milton Z. *Director, Product Marketing & Business Integration, Atomic Energy of Canada, Ltd. (AECL).* ADVANCED CANDU REACTOR (ACR)—coauthored.

Castro, Dr. Joaquin H. *Pratt and Whitney, West Palm Beach, Florida.* SCRAMJET TECHNOLOGY—coauthored.

Chen, Dr. Xiangxu. *Department of Chemistry, University of California, Irvine.* SELF-HEALING POLYMERS—coauthored.

Chong, Prof. Ken P. *Director of Mechanics and Materials, National Science Foundation, Arlington, Virginia.* DETERIORATION RESEARCH.

Chupp, Prof. Timothy E. *Department of Physics, University of Michigan, Ann Arbor.* NOBLE GAS MRI—coauthored.

Collins, Dr. Francis S. *Director, National Human Genome Research Institute, Foundation for the National Institutes of Health, Bethesda, Maryland.* HUMAN GENOME PROJECT—coauthored.

D

Dahlman-Wright, Dr. Karin. *Department of Biosciences, Karolinska Institutet, Huddinge University Hospital, Huddinge, Sweden.* ESTROGEN ACTIONS AND CARDIOVASCULAR RISK—coauthored.

Davidson, Mark R. *Department of Materials Science and Engineering, University of Florida, Gainesville.* PHOSPHOR TECHNOLOGY—coauthored.

Davidhazy, Prof. Andrew. *Imaging and Photographic Technology, School of Photographic Arts and Sciences, Rochester Institute of Technology, New York.* DIGITAL STROBOSCOPIC PHOTOGRAPHY.

Davies, Dr. Sam T. *Director, Centre for Nanotechnology and Microengineering, School of Engineering, University of Warwick, United Kingdom.* FOCUSED ION BEAM MACHINING.

De Laet, Jan. *Royal Belgian Institute for Natural Sciences, Brussels, Belgium.* PARALLEL COMPUTER PROCESSING IN SYSTEMATICS—coauthored.

De Micheli, Prof. Giovanni. *Department of Electrical Engineering, Stanford University, Stanford, California.* NETWORKING CHIP SUBSYSTEMS.

Dickie, Dr. John. *Seed Conservation Department, Royal Botanic Gardens, Kew, United Kingdom.* SEED MORPHOLOGICAL RESEARCH—coauthored.

Di Fiore, Dr. Anthony. *Department of Anthropology, New York University, New York.* PRIMATE CONSERVATION.

Dorsett, Yair. *Laboratory of RNA Molecular Biology, The Rockefeller University, New York.* RNA INTERFERENCE—coauthored.

Dowling, David R. *Department of Mechanical Engineering, University of Michigan, Ann Arbor.* TIME-REVERSED SIGNAL PROCESSING.

E

Einarson, Adrienne. *The Motherisk Program and the Division of Clinical Pharmacology & Toxicology, The Hospital for Sick Children, and the University of Toronto, Ontario, Canada.* ANTENATAL DRUG RISK—coauthored.

Elliot, Dr. James L. *Department of Earth and Planetary Sciences, Massachusetts Institute of Technology, Cambridge.* PLUTO—coauthored.

Elisseeff, Dr. Jennifer H. *Johns Hopkins University, Baltimore, Maryland.* TISSUE ENGINEERING.

F

Fellows, Dr. Christopher M. *Key Centre for Polymer Colloids, School of Chemistry, University of Sydney, Australia.* EMULSION POLYMERIZATION—coauthored.

Felthous, Prof. Alan R. *Department of Psychiatry, Southern Illinois University School of Medicine and School of Law, Chester Mental Health Center, Chester, Illinois.* IMPULSIVE AGGRESSION (FORENSIC PSYCHIATRY).

Fielding, Dr. Christopher J. *Cardiovascular Research Institute, University of California Medical Center, San Francisco.* MEMBRANE LIPID RAFTS.

Fox, Prof. Kevin. *School of Biosciences, Cardiff University, Cardiff, United Kingdom.* KNOCKOUT GENE TECHNOLOGY (NEUROSCIENCE).

Franklin, William E. *Franklin Associates, Ltd., Prairie Village, Kansas.* WOOD REUSE AND RECYCLING—coauthored.

G

Gannon, Dr. Patrick J. *Department of Otolaryngology, Mount Sinai School of Medicine, New York.* BRAIN EVOLUTION.

Gamota, Daniel. *Motorola, Inc., Schaumburg, Illinois.* NANOPRINT LITHOGRAPHY—coauthored.

Gasson, Dr. Peter. *Jodrell Laboratory, Royal Botanic Gardens, Kew, United Kingdom.* WOOD ANATOMY.

Gates, Dr. Byron D. *Department of Chemistry and Chemical Biology, Harvard University, Cambridge, Massachusetts.* SOFT LITHOGRAPHY—coauthored.

Gaynor, Thomas Macaulay. *Product Manager, Drilling Tools, Halliburton-Sperry Sun, Tay Facility, United Kingdom.* ROBOTIC CONTROLLED DRILLING.

Gilbert, Prof. Robert G. *Key Centre for Polymer Colloids, School of Chemistry, University of Sydney, Australia.* EMULSION POLYMERIZATION—coauthored.

Gingerich, Dr. Philip D. *Professor, University of Michigan, Ann Arbor.* WHALE EVOLUTION.

Giribet, Prof. Gonzalo. *Harvard University, Cambridge, Massachusetts.* TREE OF LIFE—coauthored.

Gordon-Wylie, Dr. Scott W. *Assistant Professor, Department of Chemistry, University of Vermont, Burlington.* GREEN CHEMISTRY.

Green, Owen R. *Department of Earth Sciences, University of Oxford, United Kingdom.* CYANOBACTERIAL FOSSILS—coauthored.

Greenstein, Dr. Julia L. *Immerge BioTherapeutics, Inc., Cambridge, Massachusetts.* XENOTRANSPLANTATION—coauthored.

Graham, Prof. Terrence L. *Department of Plant Pathology, Ohio State University, Columbus.* PATHOGEN RESISTANCE IN PLANTS—coauthored.

Gustafsson, Dr. Jan-Åke. *Department of Biosciences, Karolinska Institutet, Huddinge University Hospital, Huddinge, Sweden.* ESTROGEN ACTIONS AND CARDIOVASCULAR RISK—coauthored.

Gwyn, Dr. W. Charles. *Intel Corporation, Santa Clara, California.* EXTREME-ULTRAVIOLET LITHOGRAPHY.

H

Halpin, Prof. Daniel W. *Purdue University, Division of Construction Engineering and Management, West Lafayette, Indiana.* EMERGING CONSTRUCTION TECHNOLOGY.

Harley, Dr. John P. *Foundation Professor, Department of Biology, Eastern Kentucky University, Richmond.* LEPTOSPIROSIS.

Härlin, Dr. Mikael. *Department of Natural Sciences, Sodertorn University College, Huddinge, Sweden.* CLASSIFICATION NOMENCLATURE.

Harrop, Dr. Peter J. *Chairman, ID TechEx Ltd, Cambridge, United Kingdom.* SMART LABELS.

Henry, Prof. Charles S. *Department of Chemistry, Colorado State University, Fort Collins.* MINIATURIZED ANALYSIS SYSTEMS.

Henry, Dr. James L. *Department of Physiology and Pharmacology, University of Western Ontario, London, Ontario, Canada.* PAIN.

Hollenberg, Prof. Norman K. *Departments of Medicine and Radiology, Harvard Medical School and Brigham and Women's Hospital, Boston, Massachusetts.* ANGIOTENSIN RECEPTOR BLOCKERS.

Holloway, Prof. Paul H. *Department of Materials Science and Engineering, University of Florida, Gainesville.* PHOSPHOR TECHNOLOGY—coauthored.

Hopwood, Jerry. *Director, ACR Business Development, Atomic Energy of Canada, Ltd. (AECL).* ADVANCED CANDU REACTOR (ACR)—coauthored.

Hulet, Prof. Randall G. *Department of Physics and Astronomy, Rice University, Houston, Texas.* ATOMIC FERMI GASES.

Hurd, Dr. H. Scott. *Analytical Epidemiologist, Preharvest Food Safety Research Unit, National Animal Disease Center, U.S. Department of Agriculture—Agricultural Research Service, Ames, Iowa.* SALMONELLOSIS IN LIVESTOCK.

J

Jabbour, Dr. Ghassan E. *Optical Sciences Center, University of Arizona, Tucson.* ORGANIC LIGHT-EMITTING DEVICES (OLEDS).

Janies, Daniel. *Assistant Professor, Department of Biomedical Informatics, Ohio State University, Columbus.* PARALLEL COMPUTER PROCESSING IN SYSTEMATICS—coauthored.

Jordan, Dr. Elke. *National Human Genome Research Institute, Foundation for the National Institutes of Health, Bethesda, Maryland.* HUMAN GENOME PROJECT—coauthored.

K

Katz, Prof. Eugenii. *Department of Organic Chemistry, The Hebrew University of Jerusalem, Israel.* BIOFUEL CELLS.

Kern, Susan. *Department of Earth and Planetary Sciences, Massachusetts Institute of Technology, Cambridge.* PLUTO—coauthored.

Kim, Prof. Tschangho John. *University of Illinois at Urbana-Champaign.* GIS/GPS IN TRANSPORTATION.

Klass, Dr. Klaus-Dieter. *Museum für Tierkunde, Dresden, Germany.* MANTOPHASMATODEA (INSECT SYSTEMATICS).

Klionsky, Prof. Daniel J. *Department of Molecular, Cellular and Developmental Biology, University of Michigan, Ann Arbor.* AUTOPHAGY.

Kon, Fabio. *Assistant Professor, Department of Computer Science, University of São Paulo, Brazil.* ADAPTIVE MIDDLEWARE.

Koren, Dr. Gideon. *The Motherisk Program and the Division of Clinical Pharmacology & Toxicology, The Hospital for Sick Children, and the University of Toronto, Ontario, Canada.* ANTENATAL DRUG RISK—coauthored.

Kovats, James Allen. *NMR Product Champion, Sugar Land Product Center, Sugarland, Texas.* NUCLEAR MAGNETIC RESONANCE LOGGING (PETROLEUM ENGINEERING).

Kowalewski, Dr. Michal. *Associate Professor of Geobiology, Department of Geological Sciences, Virginia Polytechnic Institute and State University, Blacksburg.* CONSERVATION PALEOBIOLOGY.

L

LaBar, Malcolm P. *Program Manager, Vista, California.* MODULAR HELIUM REACTORS—coauthored.

Lawrence, Dr. Stephen S. *Department of Physics and Astronomy, Hofstra University, Hempstead, New York.* SUPERNOVA REMNANT 1987A.

Lent, Prof. Craig S. *University of Notre Dame, South Bend, Indiana.* QUANTUM-DOT CELLULAR AUTOMATA.

Letcher, Prof. Trevor M. *Professor of Chemistry, Natal University, Durban, South Africa.* HEAT (BIOLOGY).

LeVan, Dr. Susan L. *USDA Forest Service, Forest Products Laboratory, Madison, Wisconsin.* SMALL TREE UTILIZATION (FORESTRY).

Levizzani, Dr. Vincenzo. *Institute of Atmospheric Sciences and Climate, National Research Council, Bologna, Italy.* SATELLITE OBSERVATION OF CLOUDS.

Long, Dr. Maureen. *University of Florida, College of Veterinary Medicine, Department of Large Animal Clinical Sciences, Gainesville, Florida.* WEST NILE VIRUS (EQUINE).

Love, J. Christopher. *Department of Chemistry and Chemical Biology, Harvard University, Cambridge, Massachusetts.* SOFT LITHOGRAPHY—coauthored.

Luo, Dr. Hui. *AT &T Labs—Research, Middletown, New Jersey.* WIRELESS FIDELITY TECHNOLOGY.

M

MacLeod, Dr. Norman. *Keeper of Palaeontology, The Natural History Museum, London, United Kingdom.* PUNCTUATED EQUILIBRIA (EVOLUTIONARY THEORY).

Maden, Prof. Malcolm. *MRC Centre for Developmental Neurobiology, King's College, London, United Kingdom.* RETINOIDS AND CNS DEVELOPMENT.

Magallón, Dr. Susana. *Department of Botany, Instituto de Biología, Universidad Nacional Autónoma de México, Mexico City.* Paleobotany.

Marcus, Prof. W. Andrew. *Department of Geography, University of Oregon, Eugene.* Hyperspectral remote sensing (mapping).

Martineau, Dr. Daniel. *Professor, College of Veterinary Medicine, Department of Pathology and Microbiology, Université de Montréal, Quebec, Canada.* Wildlife forensic science.

Marvier, Dr. Michelle A. *Biology Department and Environmental Studies Institute, Santa Clara University, Santa Clara, California.* Biodiversity hotspots.

Matsuoka, Prof. Yoky. *Robotics Institute and Mechanical Engineering, Carnegie Mellon University, Pittsburgh, Pennsylvania.* Humanoid robots.

McConnochie, Timothy. *Department of Astronomy, Cornell University, Ithaca, New York.* Mars Global Surveyor.

McDonald, Keith D. *Navtech Consulting, Alexandria, Virginia.* Galileo Navigation Satellite System.

Meschede, Dieter. *Professor of Experimental Physics, Institute for Applied Physics, University of Bonn, Germany.* Atom lithography.

Meyers, Dr. Blake C. *Assistant Professor, Department of Plant and Soil Sciences, University of Delaware, Newark.* Plant pathogen genomics.

Michelsen, Dr. Axel. *Center for Sound Communication, Institute of Biology, University of Southern Denmark, Odense.* Dance language of bees.

Munson, Prof. Eric J. *Department of Pharmaceutical Chemistry, The University of Kansas, Lawrence.* Solid-state NMR spectroscopy—coauthored.

N

Naicker, Dr. Pavan K. *Department of Chemical Engineering, Vanderbilt University, Nashville, Tennessee.* Chemical thermodynamics modeling.

Newhouse, Dr. Michael T. *Inhale Therapeutic Systems, Inc., San Carlos, California.* Inhalation drug therapy.

Novotny, Dr. Eric. *International Launch Services, McLean, Virginia.* Satellite launch vehicles.

Nulman, Irena. *The Motherisk Program and the Division of Clinical Pharmacology & Toxicology, The Hospital for Sick Children, and the University of Toronto, Ontario, Canada.* Antenatal drug risk—coauthored.

O

Offerdahl, Dr. Thomas J. *Postdoctoral Fellow, The University of Kansas, Lawrence.* Solid-state NMR spectroscopy—coauthored.

O'Neil, Dr. Michael A. *Natural Resources Institute, University of Greenwich at Medway, United Kingdom.* Microcalorimetry.

Otsuki, Dr. Michio. *Department of Biosciences, Karolinska Institutet, Huddinge University Hospital, Huddinge, Sweden.* Estrogen actions and cardiovascular risk—coauthored.

P

Patrono, Prof. Carlo. *University of Rome La Sapienza, Rome, Italy.* COX-2 inhibitors.

Pawson, Dr. David L. *Senior Scientist, National Museum of Natural History, Washington, DC.* Brittle star optics.

Peirce, Prof. James Jeffrey. *Department of Civil and Environmental Engineering, Duke University, Durham, North Carolina.* Geotropospheric interactions.

Penczek, Prof. Stanislaw. *Polish Academy of Sciences, Lódź, Poland.* Ring-opening polymerization.

Platt, Donald. *Melbourne, Florida.* Nanosatellites.

Poggi, Prof. Agostino. *Dipartimento di Ingegneria dell'Informazione, University of Parma, Italy.* Collaborative fixed and mobile networks—coauthored.

Popel, Dr. Aleksander S. *Department of Biomedical Engineering, Whitaker Biomedical Engineering Institute, School of Medicine, Johns Hopkins University, Baltimore, Maryland.* Biomechanical measurement.

Price, Prof. H. James. *Department of Soil & Crop Sciences, Texas A & M University, College Station.* Angiosperm genomes (plant phylogeny).

Q

Quigg, Prof. Chris. *Senior Scientist, Theoretical Physics Department, Fermi National Accelerator Laboratory, Batavia, Illinois.* Higgs boson.

R

Rahman, Prof. Saifur. *Alexandria Research Institute, Virginia Polytechnic Institute & State University, Alexandria.* Renewable energy sources.

Ramos, Juan Luis. *Consejo Superior de Investigaciones Científicas, Estación Experimental del Zaidín, Granada, Spain.* Biodegradation of pollutants—coauthored.

Rapp, Dr. George (Rip). *Regents Professor of Geoarchaeology, Department of Geological Sciences, University of Minnesota, Duluth.* Geoarchaeology.

Rayfield, Dr. Emily. *Emmanuel College Research Fellow, Department of Earth Sciences, University of Cambridge, United Kingdom.* Finite element analysis (paleontology).

Reigber, Prof. Christoph. *GeoForschungsZentrum Potsdam, Germany.* GRACE (Gravity Recovery and Climate Experiment)—coauthored.

Ren, Dr. Mindong. *Assistant Professor, Department of Cell Biology, New York University School of Medicine, New York.* Lysosome-related organelles.

Rightmire, Prof. G. Philip. *Department of Anthropology, Binghamton University (SUNY), Binghamton, New York.* Human evolution in Eurasia.

Ronquist, Dr. Fredrik. *Department of Systemic Zoology, Evolutionary Biology Centre, Uppsala Unversity, Sweden.* BAYESIAN INFERENCE (PHYLOGENY).

Rowell, Dr. Roger M. *Professor, University of Wisconsin-Madison.* WATER DECONTAMINATION.

S

Sage, Dr. Andrew P. *Founding Dean Emeritus and First American Bank Professor, University Professor, School of Information Technology and Engineering, George Mason University, Fairfax, Virginia.* SYSTEM FAMILIES.

Sahazizian, Anne-Marie. *P. Eng, Engineering Services, Hydro One Inc., Toronto, Ontario, Canada.* ELECTRIC POWER SUBSTATION DESIGN.

Salminen, Dr. Justin. *Helsinski University of Technology, Department of Chemical Technology, Helsinki, Finland.* PAPER MANUFACTURING (THERMODYNAMICS).

Samuelsen, Prof. Scott. *National Fuel Cell Research Center, University of California, Irvine.* FUEL CELLS.

Sauer, Beverly J. *Senior Chemical Engineer, Franklin Associates, A Division of ERG, Prairie Village, Kansas.* WOOD REUSE AND RECYCLING—coauthored.

Savolainen, Dr. Vincent. *Royal Botanic Gardens, Kew, United Kingdom.* TREE OF LIFE—coauthored.

Schafer, Dr. William M. *Schafer Ltd, Bozeman, Montana.* MINE WASTE ACIDITY CONTROL.

Schick, Kathy. *Craft Research Center, Bloomington, Indiana.* STONE TOOL ORIGINS—coauthored.

Schlenoff, Prof. Joseph B. *Department of Chemistry & Biochemistry, The Florida State University, Tallahassee.* POLYELECTROLYTE MULTILAYERS.

Schuurman, Dr. Henk-Jan. *Immerge BioTherapeutics, Inc., Cambridge, Massachusetts.* XENOTRANSPLANTATION—coauthored.

Scowen, Dr. Paul. *Department of Physics and Astronomy, Arizona State University, Tempe.* CRAB NEBULA.

Seery, Dr. Bernard D. *NASA Goddard Space Flight Center, Greenbelt, Maryland.* JAMES WEBB SPACE TELESCOPE (JWST).

Shelton, Dr. Brian G. *MPH, PathCon Laboratories, Norcross, Georgia.* INDOOR FUNGI.

Simon, Mr. Walter A. *General Atomics, San Diego, California.* MODULAR HELIUM REACTORS—coauthored.

Snieder, Prof. Roel. *Department of Geophysics, Colorado School of Mines, Golden.* CODA WAVE INTERFEROMETRY.

Solie, Stacey. *Seattle, Washington.* MARINE HABITAT PROTECTION.

Soller, Dr. David R. *USGS/Geologic Division, National Geologic Map Database, Reston, Virginia.* GEOLOGIC MAPPING.

Stankovic, Prof. John A. *Department of Computer Science, University of Virginia, Charlottesville.* NETWORKED MICROCONTROLLERS.

Stephenson, Dr. Edward J. *Indiana University Cyclotron Facility, Bloomington, Indiana.* CHARGE SYMMETRY BREAKING.

Stroock, Prof. Abraham D. *Department of Chemical and Biomolecular Engineering, Cornell University, Ithaca, New York.* MICROFLUIDICS.

Stuppy, Dr. Wolfgang. *Seed Conservation Department, Royal Botanic Gardens, Kew, United Kingdom.* SEED MORPHOLOGICAL RESEARCH—coauthored.

Sukumar, N. *Department of Chemistry, Rensselaer Polytechnic Institute, Troy, New York.* MOLECULAR MODELING—coauthored.

Sumida, Dr. Stuart S. *Professor, Department of Biology, California State University, San Bernardino.* ORIGIN OF AMNIOTES.

Swanson, Scott D. *Department of Physics, University of Michigan, Ann Arbor.* NOBLE GAS MRI—coauthored.

T

Tapley, Prof. Byron D. *Center for Space Research, The University of Texas at Austin.* GRACE (GRAVITY RECOVERY AND CLIMATE EXPERIMENT)—coauthored.

Taylor, Dr. Shelley E. *Professor, Department of Psychology, University of California, Los Angeles.* SEX DIFFERENCES IN STRESS RESPONSE.

Thomas, Prof. Brian G. *Department of Mechanical and Industrial Engineering, University of Illinois at Urbana-Champaign.* CONTINUOUS CASTING (METALLURGY).

Toth, Dr. Nicholas. *Craft Research Center, Bloomington, Indiana.* STONE TOOL ORIGINS—coauthored.

Tuckermann, Dr. Rudolf. *Institute for Coastal Research, Physical and Chemical Analysis, GKSS—Research Centre, Germany.* ULTRASONIC TRAPPING OF GASES—coauthored.

Tuschl, Dr. Thomas. *Laboratory of RNA Molecular Biology, The Rockefeller University, New York.* RNA INTERFERENCE—coauthored.

V

Vallero, Dr. Daniel A. *Department of Environmental Science, North Carolina Central University, Durham.* ENVIRONMENTAL ENDOCRINE DISRUPTORS.

Van Cott, Kevin. *Department of Chemical Engineering, Virginia Polytechnic Institute & State University, Blacksburg.* THERAPEUTIC PROTEIN SYNTHESIS—coauthored.

van Dillewijn, Pieter. *Consejo Superior de Investigaciones Científicas, Estación Experimental del Zaidín, Granada, Spain.* BIODEGRADATION OF POLLUTANTS—coauthored.

Vardaxoglou, Dr. J. (Yiannis) C. *Professor of Wireless Communications, Department of Electronic and Electrical Engineering, Loughborough University, United Kingdom.* MOBILE-PHONE ANTENNAS.

Velander, Prof. William H. *Department of Chemical Engineering, Virginia Polytechnic Institute & State University, Blacksburg.* THERAPEUTIC PROTEIN SYNTHESIS—coauthored.

von Puttkamer, Dr. Jesco. *NASA Headquarters, Office of Space Flight, Washington, DC.* SPACE FLIGHT.

W

Wang, Prof. Pao K. *Department of Atmospheric and Oceanic Sciences, University of Wisconsin–Madison.* ATMOSPHERIC WATER VAPOR.

Weil, Dr. Anne. *Research Associate, Duke University, Department of Biological Anthroplogy and Anatomy, Durham, North Carolina.* JEHOL BIOTA MAMMALS (PALEONTOLOGY).

Weinstein, Dr. Brant M. *National Institute of Child Health and Human Development, Bethesda, Maryland.* ANGIOGENESIS.

Welcsh, Dr. Piri L. *Research Assistant Professor, University of Washington School of Medicine, Division of Medical Genetics, Seattle, Washington.* PREDICTIVE GENETICS OF CANCER.

West, Sabrina. *Department of Biology, Santa Clara University, Santa Clara, California.* POPULATION GROWTH AND SUSTAINABLE DEVELOPMENT.

Wheeler, David M. *Security and Software Architect, Group Architecture, Wireless Communications, and Computing Group, Intel Corporation, Chandler, Arizona.* TELECOMMUNICATIONS NETWORK SECURITY.

Wheeler, Dr. Ward C. *Curator of Invertebrates, American Museum of Natural History, New York.* PARALLEL COMPUTER PROCESSING IN SYSTEMATICS—coauthored.

Whitcomb, Carrie Morgan. *Director, National Center for Forensic Science, University of Central Florida, Orlando.* DIGITAL EVIDENCE.

Whitesides, Prof. George M. *Department of Chemistry and Chemical Biology, Harvard University, Cambridge, Massachusetts.* SOFT LITHOGRAPHY—coauthored.

Wignall, Dr. Paul B. *School of Earth Sciences, University of Leeds, United Kingdom.* BIOTIC RECOVERY AFTER MASS EXTINCTIONS.

Williams, Dr. Neil M. *Associate Professor, Livestock Disease Diagnostic Center, University of Kentucky, Lexington.* EQUINE ABORTION EPIDEMIC.

Wolfe, Daniel B. *Department of Chemistry and Chemical Biology, Harvard University, Cambridge, Massachusetts.* SOFT LITHOGRAPHY—coauthored.

Wright, Prof. Edward L. *Department of Astronomy, University of California, Los Angeles.* WILKINSON MICROWAVE ANISOTROPY PROBE.

Wudl, Prof. Fred. *Department of Chemistry and Biochemistry, University of California, Los Angeles.* SELF-HEALING POLYMERS—coauthored.

Z

Zanzerkia, Eva E. *Department of Geophysics, Stanford University, Stanford, California.* PRECISE EARTHQUAKE LOCATION—coauthored.

Zhang, Dr. Jie. *Motorola, Inc., Schaumburg, Illinois.* NANOPRINT LITHOGRAPHY—coauthored.

Index

Asterisks indicate page references to article titles.